NanoScience and Technology

NANOSCIENCE AND TECHNOLOGY

Series Editors:
P. Avouris B. Bhushan D. Bimberg K. von Klitzing H. Sakaki R. Wiesendanger

The series NanoScience and Technology is focused on the fascinating nano-world, mesoscopic physics, analysis with atomic resolution, nano and quantum-effect devices, nanomechanics and atomic-scale processes. All the basic aspects and technology-oriented developments in this emerging discipline are covered by comprehensive and timely books. The series constitutes a survey of the relevant special topics, which are presented by leading experts in the field. These books will appeal to researchers, engineers, and advanced students.

Atomic Force Microscopy, Scanning Nearfield Optical Microscopy and Nanoscratching
Application
to Rough and Natural Surfaces
By G. Kaupp

Applied Scanning Probe Methods VI
Characterization
Editors: B. Bhushan and S. Kawata

Applied Scanning Probe Methods VII
Biomimetics
and Industrial Applications
Editors: B. Bhushan and H. Fuchs

**Roadmap
of Scanning Probe Microscopy**
Editors: S. Morita

Nanocatalysis
Editors: U. Heiz and U. Landman

Nanostructures
Fabrication and Analysis
Editor: H. Nejo

Fundamentals of Friction and Wear on the Nanoscale
Editors: E. Gnecco and E. Meyer

Nanostructured Soft Matter
Experiment, Theory, Simulation
and Perspectives
Editor: A.V. Zvelindovsky

Charge Migration in DNA
Perspectives from Physics, Chemistry,
and Biology
Editor: T. Chakraborty

**Lateral Alignment
of Epitaxial Quantum Dots**
Editor: O. Schmidt

Applied Scanning Probe Methods VIII
Scanning Probe Microscopy
Techniques
Editors: B. Bhushan, H. Fuchs,
and M. Tomitori

Applied Scanning Probe Methods IX
Characterization
Editors: B. Bhushan, H. Fuchs,
and M. Tomitori

Applied Scanning Probe Methods X
Biomimetics
and Industrial Applications
Editors: B. Bhushan, H. Fuchs,
and M. Tomitori

Semiconductor Nanostructures
Editor: D. Bimberg

**Multiscale Dissipative Mechanisms
and Hierarchical Surfaces**
Friction, Superhydrophobicity,
and Biomimetics
By M. Nosonovsky and B. Bhushan

Dieter Bimberg (Ed.)

Semiconductor Nanostructures

With 245 Figures

 Springer

Professor Dr. Dieter Bimberg
TU Berlin, Fakultät Mathematik/
Naturwissenschaften
Institut für Festkörperphysik
Hardenbergstr. 36
10623 Berlin, Germany
E-mail: bimberg@physik.tu-berlin.de

Series Editors:

Professor Dr. Phaedon Avouris
IBM Research Division
Nanometer Scale Science & Technology
Thomas J. Watson Research Center
P.O. Box 218
Yorktown Heights, NY 10598, USA

Professor Dr. Bharat Bhushan
Ohio State University
Nanotribology Laboratory
for Information Storage
and MEMS/NEMS (NLIM)
Suite 255, Ackerman Road 650
Columbus, Ohio 43210, USA

Professor Dr. Dieter Bimberg
TU Berlin, Fakultät Mathematik/
Naturwissenschaften
Institut für Festkörperphysik
Hardenbergstr. 36
10623 Berlin, Germany

Professor Dr., Dres. h.c. Klaus von Klitzing
Max-Planck-Institut
für Festkörperforschung
Heisenbergstr. 1
70569 Stuttgart, Germany

Professor Hiroyuki Sakaki
University of Tokyo
Institute of Industrial Science
4-6-1 Komaba, Meguro-ku
Tokyo 153-8505, Japan

Professor Dr. Roland Wiesendanger
Institut für Angewandte Physik
Universität Hamburg
Jungiusstr. 11
20355 Hamburg, Germany

ISSN 1434-4904
ISBN 978-3-540-77898-1 Springer Berlin Heidelberg New York

Library of Congress Control Number: 2008923741

This work is subject to copyright. All rights are reserved, whether the whole or part of the material is concerned, specifically the rights of translation, reprinting, reuse of illustrations, recitation, broadcasting, reproduction on microfilm or in any other way, and storage in data banks. Duplication of this publication or parts thereof is permitted only under the provisions of the German Copyright Law of September 9, 1965, in its current version, and permission for use must always be obtained from Springer. Violations are liable to prosecution under the German Copyright Law.

Springer is a part of Springer Science+Business Media.

springeronline.com

© Springer-Verlag Berlin Heidelberg 2008

The use of general descriptive names, registered names, trademarks, etc. in this publication does not imply, even in the absence of a specific statement, that such names are exempt from the relevant protective laws and regulations and therefore free for general use.

Typesetting: Data prepared by VTEX using a Springer LATEX macro
Cover: eStudio Calamar Steinen

Printed on acid-free paper   SPIN: 11951438   57/3180/VTEX - 5 4 3 2 1 0

Preface

Let me start by defining the subject of this book, semiconductor nanostructures, using a strictly physical argument.

If the geometrical extent of a semiconductor, typically embedded in the matrix of another semiconductor, is reduced in one, two or three directions of space below the size of the "de Broglie wavelength" of a charge carrier—in other words, if it is reduced in size to only a few nanometers—it is a nanostructure. Reduction of dimensionality in one, two or three directions leads to quantum wells, wires or dots, respectively. The focus here is mainly on the ultimate limit, quantum dots.

The electronic—and to a lesser extent the vibronic—properties of low-dimensional structures (including their interactions such as electron–electron, electron–hole and electron–phonon interaction) depend qualitatively on the dimensionality of the structure and quantitatively on details of the geometry of the structure (its size and shape) and of the distribution of atoms inside. The electronic properties in turn control the linear and nonlinear optical and transport properties. Thus "geometrical architecture" opens enormous opportunities for designing completely novel materials or heterostructures. These opportunities extend far beyond the well-known "chemical architecture", where properties are modified by varying the chemical composition. Radically different from three-, two-, and one-dimensional structures in their electronic properties are zero-dimensional structures: quantum dots. Their "density of electronic states" is described by delta-functions and they show no dispersion of energy, thus resembling an atom in a dielectric matrix instead of a classical semiconductor. The Hamilton and momentum operators of quantum dots do not commute, leading again to novel physical properties. Figure 1 shows the variation of the density of states in structures of varying dimensionality.

The geometrical properties of quantum dots are controlled by the thermodynamic and kinetic parameters of the growth of strongly strained heterostructures, as long as we focus on the various fully epitaxial modes of their fabrication. Such

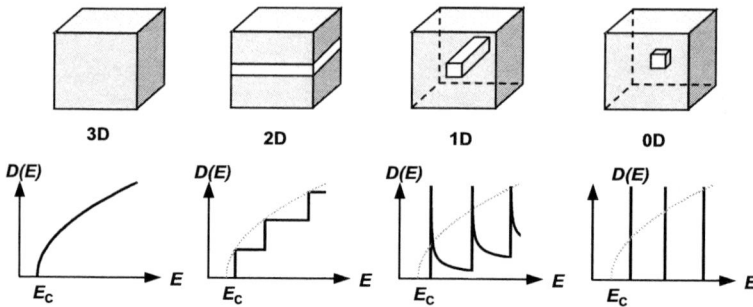

Fig. 1. Top: Schematic representations of three-, two-, one- and zero-dimensional heterostructures. Bottom: Densities of electronic states for the given case of dimensionality

growth modes lead to completely defect-free structures and interfaces. Fabrication of "classical" defect-free semiconductor heterostructures is based on the few existing lattice-matched or close to lattice-matched pseudomorphic structure families like GaAs/AlAs, or InP/InGaAsP. In contrast, quantum dot growth inverts this paradigm because a minimum difference of lattice constants is necessary to initiate self-organization effects, suddenly yielding an enormous wealth of "possible" combinations of material systems, complementary to the "classical" case.

Theoretical and experimental studies of growth mechanisms leading to the formation of nanostructures on different types of surfaces and for different semiconductors were one of the goals of the "Sonderforschungsbereich (SFB)" on "Growth Related Properties of Semiconductor Nanostructures", a "Center of Excellence", using a nonliteral translation, of the Deutsche Forschungsgemeinschaft (DFG), the German National Science Foundation. Self-ordering and self-organization effects play a decisive role for these newly discovered growth modes. Experimental investigations of nanostructures by structural, optical and transport methods, in addition to theoretical modeling of their electronic, linear and nonlinear optical properties were another objective. Novel experimental methods, like cathodoluminescence, cross-section tunneling microscopy, calorimetric absorption spectroscopy, near-field microscopy, to mention a few, had to be developed to meet the experimental challenges of gathering information on such ultra small structures. Discovery of many absolutely unique properties of nanostructures turned out to be of fundamental importance and present the starting point for a novel generation of photonic and electronic devices.

The reunification of Berlin brought together the members of this SFB, centered at the Technical University of Berlin. The principal scientists of the SFB came in almost equal numbers from some of the strongest research groups of the former West Berlin—the TU Berlin and the Fritz-Haber-Institute of the Max-Planck-Society—and from the former East-Berlin—the Humboldt University, the Paul-Drude-Institute and the Max-Born-Institute. In addition, guest scientists from all over the world contributed to its success, in particular those from the Ioffe Institute of the Russian Academy of Sciences.

The start of the SFB in the summer of 1994, after more than two years of preparation, was just in time. The international nano-wave then began to roll, triggered partly by the work of the SFB. After more than 12 years of operation, the papers published in 2006 by its researchers in international journals and at international conferences led them to the "top of the chart". The international citation index shows that many of the publications from the SFB belong to the most cited ones of the world in this area.

Three books [1–3] published in 1999, 2002 and 2004 summarize the initial worldwide fundamental work on quantum dots [1], aspects of epitaxial growth [2] and applications of quantum dots for novel nano-optoelectronics [3]. This new book summarizes what we believe represents the best, most important works on nanostructures of the last years, presented to the scientific community in a compact thus readable manner. Chapters 1–6 cover theoretical and experimental aspects of growth and structural studies. Chapters 7–10 are devoted to the theory of electronic and optical properties of quantum dots. Sections 11–16 finally focus on experimental studies of optical properties of nanostructures including ultra-high magnetic fields, ultra fast spectroscopy and emerging applications for novel nano-memories, quantum computing and cryptography. The authors of the contributions are the work package leaders of the SFB (Chaps. 1, 3–6, 8–10, 12, 13, 15 and 16) and of some of my coworkers (Chaps. 1, 2, 7, 11 and 14).

I thank my colleagues for the joy of working together for one and a half decades, lending me their support as a chairman of the SFB. I also thank the many reviewers for their constructive advice all along our way; the staff of the DFG for their never-ending patience and always-positive thinking; Claus Ascheron and the staff of Springer for their enthusiasm; Doreen Nitzsche for her energy reminding everybody to deliver and for completing the technical part of the book; and to my wife, Sigrun, for her patience when I looked overworked.

Berlin, March 2008   Dieter Bimberg

References

1. D. Bimberg, M. Grundmann, N.N. Ledentsov, *Quantum Dot Heterostructures* (Wiley, New York, 1999)
2. V.A. Shchukin, N. Ledentsov, D. Bimberg, *Epitaxy of Nanostructures* (Springer, Berlin, 2003)
3. M. Grundmann (ed.), *Nano-Optoelectronics* (Springer, Berlin, 2002)

Contents

Preface v

List of Contributors xvii

1 Thermodynamics and Kinetics of Quantum Dot Growth
Vitaly Shchukin, Eckehard Schöll and Peter Kratzer 1
1.1 Introduction 2
    1.1.1 Length and Time Scales 3
    1.1.2 Multiscale Approach to the Modeling of Nanostructures 4
1.2 Atomistic Aspects of Growth 5
    1.2.1 Diffusion of Ga Atoms on GaAs(001) 5
    1.2.2 Energetics of As_2 Incorporation During Growth 5
    1.2.3 Kinetic Monte Carlo Simulation of GaAs Homoepitaxy 6
    1.2.4 Wetting Layer Evolution 9
1.3 Size and Shapes of Individual Quantum Dots 11
    1.3.1 Hybrid Approach to Calculation of the Equilibrium Shape of Individual Quantum Dots 11
    1.3.2 Role of High-Index Facets in the Shape of Quantum Dots 13
    1.3.3 Shape Transition During Quantum Dot Growth 14
    1.3.4 Constraint Equilibrium of Quantum Dots with a Wetting Layer .. 15
1.4 Thermodynamics and Kinetics of Quantum Dot Ensembles 19
    1.4.1 Equilibrium Volume of Strained Islands versus Ostwald Ripening 19
    1.4.2 Crossover from Kinetically Controlled to Thermodynamically Controlled Growth of Quantum Dots 22
    1.4.3 Tunable Metastability of Quantum Dot Arrays 25
    1.4.4 Evolution Mechanisms in Dense Arrays of Elastically Interacting Quantum Dots 27

1.5	Quantum Dot Stacks	29
	1.5.1 Transition between Vertically Correlated and Vertically Anticorrelated Quantum Dot Growth	29
	1.5.2 Finite Size Effect: Abrupt Transitions between Correlated and Anticorrelated Growth	31
	1.5.3 Reduction of a Size of a Critical Nucleus in the Second Quantum Dot Layer	32
1.6	Summary and Outlook	34
	References	35

2 Control of Self-Organized In(Ga)As/GaAs Quantum Dot Growth
Udo W. Pohl and André Strittmatter 41

2.1	Introduction	41
2.2	Evolution and Strain Engineering of InGaAs/GaAs Quantum Dots	42
	2.2.1 Evolution of InGaAs Dots	42
	2.2.2 Engineering of Single and Stacked InGaAs QD Layers	46
2.3	Growth Control of Equally Shaped InAs/GaAs Quantum Dots	50
	2.3.1 Formation of Self-Similar Dots with a Multimodal Size Distribution	51
	2.3.2 Kinetic Description of Multimodal Dot-Ensemble Formation	54
2.4	Epitaxy of GaSb/GaAs Quantum Dots	56
	2.4.1 Onset and Dynamics of GaSb/GaAs Quantum-Dot Formation	56
	2.4.2 Structure of GaSb/GaAs Quantum Dots	58
2.5	Device Applications of InGaAs Quantum Dots	60
	2.5.1 Edge-Emitting Lasers	60
	2.5.2 Surface-Emitting Lasers	61
2.6	Conclusion	62
	References	63

3 In-Situ Monitoring for Nano-Structure Growth in MOVPE
Markus Pristovsek and Wolfgang Richter 67

3.1	Introduction	67
3.2	Reflectance	69
3.3	Reflectance Anisotropy Spectroscopy (RAS)	71
	3.3.1 RAS Spectra and Surface Reconstruction	72
	3.3.2 Monolayer Oscillations	74
	3.3.3 Monitoring of Carrier Concentration	79
3.4	Scanning Tunneling Microscopy (STM)	82
3.5	Conclusion	84
	References	85

4 Bottom-up Approach to the Nanopatterning of Si(001)
R. Koch 87

4.1	Quantum Dot Growth on Semiconductor Templates	87
4.2	$(2 \times n)$ Reconstruction of Si(001)	88

4.3	Monte Carlo Simulations on the (2 × n) Formation	90
4.4	Scanning Tunneling Microscopy Results	92
4.5	Summary and Outlook	94
References		95

5 Structural Characterisation of Quantum Dots by X-Ray Diffraction and TEM
R. Köhler, W. Neumann, M. Schmidbauer, M. Hanke, D. Grigoriev, P. Schäfer, H. Kirmse, I. Häusler and R. Schneider 97

5.1	Introduction	97
5.2	Liquid Phase Epitaxy of SiGe/Si: A Model System for the Stranski–Krastanow Process	99
	5.2.1 Dot Evolution in a Close-to-Equilibrium Regime	99
5.3	(In,Ga)As Quantum Dots on GaAs	103
	5.3.1 Shape, Size, Strain and Composition Gradient in InGaAs QD Arrays	103
	5.3.2 Chemical Composition of (In,Ga)As QDs Determined by TEM	107
	5.3.3 Controlling 3D Ordering in (In,Ga)As QD Arrays through GaAs Surface Orientation	109
5.4	Ga(Sb,As) Quantum Dots on GaAs	113
	5.4.1 Structural Characterisation of Ga(Sb,As) QDs by High-Resolution TEM Imaging	117
	5.4.2 Chemical Characterisation of Ga(Sb,As) QDs by HAADF STEM Imaging	118
References		119

6 The Atomic Structure of Quantum Dots
Mario Dähne, Holger Eisele and Karl Jacobi 123

6.1	Introduction	123
6.2	Experimental Details	124
6.3	STM Studies of InAs Quantum Dots on the Growth Surface	124
6.4	XSTM Studies of Buried Nanostructures	127
	6.4.1 InAs Quantum Dots	127
	6.4.2 InGaAs Quantum Dots	131
	6.4.3 GaSb Quantum Dots	134
6.5	Conclusion	135
References		136

7 Theory of Excitons in InGaAs/GaAs Quantum Dots
Andrei Schliwa and Momme Winkelnkemper 139

7.1	Introduction	139
7.2	Interrelation of QD-Structure, Strain and Piezoelectricity, and Coulomb Interaction	140
	7.2.1 The Binding Energies of the Few Particle Complexes	140
7.3	Method of Calculation	143
	7.3.1 Calculation of Strain	144

		7.3.2 Piezoelectricity and the Reduction of Lateral Symmetry	145

- 7.3.2 Piezoelectricity and the Reduction of Lateral Symmetry ... 145
- 7.3.3 Single Particle States ... 147
- 7.3.4 Many-Particle States ... 148
- 7.3.5 The Configuration Interaction Model ... 148
- 7.3.6 Interband Spectra ... 150
- 7.4 The Investigated Structures: Variation of Size, Shape and Composition ... 150
- 7.5 The Impact of QD Size ... 151
  - 7.5.1 The Role of the Piezoelectric Field ... 153
- 7.6 The Aspect Ratio ... 155
  - 7.6.1 Vertical Aspect Ratio ... 155
  - 7.6.2 Lateral Aspect Ratio ... 157
- 7.7 Different Composition Profiles ... 157
  - 7.7.1 Inverted Cone-Like Composition Profile ... 157
  - 7.7.2 Annealed QDs ... 159
  - 7.7.3 InGaAs QDs with Uniform Composition ... 159
- 7.8 Correlation vs. QD Size, Shape and Particle Type ... 159
- 7.9 Conclusions ... 162
- References ... 163

8 Phonons in Quantum Dots and Their Role in Exciton Dephasing
F. Grosse, E.A. Muljarov and R. Zimmermann ... 165

- 8.1 Introduction ... 165
- 8.2 Structural Properties of Semiconductor Nanostructures ... 166
- 8.3 Theory of Acoustic Phonons in Quantum Dots ... 166
  - 8.3.1 Continuum Elasticity Model of Phonons ... 167
  - 8.3.2 Phonons in Quantum Dots ... 170
- 8.4 Exciton-Acoustic Phonon Coupling in Quantum Dots ... 171
- 8.5 Dephasing of the Exciton Polarization in Quantum Dots ... 173
  - 8.5.1 Single Exciton Level: Independent Boson Model ... 174
  - 8.5.2 Multilevel System: Real and Virtual Phonon-Assisted Transitions ... 176
  - 8.5.3 Application to Coupled Quantum Dots ... 182
- 8.6 Summary ... 184
- References ... 185

9 Theory of the Optical Response of Single and Coupled Semiconductor Quantum Dots
C. Weber, M. Richter, S. Ritter and A. Knorr ... 189

- 9.1 Introduction ... 189
- 9.2 Theory ... 190
  - 9.2.1 Quantum Dot Model ... 190
  - 9.2.2 Hamiltonian ... 191
  - 9.2.3 Mathematical Formalisms ... 193
- 9.3 Single Quantum Dot Response ... 196
  - 9.3.1 Linear Absorption Spectra and Quantum Optics ... 196
  - 9.3.2 Semiclassical Nonlinear Dynamics ... 199

9.4 Two Coupled Quantum Dots 201
    9.4.1 Absorption Spectra 202
    9.4.2 Excitation Transfer 202
    9.4.3 Rabi Oscillations 203
    9.4.4 Pump-Probe/Differential Transmission Spectra 204
9.5 Multiple Quantum Dots 205
    9.5.1 Four-Wave-Mixing: Photon Echo in Quantum Dot Ensembles ... 205
    9.5.2 Absorption of Multiple Coupled Quantum Dots 205
    9.5.3 Energy Transfer of Multiple Coupled Quantum Dots 206
9.6 Conclusion 206
References 207

10 Theory of Nonlinear Transport for Ensembles of Quantum Dots
G. Kießlich, A. Wacker and E. Schöll 211
10.1 Introduction 211
10.2 Coulomb Interaction within a Quantum Dot Layer 211
10.3 Transport in Quantum Dot Stacks 213
10.4 Current Fluctuations and Shot Noise 214
10.5 Full Counting Statistics and Decoherence in Coupled Quantum Dots 216
10.6 Conclusion 218
References 219

11 Quantum Dots for Memories
M. Geller and A. Marent 221
11.1 Introduction 221
11.2 Semiconductor Memories 222
    11.2.1 Dynamic Random Access Memory (DRAM) 222
    11.2.2 Nonvolatile Semiconductor Memories (Flash) 223
    11.2.3 A QD-based Memory Cell 224
11.3 Charge Carrier Storage in Quantum Dots 226
    11.3.1 Experimental Technique 226
    11.3.2 Carrier Storage in InGaAs/GaAs Quantum Dots 228
    11.3.3 Hole Storage in GaSb/GaAs Quantum Dots 229
    11.3.4 InGaAs/GaAs Quantum Dots with Additional AlGaAs Barrier ... 230
11.4 Conclusion and Outlook 233
References 235

12 Visible-Bandgap II–VI Quantum Dot Heterostructures
Ilya Akimov, Joachim Puls, Michael Rabe and Fritz Henneberger 237
12.1 Introduction 237
12.2 Epitaxial Growth 238
12.3 Few-Particles States and Their Fine Structure 241
    12.3.1 Excitons and Biexcitons 241
    12.3.2 Trions in Charged Quantum Dots 243

12.4 Coherent Control of the Exciton–Biexciton System 245
12.5 Spin Relaxation of Excitons, Holes, and Electrons 247
    12.5.1 Exciton Quantum Coherence 247
    12.5.2 Hole Spin Lifetime 248
    12.5.3 Spin Dynamics of the Resident Electron 249
12.6 Diluted Magnetic Quantum Dots 251
References 253

13 Narrow-Gap Nanostructures in Strong Magnetic Fields
T. Tran-Anh and M. Ortenberg 255
13.1 Introduction 255
13.2 Materials: HgSe/HgSe:Fe 256
13.3 Fabrication of HgSe/HgSe:Fe Nanostructures 256
    13.3.1 Quantum Wells 257
    13.3.2 Roof-Ridge Quantum Wires 258
    13.3.3 Quantum Dots 259
13.4 Electronic Characterization of the HgSe/HgSe:Fe Nano-Structures in Strong Magnetic Fields 262
    13.4.1 High-Field Magneto Transport 262
    13.4.2 Infrared Magneto-Resonance Spectroscopy 263
13.5 Summary 267
References 267

14 Optical Properties of III–V Quantum Dots
Udo W. Pohl, Sven Rodt and Axel Hoffmann 269
14.1 Introduction 269
14.2 Confined States and Many-Particle Effects 270
    14.2.1 Renormalization 270
    14.2.2 Phonon Interaction 274
    14.2.3 Electronic Tuning by Strain Engineering 276
    14.2.4 Multimodal InAs/GaAs Quantum Dots 278
14.3 Single InAs/GaAs Quantum Dots 281
    14.3.1 Spectral Diffusion 281
    14.3.2 Size-Dependent Anisotropic Exchange Interaction ... 282
    14.3.3 Binding Energies of Excitonic Complexes 285
    14.3.4 Data Storage Using Confined Trions 286
    14.3.5 Electronic Tuning by Annealing 287
14.4 Optical Properties of InGaN/GaN Quantum Dots 288
    14.4.1 Time-Resolved Studies on Quantum Dot Ensembles .. 289
    14.4.2 Single-Dot Spectroscopy 292
14.5 Summary 296
References 298

15 Ultrafast Coherent Spectroscopy of Single Semiconductor Quantum Dots
Christoph Lienau and Thomas Elsaesser 301
15.1 Introduction 301
15.2 Interface Quantum Dots 303
15.3 Coherent Spectroscopy of Interface Quantum Dots: Experimental
       Technique 305
15.4 Coherent Control in Single Interface Quantum Dots 308
       15.4.1 Ultrafast Optical Nonlinearities of Single Interface Quantum Dots 308
       15.4.2 Rabi Oscillations in a Quantum Dot 312
       15.4.3 Optical Stark Effect: Ultrafast Control of Single Exciton
              Polarizations 315
15.5 Coupling Two Quantum Dots via the Dipole–Dipole Interaction 319
15.6 Summary and Conclusions 323
References 325

16 Single-Photon Generation from Single Quantum Dots
Matthias Scholz, Thomas Aichele and Oliver Benson 329
16.1 Introduction 329
16.2 Single Quantum Dots as Single-Photon Emitters 331
       16.2.1 Photon Statistics of Single-Photon Emitters 331
       16.2.2 Micro-Photoluminescence 332
       16.2.3 Single Photons from InP Quantum Dots 333
16.3 Multiphoton Emission from Single Quantum Dots 334
16.4 Realization of the Ultimate Limit of a Light Emitting Diode 339
16.5 Applications in Quantum Information Processing 343
       16.5.1 Quantum Key Distribution 343
       16.5.2 Quantum Computing 344
16.6 Outlook 346
References 347

Index 351

List of Contributors

Thomas Aichele
Institut für Physik
Humboldt-Universität zu Berlin
Hausvogteiplatz 5-7
D-10117 Berlin, Germany
thomas.aichele@
physik.hu-berlin.de

Ilya Akimov
Institut für Physik
Humboldt-Universität zu Berlin
Newtonstr. 15
D-12489 Berlin, Germany
ilya.akimov@
physik.hu-berlin.de

Oliver Benson
Institut für Physik
Humboldt-Universität zu Berlin
Hausvogteiplatz 5-7
D-10117 Berlin, Germany
oliver.benson@
physik.hu-berlin.de

Dieter Bimberg
Institut für Festkörperphysik
Technische Universität Berlin
Hardenbergstr. 36
D-10623 Berlin, Germany
bimberg@physik.tu-berlin.de

Mario Dähne
Institut für Festkörperphysik
Technische Universität Berlin
Hardenbergstr. 36
D-10623 Berlin, Germany
daehne@physik.tu-berlin.de

Holger Eisele
Institut für Festkörperphysik
Technische Universität Berlin
Hardenbergstr. 36
D-10623 Berlin, Germany
ak@physik.tu-berlin.de

Thomas Elsaesser
Max-Born-Institut für Nichtlineare Optik
und Kurzzeitspektroskopie
Max-Born-Str. 2a
D-12489 Berlin, Germany
elsasser@mbi-berlin.de

Martin Geller
Institut für Festkörperphysik
Technische Universität Berlin
Hardenbergstr. 36
D-10623 Berlin, Germany
geller@
sol.physik.tu-berlin.de

Daniil Gregoriev
Institut für Physik
Humboldt-Universität zu Berlin
Newtonstr. 15
D-12489 Berlin, Germany
grigoriev.daniil@
physik.hu-berlin.de

Frank Grosse
Institut für Physik
Humboldt-Universität zu Berlin
Newtonstr. 15
D-12489 Berlin, Germany
frank.grosse@
physik.hu-berlin.de

Michael Hanke
Institut für Physik
Martin-Luther-Universität
  Halle-Wittenberg
Hoher Weg 8
D-06120 Halle/Saale, Germany
michael.hanke@
physik.uni-halle.de

Ines Häusler
Institut für Physik
Humboldt-Universität zu Berlin
Newtonstr. 15
D-12489 Berlin, Germany
haeusler@
physik.hu-berlin.de

Fritz Henneberger
Institut für Physik
Humboldt-Universität zu Berlin
Newtonstr. 15
D-12489 Berlin, Germany
henne@physik.hu-berlin.de

Axel Hoffmann
Institut für Festkörperphysik
Technische Universität Berlin
Hardenbergstr. 36
D-10623 Berlin, Germany
hoffmann@
physik.tu-berlin.de

Karl Jacobi
Fritz-Haber-Institut der
  Max-Planck-Gesellschaft
Faradayweg 4-6
D-14195 Berlin, Germany
jacobi@fhi-berlin.mpg.de

Gerold Kießlich
Institut für Theoretische Physik
Technische Universität Berlin
Hardenbergstr. 36
D-10623 Berlin, Germany
kieslich@
physik.tu-berlin.de

Holm Kirmse
Institut für Physik
Humboldt-Universität zu Berlin
Newtonstr. 15
D-12489 Berlin, Germany
holm.kirmse@
physik.hu-berlin.de

Andreas Knorr
Institut für Theoretische Physik
Technische Universität Berlin
Hardenbergstr. 36
D-10623 Berlin, Germany
knorr@
itp.physik.tu-berlin.de

Reinhold Koch
Paul-Drude-Institut für
  Festkörperelektronik
Hausvogteiplatz 5-7
D-10117 Berlin, Germany
koch@pdi-berlin.de

Rolf Köhler
Institut für Physik
Humboldt-Universität zu Berlin
Newtonstr. 15
D-12489 Berlin, Germany
rolf.koehler@
physik.hu-berlin.de

Peter Kratzer
Fachbereich Physik
Universität Duisburg-Essen
Lotharstr. 1
D-47048 Duisburg, Germany
peter.kratzer@uni-due.de

Christoph Lienau
Institut für Physik
Carl von Ossietzky Universität
D-26111 Oldenburg, Germany
christoph.lienau@
uni-oldenburg.de

Andreas Marent
Institut für Festkörperphysik
Technische Universität Berlin
Hardenbergstr. 36
D-10623 Berlin, Germany
marent@
sol.physik.tu-berlin.de

Egor A. Muljarov
Institut für Physik
Humboldt-Universität zu Berlin
Newtonstr. 15
D-12489 Berlin, Germany
muljarov@gpi.ru

Wolfgang Neumann
Institut für Physik
Humboldt-Universität zu Berlin
Newtonstr. 15
D-12489 Berlin, Germany
wolfgang.neumann@
physik.hu-berlin.de

Michael von Ortenberg
Institut für Physik
Humboldt-Universität zu Berlin
Newtonstr. 15
D-12489 Berlin, Germany
michael.von.ortenberg@
physik.hu-berlin.de

Udo W. Pohl
Institut für Festkörperphysik
Technische Universität Berlin
Hardenbergstr. 36
D-10623 Berlin, Germany
pohl@physik.tu-berlin.de

Markus Pristovsek
Institut für Festkörperphysik
Technische Universität Berlin
Hardenbergstr. 36
D-10623 Berlin, Germany
prissi@
gift.physik.tu-berlin.de

Joachim Puls
Institut für Physik
Humboldt-Universität zu Berlin
Newtonstr. 15
D-12489 Berlin, Germany
puls@physik.hu-berlin.de

Michael Rabe
Institut für Physik
Humboldt-Universität zu Berlin
Newtonstr. 15
D-12489 Berlin, Germany
michael.rabe@dca.fi

Marten Richter
Institut für Theoretische Physik
Technische Universität Berlin
Hardenbergstr. 36
D-10623 Berlin, Germany
mrichter@
itp.physik.tu-berlin.de

Wolfgang Richter
Dipartimento di Fisica
Università di Roma 'Tor Vergata'
Via della Ricerca Scientifica
I-00133 Roma, Italy
wolfgang.richter@
tu-berlin.de

Sandra Ritter
Institut für Theoretische Physik
Technische Universität Berlin
Hardenbergstr. 36
D-10623 Berlin, Germany
sandra@
itp.physik.tu-berlin.de

Sven Rodt
Institut für Festkörperphysik
Technische Universität Berlin
Hardenbergstr. 36
D-10623 Berlin, Germany
srodt@physik.tu-berlin.de

Peter Schäfer
Institut für Physik
Humboldt-Universität zu Berlin
Newtonstr. 15
D-12489 Berlin, Germany
peter.schaefer@physik.hu-berlin.de

Andrei Schliwa
Institut für Festkörperphysik
Technische Universität Berlin
Hardenbergstr. 36
D-10623 Berlin, Germany
andrei@
sol.physik.tu-berlin.de

Martin Schmidbauer
Institut für Kristallzüchtung
Max-Born-Str. 2
D-12489 Berlin, Germany
schmidbauer@ikz-berlin.de

Reinhard Schneider
Laboratorium für
  Elektronenmikroskopie
Universität Karlsruhe
D-76128 Karlsruhe, Germany
schneider@
lem.uni-karlsruhe.de

Eckehard Schöll
Institut für Theoretische Physik
Technische Universität Berlin
Hardenbergstr. 36
D-10623 Berlin, Germany
schoell@physik.tu-berlin.de

Matthias Scholz
Institut für Physik
Humboldt-Universität zu Berlin
Hausvogteiplatz 5-7
D-10117 Berlin, Germany
matthias.scholz@
physik.hu-berlin.de

Vitaly Shchukin
Institut für Festkörperphysik
Technische Universität Berlin
Hardenbergstr. 36
D-10623 Berlin, Germany
shchukin@
sol.physik.tu-berlin.de

André Strittmatter
Institut für Festkörperphysik
Technische Universität Berlin
Hardenbergstr. 36
D-10623 Berlin, Germany
strittma@
sol.physik.tu-berlin.de

Tuan Tran-Anh
Institut für Theoretische Physik
Technische Universität Berlin
Hardenbergstr. 36
D-10623 Berlin, Germany
tuan@physik.hu-berlin.de

Andreas Wacker
Department of Physics
University of Lund
Box 118
SE-22100 Lund, Sweden
andreas.wacker@fysik.lu.se

Carsten Weber
Institut für Theoretische Physik
Technische Universität Berlin
Hardenbergstr. 36
D-10623 Berlin, Germany
cw@itp.physik.tu-berlin.de

Momme Winkelnkemper
Institut für Festkörperphysik
Technische Universität Berlin
Hardenbergstr. 36
D-10623 Berlin, Germany
momme@
sol.physik.tu-berlin.de

Roland Zimmermann
Institut für Physik
Humboldt-Universität zu Berlin
Newtonstr. 15
D-12489 Berlin, Germany
zim@physik.hu-berlin.de

1

Thermodynamics and Kinetics of Quantum Dot Growth

Vitaly Shchukin, Eckehard Schöll, and Peter Kratzer

Abstract. Basic processes responsible for the formation of quantum dot (QD) nanostructures occur on a large range of length and time scales. Understanding this complex phenomenon requires theoretical tools that span both the atomic-scale details of the first-principles methods and the more coarse-scale continuum approach. By discussing the time scale hierarchy of different elementary kinetic processes we emphasize several levels of constraint equilibrium of the system and elucidate pathways to reach corresponding stable or metastable states. Main focus is given to the InAs/GaAs material system which is the most advanced one for applying QDs in optoelectronics. First principles calculations of the potential energy surfaces by the density functional theory (DFT) gain the knowledge about potential minima corresponding to the preferred adsorption sites and barriers that govern the rates of diffusion, desorption, and island nucleation in both unstrained and strained systems. Based on these ab initio parameters, kinetic Monte Carlo (kMC) simulations have allowed a detailed theoretical description of GaAs/GaAs and InAs/InAs homoepitaxial growth and elucidated the nucleation and evolution of InAs islands on GaAs. A hybrid approach combining DFT calculations of the surface energies and continuum elasticity theory for the strain relaxation energy has given the equilibrium shape of InAs/GaAs QDs as a function of volume and explained the observed shape transitions. For the ensembles of strained QDs, the Fokker-Planck evolution equation has explained the formation of different types of metastable states in sparse and dense arrays, and the kMC simulations have proposed a tool to distinguish kinetically controlled and thermodynamically controlled QD growth. By continuum elasticity theory in elastically anisotropic semiconductor systems, transitions between vertically correlated and vertically anticorrelated growth of QD stacks has been explained, and yet another approach has been proposed to control the formation of complex nanoworlds.

1.1 Introduction

Data transmission, processing and storage form the backbone of the modern information society. At the beginning of 1990s, a few outstanding discoveries concerning self-organization phenomena on crystal surfaces marked a change of major paradigms in semiconductor physics and technology. The new approach in epitaxy enables fast parallel fabrication of large densities of quantum dots and wires for almost unlimited material combinations and has become the basis for the powerful new branch of nanotechnology [1, 2].

Quantum dots (QDs), nanometer-scale coherent inclusions showing a discrete atom-like electronic spectrum up to room temperature and above, have major advantages for applications as an active medium of optoelectronic devices. These properties have already resulted in many breakthroughs in the field of semiconductor lasers. QD lasers, based on the most-studied three-dimensional In(Ga)As/GaAs QDs, have shown superior performance with respect to conventional quantum well lasers regarding the following parameters [3, 4]: *i*) higher characteristic temperature T_0 referring to the thermal stability of the threshold current density; *ii*) higher robustness against structural defects; *iii*) better beam quality and effective suppression of filamentation, etc. Combining self-organization phenomena and nanoengineering, including a defect-reduction technique that enables selective elimination of dislocated QDs [4, 5], has allowed fabrication of top performance GaAs-based QD lasers for the practical 1300 nm spectral range. Further advances in nanoengineering include a defect-reduction technique in thick metamorphic layers that has allowed blocking the propagation of extended defects in the GaInAs films grown on GaAs substrates to the upper layers. This enables fabrication of high-performance degradation-robust InAs/GaInAs QD lasers for the spectral range of 1460–1500 nm [6–8]. Using two-dimensional InAs/GaAs QDs obtained in the submonolayer deposition mode results in an ultra-high volume density of QDs. This method enables the ultrahigh frequency (20 GB/s) thermally insensitive (up to 85°C) operation of a vertical cavity surface emitting laser (VCSEL) [9–11].

A complementary field of exciting QD applications requires, on the contrary, an ultra-low QD density which allows access to single dots. The QD-based sources of linearly polarized single photons [12, 13] as well as a source of entangled photons [14] have been realized. These sources form an element basis for the emerging field of quantum cryptography and have potential use in future quantum computing systems.

The progress in the area of epitaxial nanostructures and, in particular, the development of quantum dot- and quantum wire-semiconductor technology employing self-organization phenomena requires a profound understanding of the basic physics behind the spontaneous formation of nanostructures. The progress can only be reached and has been actually reached by combined efforts in

(i) Designing growth experiments
(ii) Developing theory of spontaneous nanostructuring
(iii) Performing precise structural and optical characterization of the grown objects

(iv) Developing experimental tools that allow for controlling and tuning of geometrical parameters and electronic spectra of the nanostructures
(v) Optimizing growth techniques to meet device requirements; and
(vi) Fabricating novel nanostructure-based devices that, in fact, fuel the research area

1.1.1 Length and Time Scales

This chapter focuses on the theoretical modeling of the basic processes responsible for the formation of QD nanostructures. The processes underlying nanostructure formation occur over a large range of length and time scales. One can mention deposition/evaporation of atoms on/from the surface, chemical reactions, diffusion hopping of adatoms over the surface, nucleation of islands, attachment/detachment of adatoms to/from islands, evolution of an island ensemble towards equilibrium/ripening, intermixing/segregation, etc.

A complete understanding of the behavior of materials requires theoretical tools that span both the atomic-scale details of the first-principles methods and the more coarse-grained description by a continuum approach. An overview of various computational strategies focused on combining traditional methods—density functional theory,
molecular dynamics, Monte Carlo (MC) methods and continuum description—within a unified multiscale approach is given in [15].

Figure 1.1(a) illustrates physical times and sizes of QD nanostructures that are accessible by different methods within $t_c = 24$ h computational time of a central processor unit of a computer (CPU time). Within an overall CPU time of t_c one can perform n steps of CPU time τ_c, and thereby advance the physical time by t_p in steps of the physical time increment τ_p, i.e. $t_c = n\tau_c$ and $t_p = n\tau_p$. The best known scaling of the numerical effort with the number of atoms N is linear (see, e.g. [16]). The physical time reachable in a given CPU time for a system with characteristic length L and characteristic density $\rho = N/V = NL^{-3}$ is then

$$t_p = Ct_c\tau_p \cdot L^{-3}, \qquad (1.1)$$

with a prefactor C depending on the underlying method for the force evaluation. Different gray levels in Fig. 1.1(a) indicate accessible time scales and sizes that can be modeled by *ab initio* molecular dynamics, or many-body potential (MBP) molecular dynamics, or by *ab initio* or MBP kinetic Monte Carlo (kMC) simulations. The regime of continuum-elasticity theory (CET) is shown as being independent from the physical time as it is primarily a static approach.

Apart from computational feasibility, one should emphasize that each of the theoretical tools has its own advantages. Thus, *ab initio* methods give an exact ultimate answer for a given materials system. On the other hand, continuum theory is able to describe different material systems on a coarse scale, predicting different scenarios that depend on the material parameters. A multiscale modeling of nanostructure formation allows us to combine the advantages of several of the more traditional methods.

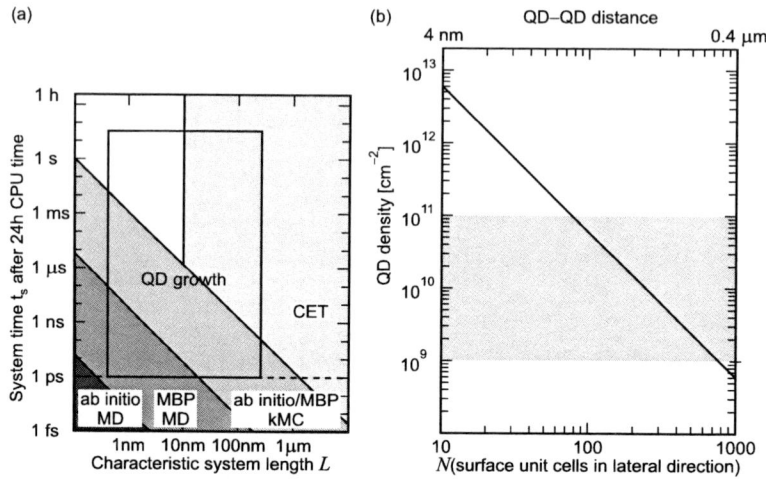

Fig. 1.1. Physical times and sizes of QD nanostructures that are accessible by different methods within a CPU time of $t_c = 24$ h and a time increment of $\tau_p = 1$ fs (**a**). The computational effort of a typical relaxation with a many-body potential is indicated by a dashed line. The regions of QD growth (rectangle) can be deduced from the experimentally observed QD densities (*gray area* in (**b**)), the relation between QD density and QD–QD distance is shown by a solid line in (**b**), and the reported typical formation times of a few seconds. From [16], with permission

1.1.2 Multiscale Approach to the Modeling of Nanostructures

For assessing the thermodynamic stability of semiconductor nanostructures in various stages during their preparation, the thermodynamics of semiconductor surfaces is obviously a very important factor. In this context it is important to emphasize several levels of partial, or *constraint equilibrium*. For example, if the diffusion of atoms over the island facets is faster than the attachment/detachment of adatoms to/from an island, an island can form an *equilibrium island shape* at a given volume. On the time scale where attachment/detachment processes are fast, and migration of adatoms between the islands is slow, a *local surface equilibrium* between an island and the wetting layer (adatom sea) can form. On the time scale where material exchange between islands mediated by the wetting layer occurs, the system evolves towards a *global surface equilibrium* provided the deposition is stopped (growth interruption); evaporation of atoms and intermixing/segregation processes which would occur via bulk migration are negligible.

Most semiconductor surfaces are reconstructed, therefore first-principles calculations using realistic atomic structures are crucial to gain knowledge about the surface free energies and surface stresses of the surfaces and interfaces that are being formed. In the preparation of self-assembled quantum dots by Stranski–Krastanow growth, examples are the wetting layer, the interfaces between a quantum dot and the capping layer, or the side facets of free-standing quantum dots. By combining density-functional theory (DFT) calculations using detailed surface atomic structure with thermodynamic considerations, i.e. by *ab initio* thermodynamics, valuable insight can be gained into the thermodynamics of surface and interface formation.

As a basis for the understanding of possible kinetic limitations that could occur during the growth of quantum dots, it is important to have detailed information not only about thermodynamics, but also about the growth kinetics, both of pure semiconductors and heterostructures. For applying quantum dots in optoelectronic devices, InAs/GaAs is the most advanced material system where numerous experiments have been performed and suitable growth conditions are best understood. Therefore most of the theoretical work presented in the following focuses on this system. Understanding growth kinetics from an atomistic perspective starts with DFT calculations of potential-energy surfaces (PESs) for the relevant molecular processes. From the PES, knowledge can be gained both about the energy minima corresponding to the preferred adsorption sites of Ga or In atoms or arsenic molecules, as well as knowledge about the energy barriers that govern the rates of diffusion, desorption and island nucleation. For homoepitaxy of GaAs on a GaAs(001) substrate, an extensive set of DFT calculations has been performed in order to obtain these data. Although the heteroepitaxial system InAs/GaAs is clearly much more complicated to describe on the atomic level, in this case the atomistic approach using DFT calculations has succeeded in elucidating several structures and processes relevant to quantum dot growth.

1.2 Atomistic Aspects of Growth

1.2.1 Diffusion of Ga Atoms on GaAs(001)

For epitaxy on the frequently used GaAs(001) substrate, the most important surface reconstructions are the $\beta 2(2 \times 4)$ and the $c(4 \times 4)$ reconstructions. These are most stable under the moderately arsenic-rich or very arsenic-rich conditions typical for molecular beam epitaxy at 500–600°C and 400–500°C, respectively. For Ga adatom diffusion on the GaAs(001)$\beta 2(2 \times 4)$ surface, density-functional theory calculations [17] have shown that the surface diffusion is highly anisotropic, with energy barriers of 1.2 eV and 1.5 eV in the [$\bar{1}10$] direction (along the trenches) and the [110] direction, respectively. Remarkably, this study showed that Ga adatoms are able to split the surface arsenic dimers that are part of the $\beta 2(2 \times 4)$ reconstruction, and find their most favorable binding sites in this position. The same holds true for In adatoms on various (001) surfaces [18, 19]. However, other binding sites for In (outside As dimers) exist that are more favorable. For both Ga or In deposition, the ability of these adatoms to break up As dimers is an important atomistic step for growth on arsenic-rich (001) surfaces. Figure 1.2 shows, for the case of In/GaAs(001)$c(4 \times 4)$, the potential energy surface that governs the insertion of the In atom into an As dimer.

1.2.2 Energetics of As$_2$ Incorporation During Growth

In molecular-beam epitaxy (MBE), both As$_4$ or As$_2$ molecules (obtained by cracking As$_4$) can be used as sources of arsenic. Density-functional theory calculations

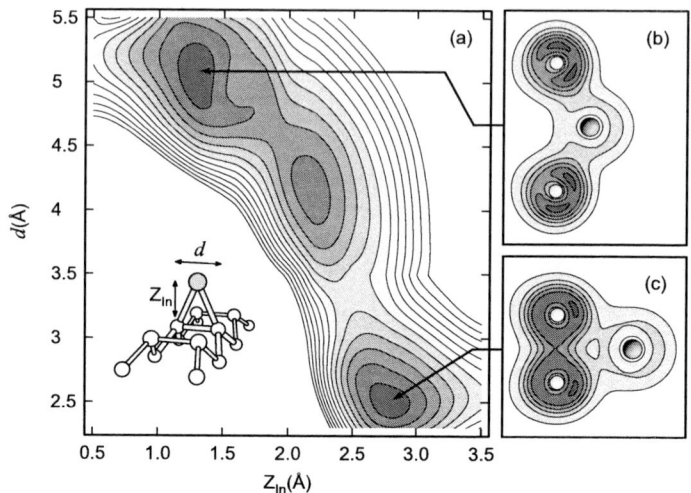

Fig. 1.2. (a) Binding energy of an In adatom interacting with the center As dimer in the GaAs(001)$c(4 \times 4)$ reconstruction as a function of the As–As distance d and the In height above the midpoint of the dimer z_{In}, as indicated in the inset. (**b, c**) Bonding configuration and valence electron density in the plane containing the In adatom and the As dimer for two minima of E_b

performed for the case of As_2 have shown that As_2 molecules bind only weakly on perfect GasAs(001)$\beta 2(2 \times 4)$ or GaAs(001)$c(4 \times 4)$ surfaces. A similar behavior is expected for As_4 molecules, since they are less reactive than As_2. According to these DFT calculations [20, 21], the As_2 molecules find sufficiently strong binding sites (binding energy > 1.5 eV) only after overcoming high energy barriers, or if additional Ga adatoms are already present on the surface, e.g. from previous Ga deposition. The latter finding is in accordance with the experimental observation that the sticking coefficient of an As_2 molecular beam on GaAs(001)$\beta 2(2 \times 4)$ is measurably different from zero only if a Ga evaporation source is operative at the same time [22].

1.2.3 Kinetic Monte Carlo Simulation of GaAs Homoepitaxy

The data about binding energies of Ga adatoms and As_2 molecules and the energy barriers for desorption, diffusion and island nucleation obtained from the DFT calculations were combined in a kinetic Monte Carlo (kMC) study of GaAs homoepitaxy [23]. With this technique, it is possible to overcome the huge gap between the time scales of molecular processes (picoseconds) and the relevant time scales for growth experiments (several seconds up to minutes). As a result of these simulations, it was possible to predict the GaAs island density as a function of temperature and flux [23, 25, 26]. Particularly advantageous is the adatom-density kinetic Monte Carlo method [27] that speeds up the simulations for large sample areas.

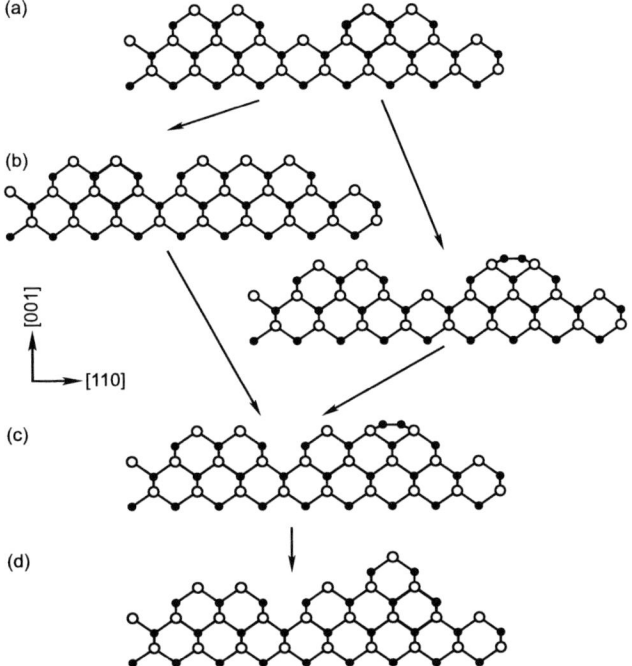

Fig. 1.3. Schematic representation of the growth mechanism in GaAs homoepitaxy on the GaAs(001)-$\beta 2$ surface (*side view*). (**a**) The substrate is corrugated on the atomic scale, with "hills" of As dimers (*open circles, dimer axis perpendicular to the plane of the graph*) and "trenches". Ga atoms (*filled circles*) with dangling bonds appear at the "sidewalls" of the trenches. (**b**) Left: Attachment of material in the trench, yielding a local $\beta 1$-reconstruction; right: Ga dimer as a metastable growth intermediate. (**c**) Formation of a Ga dimer adjacent to locally filled trench. (**d**) The island extends into a new layer after As_2 adsorption

With the kMC simulations, we managed to pinpoint a sequence of processes that give rise to island growth on the GaAs(001)-$\beta 2$ substrate [23]. Some metastable intermediates are shown in Fig. 1.3. By attachment of material in the trenches of the $\beta 2$-reconstruction, two Ga atoms are incorporated into the surface, and a third As dimer is added to the two top-level As dimers of the $\beta 2$-reconstruction, yielding a unit cell with local $\beta 1$-reconstruction (Fig. 1.3(b), left part). If the rate of desorption of arsenic exceeds the rate of adsorption (i.e., when growing at high temperatures and under low As fluxes), mobile Ga adatoms may agglomerate into Ga dimers (Fig. 1.3(b), right part). The formation of these Ga dimers occurs preferentially near sites where material has already filled the trenches, i.e., where the local $\beta 1$-reconstruction has appeared (Fig. 1.3(c)). Finally, if two or more such Ga dimers have formed at sites adjacent in the [$\bar{1}10$] direction, these offer favorable adsorption sites for As_2 molecules. Adsorption of As_2 on these sites results in a small island that extends into a new layer (Fig. 1.3(d)). Experimental studies in combination with modeling have demonstrated that these islands, once they grow larger, redevelop the

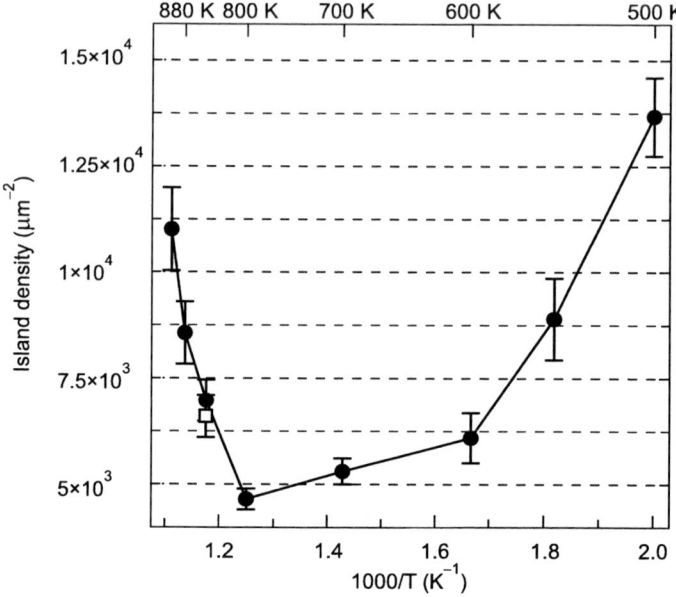

Fig. 1.4. Island number density for GaAs homoepitaxy after deposition for 1 s, using a Ga flux of 0.1 ML/s and an effective As_2 flux of 100 ML/s. (This rather high value is used to take into account the effect of multiple bounces of As_2 molecules on the surface.) The filled circles are results of kMC simulations ([23]), the open square symbol is the measured island density from [24], at a Ga flux of 0.1 ML/s and a (direct) As_2 flux of 0.8 ML/s

well-known $\beta 2$-reconstruction pattern [28]. As a result of this growth mechanism, such islands are elongated along the $[\bar{1}10]$ direction. One can get a good idea about a typical growth sequence by looking at animated snapshots taken in millisecond time intervals of the atomic configurations obtained from the kMC simulations. Such a "movie" is available from the Web [29].

Already after depositing a fraction of a monolayer of GaAs, the island density reaches a saturation value. At the most frequently used growth temperature of 580°C, the kMC simulations predict an island density of 7×10^{11} cm^{-2}, in good agreement with experiments using molecular beam epitaxy followed by an analysis of the island density in scanning tunneling microscopy images, yielding 6.6×10^{11} cm^{-2}, see [24, 30]. Simulations of hypothetical growth at widely different temperatures, 500 K < T < 900 K, assuming that the $\beta 2$-reconstruction could be stabilized over the whole temperature range, show a remarkable temperature dependence of the island density (see Fig. 1.4). While at low temperature the island density decreases with increasing temperature, as expected from nucleation theory, the experimentally relevant growth regime is on the rising slope of the island density at high growth temperatures. This growth regime is a unique feature of a two-component system: At $T \sim 800$ K, desorption of As_2 sets in; therefore the attachment of material to the edges of existing islands (that requires permanent arsenic incorporation) becomes reversible above this temperature. This leads to the observed change in growth kinetics and to an increase in the nucleation of new islands with increasing temperature [23, 26]. Around its minimum at ~800 K, the island density in the kMC simulations is found to be an increasing function of As$_2$ flux, again in agreement with experiment [24].

In a similar spirit, kMC simulations were performed for the interaction of As$_2$ with the InAs(001) surface, and for island growth in homoepitaxy on InAs(001). Interestingly, the (2 × 4)-reconstructed InAs(001) surface undergoes a reversible phase transition between the $\beta2(2 \times 4)$ and the $\alpha2(2 \times 4)$ reconstruction as a function of both temperature and arsenic flux [31, 32]. Both experiment and kMC simulations find that the saturation island density after deposition of a fraction of a monolayer is slightly lower for InAs as compared to GaAs. While InAs epitaxy is done at a lower substrate temperature compared to GaAs epitaxy, the effect of temperature is overcompensated by the generally lower energy barriers for In atoms, and higher rate constants for both In diffusion and As$_2$ desorption on InAs(001). Despite the strong similarities between InAs and GaAs, there are some differences in details of the growth scenario: For InAs, the island density in InAs homoepitaxy was found to be a decreasing function of As$_2$ flux (from 5.6×10^{11} cm^{-2} to 2.5×10^{11} cm^{-2}), both in experiments and in kMC simulations [33].

For the self-assembled growth of quantum dots, heteroepitaxy of InAs on GaAs is most relevant. Therefore, it is important to discuss whether the findings for homoepitaxy are transferable to the situation in heteroepitaxy. Experimentally, it has been demonstrated that MBE growth of InAs on the GaAs(001)-$\beta2(2 \times 4)$ substrate can be performed in such a way that the resulting morphology of two-dimensional (2D) islands is qualitatively similar to homoepitaxy of GaAs [34, 35]. However, this submonolayer growth regime is not typical for quantum dot growth, because the latter requires deposition of 1.5 to 3 monolayers (ML) of InAs. For deposition of 1 ML of InAs or more, a wetting layer with a surface reconstruction substantially different from the substrate is formed (see Sect. 1.2.4). Furthermore, not all of the 2D islands develop into quantum dots, since typically observed quantum dot densities (between 10^{10} cm^{-2} and 10^{11} cm^{-2}) are about one order of magnitude lower than the nucleation density of the 2D islands (between 10^{11} cm^{-2} and 10^{12} cm^{-2}). Nevertheless, it has been found experimentally that the scaling properties of island size distributions carry over from 2D to 3D island formation driven by heteroepitaxial strain [36–38]. While such a scaling behavior is supported by theoretical arguments for 2D islands, the reason for its validity for the 3D case is not yet fully understood by theory.

1.2.4 Wetting Layer Evolution

For MBE growth of InAs quantum dots on GaAs, typically somewhat lower temperatures (400–550°C) are used than for homoepitaxy of GaAs (550–600°C). Under the commonly used As flux, the $c(4 \times 4)$ reconstruction of GaAs(001) appears. The energy barriers for diffusion of single Ga [39] or In [18] atoms on this surface have been determined by density-functional theory. It is found that surface diffusion is close to being isotropic, with energy barriers of 0.94 eV for Ga and 0.67 eV for In atoms.

However, small amounts of In deposited on GaAs(001)$c(4\times 4)$ strongly affect the surface morphology. First "incomplete" $c(4 \times 4)$ reconstruction patterns develop, in which part of the As atoms in the surface As dimers are replaced by cations [40–42], followed by a new phase that starts developing at surface steps [43]. Upon further In deposition, the formation of a wetting layer with both commensurate and incommensurate (1×3) and (2×3) reconstruction patterns has been observed by reflection high-energy electron diffraction (RHEED) measurements [44].

Since growth of InAs quantum dots on GaAs(001) proceeds in the presence of this wetting layer, it is important to understand its atomic structure and the thermodynamic driving force for its formation. With the help of DFT calculations, it is possible to investigate the formation energy of the wetting layer as a function of its thickness (= indium deposition) and of variations of the growth conditions. The latter are reflected in the calculations by the choice of the arsenic chemical potential μ_{As}. Both a segregated film of pure InAs, with either $\beta 2(2 \times 4)$ or $\alpha 2(2 \times 4)$ reconstruction, and formation of an alloyed wetting layer have been considered. For the latter alternative, an As-terminated (2×3) reconstruction was selected, motivated by X-ray diffraction experiments [45] on $In_xGa_{1-x}As$ surface alloy films that observe this reconstruction at $x = 2/3$, along with a triple-period ordering of In and Ga on the cation sublattice sites. For very As-rich conditions, as shown in Fig. 1.5, the alloyed (2×3)-reconstructed wetting layer is found to be most favorable. Both for less As-rich conditions and thicker films, the $\alpha 2(2 \times 4)$ reconstruction is found to be lower in energy, and to extend its range of stability at the expense of the (2×3) reconstruction. The $\beta 2(2 \times 4)$ reconstruction that had been studied previously [46, 47]

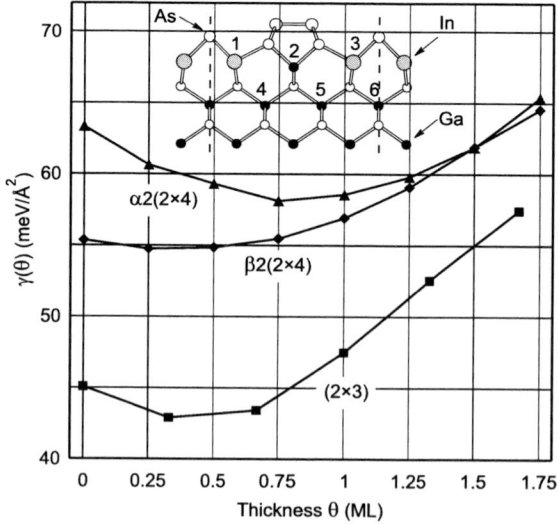

Fig. 1.5. Formation energy γ_f of the InAs wetting layer on GaAs as a function of indium deposition θ at very arsenic-rich conditions, $\mu_{As} = \mu_{As(bulk)}$. The inset shows the cation sites included in the calculation of $\gamma_f(\theta)$ for the (2×3) structure

is found to occur in an intermediate regime of moderately As-rich conditions. The strain energy stored in the wetting layer in all cases shows up as a linear increase of the formation energy with the thickness, as seen in Fig. 1.5.

1.3 Size and Shapes of Individual Quantum Dots

1.3.1 Hybrid Approach to Calculation of the Equilibrium Shape of Individual Quantum Dots

While quantum dots display electronic properties different from bulk materials due to the confinement of electrons and holes on the nanometer scale, their structural and elastic properties are less affected by the physics on the nanoscale. Typical InAs quantum dots of 10 nm to 20 nm base length consist of the order of 10^4 atoms. If an InAs island of this size is grown pseudomorphically on GaAs, the elastic energy introduced both in the island and in the substrate due to the mismatch of the lattice constants of the two materials can be well described by classical elasticity theory. Deviations from the classically expected behavior, e.g. due to surface stress, are limited to relatively small regions of space and can be treated separately if required. This can be done, e.g., by taking into account modifications of the surface energies due to surface stress. Therefore it is possible to separate the total energy associated with the formation of strained free-standing islands into the energy gain from partial strain relief, and the energy cost due to the formation of side facets and edges. Knowing the total formation energy, the shape of such an island can be determined under the assumption of thermal equilibrium, by minimizing the formation energy for a fixed amount of material in the island.

In order to calculate the contributions to the formation energy, a hybrid approach has been devised [48, 49] that allows us to employ specific methods most suitable for calculating each contribution separately. The energy gain due to strain relaxation is calculated according to continuum elasticity theory, e.g. by a finite-element method, while surface energies and surface stresses are calculated using density functional theory. The edge energies can be estimated from DFT calculations as well. They become less important for larger islands, and have therefore been neglected in most studies.

Applying the hybrid approach, the equilibrium quantum dot shape has been determined for InAs on GaAs(001) [48, 49], using low-index facets as boundaries of the InAs islands (see Fig. 1.6(a)). A similar study was performed for InP quantum dots on GaP(001) [50]. These studies showed that the equilibrium shape of quantum dots depends both on the chemical environment during quantum dot growth, and on the size of the dots. For InAs quantum dots grown on GaAs, the shapes predicted by the hybrid approach for very arsenic-rich growth conditions are shown in Fig. 1.6(c,d). For example, the InAs($\bar{1}\bar{1}1$) facet, for which a low-energy As-rich reconstruction exists, becomes very prominent in the quantum dot shape for As-rich conditions (see Fig. 1.6(c,d)). The equilibrium shape results from an optimization process that allows us to continuously change the shape (and thus the fraction of total surface area

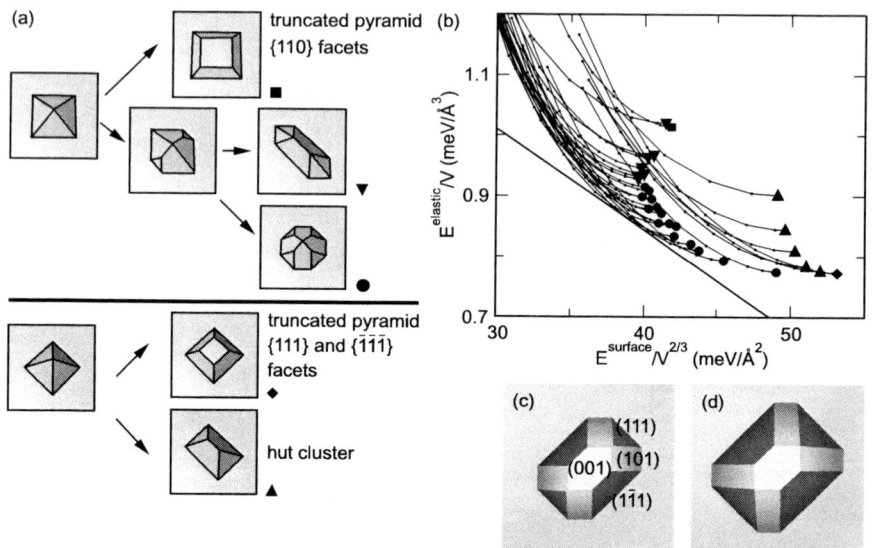

Fig. 1.6. (a) Island shapes for InAs quantum dots on GaAs considered in [49], consisting of low-index facets. The basic shapes are pyramids oriented in two different ways relative to the substrate. Further shapes are generated by cutting off the top of the pyramid, or by cutting off the apices at its base, leading to multifaceted islands. (b) The elastic energy per volume E^{elastic}/V versus the surface energy per area $E^{\text{surface}}/V^{2/3}$ for InAs islands. The symbols refer to the shapes displayed in (a): Square: square-based pyramid with four $\{101\}$ facets. Diamond: square-based pyramid with two $\{111\}$ and two $\{\bar{1}\bar{1}\bar{1}\}$ facets. Triangles up: huts with two $\{111\}$ and two $\{\bar{1}\bar{1}\bar{1}\}$ facets. Triangles down: square-based $\{101\}$ pyramids with $\{\bar{1}\bar{1}\bar{1}\}$ truncated edges. Circles: islands with four $\{101\}$, two $\{111\}$, and two $\{\bar{1}\bar{1}\bar{1}\}$ facets. The small dots denote the corresponding truncated islands that are connected by the full lines. The dashed line is the curve of constant total energy $E^{\text{elastic}} + E^{\text{surface}}$ that selects the equilibrium shape for the volume $V = 2.14 \times 10^5$ Å^3. (c)–(d) The equilibrium shape of a strained coherent InAs island in an As-rich environment at two different volumes, (c) $V \approx 2 \times 10^5$ Å^3 ($\sim 10\,000$ atoms), (d) $V \approx 4 \times 10^5$ Å^3 ($\sim 20\,000$ atoms). From [49], with permission

contributed by each side facet) as a function of quantum dot size. The principle of the optimization process is illustrated in Fig. 1.6(b). For the balance between the energy gain due to elastic relaxation, and the energy cost of forming the side facets, the size of the quantum dot is crucial: Each quantum dot size corresponds to a straight line with a specific slope in the diagram of Fig. 1.6(b), where the dimensionless elastic energy and the surface energy (both normalized with the appropriate power of the quantum dot volume) are plotted on the axes. Varying the size of a quantum dot with given shape shows up in Fig. 1.6(b) as a curve labeled by a specific symbol. The locus where the straight line becomes tangent to the lower envelope of all curves defines the optimum island shape for a given volume. Larger quantum dots correspond to a smaller slope, and therefore touch the envelope at a different point. While the quantum dots grow, the magnitude of the elastic energy relief (which is proportional to the quantum dot volume) increases more strongly, compared to the surface energy contribution. Therefore, large quantum dots show a higher aspect ratio (defined as height-to-base ratio) than small ones. The top facet is predicted to contribute a smaller fraction of the overall surface area for the bigger quantum dots. Again, this can be seen by inspecting the equilibrium shapes in Fig. 1.6(c,d).

For InAs quantum dots grown on GaAs, the typically observed shapes have a smaller aspect ratio (between 0.2 and 0.3) compared to the prediction of the hybrid approach for quantum dots bounded exclusively by low-index facets (aspect ratio 0.3 to 0.5). Only for quantum dots grown with an unusually small growth rate at rather high temperatures [51], a shape has been observed that is qualitatively similar to the one theoretically predicted. This can be interpreted as a hint that even the shape of quantum dots is not fully determined by thermal equilibrium for MBE growth of InAs quantum dots under typical conditions. Another indication for a kinetically determined shape comes from InAs quantum dots grown on the GaAs(113) substrate, where very elongated islands have been observed in STM studies [52]. From the (calculated) surface energies of various InAs surface orientations, which deviate only in a range of $\pm 10\%$ from their mean value of ~ 40 meV/Å^2, one would not expect the equilibrium shape to be so strongly elongated. On the other hand, the shapes predicted by the hybrid approach in [50] for InP quantum dots were found to be in good agreement with those observed for large MOCVD-grown dots [53].

1.3.2 Role of High-Index Facets in the Shape of Quantum Dots

For a more refined understanding of the shape of quantum dots, knowledge about the atomic structure and surface energies of high-index surfaces turned out to be important, since they play a role as side facets of the quantum dots. With this motivation, both the GaAs(114) [54, 55] and the GaAs(2 5 11) [56, 57] surfaces were investigated in combined experimental and theoretical studies. By growing thin films of GaAs on specially cut wafers, it has been demonstrated that both the (114) and the (2 5 11) surfaces are stable under the conditions of molecular beam epitaxy. Structural models for the As-rich GaAs(2 5 11) reconstruction, and for both the Ga-rich $\omega(2 \times 4)$ [54] and As-rich $\alpha(2 \times 4)$ reconstructions [55] of GaAs(114) were proposed on the basis of STM images. Density-functional theory calculations of the surface energies showed that these reconstructions are indeed low in energy (within their respective range of stability). Analogous results were obtained for InAs by calculating the surface energies for these high-index orientations, see Fig. 1.7(a). Comparing the observed STM images for GaAs with calculated images using the Tersoff–Hamann approach, further support for the proposed structural models could be obtained. While the newly proposed surface reconstructions for the (114) and (2 5 11) surfaces fulfill the electron-counting rule, the calculations also demonstrated that this rule becomes less important if the size of the unit cell of the reconstruction is large. As a result, the (137) and (3 7 15) reconstructions of GaAs and InAs, which are built up from similar structural elements (As dimers, three-fold coordinated As and Ga surface atoms) as the (2 5 11) surface, but do not comply with the electron counting rule, are similarly low in surface energy (see Fig. 1.7(a)).

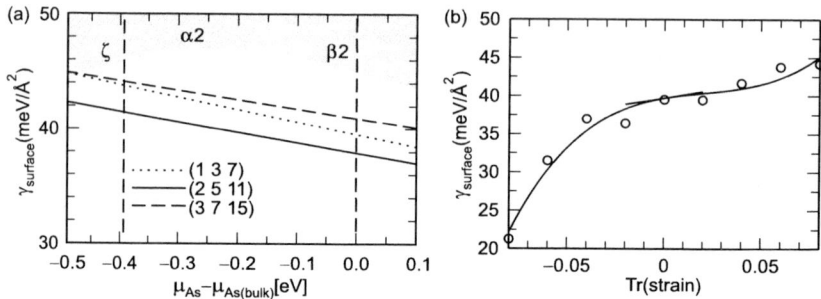

Fig. 1.7. (a) Calculated surface energies as a function of the chemical potential of arsenic for the InAs(2 5 11) surface with three As dimers per unit cell (*solid line*), for the InAs(3 7 15) surface with two As dimers per unit cell (*dashed line*), and for the InAs(137) surface with one As dimer per unit cell (*dotted line*). For comparison, surface energies of the InAs(001) $\beta 2(2 \times 4)$, $\alpha 2(2 \times 4)$ and $\zeta(4 \times 2)$, which have the lowest energies of the presently known reconstructions of InAs(001), are indicated by the shaded region. (b) Surface energy γ of the InAs(137) surface as a function of strain, ε, per unit surface area (of the unstrained material)

1.3.3 Shape Transition During Quantum Dot Growth

The improved understanding of high-index facets, together with the concept of constrained equilibrium in which the base area occupied by a quantum dots remains fixed during the advanced stages of growth, gave rise to a refined scenario for shape evolution: The quantum dots undergo a shape transition after exceeding a specific size, accompanied by an abrupt drop in the chemical potential for the In adatoms to be incorporated in the dot [58]. In theoretical support of this growth scenario, a detailed discussion of the energetic contributions from elastic relaxation and from the surface energy terms within the hybrid approach has been worked out for the sequence of shapes shown in Fig. 1.8. First, flat islands are formed that are bounded predominantly by high-index facets from the families of {137}, {2 5 11} or {3 7 15} orientations. In accordance with the experimental observations [59], the simplest case of {137} facets is considered in the following. The initial formation of flat islands is also supported by theoretical considerations, since it was shown by a DFT calculation [58] (cf. Fig. 1.7(b)) that the surface energy of InAs(137) is lowered considerably by compressive strain, as is present on the facets of small flat quantum dots. In the course of growth, the occurrence of low-index facets from the {110} and {111} families (that form a steeper angle with the substrate) may become energetically favorable at some point. This is plausible because the steeper quantum dot shape allows for an improved strain relaxation. However, the cost of creating the extra low-index facets pays off only if the quantum dot has already reached a certain minimum size. (Recall that strain relaxation dominates over the surface contributions for large quantum dots.) If we consider the base area of the islands as fixed (which is reasonable since it is a slowly varying quantity), it can be shown that the transition between the island shapes is abrupt. Considering equilibrium among the atoms of an island with the constraint of a fixed base area allows us to define a chemical potential

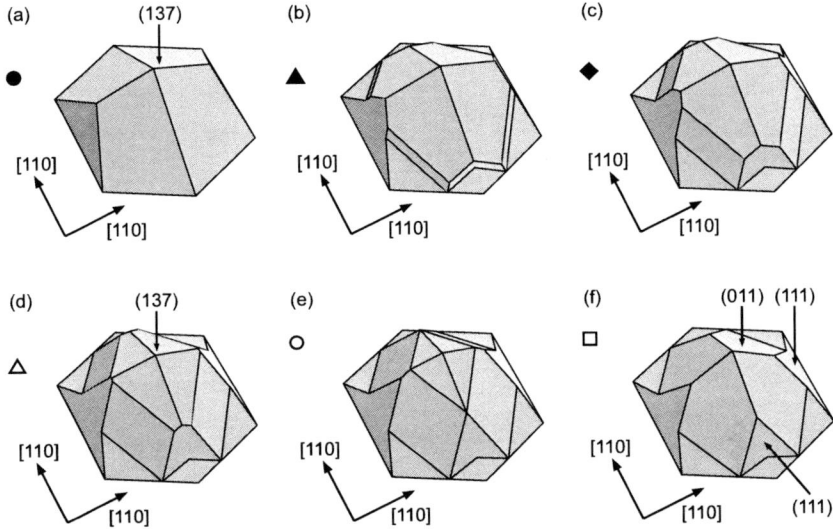

Fig. 1.8. Proposed sequence of shapes for the growth of InAs quantum dots on GaAs(001). Small quantum dots, (**a**), are bounded by {137} and {$\bar{1}\bar{1}1$} facets. Growth proceeds mostly through layer-by-layer growth on the {137} facets; however, the newly grown layers do not make contact with the (001) substrate (**b**). As a result, {110} and {111} facets develop at the lower end of the added layers, giving the quantum dot an increasingly steeper appearance (**c**)–(**e**). Eventually, a sharp tip could possibly develop if growth of the {110} facets extends to the top (**f**)

for the In atoms in the island. Since the free energy of the island as a function of island size has a cusp at the volume of the shape transition, the so-defined chemical potential is discontinuous (see Fig. 1.9). The influx of In adatoms from the wetting layer into the quantum dot is driven by the chemical potential difference between the outside and the inside region. For this reason, the discontinuity of the chemical potential has important consequences for the growth kinetics: it has been shown that it leads to so-called anomalous coarsening [60, 61]. As a result, bimodal island size distributions are to be expected, and the width of the smaller size component may be rather narrow. Hence the anomalous coarsening, in addition to the delaying effect of strain on the growth of larger quantum dots (to be described in the following), could explain the frequently obtained narrow size distribution that is desirable for many applications of quantum dots. Moreover, it is noteworthy that the shape evolution suggested by Fig. 1.8 results in an aspect ratio of the islands of ~ 0.3, in satisfactory agreement with the majority of experiments on InAs quantum dots.

1.3.4 Constraint Equilibrium of Quantum Dots with a Wetting Layer

Another important outcome of first-principles calculations was the active role played by the wetting layer during quantum dot growth. The growing quantum dots consume

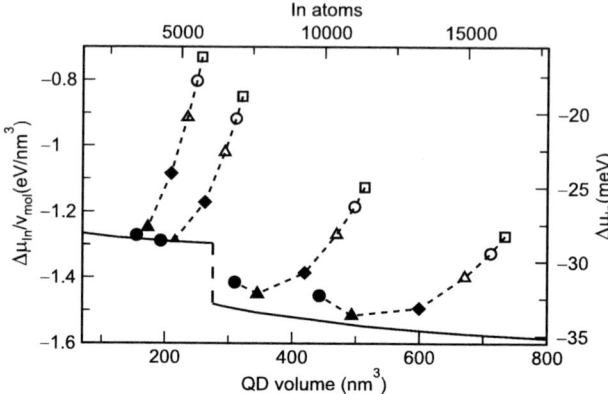

Fig. 1.9. Chemical potential of In atoms in quantum dots of various fixed base areas, as a function of the dot volume. The curves, from upper left to lower right, correspond to quantum dots with a base diameter in [110] direction of 19.8, 24.7, 39.6, and 56.6 nm, respectively. The symbols along the curves refer to the different shapes shown in Fig. 1.8. For quantum dots of small base area, adding material on top of the pyramid (Fig. 1.8(a)) would result in an increase of chemical potential and hence does not occur spontaneously. For quantum dots with a base length larger than 30 nm, however, a transition from the shape Fig. 1.8(a) to Fig. 1.8(b) becomes a spontaneous process accompanied by a lowering of $\Delta\mu_{In}$ (*dotted lines*). The unconstrained chemical potential (*solid line*) shows an abrupt drop at the growth transition

material from the wetting layer. Consequently, the thickness of the wetting layer after completed quantum dot growth is typically smaller than its critical thickness required for the onset of quantum dot nucleation [46, 47]. This phenomenon is clearly observable for SiGe quantum dots on Si, where STM images show a ring-shaped depleted zone around the quantum dots [62]. For InAs quantum dots on GaAs, experimental evidence for material uptake has been found as well (following its theoretical prediction in [47]), both indirectly (from an analysis of the total volume of the islands formed [63]), and directly from inspection of STM images of quantum dots grown near steps [64–66]. For the modeling, this means that another energy contribution stemming from the thinning of the wetting layer has to be added to the energy balance of the hybrid approach, if we are to correctly describe the evolution of a quantum dot ensemble. This third contribution—that adds to the elastic and the surface energy contribution—rises as the quantum dots grow larger (see Fig. 1.10(a)). Consequently, there is a minimum in the energy per volume for an ensemble of identical quantum dots in equilibrium with the wetting layer. In [47], an identical shape (a truncated pyramid) and size of the quantum dots has been assumed for the sake of simplicity. With the density of 3D island nucleation n_{3D} as input, this theory enables us to predict the size reached by the quantum dots for any given amount θ_0 of deposited material. As shown in Fig. 1.10(b), the results are in good agreement with experimental observations.

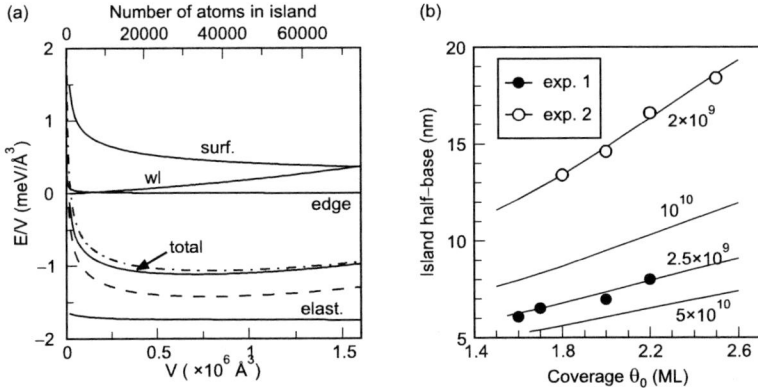

Fig. 1.10. (a) Total energy gain per volume due to island formation for an ensemble of identical islands in equilibrium with a wetting layer, according to the hybrid approach. Various energy contributions (*solid lines*) are shown for an island density of $n_{3D} = 10^{10}$ cm^{-2} and indium deposition $\theta_0 = 1.8$ ML. The dashed line is the total energy gain for $n = 10^{10}$ cm^{-2}, $\theta_0 = 1.5$ ML. (b) Dependence of the final size of the quantum dots, expressed by half of their base length, on the amount of InAs deposited (coverage θ_0) assuming various island densities (labels in cm^{-2}). The experimental values are taken from [67] (●) and estimated from [68] (○)

For unequal sizes of the quantum dot nuclei, the theory needs to be extended to include the kinetics of quantum dot growth. It can be shown (see Sect. 1.4.4) that dense metastable arrays of quantum dots kinetically evolve into a sharply peaked size distribution due to the repulsive interaction between quantum dots. Direct insight into the temporal evolution of quantum dots can be obtained from kMC simulations. This requires us to include the effect of strain on the interaction between islands and adatoms in the energy functional governing the kMC simulations. In this way, it has been possible to follow the evolution of 2D platelets that can be considered the starting point of subsequent 3D island growth [69, 70]. As a result of these simulations, it has been observed that the narrow size distribution of the platelets is accompanied by spatial ordering, where platelets align either in chains [71] or arrays [72] oriented along elastically soft directions. These growth simulations have been used as input for a realistic modeling of transport features like capacitance-voltage characteristics [73]. Another example for the use of kMC simulations is the investigation of Si micro-crystal nucleation inside droplets of liquid indium [74] where micro-crystals of pyramidal shape develop. Transitions between vertically correlated and anticorrelated growth of self-organized quantum dot stacks have also been studied by kMC simulations [75].

Studies of the growth of quantum dots by material transport on the wetting layer require detailed knowledge about the underlying microscopic diffusion processes. With this motivation, the energy barriers for hopping of In adatoms have been investigated by DFT calculations, both for the GaAs(001)$c(4 \times 4)$ surface [18] initially present before In deposition, as well as for an alloyed wetting layer surface, terminated by an In$_{2/3}$Ga$_{1/3}$As layer in (2×3) or (1×3) reconstruction [19].

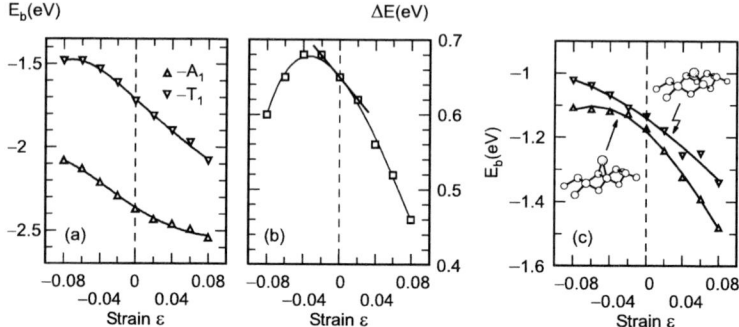

Fig. 1.11. (a) Binding energy E_b as a function of isotropic strain ϵ for an indium adatom at the stable binding site A_1, and at the energy barrier for surface hopping, T_1. (b) Diffusion barrier $\Delta E \equiv E_b(T_1) - E_b(A_1)$ as a function of ε. Full curves on both panels represent least-squares polynomial fits to the calculated points. (c) Binding energy of an indium adatom for the depicted bonding configurations, inside or outside As dimers, as a function of strain

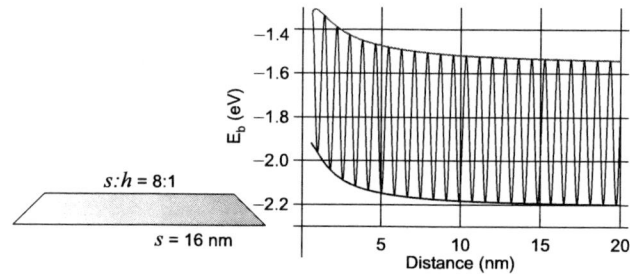

Fig. 1.12. Energy profile (*oscillating curve*) for an In adatom approaching perpendicular to a very long, coherently strained InAs island (of width s and height h) on the $c(4 \times 4)$-reconstructed GaAs(001) surface. Superimposed on the diffusion potential due to the atomic structure of the surface, the strain field in the substrate induced by the island has a repulsive effect that lifts both the binding energies (*thick lower line*) and transition state energies (*thick upper line*) close to the island

It was found that even at low island density, when the quantum dots are far apart and do not interact directly, the strain field induced in the substrate may affect material transport. This, in turn, leads to an indirect interaction between growing quantum dots competing for deposited adatoms. For the example of In diffusion on the GaAs(001)$c(4 \times 4)$ surface, DFT calculations could show that the compressive strain induced by the quantum dot in the surrounding substrate may hinder material transport considerably [18, 25], because compressive strain lowers the binding energy of In adatoms and raises the energy barrier for diffusion (at least for moderately negative strain values, see Fig. 1.11(b)). By combining the DFT results with information about the local strain field around an island obtained from elasticity theory within the flat-island approximation, it is possible to map out the potential energy profile for In diffusion on length scales that are large compared to the atomic scale. As shown in Fig. 1.12, a repulsive wall is built up around the island. If two islands compete for deposited material, this strain effect on the diffusion will allow the smaller island to collect more adatoms and thus to catch up with the larger island when further material is deposited. This mechanism offers an explanation for the narrow island size distributions attainable experimentally even for low island densities that preclude direct elastic interactions between the islands.

1.4 Thermodynamics and Kinetics of Quantum Dot Ensembles

1.4.1 Equilibrium Volume of Strained Islands versus Ostwald Ripening

Growth interruption or annealing is frequently used as a part of the technological process to let a system come to—or at least closer to—equilibrium. Under typical experimental conditions, evaporation of atoms and intermixing/segregation are negligible, and the heteroepitaxial system evolves toward a constraint *surface equilibrium*. This is the equilibrium theory of heteroepitaxial growth. The latter traditionally distinguishes three growth modes: *i*) Frank–van-der-Merwe, or layer-by-layer growth; *ii*) Volmer–Weber, or three-dimensional (3D) island growth; and *iii*) Stranski–Krastanov growth, where a flat wetting layer is formed first and 3D islands are formed when the wetting layer reaches a critical thickness.

Here we focus on the Stranski–Krastanow growth regime, which is realized in semiconductor systems of major interest, like Ge/Si and InAs/GaAs. If the amount of deposited material is below 1 monolayer, an array of monolayer-high, two-dimensional islands forms. For a dilute array of islands, the energy of formation of a single island consisting of N atoms is given by

$$E(N) = -WN + C_1\sqrt{N} - C_2\sqrt{N}\ln(\sqrt{N}). \qquad (1.2)$$

The first term is the binding energy between the atoms in the adsorbate layer, and the second term is the island edge energy due to broken chemical bonds. The third term is the elastic relaxation energy associated with the discontinuity of the surface stress tensor at the island boundaries [76–79].

The energy per atom, $\varepsilon(N) = E(N)/N$ is plotted in Fig. 1.13(a). It always has a minimum at the optimum size $N_0 = \exp[2(C_1/C_2) + 1]$ at which the energy per atom is lower than in a fully ripened island ($N \to \infty$) by the quantity $\varepsilon_0 = C_2 N_0^{-1/2}$, which demonstrates the existence of an equilibrium volume of a 2D island.

Direct experimental proofs of quantum-dot behavior of flat monolayer-high islands have been obtained for 1–2 monolayer-high CdSe insertions in ZnSSe matrix from spot-focused cathodoluminescence studies [80] as well as from photoluminescence studies of the islands in small etched mesas [81]. One of the key advantages of submonolayer QDs is a possibility to obtain arrays of islands with an ultra-small size and ultra-high density. Thus, an array of submonolayer InAs/GaAs QDs has been used as an active medium in vertical-cavity surface emitting lasers (VCSELs) allowing the ultrahigh frequency (20 GB/s) thermally insensitive (up to 85°C) operation [9–11].

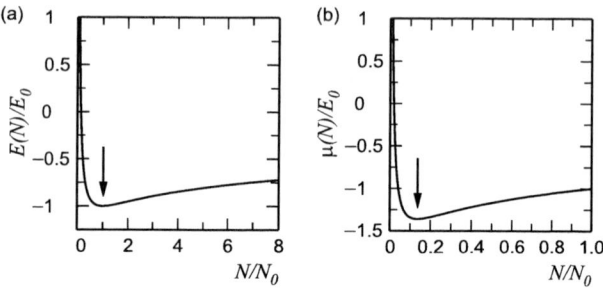

Fig. 1.13. (a) The energy per atom versus the volume of a 2D strained island, $E(N)/E_0$. (b) The chemical potential of a 2D strained island, $\mu(N) = dE/dN$. Arrows point on the characteristic volumes of islands, N_0, and N_1, respectively

If the amount of the deposited material exceeds the critical thickness of the wetting layer, three-dimensional islands form. Let N atoms from the wetting layer form an islands. Then the formation energy of the island can be written as follows [82]

$$E(N) = -\Delta E_{\text{elast}} N + \Delta E_{\text{surf}} N^{2/3} + \Delta E_{\text{edges}} N^{1/3} \\ - \Delta E_{\text{elast}}^{\text{edges}} N^{1/3} \ln(N^{1/3}). \quad (1.3)$$

The first term in (1.3) is the energy of the volume elastic relaxation due to the transition of the material from a highly strained flat wetting layer to a partially relaxed 3D strained island. The second term is the change of the total surface energy of the system due to the formation of tilted facets of the island and disappearance of the wetting layer surface beneath the island. The third term is the short-range contribution to the energy of the edges. The last, fourth term is the energy of the elastic relaxation due to surface stress tensor discontinuity at the edges [83, 84].

The first and the fourth terms in (1.3) are always negative, and the third term is always positive. The second term, the change of the surface energy given by the second term, can be written for a particular square-base pyramidal shape as follows [85]

$$\Delta E_{\text{surf}} = (6 \cot \theta_0 v)^{2/3} \left[\gamma(\theta_0) \sec \theta_0 - \gamma_{\text{WL}} - g_1(\theta_0) \tau \varepsilon_0 - g_2(\theta_0) S \varepsilon_0^2 \right]. \quad (1.4)$$

Here ϑ_0 is the tilt angle of the side facets, v is the unit cell volume, $\gamma(\theta_0)$ is the surface energy of the side facets, γ_{WL} is the surface energy of the wetting layer, ε_0 is the lattice mismatch between the deposited material and the substrate, the third and the fourth terms are strain-induced renormalization terms in the surface energy, τ is a typical value of the surface stress, S is a typical value of the second-order elastic moduli of the surface, and g_1 and g_2 are geometrical factors. A key point is that ΔE_{surf} can be both positive and negative.

In an array of 3D islands, the substrate-mediated elastic interaction energy contributes to the total energy of the island array. To address the question of the minimum energy state of an array of 3D islands on a wetting layer surface, it is convenient to consider an array of equal-volume and equal-shape pyramids. Assume the total number of atoms in the islands is fixed and the island shape is fixed, then seek the minimum energy per atom. The energy per atom versus the island volume N equals

$$\frac{E(N)}{N} = -\Delta E_{\text{elast}} + E_0 \left[\frac{2\alpha}{e^{1/2}} \left(\frac{N_0}{N} \right)^{1/3} - \frac{2}{3} \left(\frac{N_0}{N} \right)^{2/3} \ln\left(\frac{e^{3/2} N_0}{N} \right) \right.$$
$$\left. + \frac{4\beta}{e^{3/4}} \left(\frac{N_0}{N} \right)^{1/2} \right]. \tag{1.5}$$

Here $N_0 = \exp[3(\Delta E_{\text{edges}}/\Delta E_{\text{elast}}^{\text{edges}} + 1/2)]$ is the equilibrium island volume for the particular case $\Delta E_{\text{surf}} = 0$. The value $E_0 = (1/2)\Delta E_{\text{elast}}^{\text{edges}} N_0^{-2/3}$. The value $\alpha = e^{1/2} \Delta E_{\text{surf}}/\Delta E_{\text{elast}}^{\text{edges}}$ is the ratio of the change of the surface energy due to the formation of islands, ΔE_{surf}, and of the contribution of the edges to the elastic relaxation energy. The value β is the ratio of the average interaction energy between the islands, and $\Delta E_{\text{elast}}^{\text{edges}}$, it increases upon coverage q as $q^{3/2}$ [82].

Figure 1.14(a) shows the energy per atom, $E'(N) = E(N)/N + \Delta E_{\text{elast}}$ versus the island volume N in a dilute limit, where the elastic interaction between islands is neglected. For clarity, the energy versus $N^{1/3}$ is plotted. It can be seen that, if $\alpha < 0$, there exists a finite island size corresponding to the minimum of the energy per atom; the minimum is governed by the strain-induced renormalization of the surface energy. An interval of α where $0 < \alpha < 1$, the energy per atom minimum at a finite island size is due to the surface stress relaxation at the island edges. If $\alpha > 1$,

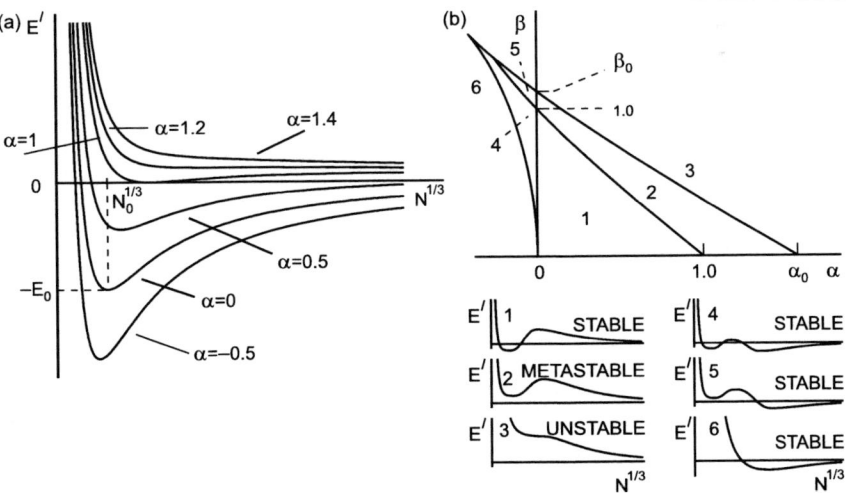

Fig. 1.14. (a) Energy per atom versus island volume in a dilute array of 3D strained islands. (b) Phase diagram of an array of elastically interacting 3D strained islands. The parameter α is the ratio of the change of the surface energy due to the formation of islands, ΔE_{surf}, and of the contribution of the edges to the elastic relaxation energy, $|\Delta E_{\text{elast}}^{\text{edges}}|$. The parameter β is the ratio $E_{\text{inter}}/|\Delta E_{\text{elast}}^{\text{edges}}|$

the island array exhibits a thermodynamic driving force to Ostwald ripening, where all material should be ultimately collected in one huge island. Practically ripening results in the formation of huge islands, where defects form. If the density of islands increases, the elastic interaction between islands mediated by the substrate becomes an important contribution to the energy per atom in the array of islands. This is a positive energy of elastic repulsion that favors ripening. Figure 1.14(b) shows an equilibrium phase diagram. A stability region implies that an array of equal-size islands is the minimum energy state of the system. In a region of metastability a thermodynamic driving force to ripening exists, but an array of equal-size islands may occur as a metastable state. In a region of instability no metastable state is possible, and an array of islands will ripen.

In a real system, the constraint surface equilibrium implies, that islands and the wetting layer can exchange atoms, and the total number of atoms is fixed in the combined system "islands plus wetting layer". The equilibrium phase diagram is more complex but still contains parameter regions of equal-volume islands, where ripening is not thermodynamically favorable [86]. This conclusion persists [87] if one also takes into account the change of the island shape upon volume increase, similar to one discussed earlier in Sects. 1.3.1 and 1.3.2.

By changing control parameters of a system, it is possible to drive an array of islands from a stable state to an unstable one which results in ripening. For III–V heteroepitaxial systems, such control can be realized, e.g., by changing the arsenic pressure in the vapor [88–90]. Existence of thermodynamically stable ensembles of the islands can be confirmed by reversible changes of islands density, volume and the wetting layer thickness upon cycled temperature ramping and cooling [91, 92]. At the same time an array of islands can also be formed as an intermediate state of the ripening process. Experimental tools that allow us to distinguish kinetically dominated arrays of islands from thermodynamically dominated ones are discussed in detail in [87]. One of the tools is described in Sect. 1.4.2.

The minimum energy per atom (MEA) attained for islands of a certain finite volume corresponds to the equilibrium volume at $T = 0$. Upon temperature increase, the equilibrium distribution of islands shows broadening and a shift of the distribution function maximum toward smaller volumes [93]. At higher temperatures, a second maximum evolves in the distribution function due to the gas of single adatoms. Finally, the distribution function maximum corresponding to nanoscale islands disappears and the distribution function becomes monotonically decreasing. A key feature of the equilibrium island distribution that can be observed experimentally is a decrease of the average island volume upon temperature increase.

1.4.2 Crossover from Kinetically Controlled to Thermodynamically Controlled Growth of Quantum Dots

As a relatively narrow volume distribution of the islands can be both a thermodynamically stable state of the system and an intermediate kinetically controlled state, it is of major interest to establish experimental tools that would allow us to distinguish between the two cases.

To address this issue, a kinetic Monte Carlo (kMC) simulation of the formation of two-dimensional strained islands upon growth interruption has been carried out. The growth simulations of [94] use an event-based algorithm applied to a solid-on-solid model with deposition and diffusion as the relevant processes. Diffusion of adatoms occurs on a square lattice by nearest neighbor hopping. Atoms can cross island edges by surmounting a Schwöbel barrier. The relevant energies in our simulations are the binding energy to the surface $E_s = 0.7\,\mathrm{eV}$ and the strength of the $n \leq 4$ nearest neighbor bonds $E_b = 0.3\,\mathrm{eV}$ that influence the time scale for diffusion and island formation, respectively. Existing islands generate an elastic strain field caused by the lattice mismatch. This strain field influences detachment from island boundaries and the motion of adatoms in the vicinity of islands through a position-dependent energy correction term E_{str}.

The hopping rate for a single atom is then given by an Arrhenius law

$$p = \nu \exp\left[-\frac{E_s + nE_b - E_{\mathrm{str}}}{k_B T}\right], \quad (1.6)$$

with the attempt frequency $\nu = 10^{13}\,\mathrm{s}^{-1}$, and the strain energy density has been calculated using Green's function approach and the normalized per atomic bond.

The simulations have been performed on a lattice of 250×250 atomic sites. As an initial step a coverage of 4% was deposited randomly on the surface at a flux of 1 ML/s. Every 0.01 s a histogram of the island size distribution is recorded. To reduce the noise, ten simulations with different initial conditions have been used to calculate an average. Figure 1.15 displays the simulation results for the temporal evolution of an average island size $< \sqrt{N} >$ for temperatures of $T = 675\,\mathrm{K}$, $700\,\mathrm{K}$ and $725\,\mathrm{K}$.

From Fig. 1.15 it is evident that in the initial stages of island growth the size distribution is clearly kinetically controlled. At lower temperatures many small islands are formed whereas at higher temperatures fewer and larger islands emerge.

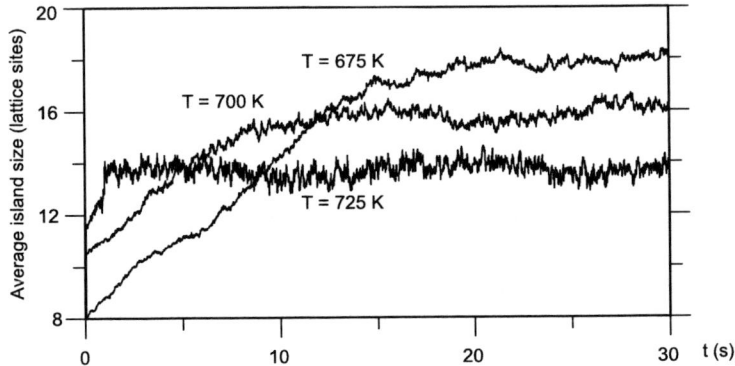

Fig. 1.15. Temporal evolution of the average island size for $T = 675\,\mathrm{K}$, $T = 700\,\mathrm{K}$, and $T = 725\,\mathrm{K}$. Monte Carlo simulations have been performed on a 250×250 grid and averaged over ten runs with the same set of parameters

On short time scales of a few seconds the islands do not grow by a considerable amount and the scaling of the island size with temperature is still kinetically controlled.

At lower temperatures the nucleation of islands is the dominant process. Since the adatom mobility is low, the density of single adatoms increases fast during the deposition and pairs of atoms are formed randomly. Those act as nuclei for islands. Consequently, one observes many small islands for low temperatures.

With increasing temperature the adatoms become more and more mobile. A single adatom in a hot system can travel a long distance until it finds an existing island to which it will attach. The adatom density therefore decreases and nucleation of new islands is suppressed. The final spatial configuration in the kinetically controlled regime exhibits few large islands.

Right after the deposition, however, the islands begin to equilibrate. The system is now in an intermediate state between kinetically and thermodynamically controlled growth conditions. The slow increase of island sizes and a crossover of the average island size for systems of different temperatures is a characteristic of this regime.

For low temperatures the growth process is the slowest and the higher the temperature becomes the faster the islands approach their average equilibrium size. Once the equilibrium size distribution is reached, the average island diameter remains constant. In the course of equilibration the islands in the low temperature systems continue to grow until they reach their equilibrium size at an average diameter above that of the islands of the hotter systems, as is expected for islands grown under equilibrium conditions.

From the results of the thermodynamic theory and of the kinetic simulations, an experimental tool emerges that allows us to distinguish between kinetically controlled islands and thermodynamically controlled islands. If, upon increase of the substrate temperature, an average number of atoms in the islands, or the average island volume increases, the island formation is controlled predominantly by the growth kinetics. If, with increasing of the substrate temperature, the average island volume decreases, the island formation is controlled predominantly by thermodynamics. For submonolayer islands, the height is fixed, and the island volume is proportional to the square of the island lateral size, thus the above arguments apply to the dependence of the lateral size vs. temperature.

This developed method has been used to analyze an array of submonolayer (0.3 ML) InAs islands on GaAs substrate. Two structures of InAs islands deposited at two different substrate temperatures, 350°C and 480°C have been capped by GaAs and studied by cross-section high-resolution transmission electron microscopy (HRTEM) and photoluminescence (PL) [95], see also [2]. Cross-sectional HRTEM images processed by using DALI (digital analysis of lattice images) evaluation program (see, e.g. [96]) have revealed local map of the vertical lattice parameter, showing higher values of the latter in the regions with higher In content. HRTEM has indicated smaller islands in the sample grown at the higher temperature. PL spectra referring the entire ensemble of islands have revealed a blue shift of the QD peak in the sample grown at a higher temperature confirming a larger volume of the islands. Thus, the higher the substrate temperature during island formation, the smaller the average volume, which indicates that the array of strained islands is predominantly thermodynamically controlled.

1.4.3 Tunable Metastability of Quantum Dot Arrays

The energetics of an array of strained islands, namely the existence of an energetically preferred island volume, makes a strong impact on the ensemble evolution kinetics even at island volumes far below the preferred one. The behavior is similar for 2D (monolayer-high) islands and for 3D islands with a strain-renormalized surface energy. For simplicity 2D islands have been considered in detail. For a dilute array of islands, the energy of formation of a single island consisting of N atoms is given by (1.2). The chemical potential of the island is given by $\mu = \mathrm{d}E(N)/\mathrm{d}N$, which is displayed in Fig. 1.13(b) and has a minimum at $N_1 = N_0/e^2 \approx 0.14 N_0$.

The time evolution of the island volume distribution function $f(t, N)$ can be described by a Fokker–Planck equation,

$$\frac{\partial}{\partial t} f(t, N) = -\frac{\partial}{\partial N} J(t, N), \tag{1.7}$$

where the flux in the configurational space of island volume is

$$J(t, N) = \omega(N) \left[\frac{\bar{\mu} - \mu(N)}{k_\mathrm{B} T} f(t, N) - \frac{\partial}{\partial N} f(t, N) \right]. \tag{1.8}$$

Here the typical case is considered where the kinetics are limited by attachment (detachment) process to (from) island perimeter and the kinetic factor $\omega(N) = N^{1/2}$. The first term in (1.8) is conventionally referred to as the drift contribution and is proportional to the difference between the chemical potential $\mu(N)$ of an island having N atoms and that of the adatom sea, $\bar{\mu}$. The second term is known as the diffusion contribution. The time-dependent, mean field chemical potential $\bar{\mu}$ is determined by the mass conservation law which, in the absence of nucleation of new islands, yields the relationship between the island flux $J(t, N)$ integrated over all islands, and the deposition flux Φ,

$$\int_0^\infty J(t, N) \, \mathrm{d}N = \Phi. \tag{1.9}$$

Figures 1.16(a) and (b) show the results [61, 97] of the numerical solution of (1.7)–(1.9) under conditions of annealing, or growth interruption. The temperature is defined in units of $\Theta = C_2 \sqrt{N_0}/k_\mathrm{B}$ corresponding to the energy of an island containing N_0 atoms of energy per atom ε_0. In all calculations we use $T/\Theta = 10^{-3}$ and $C_1/C_2 = 3.27$ diving $N_0 = 5.1 \times 10^3$. The initial distribution of islands is below the value N_1, and the initial evolution shown in Fig. 1.16 is governed by the negative gradient in $\mu(N)$ and, as such, is similar to conventional capillarity-driven ripening, in which the chemical potential decreases monotonically. Small islands with a chemical potential above $\bar{\mu}$ shrink, and large islands, with a chemical potential below $\bar{\mu}$, grow. The island distribution broadens and evolves to larger volumes as shown in Fig. 1.16(b).

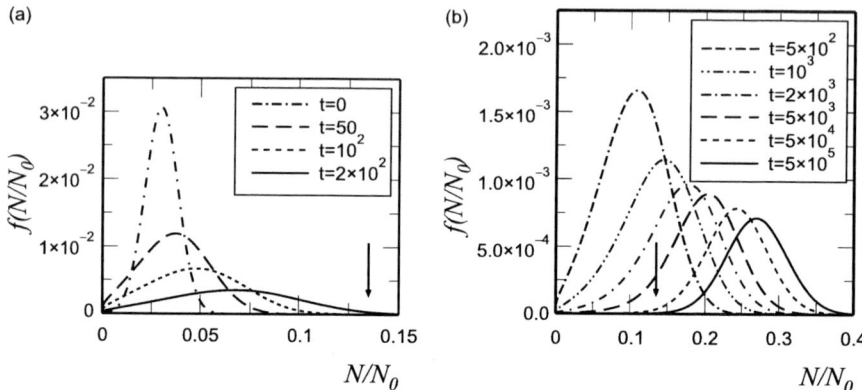

Fig. 1.16. (a) Early and (b) late evolution of the island volume distribution function $f(t, N)$ with time (scaled units) for a Gaussian initial distribution located below N_1. The solid arrow indicates the chemical potential minimum at N_1

Figure 1.16(b) shows the same island distribution evolved to larger times. The distribution passes the volume of the minimum chemical potential N_1 and at later times (say, at $t = 5 \times 10^5$) lies essentially above N_1. The following evolution of the island distribution is governed by a positive chemical potential gradient, $d\mu/dN > 0$. Then small islands with a chemical potential below $\bar{\mu}$ grow, and large islands, with a chemical potential above $\bar{\mu}$, shrink. The drift term in (1.7) is responsible to *inverse ripening* narrowing the island distribution. The narrowing is opposed by the diffusion term, and the two contributions to the flux J nearly cancel each other. This cancellation of terms produces a metastable state which can effectively suppress the evolution of $f(t, N)$ on experimentally relevant time scales.

Metastable states should play an important role in any kinetics of the surface nanostructures provided that positive gradients in chemical potential exist with respect to island size. Therefore, if material is deposited such that $\bar{\mu}$ is only slightly enhanced by the deposition flux, the island size distribution will be dominated by the metastable state at that particular coverage. In regions of positive chemical potential gradient, the size distribution can then be tuned to a desired size by depositing material for a required time. This "close-to-equilibrium" procedure is illustrated in Fig. 1.17 in which a uniform initial distribution of the islands between $N = 0$ and $N = 0.07N_0$ ($< N_1$) is chosen to mimic the early stages of island nucleation. For times smaller than $t = 10^3$, the deposition has little effect and the evolution is similar to the annealing case (Fig. 1.17(a)). The distribution becomes metastable as it passes above N_1 and further flux causes the Gaussian-like state to drift to higher volumes with only a slight broadening of the profile (Fig. 1.17(b)).

Within the interval of island volumes with a positive chemical potential gradient leads, many metastable states are possible with a different average volume. Depositing material by a small flux or using a two-stage (growth and annealing) procedure, it is possible to effectively tune metastable states. Similar behavior occurs for 3D

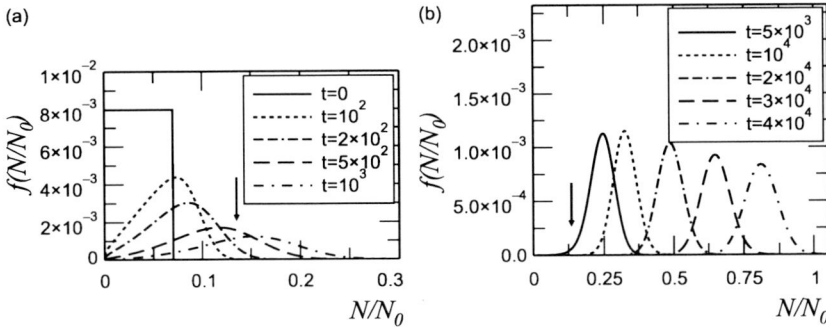

Fig. 1.17. (a) Early and (b) late evolution of the island volume distribution function $f(t, N)$ with time (scaled units) for a uniform initial distribution located below N_1 in the presence of a small deposition flux. The solid arrow indicates the chemical potential minimum at N_1

strained islands with the strain-renormalized surface energy, which have the minimum energy per atom volume [98]. The predicted kinetics of the ensemble evolution is in agreement with the experimental observations of Fe islands on NaCl(001) substrate (magnetic quantum dots) performed by atomic force microscopy (AFM) [99].

1.4.4 Evolution Mechanisms in Dense Arrays of Elastically Interacting Quantum Dots

In dense arrays of islands, where the average distance between islands is comparable with their base length, the elastic interaction energy between the islands becomes important. As discussed Sect. 1.4.1, the elastic interaction energy on the average is the positive energy of elastic repulsion. Therefore it reduces the domain of the phase diagram [82] corresponding to a stable array of the equal-size islands and favors ripening, or coarsening.

Elastic interaction takes on a completely different role with local variations in the strain field in dense arrays of islands. This role can be most easily elucidated on a model example of an array of strained islands having a conical shape [61, 100]. The total energy of the island array is then given by the formula

$$E_{\text{total}} = \frac{3}{2}\beta V_a^{2/3} - wJ\tan\vartheta\, V_a + \frac{w}{\pi}\sum_{b\neq a}\frac{V_a V_b}{R_{ab}^3} F\left(\frac{\rho_a}{R_{ab}}, \frac{\rho_b}{R_{ab}}\right), \qquad (1.10)$$

where the first term represents the additional surface energy associated with the island formation. The second term is the elastic self-relaxation energy of the island, and the third term represents the elastic interaction energy between the ath island and all other islands. Here V_a and ρ_a are the respective volume and the base radius of the ath cone and R_{ab} is the distance between the basal centers of islands a and b. In the coefficient $w = (1+\nu)(1-\nu)^{-1}Y\varepsilon_0^2$, where ε_0 is the lattice mismatch between the deposit and the substrate, Y and ν are Young's modulus and Poisson's ratio, respectively, assuming both materials are equal, and the numerical factor $J = 1.059$.

The coefficient $\beta = 2\pi^{1/3}3^{-1/3}(\cot\vartheta)^{2/3}(\Delta\Gamma)$, where $\Delta\Gamma = \gamma(\vartheta)\sec\vartheta - \gamma(0)$ and $\gamma(\vartheta)$ and $\gamma(0)$ are the surface energies of the tilted surface of the island and of the flat surface of the wetting layer, respectively. The role of local elastic interactions is more pronounced in the case where $\Delta\Gamma > 0$ so that, even without elastic interactions, islands would tend to ripen to reduce the overall surface energy.

Now we can discuss the evolution of an array of islands in the regime of attachment-limited kinetics. Islands can attach atoms from the adatom sea and detach atoms which go to the adatom sea. The local flux of atoms to/from each island is governed by the local difference between the chemical potential of an atom, μ_a and the adatom sea, $\bar{\mu}$,

$$\frac{dV_a}{dt} = V_a^{1/3}[\bar{\mu} - \mu_a], \quad (1.11)$$

where the chemical potential of an island is defined as $\mu_a = \partial E_{\text{total}}/\partial V_a$. The elastic interaction energy between the two conical islands has been calculated exactly in [102]. The key feature of this energy is that its contribution to the chemical potential *diverges* as two islands nearly contact each other.

To emphasize the impact of the elastic interaction on the evolution of a dense array of islands, an initially hexagonal array of identical conical islands was considered, and an initial perturbation in island volumes and positions was introduced. Figures 1.18(a) and 1.18(b) compare the evolution of island radii without strain and with strain included. The evolution of a dense array of islands without strain is dominated by the coalescence events, when two islands touch each other and form a single island by adding up their volumes. The coalescence events manifest themselves as abrupt jumps in island radii. When the strain is included, no abrupt jumps occur. The latter means that no coalescence on impact occur in an array of strained islands, and the coarsening proceeds via the Ostwald ripening mechanism. Thus,

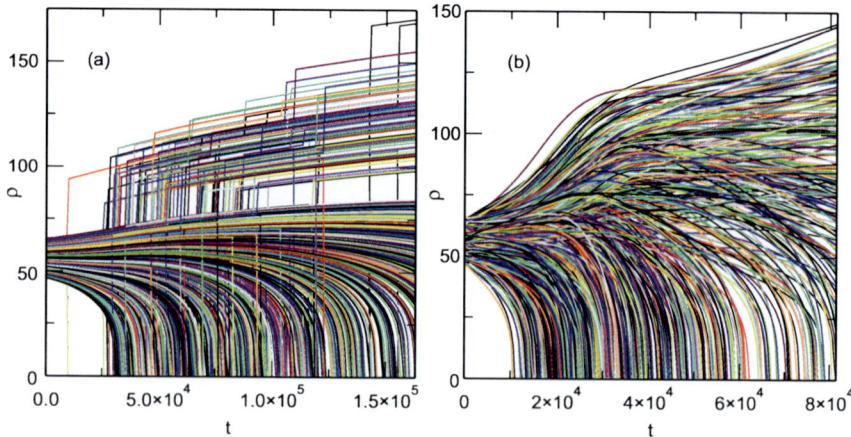

Fig. 1.18. Time evolution for the scaled radii ρ for the initial arrays shown in (a) for zero strain and (b) with strain included

a dramatic increase of the chemical potential of a strained island in the vicinity of another strained island due to local strain fields suppresses the coalescence. In addition, the elastic interaction alters the temporal behavior of the average density and radius of the islands, which is in agreement with the experimental data on dense arrays of GeSi/Si islands [103].

In ultra-dense arrays, where islands nearly touch each other, a metastable state may occur, which is stable against small perturbations in island volumes and positions [100, 104]. Thus, surprisingly, a positive elastic energy of elastic repulsion between islands can stabilize ultra-dense arrays.

1.5 Quantum Dot Stacks

Combining on the one hand self-organization phenomena providing quantum dots and quantum wires, and nanoengineering on the other, allows us to extend substantially the variety of nanostructures as well as to control their geometry and electronic spectrum. A straightforward way is to grow a multisheet array of quantum dots separated by spacers. This eventually results in the formation of multisheet arrays of quantum dots.

The growth of typical multilayer arrays of quantum dots, e.g. InAs/GaAs, Ge/Si, etc. often exhibits the formation of vertically correlated arrays, wherein the dots of the next layer are located above the dots of the previous layer forming vertical columns (see, e.g. [2] and the references therein). The vertical columns with a thin spacer result in electronic coupling of the neighboring QDs.

The advantages of using QD stacks include *i*) an enhanced volume density of QDs, resulting in a higher performance of optoelectronic devices; *ii*) engineering of electronic states in coupled QDs; and *iii*) enhancement of size homogeneity and spatial ordering.

The theoretical understanding of spatial correlation in QD stacks is based on the arguments of constraint thermodynamics. As bulk diffusion in semiconductors at typical growth temperatures is negligibly slow, the structure of the buried islands does not change during the formation of every next sheet of the islands. The islands on the surface are formed in the static strain field created by the buried islands. It has been conventionally believed that the islands of the next layer are formed at the positions of the minimum elastic energy density. This approach has explained the formation of vertical columns of QDs [105] and the enhancement of ordering in the upper layers [106].

1.5.1 Transition between Vertically Correlated and Vertically Anticorrelated Quantum Dot Growth

Surprisingly, the growth of multisheet arrays of CdSe submonolayer islands separated by ZnSe spacers has revealed anticorrelated growth, wherein the islands of the next sheet form over the spacings between the islands of a previous sheet [107].

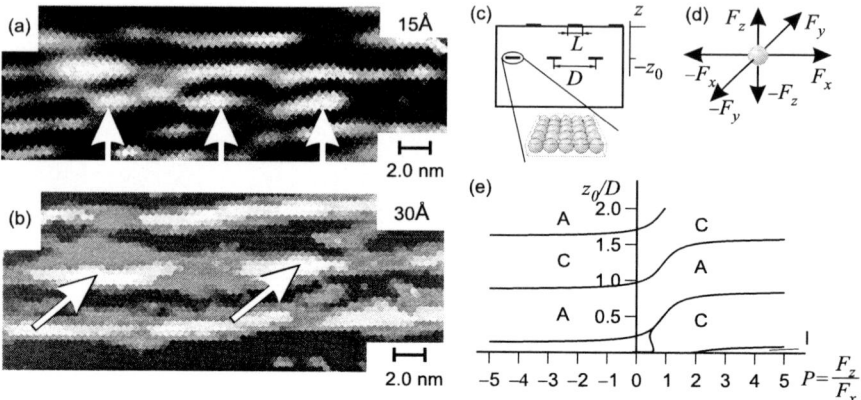

Fig. 1.19. Structure of multilayer arrays of 2D islands. (**a**)–(**b**) Processed cross-sectional high-resolution transmission electron microscopy (HRTEM) images of CdSe/ZnSe multilayer island structure. White arrows: a guide to the eye. (**a**) Spacer thickness 15 Å, vertically correlated array. (**b**) Spacer thickness 30 Å, vertically anticorrelated array. (**c**) Schematic structure of a two-layer array. (**d**) A single atom in a buried island as a dipole force source. (**e**) Diagram of a two-layer structure of 2D islands, "C" refers to a correlated array, "A" stands for an anticorrelated array, and "I" means an intermediate one

This phenomena has been explained and the theory of the formation of a multi-sheet array of the islands has been developed [108]. The key point is the elastic anisotropy of the semiconductors, characterized by the dimensionless parameter $\xi = (c_{11} - c_{12} - 2c_{44})/c_{44}$, where c_{11}, c_{12}, and c_{44} are elastic moduli in the Voigt notation. In Si, Ge and III–V semiconductors having the zinc-blend, structures are cubically anisotropic medium, with $\xi < 0$ and elastically soft directions $\langle 100 \rangle$. As a consequence of the anisotropy, the elastic strain field created on the surface by a periodic array of buried islands (Fig. 1.19(c)) on the surface exhibits an oscillatory decay as a function of the spacer thickness, z_0. If each of the atoms constituting a buried island is presented as an elastic dipole (Fig. 1.19(d)), the resulting structure of the surface islands depends on two parameters, the ratio of the spacer thickness to the lateral period, z_0/D, and the parameter F_z/F_x which characterizes an elementary elastic dipole. Figure 1.19(e) shows a diagram of the parameter regions, in which the two layers are either correlated or anticorrelated. For a small spacer thickness an intermediate structure is also possible. The main feature of Fig. 1.19(e) is that, with an increase of the spacer thickness, the relative structure of a double-sheet array changes from correlated to anticorrelated and back.

After the theory of transition between correlated and anticorrelated arrangement had been developed, the same material system of CdSe submonolayer islands in ZnSe matrix has revealed vertical correlation at a thinner spacer (Fig. 1.19(a)) [109]. Transition between vertically correlated and vertically anticorrelated arrangements has an important implication on the optical properties of the nanostructures. In the case of a small spacer thickness and vertically correlated arrangement, the wave function of a localized exciton is cigar-like extended in the vertical direction; the photoluminescence from the coupled quantum dots is TM-polarized [110]. If the islands in the neighboring sheets are anticorrelated, no electronic coupling between the dots occurs and the excitons are localized in separate dots. The wave function is then extended in the lateral direction, and the photoluminescence is TE-polarized.

1.5.2 Finite Size Effect: Abrupt Transitions between Correlated and Anticorrelated Growth

Further theoretical studies—focused on the elastic interaction of buried and surface point inclusions via an elastically anisotropic matrix [111]—have shown that the minimum interaction energy occurs at a certain angle of inclination α with respect to the vertical direction, whereas $\alpha = 19°$ for Si, $\alpha = 25°$ for GaAs, and $\alpha = 33°$ for ZnSe. However, these values do not explain the experimental observations of anticorrelated arrangement of GaAlAs QDs in GaAs matrix at much larger angles of inclinations [112].

The exact theoretical consideration of the arrangements in stacks of 3D QDs was carried out in [113]. Figure 1.20 emphasizes a drastic difference in the elastic strain fields created on the surface by an array of buried point-like QDs, on the one hand, and an array of finite-size QDs on the other hand, as opposed to a single point-like QD [111]. Figure 1.20(a) shows the minima (black) and the maxima (white) of the elastic interaction energy E in the relaxed surface for a single buried point-like QD. Figures 1.20(b)–(d) refer to a periodic quadratic array of QDs with a lateral spacing of $l = 25$ l.s. (lattice sites). In a neighborhood of approximately $l/2$ around each QD the elastic properties are governed by the single point defect. Outside this region, the strain fields are overlapping. Starting with a spacer thickness of $d \approx 14$ l.s., the inclination angle α increases more and more until at $d \approx 17$ l.s. the two minima induced by two neighboring QDs meet and form a flat double minimum exactly in the middle between these QDs (Fig. 1.20(c)). With further increasing spacer thickness, the position of the double minimum remains stable and unchanged for a considerable range of d values. Finally, at $d = 31$ l.s. the minimum starts moving back to a position vertically above the QDs.

The transition of the minima from a position close to vertically above the QDs to in between the QDs is also visible in the elastic energy density profiles at $d = 10$ l.s. and $d = 25$ l.s. in Fig. 1.20(d). This indicates a transition from correlated to anticorrelated growth. From Fig. 1.20(c) one infers an inclination angle for the alignment of QDs of about $\alpha \approx 45°$ for GaAs which is in reasonable agreement with the experimentally observed value of 50° [112].

Figures 1.20(e)–(g) illustrate the influence of the shape and finite volume of the QDs. A square array of pyramids, each with a base length of 20 l.s. × 20 l.s. and a height of 5. l.s. positioned with a lateral distance of $l = 40$ l.s. is considered. Compared with an array of point-like elastic defects, there are significant differences. The inclination angle α remains unchanged. But the finite volume of the Qds results in a large range of spacer thicknesses d where the minimum is directly located above the buried structures. At a spacer thickness of about 20 l.s. the energy minimum

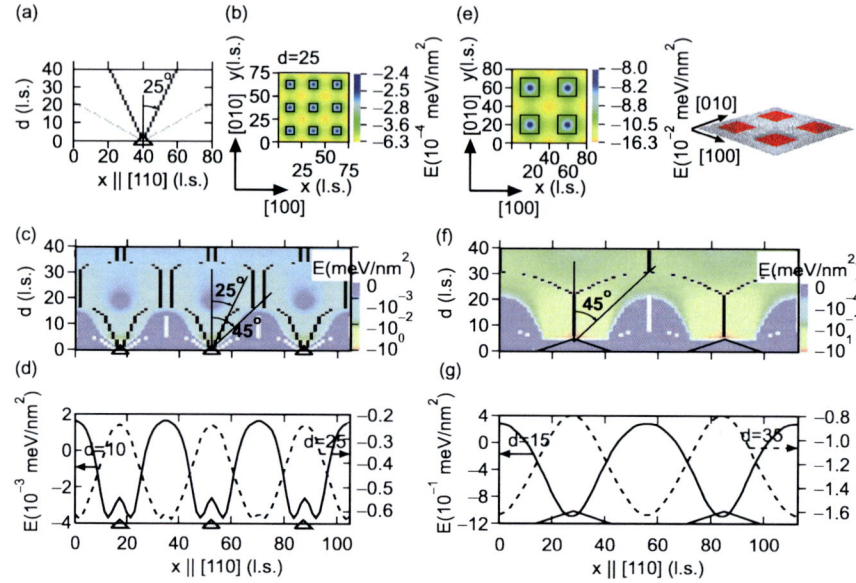

Fig. 1.20. Elastic strain field of buried strained islands in a GaAs matrix. The figure represents the strain field of one single point-like inclusion (**a**), a periodic array of point defects (**b**)–(**d**), and a periodic array of finite-size pyramids (**e**)–(**g**). (**a**), (**c**) and (**f**) show the minima (*black*) and maxima (*white*) of the elastic energy E per unit area in the [110] direction of the relaxed surface vs. spacer thickness d. (**c**) and (**f**) show $E < 0$ additionally in gray scale coding. (**b**) The density plot E in the [100]–[010] surface plane at $d = 25$. (**d**) E for $d = 10$ and $d = 25$. (**e**) The density plot at $d = 35$ (inset: 3D view). (**g**) E for $d = 10$ and $d = 35$. The positions and sizes of the buried islands are indicated by triangles [(**a**), (**c**), (**d**), (**f**), and (**g**)] and by boxes [(**b**) and (**e**)], respectively. All lengths are in units of the lattice parameter $a_{GaAs} = 0.565$ nm.

splits and moves towards the anticorrelated positions between the QDs. Due to the finite size effects, the transition from the correlated to anticorrelated growth happens much more abruptly than in an array of point defects (Fig. 1.20(c)). Such an abrupt transition is indeed observed experimentally [112]. Also the double minimum of the energy profile is replaced by a single minimum (Fig. 1.20(g)). Note that the transition occurs at larger spacer thickness d for the array of QDs with a larger lateral distance of $l = 40$ l.s., and this ratio should scale for arrays with even larger lateral distance at fixed angle α. This is consistent with the experiment [112], where the transition is observed at a spacer thickness of 150 monolayers for an in-plane nearest-neighbor distance of 80–100 nm.

1.5.3 Reduction of a Size of a Critical Nucleus in the Second Quantum Dot Layer

The arguments of constraint thermodynamics that favor the formation of the QDs in the minima of the elastic energy density have allowed us to explain the spatial

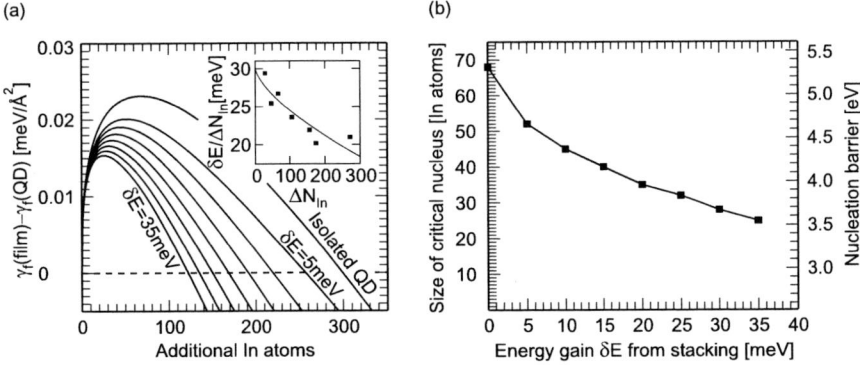

Fig. 1.21. The energy gain δE due to stacking arrangement reduces both the nucleation barriers and the size of the critical nucleus and leads to preferred nucleation above the overgrown QDs. The value of δE depends on QD shape, spacer thickness, and lateral positions with respect to overgrown QD (Fig. 1.19, Fig. 1.20)

correlations and anticorrelations in QD stacks. This approach, however, does not quantify the energetic gain for nucleation in the elastic energy density minima as compared to other nonpreferred sites. To judge the impact of the strain tensor on the surface arising from the overgrown QD as compared to possible kinetic effects, a quantitative calculation has been undertaken [114, 115] of the size reduction of the critical nucleus and lowering the nucleation barrier in the preferred nucleation site.

To consider the nucleation barrier, we focus again on the InAs/GaAs system and compare the formation energies γ_f of homogeneous InAs films of increasing thickness with QDs of increasing volume on a wetting layer of a constant thickness. The formation energy of the upper QD is lowered in the strain field

$$\gamma_f(\text{QD}, N, x, y) \approx \frac{E(N)}{A} + \delta E_{\text{elast}}(N, x, y) \cdot \frac{N}{A}. \qquad (1.12)$$

Here $\Delta E(N)$ is the energy of the free-standing QD with no overgrown QD beneath given by (1.3), N is the number of In atoms in a QD, A is the area of the referred simulation cell, where one QD per cell is assumed. The volume dependence of $\delta E_{\text{elast}} \cdot N$ is well described by a proportionality to the base area of the upper QD (inset in Fig. 1.21(a)) that senses the strain due to the overgrown QD beneath. With this result, the formation energies of hut-shaped QDs for different energy gains due to the stacking arrangement can be calculated. The obtained difference between the formation energy of a homogeneous film and a hut-shaped QD for different energy gains due to different stacking arrangements is shown in Fig. 1.21(a). The energy gain reduces the energy barrier for nucleation $\Delta \gamma_c$ and the size of the critical nucleus N_c.

Numerical calculations have revealed a linear relationship between the two y axes in Fig. 1.21(b). The critical nucleus of a free-standing QD of approximately 70 In atoms is reduced by the elastic interaction in the QD stack by up to a factor of 3 to approximately 25 In atoms (Fig. 1.21(b)) in the investigated range of energy gains δE. The energy barrier for forming the critical nucleus is lowered from $\approx 5.3 \, \text{eV}$

to ≈ 3.5 eV. The resulting increase in the nucleation rate above the overgrown QDs can thus quantitatively explain the strong tendency of the experimentally observed correlations in QD stacks. Note that the above change of $\Delta\gamma_c$ by ≈ 2 eV would increase the nucleation rate at a temperature of 750 K by several orders of magnitude if we assume that the rate is proportional to $\exp(-\Delta\gamma_c/k_B T)$. A weakening of the preference for perfect correlations can be due to a reduced energy gain because of, e.g., intermixing of Ga and In, or the formation of defects.

1.6 Summary and Outlook

Multiscale modeling of semiconductor nanostructures' self-organized formation has provided us with a deep insight into this broad class of exciting phenomena. Based on combination of density functional theory, multibody potentials, kinetic Monte Carlo simulations, and static continuum elasticity theory, an understanding of key aspects of the quantum dot growth has been gained.

Conclusions have been theoretically established and confirmed by careful analysis of experimental data in the following aspects of quantum dot growth.
Our investigations into quantum dots as self-organized nanostructures have revealed that:

- Gross features of quantum dot evolution are captured theoretically by constraint thermodynamic equilibrium.
- Cross-over from nucleation kinetics to thermodynamics occurs in early stages of 2D island growth.

From our study of the shape of free-standing (MBE-grown) quantum dots, we learned that:

- Shape is dominated by high-index (low-energy) facets.
- Transition to dome shape occurs due to incomplete facet growth.

Finally, our look at in-plane ordering of quantum dots led us to conclude that:

- Substrate-mediated elastic repulsion between QDs leads to ordering.
- Kinetics in dense arrays of elastically interacting QDs hinders coalescence on impact in favor of Ostwald ripening.
- Kinetics in ultra-dense arrays of elastically interacting QDs defies ripening.

Regarding 3D ordering in QD stacks:

- Alternation of vertical correlation and anti-correlation can be explained by elastic anisotropy of the matrix.

A few aspects of QD growth are still poorly understood and represent real challenges for further theoretical investigations. First, these are alloy-based QDs, like GaInAs/GaAs, SiGe/Si, etc. An alloy manifests itself not only as a material with a reduced lattice mismatch with respect to the substrate, but also represents an additional degree of freedom connected with a possible strongly inhomogeneous alloy composition profile. The impact of possible alloy phase separation on QD formation has been pointed out in a few theoretical works (see, e.g. [116, 117]) though the understanding is far from being complete.

Second, the overgrowth (capping) of strained islands that can impose a strong change on island volume, shape, and alloy composition, is very poorly understood. Atomistic simulations of strain relaxation at different stages of InAs/GaAs QD overgrowth [118] may be considered as a first step, but more extended studies are needed.

Third, and of great technological importance, is the physics behind the defect-reduction technique, allowing efficient manipulation with threading dislocations, allowing their bending and complete blocking thus enabling growth of practically dislocation-free layers on top of heavily dislocated films [4]. Development of an adequate theory is mandatory for further technology optimization and for bringing the first examples of optoelectronic devices for the spectral region around 1550 nm on GaAs substrate to a required level for practical applications.

The authors acknowledge support from the Deutsche Forschungsgemeinschaft in the frame of the Sonderforschungsbereich 296 "Wachstumskorrelierte Eigenschaften niederdimensionaler Halbleiterstrukturen".

References

1. D. Bimberg, M. Grundmann, N.N. Ledentsov, *Quantum Dot Heterostructures* (Wiley, Chichester, 1998)
2. V.A. Shchukin, N.N. Ledentsov, D. Bimberg, *Epitaxy of Nanostructures* (Springer, Berlin, 2003)
3. N.N. Ledentsov, in *Future Trends in Microelectronics: The Nano Millennium*, ed. by S. Luryi, J. Xu, A. Zaslavsky (Wiley, New York, 2002), p. 195
4. N.N. Ledentsov, A.R. Kovsh, V.A. Shchukin, S.S. Mikhrin, I.L. Krestnikov, A.V. Kozhukhov, L.Y. Karachinsky, M.V. Maximov, I.I. Novikov, Y.M. Shernyakov, I.P. Soshnikov, A.E. Zhukov, E.L. Portnoi, V.M. Ustinov, D. Gerthsen, P. Bhattacharya, N.D. Zakharov, P. Werner, F. Hopfer, M. Kuntz, D. Bimberg, in *Semiconductor and Organic Optoelectronic Materials and Devices*, ed. by C.-E. Zah, Y. Luo, S. Tsuji. Proceedings of SPIE, vol. 5624 (SPIE, Bellingham, 2005), p. 335
5. H.Y. Liu, S.L. Lew, T. Badcock, D.J. Mowbray, M.S. Skolnick, S.K. Ray, T.L. Choi, K.M. Groom, B. Stevens, F. Hasbullah, C.Y. Jin, M. Hopkinson, R.A. Hogg, Appl. Phys. Lett. **89**, 073113 (2006)
6. N.N. Ledentsov, A.R. Kovsh, A.E. Zhukov, N.A. Maleev, S.S. Mikhrin, A.P. Vasil'ev, E.S. Semenova, M.V. Maximov, Y.M. Shemyakov, N.V. Kryzhanovskaya, V.M. Ustinov, D. Bimberg, Electron. Lett. **39**, 1126 (2003)
7. L.Y. Karachinsky, T. Kettler, N.Y. Gordeev, I.I. Novikov, M.V. Maximov, Y.M. Shernyakov, N.V. Kryzhanovskaya, A.E. Zhukov, E.S. Semenova, A.P. Vasil'ev, V.M. Ustinov, N.N. Ledentsov, A.R. Kovsh, V.A. Shchukin, S.S. Mikhrin, A. Lochmann, O. Schulz, L. Reissmann, D. Bimberg, Electron. Lett. **41**, 478 (2005)
8. Z. Mi, P. Bhattacharya, J. Yang, Appl. Phys. Lett. **89**, 153109 (2006)
9. F. Hopfer, A. Mutig, M. Kuntz, G. Fiol, D. Bimberg, N.N. Ledentsov, V.A. Shchukin, S.S. Mikhrin, D.L. Livshits, I.L. Krestnikov, A.R. Kovsh, N.D. Zakharov, P. Werner, Appl. Phys. Lett. **89**, 141106 (2006)

10. N.N. Ledentsov, F. Hopfer, A. Mutig, V.A. Shchukin, A.V. Savel'ev, G. Fiol, M. Kuntz, V.A. Haisler, T. Warming, E. Stock, S.S. Mikhrin, A.R. Kovsh, C. Bornholdt, A. Lenz, H. Eisele, M. Dähne, N.D. Zakharov, P. Werner, D. Bimberg, Proc. SPIE **6468**, 47 (2007); Proc. SPIE Photonics West, January 21–26, 2007, San Jose, CA
11. N.N. Ledentsov, V.A. Shchukin, D. Bimberg, in *Future Trends in Microelectronics: Up the Nano Creek*, ed. by S. Luryi, J. Xu, A. Zaslavsky (Wiley, New York, 2007)
12. A.J. Bennett, D.C. Unitt, P. Atkinson, D. Ritchie, A.J. Shields, Opt. Express **13**, 50 (2005)
13. A. Lochmann, E. Stock, O. Schulz, F. Hopfer, D. Bimberg, V.A. Haisler, A.I. Toropov, A.K. Bakarov, A.K. Kalagin, Electron. Lett. **42** (2006)
14. R.M. Stevenson, R.J. Young, P. Atkinson, K. Cooper, D.A. Ritchie, A.J. Shields, Nature **439**, 179 (2006)
15. D.D. Vvedensky, J. Phys.: Condens. Matter **16**, R1537 (2004)
16. T. Hammerschmidt, Ph.D. thesis, Fritz-Haber-Institut der Max-Planck Gesellschaft. http://w3.rz-berlin.mpg.de/~hammer/Diss/Diss-ThomasHammerschmidt.pdf
17. A. Kley, P. Ruggerone, M. Scheffler, Phys. Rev. Lett. **79**, 5278 (1997)
18. E. Penev, P. Kratzer, M. Scheffler, Phys. Rev. B **64**, 085401 (2001)
19. E. Penev, S. Stojković, P. Kratzer, M. Scheffler, Phys. Rev. B **69**, 115335 (2004)
20. C.G. Morgan, P. Kratzer, M. Scheffler, Phys. Rev. Lett. **82**, 4886 (1999)
21. P. Kratzer, C.G. Morgan, M. Scheffler, Phys. Rev. B **59**, 15246 (1999)
22. C.T. Foxon, B.A. Joyce, Surf. Sci. **64**, 293 (1977)
23. P. Kratzer, M. Scheffler, Phys. Rev. Lett. **88**, 036102 (2002)
24. G.R. Bell, M. Itoh, T.S. Jones, B.A. Joyce, Surf. Sci. **423**, L280 (1999)
25. P. Kratzer, E. Penev, M. Scheffler, Appl. Phys. A **75**, 79 (2002)
26. P. Kratzer, E. Penev, M. Scheffler, Appl. Surf. Sci. **216**, 436 (2002)
27. L. Mandreoli, J. Neugebauer, R. Kunert, E. Schöll, Phys. Rev. B **68**, 155429 (2003)
28. M. Itoh, G.R. Bell, A.R. Avery, T.S. Jones, B.A. Joyce, D.D. Vvedensky, Phys. Rev. Lett. **81**, 633 (1998)
29. MPEG-videos of an animated growth sequence can be downloaded, for two different viewpoints, from ftp://ftp.aip.org/epaps/phys_rev_lett/E-PRLTAO-87-031152/isl-gaas-front.mpg ftp://ftp.aip.org/epaps/phys_rev_lett/E-PRLTAO-87-031152/isl-gaas-top.mpg
30. B.A. Joyce, D.D. Vvedensky, T.S. Jones, M. Itoh, G.R. Bell, J.G. Belk, J. Cryst. Growth **201/202**, 106 (1999)
31. F. Grosse, W. Barvosa-Carter, J.J. Zinck, M. Wheeler, M.F. Gyure, Phys. Rev. Lett. **89**, 116102 (2002)
32. F. Grosse, M.F. Gyure, Phys. Rev. B **66**, 075320 (2002)
33. F. Grosse, W. Barvosa-Carter, J.J. Zinck, M.F. Gyure, Phys. Rev. B **66**, 075321 (2002)
34. G.R. Bell, T.J. Krzyzewski, P.B. Joyce, T.S. Jones, Phys. Rev. B **61**, R10551 (2000)
35. J. Krzyzewski, P.B. Joyce, G.R. Bell, T.S. Jones, Surf. Sci. **482–485**, 891 (2001)
36. Y. Ebiko, S. Muto, D. Suzuki, S. Itoh, K. Shiramine, T. Haga, Y. Nakata, N. Yokoyama, Phys. Rev. Lett. **80**, 2650 (1998)
37. T.J. Krzyzewski, P.B. Joyce, G.R. Bell, T.S. Jones, Phys. Rev. B **66**, 201302(R) (2002)
38. F. Arciprete, E. Placidi, V. Sessi, M. Fanfoni, F. Patella, A. Balzarotti, Appl. Phys. Lett. **89**, 041904 (2006)
39. J.G. LePage, M. Alouani, D.L. Dorsey, J.W. Wilkins, P.E. Blöchl, Phys. Rev. B **58**, 1499 (1998)
40. A. Ohtake, J. Nakamura, S. Tsukamoto, N. Koguchi, A. Natori, Phys. Rev. Lett. **89**, 206102 (2002)
41. A. Ohtake, N. Koguchi, Appl. Phys. Lett. **83**, 5193 (2003)

42. E. Penev, P. Kratzer, M. Scheffler, Phys. Rev. Lett. **93**, 146102 (2004)
43. M.C. Xu, Y. Temko, T. Suzuki, K. Jacobi, Surf. Sci. **580**, 30 (2005)
44. J.G. Belk, C.F. McConville, J.L. Sudijono, T.S. Jones, B.A. Joyce, Surf. Sci. **387**, 213 (1997)
45. M. Sauvage-Simkin, Y. Garreau, R. Pinchaux, M.B. Véron, J.P. Landesman, J. Nagle, Phys. Rev. Lett. **75**, 3485 (1995)
46. L.G. Wang, P. Kratzer, M. Scheffler, N. Moll, Phys. Rev. Lett. **82**, 4042 (1999)
47. L.G. Wang, P. Kratzer, N. Moll, M. Scheffler, Phys. Rev. B **62**, 1897 (2000)
48. E. Pehlke, N. Moll, A. Kley, M. Scheffler, Appl. Phys. A **65**, 525 (1997)
49. N. Moll, M. Scheffler, E. Pehlke, Phys. Rev. B **58**, 4566 (1998)
50. Q. Liu, N. Moll, M. Scheffler, E. Pehlke, Phys. Rev. B **60**, 17008 (1999)
51. G. Costantini, C. Manzano, R. Songmuang, O.G. Schmidt, K. Kern, Appl. Phys. Lett. **82**, 3194 (2003)
52. Y. Temko, T. Suzuki, P. Kratzer, K. Jacobi, Phys. Rev. B **68**, 165310 (2003)
53. K. Georgsson, N. Carlsson, L. Samuelson, W. Seifert, L.R. Wallenberg, Appl. Phys. Lett. **67**, 2981 (1995)
54. J. Márquez, P. Kratzer, L. Geelhaar, K. Jacobi, M. Scheffler, Phys. Rev. Lett. **86**, 115 (2001)
55. J. Márquez, P. Kratzer, K. Jacobi, J. Appl. Phys. **95**, 7645 (2004)
56. L. Geelhaar, J. Márquez, P. Kratzer, K. Jacobi, Phys. Rev. Lett. **86**, 3815 (2001)
57. L. Geelhaar, Y. Temko, J. Márquez, P. Kratzer, K. Jacobi, Phys. Rev. B **65**, 155308 (2002)
58. P. Kratzer, Q.K.K. Liu, P. Acosta-Diaz, C. Manzano, G. Costantini, R. Songmuang, A. Rastelli, O.G. Schmidt, K. Kern, Phys. Rev. B **73**, 205347 (2006)
59. J. Márquez, L. Geelhaar, K. Jacobi, Appl. Phys. Lett. **78**, 2309 (2001)
60. F.M. Ross, J. Tersoff, R.M. Tromp, Phys. Rev. Lett. **80**, 984 (1999)
61. D.E. Jesson, T.P. Munt, V.A. Shchukin, D. Bimberg, Phys. Rev. Lett. **92**, 115503 (2004)
62. U. Denker, O.G. Schmidt, N.Y. Jin-Phillipp, K. Eberl, Appl. Phys. Lett. **78**, 3723 (2001)
63. P.B. Joyce, T.J. Krzyzewski, G.R. Bell, T.S. Jones, S. Malik, D. Childs, R. Murray, Phys. Rev. B **62**, 10891 (2000)
64. T.J. Krzyzewski, P. Joyce, G.R. Bell, T.S. Jones, Surf. Sci. **517**, 8 (2002), cf. Fig. 6
65. E. Placidi, F. Arciprete, V. Sessi, M. Fanfoni, F. Patella, A. Balzarotti, Appl. Phys. Lett. **86**, 241913 (2005)
66. M.C. Xu, Y. Temko, T. Suzuki, K. Jacobi, Surf. Sci. **589**, 91 (2005)
67. A. Polimeni, A. Patanè, M. Capizzi, F. Martelli, L. Nasi, G. Salviati, Phys. Rev. B **53**, R4213 (1996)
68. J.M. Moison, F. Houzay, F. Barthe, L. Leprince, E. André, O. Vatel, Appl. Phys. Lett. **64**, 196 (1994)
69. E. Schöll, S. Bose, Sol. State Electron. **42**, 1587 (1998)
70. M. Meixner, R. Kunert, S. Bose, E. Schöll, V.A. Shchukin, D. Bimberg, E. Penev, P. Kratzer, in *Proc. 25th International Conference on the Physics of Semiconductors (ICPS-25)*, Osaka 2000, ed. by N. Miura, T. Ando (Springer, Berlin, 2001), p. 381
71. M. Meixner, E. Schöll, M. Schmidbauer, R. Köhler, Phys. Rev. B **64**, 245307 (2001)
72. M. Meixner, R. Kunert, E. Schöll, Phys. Rev. B **67**, 195301 (2003)
73. R. Wetzler, R. Kunert, A. Wacker, E. Schöll, New J. Phys. **6**, 81 (2004)
74. M. Block, R. Kunert, E. Schöll, T. Boeck, T. Teubner, New J. Phys. **6**, 166 (2004)
75. M. Meixner, E. Schöll, Phys. Rev. B **67**, 121202(R) (2003)
76. V.I. Marchenko, A.Y. Parshin, Zh. Eksp. Teor. Fiz. **79**, 257 (1980) [Sov. Phys. JETP **52**, 129 (1980)]
77. A.F. Andreev, Pis'ma Zh. Eksp. Teor. Fiz. **32**, 654 (1980) [JETP Lett. **32**, 640 (1980)]
78. V.I. Marchenko, Pis'ma Zh. Eksp. Teor. Fiz. **33**, 397 (1981) [JETP. Lett. **33**, 381 (1981)]

79. D. Vanderbilt, Surf. Sci. **268**, L300 (1992)
80. R. Engelhardt, V. Türck, U.W. Pohl, D. Bimberg, J. Cryst. Growth **184/185**, 311 (1998)
81. T. Kümmell, R. Weigand, G. Bacher, A. Forchel, Appl. Phys. Lett. **73**, 3106 (1998)
82. V.A. Shchukin, N.N. Ledentsov, P.S. Kop'ev, D. Bimberg, Phys. Rev. Lett. **75**, 2968 (1995)
83. A.F. Andreev, Zh. Eksp. Teor. Fiz. **80**, 2042 (1980) [Sov. Phys. JETP **53**, 1063 (1981)]
84. V.I. Marchenko, Zh. Eksp. Teor. Fiz. **81**, 1141 (1981) [Sov. Phys. JETP **54**, 605 (1981)]
85. V.A. Shchukin, N.N. Ledentsov, M. Grundmann, P.S. Kop'ev, D. Bimberg, Surf. Sci. **352–354**, 117 (1996)
86. I. Daruka, A.-L. Barabási, Phys. Rev. Lett. **79**, 3708 (1997)
87. V.A. Shchukin, D. Bimberg, Rev. Mod. Phys. **71**, 1125 (1999)
88. N.N. Ledentsov, M. Grundmann, N. Kirstaedter, O. Schmidt, R. Heitz, J. Böhrer, D. Bimberg, V.M. Ustinov, V.A. Shchukin, P.S. Kop'ev, Z.I. Alferov, S.S. Ruvimov, A.O. Kosogov, P. Werner, U. Richter, U. Gösele, J. Heydenreich, in *Proceedings of the 7th International Conference on Modulated Semiconductor Structures*, Madrid, Spain, July 1995. Solid State Electron. **40**, 785 (1996)
89. K. Ozasa, Y. Aoyagi, Y.J. Park, L. Samuelson, Appl. Phys. Lett. **71**, 797 (1997)
90. F. Heinrichsdorff, A. Krost, D. Bimberg, A.O. Kosogov, P. Werner, Appl. Surf. Sci. **123/124**, 725 (1998)
91. N.N. Ledentsov, V.A. Shchukin, D. Bimberg, V.M. Ustinov, N.A. Cherkashin, Y.G. Musikhin, B.V. Volovik, G.E. Cirlin, Z.I. Alferov, Semicond. Sci. Technol. **16**, 502 (2001)
92. R. Leon, J. Wellman, X.Z. Liao, J. Zuo, D.J.H. Cockayne, Appl. Phys. Lett. **76**, 1558 (2000)
93. V.A. Shchukin, N.N. Ledentsov, D. Bimberg, in *Self-Organized Processes in Semiconductor Alloys—Spontaneous Ordering, Composition Modulation, and 3-D Islanding*, ed. by D.M. Follstaedt, B.A. Joyce, A. Mascarenhas, T. Suzuki, Symp. Proc., vol. 583 (Mat. Res. Soc., Pittsburgh, 2000), p. 23
94. M. Meixner, E. Schöll, V.A. Shchukin, D. Bimberg, Phys. Rev. Lett. **87**, 236101 (2001)
95. V.A. Shchukin, N.N. Ledentsov, A. Hoffmann, D. Bimberg, I.P. Soshnikov, B.V. Volovik, V.M. Ustinov, D. Litvinov, D. Gerthsen, Phys. Stat. Sol. (b) **224**, 503 (2001)
96. A. Rosenauer, S. Kaiser, T. Reisinger, J. Zweck, W. Gebhardt, D. Gerthsen, Optik (Stuttgart) **102**, 63 (1996)
97. T.P. Munt, D.E. Jesson, V.A. Shchukin, D. Bimberg, Phys. Rev. B **75**, 085422 (2007)
98. T.P. Munt, D.E. Jesson, V.A. Shchukin, D. Bimberg, Appl. Phys. Lett. **85**, 1784 (2004)
99. Z. Gai, B. Wu, J.P. Pierce, G.A. Farnan, D. Shu, M. Wang, Z. Zhang, J. Shen, Phys. Rev. Lett. **89**, 235502 (2002)
100. V.A. Shchukin, D. Bimberg, T.P. Munt, D. Jesson, Phys. Rev. Lett. **90**, 076102 (2003)
101. D.E. Jesson, T.P. Munt, V.A. Shchukin, D. Bimberg, Phys. Rev. B **69**, 041302 (2004)
102. V.A. Shchukin, D. Bimberg, T.P. Munt, D.E. Jesson, Phys. Rev. B **70**, 085416 (2004)
103. J.A. Floro, M.B. Sinclair, E. Chason, L.B. Freund, R.D. Twesten, R.Q. Hwang, G.A. Lucadamo, Phys. Rev. Lett. **84**, 701 (2000)
104. V.A. Shchukin, D. Bimberg, T.P. Munt, D.E. Jesson, Ann. Phys. **320**, 237 (2005)
105. J. Tersoff, C. Teichert, M.G. Lagally, Phys. Rev. Lett. **76**, 1675 (1996)
106. F. Liu, S.E. Davenport, H.M. Evans, M.G. Lagally, Phys. Rev. Lett. **82**, 2528 (1999)
107. M. Straßburg, V. Kutzer, U.W. Pohl, A. Hoffmann, I. Broser, N.N. Ledentsov, D. Bimberg, A. Rosenauer, U. Fischer, D. Gerthsen, I.L. Krestnikov, M.V. Maximov, P.S. Kop'ev, Z.I. Alferov, Appl. Phys. Lett. **72**, 942 (1998)
108. V.A. Shchukin, D. Bimberg, V.G. Malyshkin, N.N. Ledentsov, Phys. Rev. B **57**, 12262 (1998)

109. M. Strassburg, R. Heitz, V. Türck, S. Rodt, U.W. Pohl, A. Hoffmann, D. Bimberg, I.L. Krestnikov, V.A. Shchukin, N.N. Ledentsov, Z.I. Alferov, D. Litvinov, A. Rosenauer, D. Gerthsen, J. Electron. Mater. **28**, 506 (1999)
110. I.L. Krestnikov, M. Straßburg, M. Caesar, A. Hoffmann, U.W. Pohl, D. Bimberg, N.N. Ledentsov, P.S. Kop'ev, Z.I. Alferov, D. Litvinov, A. Rosenauer, D. Gerthsen, Phys. Rev. B **60**, 8696 (1999)
111. V. Holý, G. Springholz, M. Pinczolits, G. Bauer, Phys. Rev. Lett. **83**, 356 (2000)
112. X.-D. Wang, N. Liu, C.K. Shih, S. Govindaraju, A.L. Holmes Jr., Appl. Phys. Lett. **85**, 1356 (2004)
113. R. Kunert, E. Schöll, Appl. Phys. Lett. **89**, 153103 (2006)
114. R. Kunert, E. Schöll, T. Hammerschmidt, P. Kratzer, in *Proc. 28th International Conference on Physics of Semiconductors*, Vienna, Austria, July 2006, ed. by W. Jantsch, F. Schäffler (Springer, Berlin, 2006)
115. T. Hammerschmidt, P. Kratzer, unpublished
116. J. Tersoff, Phys. Rev. Lett. **81**, 3183 (1998)
117. N. Liu, J. Tersoff, O. Baklenov, A.L. Holmes, Jr., C.K. Shih, Phys. Rev. Lett. **84**, 334 (2000)
118. T. Hammerschmidt, P. Kratzer, Am. Inst. Phys. Conf. Proc. **772**, 601 (2005)

2

Control of Self-Organized In(Ga)As/GaAs Quantum Dot Growth

Udo W. Pohl and André Strittmatter

Abstract. Self-organized growth of InAs, InGaAs, and GaSb quantum dots in GaAs matrix is investigated to establish a basis for a targeted engineering. All studied dots form in the Stranski–Krastanow mode on a two-dimensional wetting layer in a regime which is predominantly controlled by kinetics. Equilibrium-near conditions are found on a local scale. The material transfer during dot formation shows distinct differences for the three studied kind of dots. For InGaAs and GaSb dots a pronounced transfer from an intermixed wetting layer to the dots is found. Such dots have an inhomogenous composition, and dot ensembles with a bimodal size distribution may be obtained in both cases. For InAs dots material transfer is observed solely between the purely binary dots, while the wetting layer remains unaffected and is not intermixed. Dots with a multimodal size distribution are demonstrated, referring to well-defined self-similar dot shapes and size variations in steps of integral InAs monolayers.

2.1 Introduction

Semiconductor research and device development has seen a progressive reduction in dimensionality, going from bulk down to quantum wells and quantum wires, finally to quantum dots (QDs). QDs represent the ultimate limit in carrier confinement with discrete atomic-like energy states. Such states are significantly different from those of systems with higher dimensionality, resulting in potential applications in novel optoelectronic devices. The most successful approach for the fabrication of coherent, dislocation-free semiconductor QDs is the self-organized (also referred to as self-assembled) technique of Stranski–Krastanow growth. This kind of growth mode may be induced by epitaxially growing a layer of, e.g., InAs on a substrate with a significantly different lattice constant like GaAs. The unstrained lattice constant of InAs is $\approx 7\%$ larger than that of GaAs. The deposited InAs initially grows as a highly strained two-dimensional layer. However, beyond a thickness of only ≈ 1.5 InAs monolayers (MLs) growth transforms to a three-dimensional mode. The driving force for this transformation is a reduction of the elastic strain energy [1] as the material in the spontaneously formed islands is not constrained by surrounding material and can relax laterally. Some part of the InAs material does not redistribute to form islands but remains as a two-dimensional layer, because the surface free energy of InAs is lower than that of the GaAs substrate. Since this layer wets the substrate surface it is referred to as wetting layer (WL). For practical use of such islands the structure is covered by a cap layer that is typically identical to the substrate material. The size of embedded self-organized InAs/GaAs structures is in the nanometer range, i.e. of the order of the exciton Bohr radius, giving rise to fully quantized confined electron and hole states. These nanostructures are therefore generally termed quantum dots. The electronic properties of the QDs are determined by the size, shape and composition of the QD material and the composition of surrounding matrix material. This chapter addresses the control of self-organized InAs and InGaAs QD growth in GaAs matrix required to tailor electronic properties for practical applications.

2.2 Evolution and Strain Engineering of InGaAs/GaAs Quantum Dots

InGaAs/GaAs quantum dots represent the material system of choice for applications such as laser diodes and optical amplifiers. Most of the general results described in the following were obtained using samples grown by metalorganic vapor phase epitaxy (MOVPE). The evolution of $In_xGa_{1-x}As$ quantum dots follows the thermodynamics of Stranski–Krastanow growth mode addressed in Chap. 1. Consequently, a critical layer thickness in the range of 2 ML exists for the onset of QD formation for In-compositions $x_{In} > 0.3$. During growth interruption after deposition, the InGaAs QDs grow by a mass transfer of In atoms from the wetting layer to the QDs, in contrast to InAs QDs described in Sect. 2.3. The final In distribution inside the InGaAs QDs develops during overgrowth with matrix material. Strain engineering composed of both strain relaxation at the QD interface to the surrounding matrix material as well as size enlargement of QDs by activated alloy phase separation of ternary matrix material, respectively, is the common approach to shift the emission wavelength of QDs towards 1.3 µm for telecommunication applications (see Chap. 14). One important issue with respect to laser devices as well as optical amplifiers is the stacking of several QD layers in order to achieve, for example, high-speed operation of laser diodes. The most critical parameter is the spacing between individual QD planes which may have impact on electronic as well as structural properties of individual QD planes by wave-function or strain-field overlap, respectively.

2.2.1 Evolution of InGaAs Dots

The formation of $In_xGa_{1-x}As$ quantum dots has been studied for $0.3 < x < 1$. QDs have been found for $x > 0.33$ only. Below this value plastic strain relaxation by,

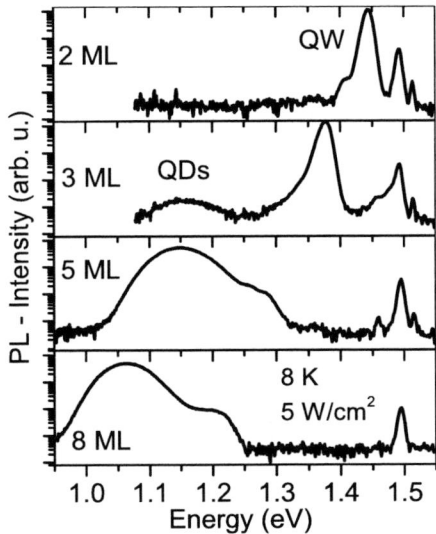

Fig. 2.1. PL spectra of $In_{0.5}Ga_{0.5}As$/GaAs samples with varied deposition thickness. From [3]

e.g., dislocations without QD formation is observed for layers exceeding the critical thickness of 5 nm [2]. Integral photoluminescence (PL) spectra are commonly recorded to monitor QD properties. By comparing intensities and energy positions of spectral features we can reveal the typical evolution of QDs upon increasing the deposition thickness. This is illustrated in Fig. 2.1 for an $In_{0.5}Ga_{0.5}As$ layer. For deposition thicknesses below the critical value (e.g., 2 ML) only InGaAs quantum well (QW) luminescence and GaAs near-band-edge luminescence is observed. The QW emission energy shows a red shift for larger thicknesses as expected, until beyond 3 ML the QD formation begins. At this point the WL luminescence rapidly decreases and the QD luminescence appears at lower energies. The strong increase of the QD luminescence intensity reflects the simultaneous formation of numerous QDs [3].

Up to 8 ML thickness the QD size increases as indicated by the shift toward lower energies and the homogeneity improves. Beyond this thickness, the luminescence intensity decreases (not shown) due to plastic strain relaxation inside the QDs. Similar trends for the QD evolution and for the layer thickness of the 2D–3D transition have been observed by other groups using both MBE [4] as well as MOVPE [5]. The density and equilibrium shape of InGaAs QDs undergoes a transition from low-density, high QDs to high-density, flat QDs as the deposition layer thickness increases [2]. Corresponding PL spectra and TEM images are displayed in Fig. 2.2. As read from the half width in the PL spectra, the ensemble of high QDs exhibits a narrower size distribution and a lower optical transition energy than the flat QDs. The corrugation of the GaAs cap layer above the high QDs observed in the TEM image indicates a size effect being responsible for the energy shift rather than a higher In content of the QDs.

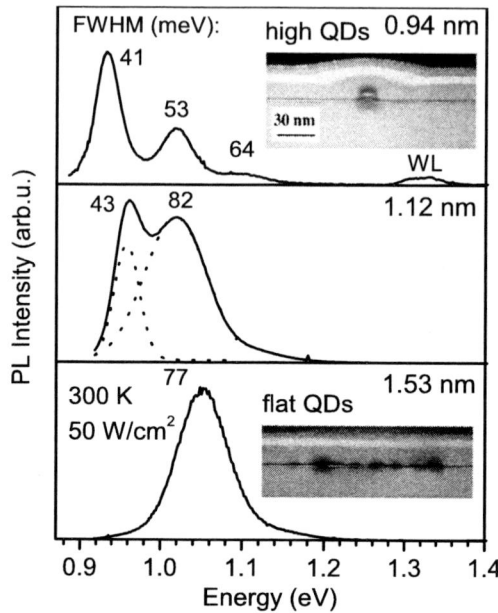

Fig. 2.2. PL spectra and TEM images of $In_{0.4}Ga_{0.6}As/GaAs$ samples with varied deposition thickness. From [2]

A bimodal ensemble composed of high and flat QDs is found for intermediate values of the deposition layer thickness as seen in the PL spectrum of a 1.12 nm thick layer. Long-range adatom diffusion can be identified as the underlying mechanism from the fact that high QDs are formed for low deposition thicknesses and long growth interruptions. Bimodal QD ensembles have also been observed for InAs QDs grown by MBE when long growth interruptions were applied [6]. However, the proposed underlying mechanism of a laterally varying thickness of the WL is different from the laterally varying In composition due to In segregation described here. Furthermore, the bimodal distribution of high and flat InGaAs QDs represents a stable state during growth interruption once it has developed.

The dependence of the critical layer thickness d_c of the 2D–3D transition on the In concentration may reflect either a constant amount of accumulated strain energy within the wetting layer being proportional to ϵ^2, i.e. $(d_c \sim \epsilon^{-2})$, or a constant amount of In incorporation in the wetting layer that is proportional to ϵ, i.e. $(d_c \sim \epsilon^{-1})$ [2]. The phase diagram Fig. 2.3 demonstrates that the incorporated In amount is a more likely cause to explain the dependence. A value of 1.67 ML (0.54 nm) InAs is derived by extrapolation to $x_{In} = 1$. This value is only slightly higher than the experimental value of 1.60 ML for pure InAs QDs grown by MOVPE [7] and is well explained by theory assuming In segregation toward the growth surface [8]. Further insight into InGaAs QD formation is gained by X-ray diffraction measurements of the In concentration of the InGaAs WL before and after the layer thickness has exceeded the critical value; cf. Fig. 2.4.

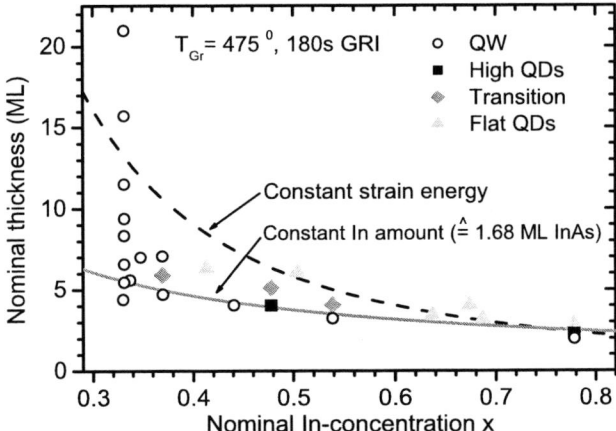

Fig. 2.3. Phase diagram of InGaAs/GaAs QD formation. Symbols denote the type of confining low-dimensional structure obtained. From [2]

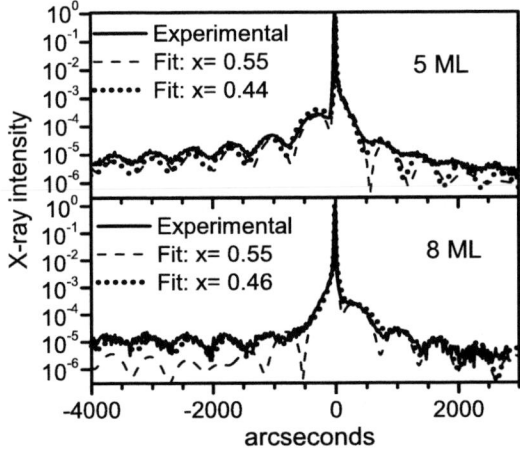

Fig. 2.4. XRD $\omega/2\Theta$ scans of the (004) reflection of $In_{0.5}Ga_{0.5}As$/GaAs samples with 5 ML and 8 ML deposition thickness. The dotted curve is a dynamical simulation. From [9]

Before QD formation starts, the reflection spectrum as caused by the WL's chemical composition and the thickness can be perfectly simulated using a unique value $x_{In} = 0.55$. However, once QDs are formed the In concentration of the WL decreases by 20% to $x_{In} = 0.44$ as evidenced by the simulations shown in Fig. 2.4 [9]. Strain relaxation is the driving force behind this effect. The strain energy of the InGaAs WL is lowered and the excess In atoms are transferred into three-dimensional QDs. Here, strain relaxation also occurs since bending of lattice planes is possible. These findings are in agreement with results obtained through many other techniques like, e.g., HRTEM and STM [10, 11].

2.2.2 Engineering of Single and Stacked InGaAs QD Layers

Single InGaAs QD Layers

A wide-spread parameter space is generally faced for the growth of InGaAs QDs by MOVPE and we are faced with making several choices. Early investigations showed that growth temperature and As partial pressure during deposition of the InGaAs wetting layer and the subsequent growth interruption are two of the most crucial parameters [2, 3]. Both have an impact on the kinetics of quantum dot evolution and on material quality. The situation becomes even more complicated if precursor decomposition depends on temperature. This applies for arsine, the common precursor for the MOVPE of GaAs-based devices. In this case, growth temperature variations affect the As partial pressure, aggravating an independent optimization of both parameters. The alternative precursor, tertiarybutylarsine (TBAs), is fully decomposed at 450°C. Moreover, it is much less toxic than arsine, thus requiring less serious safety precautions.

Using TBAs, it is possible to realize an independent optimization of the influence of the arsenic partial pressure and the growth temperature. As seen in Fig. 2.5, the emission wavelength of QDs rather depends on As partial pressure than on temperature between 470°C to 520°C. The increasing emission wavelength with an increase of the As partial pressure is most likely due to formation of larger QDs. This can be understood in terms of a lowered surface energy at higher As pressures [13, 14] allowing for larger material accumulation [12]. The highest density of up to 10^{11} QDs/cm^2 obtained for $x_{In} = 0.67$ leads to emission energies around 1.1 eV (1100–1150 nm) at room temperature [15]. Long-wavelength emission at 1.3 µm along with a QD density above 1×10^{10} cm^{-2} has been achieved only by sophisticated strain engineering techniques.

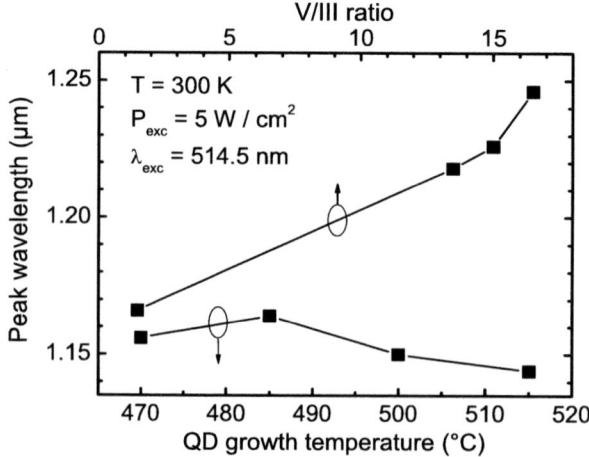

Fig. 2.5. Emission wavelength of the ground state PL peak as a function of V/III ratio and growth temperature for QDs grown using TBAs. From [12]

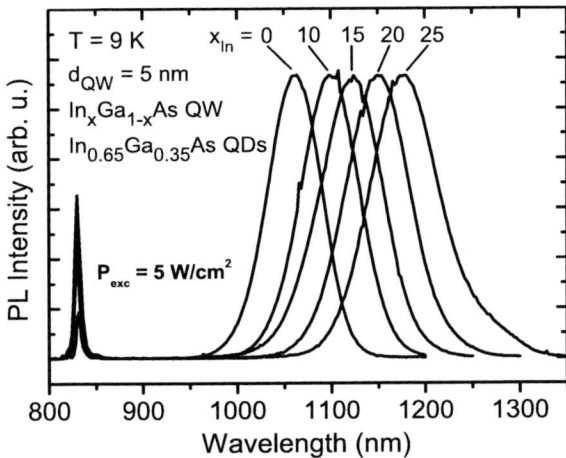

Fig. 2.6. Normalized PL spectra of InGaAs quantum dots with 80% In content, overgrown with an InGaAs quantum well of varied In content. From [16]

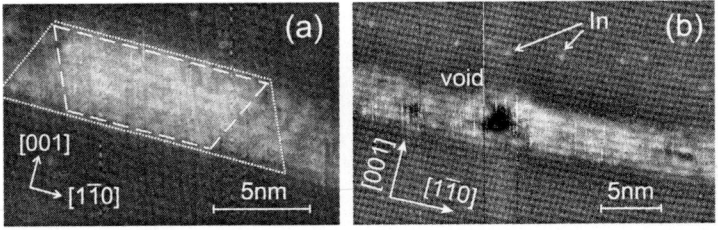

Fig. 2.7. Left: Cross-sectional STM image of empty states ($U_{tip} = +2.1$ V) of an overgrown InGaAs QD with nominally 80% In; from [21]. Right: XSTM image of a large overgrown InGaAs QD. In atoms missing in the void are found at the position of the cap surface during growth interruption. From [22]

Overgrowth of single, high-density QD layers by InGaAs has been proven to extend the emission wavelength towards 1.3 μm and beyond, still keeping a high material quality. Two effects are identified for this red shift. On one side, the overgrown InGaAs layer partially relaxes the strain of the QDs which lowers the energy of the ground state transition; cf. Chap. 14. On the other side, a process called activated alloy phase separation leads to preferential accumulation of In at the locations of QDs, thereby increasing the effective QD volume and shifting the emission towards longer wavelength [17]. The strategy of the strained material to minimize total energy is the same as for the evolution of InGaAs QDs during growth interruptions by mass transfer of In atoms. Furthermore, an In redistribution from the QD apex to the side facets was found if InAs QDs are overgrown using GaAs [18]. This leads to smaller QD dimensions. If, however, InGaAs is used to overgrow InAs QDs the size and shape of the QDs can be preserved [19]. A detailed analysis of the In distribution was obtained using XSTM measurements; see Fig. 2.7(a). The In distribution of QDs overgrown by InGaAs differs from that of InGaAs QDs capped by GaAs. The distribution of InGaAs-capped QDs can be described by an inverted truncated cone while GaAs-capped QDs are best described by an inverted truncated pyramid [20, 21]. From such measurements a preferential accumulation of In atoms at the side facets of QDs during overgrowth with InGaAs is concluded. Furthermore, the InGaAs capping layer initially grows predominantly between quantum dots.

Limitations on overgrowth of large quantum dots are given by the formation of nanovoids, as shown in Fig. 2.7(b) for a sample aimed for 1.3 µm laser diodes. The evolution of nanovoids can be explained by kinetic effects during capping, annealing and overgrowth of quantum dots as outlined in Chap. 6. Larger QDs are covered by thinner GaAs cap layers as compared to smaller QDs since the cap layer grows predominantly in the depressions between QDs. The usually applied high temperature for the growth of high-quality Ga(Al)As matrix material ($\approx 600°C$) drives the outward diffusion of atoms toward the surface, leaving a void in the center of such QDs. These voids persist upon further overgrowth for certain growth conditions, e.g. high growth rates [22].

Stacked InGaAs QD Layers

Stacking of QD layers is the key technology for realizing fast semiconductor laser diodes and optical amplifiers since the required modal gain relies on the total number of dots emitting within a certain wavelength range. Close spacing is desirable in view of an efficient overlap with the optical wave and of homogeneous electrical pumping. Ensemble broadening and defect generation upon stacking are hence the limits to be extended. Close stacking of QD planes also provides means to alter QD properties. On one hand, overlap of wave functions of two vertically correlated QDs is a possibility to create long-wavelength emission. On the other hand, high density QD seed layers can be used to increase the QD density for subsequently grown layers with large size QD.

Strain fields and surface corrugations as established by QDs in one layer have an impact on the growth of the next QD plane. For close spacing a column-like, vertical alignment of QDs in different planes is usually observed since the maximum of the strain field of a shallow buried QD provides the nucleus for subsequent QD formation. The interaction length of the strain fields depends on size and composition of the underlying QDs. For very small InAs QDs vertical alignment was found only for spacer thicknesses up to ≈ 4 nm while for larger InGaAs QDs alignment was found at spacing of up to 20 nm (Fig. 2.8) [23, 24]. At intermediate spacer layer thicknesses between perfectly correlated and completely uncorrelated stacking of QD layers clustering may be favored. In this case only the strain fields of the largest QDs are strong enough to create nucleation sites for the subsequently grown QD layer. Thus, QD material accumulates preferentially at those sites and quickly exceeds the strain limit for dislocation formation.

Nondestructive assessment of structural parameters has been demonstrated by X-ray diffraction techniques. As an example, Fig. 2.9(d) shows X-ray intensity maps around the GaAs(224) reciprocal lattice point of a five-fold stacked QD sample with

Fig. 2.8. Cross-section transmission electron micrograph of stacked $In_{0.6}Ga_{0.4}As/$ GaAs QDs with 20 nm spacers showing pronounced vertical ordering. From [23]

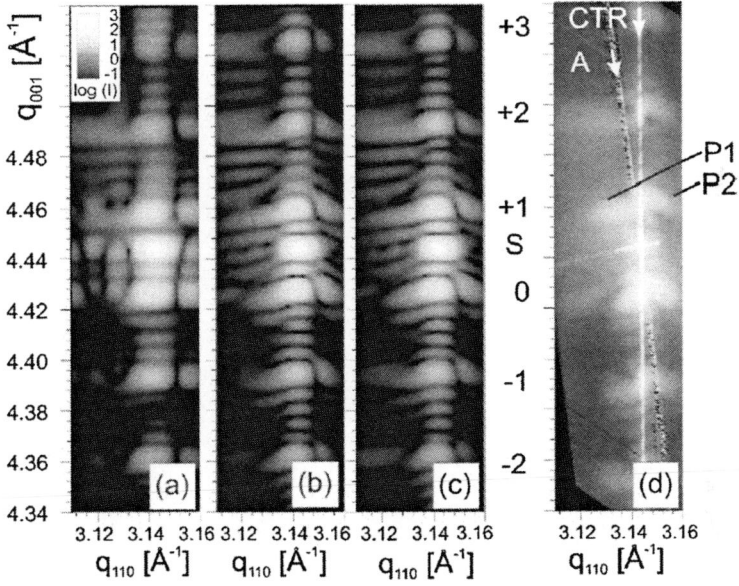

Fig. 2.9. Measured X-ray intensity distribution in the vicinity of the asymmetrical (224) reciprocal lattice point (**d**) compared to simulations of the scattered intensity calculated using different QD shapes and composition profiles (**a**)–(**c**). Feature (A) is caused by a detector artifact. From [25]

20 nm spacing between the QD planes. The nonzero lateral components of the local QD strain fields lead to diffuse scattered intensity around the crystal truncation rod (CTR) which was simulated using distorted-wave Born approximation and a 3D strain field distribution of the QDs obtained by the finite element method [23, 25, 26]. Three intensity maps were calculated assuming a lens-like QD shape of constant In content (Fig. 2.9(a)), a lens-like shape of stepwise increasing In-content (Fig. 2.9(b)), and an inverse truncated cone of stepwise increasing In content (Fig. 2.9(c)); see Sect. 3 of Chap. 5. Only the models of increasing In content allow for a correct simulation of the diffuse scattered intensity. Thus, transfer processes of In atoms from the WL towards the QD apex as already concluded from symmetric XRD measurements are confirmed by the simulation. Using grazing incidence diffraction (GID) techniques, in-plane properties of QD arrays also become accessible. For instance,

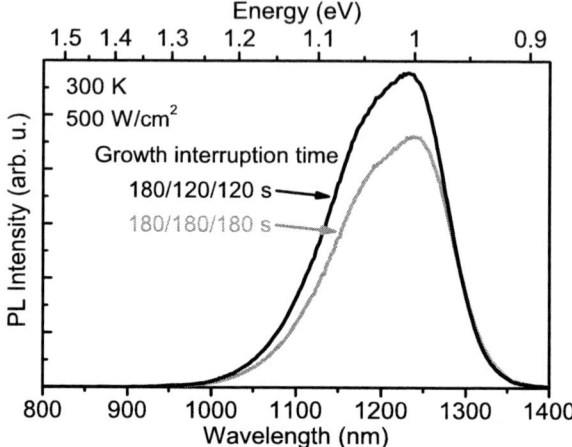

Fig. 2.10. Recovery of PL intensity by individual adjustment of growth parameters for each QD layer within a three-fold QD stack

an average in-plane dot distance of 140 nm has been measured for the given example [26]. Another issue related to stacking is segregation of In atoms toward the surface. The additional material is delivered to the following QD growth step, increasing the average QD size. Thus, the QD ensemble luminescence of stacked layers is usually broadened and red-shifted. Again, the probability for defect formation grows and the number of QD layers stacked without defect generation is limited. Individually adjusted growth parameters for each QD layer can overcome this limitation. As an example, the growth interruption time for the QD evolution was optimized for a three-layer stack; see Fig. 2.10. Hillocks on the surface disappeared and photoluminescence intensity was increased upon optimization. Simultaneously, the number of stacked QD layers could be increased from 3 up to 5 [27].

Surface corrugations induced by QDs may also act as nucleation sites for subsequent QD layers [28]. Therefore, flat surfaces are mandatory in order to keep the optimal QD growth parameters constant for each layer. Higher growth temperature, proper V/III ratio as well as growth interruptions during MOVPE growth of GaAs spacer layers drastically improve the morphology [29].

2.3 Growth Control of Equally Shaped InAs/GaAs Quantum Dots

Studies of the electronic properties of quantum dots—e.g., those reported in Chaps. 7 and 14—require well-defined structures. Fabrication of such dots by application of the widely used self-organized Stranski–Krastanow growth is hampered by the action of entropy at growth temperature, leading to ensembles of dots that vary in size, shape and composition. Since this growth mode still yields defect-free dots with superior electronic properties, the technique was recently advanced to improve the shape homogeneity within the ensemble and the integrity of the dot-matrix interfaces [30]. The development is designed to find a growth regime that simultaneously favors the formation of equilibrium-near dots and sharp interfaces. An early study on chemical-beam epitaxy of strained InAs quantum wells in InP matrix indicated the existence of respective growth parameters [31]. Introduction of a growth interruption after InAs deposition and prior to InP cap layer growth was shown to produce layers with up to eight well-separated PL lines. The lines were attributed to one-monolayer (ML) thickness fluctuations of the well, as previously also reported for lattice-matched materials GaAs/AlGaAs and InP/GaInAs. The interpretation was later improved, stating that three-dimensional islands with heights corresponding to integral numbers of InAs MLs form under the given conditions [32, 33].

2.3.1 Formation of Self-Similar Dots with a Multimodal Size Distribution

Using a GaAs matrix for InAs dots, the strain is nearly doubled with respect to an InP matrix, accordingly reducing the critical layer thickness for elastic strain relaxation in SK growth. A critical value near only 1.5 ML [34] leads to the requirement of an atomically flat GaAs surface prior to InAs deposition to obtain well-defined bottom interfaces of the dots. This is easily achieved by an appropriate process [29]. The InAs/InP islands were reported to form from a two-dimensional InAs layer during the mentioned growth interruption [31]. This means that the deposited amount of InAs was close to the critical value for the SK transition on InP. A smooth growth front is also maintained during deposition of InAs on GaAs, as proved by in situ reflection anisotropy-spectroscopy [35]. Comparable conditions encouraging growth of a multimodal QD ensemble are hence also given for the InAs/GaAs material system. To induce such QD formation, ≈ 1.9 ML thick InAs layers were deposited at typ. 490°C with a comparatively high deposition rate of 0.4 ML/s. Growth interruptions (GRIs) of variable duration were applied after InAs deposition and prior to GaAs cap layer deposition, showing a strong effect on the structural and optical properties of the layers [30, 36]. As illustrated in Fig. 2.11 a sample grown without GRI shows a narrow PL which corresponds to an e_0-hh_0 exciton recombination in a pseudomorphic InAs/GaAs quantum well of ≈ 1.8 ML thickness, in good agreement with the amount of deposited material. The asymmetry at the low-energy side is well described by an exponential function $I \propto \exp(E/E_0)$; cf. inset in Fig. 2.13. The dependence indicates QW states with a DOS tail originating from potential fluctuations due to thickness variations [36]. This proves the formation of a rough InAs/GaAs quantum well if no GRI is applied.

Samples grown with an additional growth interruption clearly show PL of a QD ensemble, peaking near 1.2 eV for a GRI of 5 s (Fig. 2.11). The QDs hence form *after* InAs deposition. The apparent modulation of the QD-ensemble emission originates from the decomposition into narrow emission lines of subensembles. The subensembles are related to InAs QDs which differ in height by one InAs ML [30, 37]. The numbers on the peaks in Fig. 2.11 mark the heights of the respective subensemble QDs in ML units. The multimodal nature of the ensemble size-distribution was con-

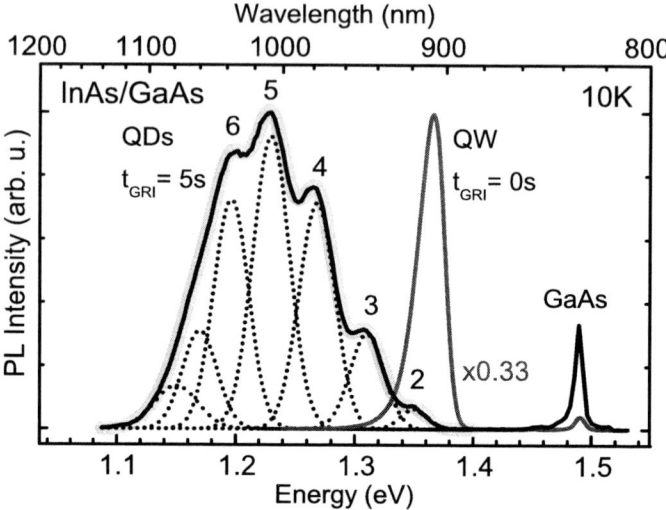

Fig. 2.11. PL spectra of InAs/GaAs samples grown without and with growth interruption (GRI). GaAs denotes matrix near band-edge emission, the thick gray curve is the sum of the dotted Gaussian fits. From [30]

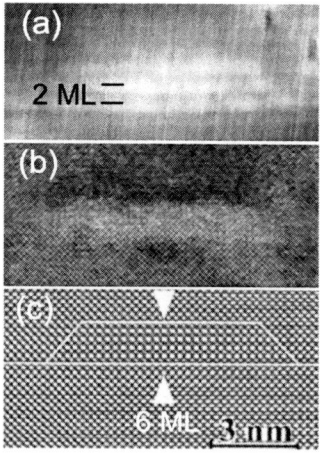

Fig. 2.12. Cross-section STM and high-resolution TEM images of InAs/GaAs QDs with 6 ML height. Bottom: Fourier-filtered image of the TEM micrograph. From [30, 38]

cluded from a combined optical and theoretical study pointed out in Chap. 14. Structural studies by transmission-electron micrographs [30] and cross-sectional scanning tunneling images [38] showed that both the base and the top layers of the QDs are flat and that the dots always consist of plain, continuous InAs; cf. Fig. 2.12. A recent STM study showed that the formation of flat InAs QD top-facets during GaAs cap-layer growth is favored by thermodynamics, leading to equilibrium-near dot shapes

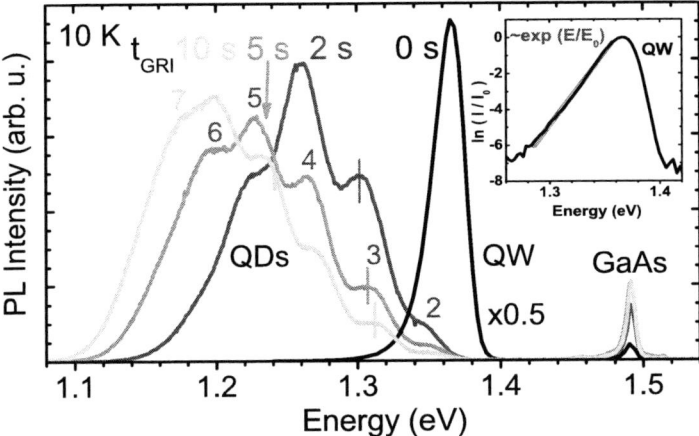

Fig. 2.13. PL spectra of InAs/GaAs QD samples grown with varied duration of the growth interruptions. From [40]

[18]. The structural analyses confirm that QDs differ in height by integer numbers of InAs monolayers. Due to steep side facets, all QDs have the shape of a truncated pyramid.

Application of short growth interruptions shows how the multimodal dot ensemble forms from the initial rough QW [39]. A very short interruption (0.5 s) induces a blue shift of the QW-related PL and leads to a weak broad emission near 1.3 eV, originating from thinning of the QW and enhanced appearance of locally thicker QW regions, respectively. The blue shift saturates for longer interruptions (0.8 s), and the emission of a multimodal QD ensemble with individual subensemble peaks evolves from the shallow localizations in the density-of-states tail of the rough QW. The material of the initial InAs deposition hence partially concentrates at some QD precursors located on a wetting layer of constant thickness. For longer GRI durations, the PL energy of the QD ensemble shifts to lower energy due to an average QD size increase. Note that the emission lines of individual subensembles do *not* shift to the red; see Fig. 2.13. They rather shift to the blue as indicated by bars on the maxima near 1.3 eV. The shift indicates a lateral dissolution of subensembles of smaller QDs and a materials redistribution to form subensembles of larger QDs. Dissolution of smaller QDs is also reflected by a decreasing QD areal density with prolonged GRI. Concomitant PLE measurements demonstrate that thickness and composition of the wetting layer remains constant during this process and can be described by 1 ML pure InAs [30]. Consequently the evolution proceeds solely by mass transport among the dots.

The dynamics of evolution are revealed by analyzing the PL intensity of individual subensembles. Calculations show that the relative oscillator strength of the QD emission is largely insensitive to the size of the QDs. The integral intensity of the subensemble peaks is hence a reasonable measure for the number of QDs with a specific corresponding height [40]. The temporal evolution of these intensities, given

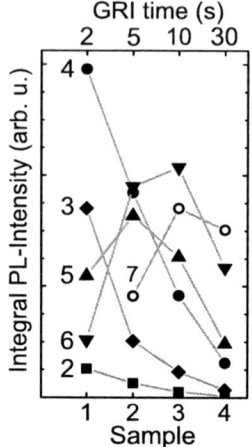

Fig. 2.14. Integral PL intensity of subensemble peaks measured from samples grown with various growth interruptions prior to GaAs capping. Numbers refer to QD heights in monolayers as indicated in Fig. 2.13. From [40]

in Fig. 2.14, directly shows the formation of subensembles with larger dots by dissolution of the subensembles built by smaller QDs. For example, the number of QDs with 5 ML height grows during the initial 5 s GRI duration at the expense of those with 2 to 4 ML height. Then these QDs disappear by feeding the 6 ML high QDs, which after 10 s GRI also start to dissolve.

Targeted adjustment of QD emission energy is a desirable feature for applications. Figure 2.15 shows that the emission energy of a selected subensemble may be varied even for a fixed height of the QDs by varying growth conditions [39]. In sample (a) lateral growth was enhanced by adding Sb during InAs deposition acting as a surfactant [41]. In samples (b) and (c) the lateral QD size was gradually reduced by changing deposition conditions. The resulting mean ground-state emission energy increases by ≈ 0.25 eV for a QD subensemble with 4 InAs ML high QDs.

2.3.2 Kinetic Description of Multimodal Dot-Ensemble Formation

Formation of a dot ensemble with a multimodal size distribution is theoretically well described by a kinetic approach [30, 40]. The model assumes strained dots of truncated pyramidal shape being surrounded by an adatom sea which represents the InAs wetting layer. Dot growth and dissolution occurs by adatom attachment and detachment with a kinetics which is basically controlled by the stress concentration at the dot's base perimeter. The elastic energy density of a dot creates a barrier for the nucleation at the side facets that increases with QD height. Growth by adding new layers on the top facet is hence favored. Material for growth of a dot is provided from dissolution of smaller dots. Each time a top layer is completed, the Gibbs free energy has local minimum [30]. For numerical modeling, adatom exchange between sea and dot is described by rate equations [40]. An Arrhenius

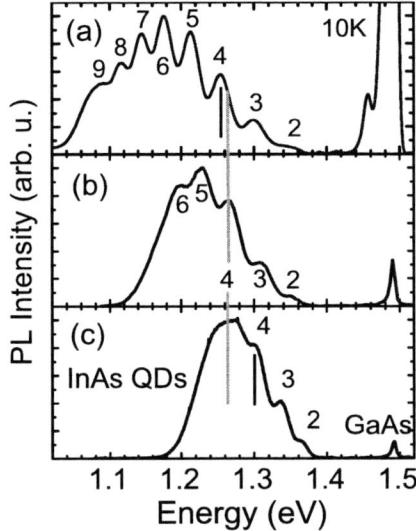

Fig. 2.15. PL spectra of InAs/GaAs QD ensembles grown under various conditions with 5 s growth interruption. The bar marks a subensemble with 4 ML high QDs. From [39]

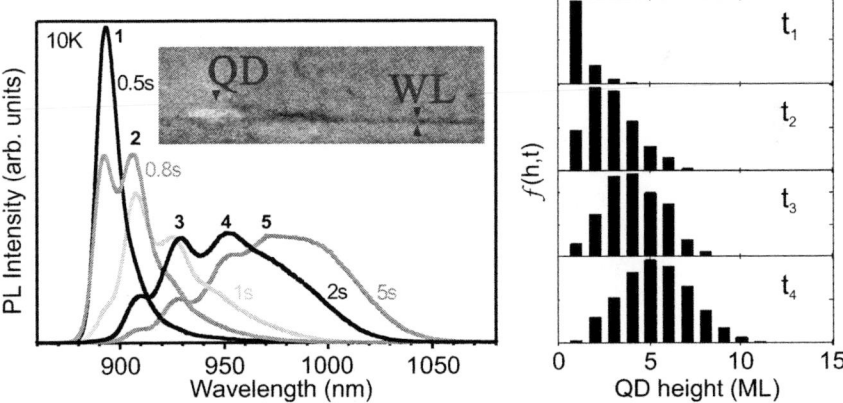

Fig. 2.16. Simulated (*left*) and measured (*right*) initial stages of the evolution of a QD ensemble with multimodal size distribution. Inset: TEM image of QD and WL evolved after 5 s growth interruption. From [42], inset from [39]

dependence in the probability of dot-height increase or decrease accounts for the strain-dependent barrier. Parameters for computing InAs/GaAs dots were adopted from continuum elasticity to account for the elastic-energy change occurring at nucleation of a new layer, and from density-functional theory to consider the step energy of the nucleus. Results of the numerical solution are given in Fig. 2.16. The simulated ensemble initially consists of 1 ML high dots and some minor contributions of higher dots. At later stages the maximum shifts towards higher dots by either dissolution or enlargement of more shallow dots. The model accounts for major features of the experimental data: Formation of a dot ensemble with multimodal size distribution, an evolution by dissolution of smaller and growth of larger QDs, dissolution of the QDs by reduction of height, and the persistence of a flat top facet.

2.4 Epitaxy of GaSb/GaAs Quantum Dots

Strained GaSb quantum dots in GaAs matrix have a staggered type II band lineup with a very large valence band offset of 400 meV, making dots of this material interesting for memory applications (discussed in Chap. 11). The lattice mismatch in the GaSb/GaAs system $\Delta a_0/a_0 = 7.8\%$ is similar to the InAs/GaAs case. Growth of GaSb/GaAs QDs in the Stranski–Krastanow mode is, however, not as straightforward as that of In(Ga)As QDs. The sticking coefficient of Sb at the growth surface is much smaller than that of As [43], leading to memory effects. Growth of GaAsSb within the miscibility gap below 620°C [44] requires careful control of growth parameters [45]. Moreover, recent studies using MBE revealed two growth modes that are controlled by the Sb/Ga ratio. Ga-rich growth conditions result in the interfacial-misfit (IMF) growth mode which is characterized by misfit dislocations confined to the interface between epilayer and substrate [46]. This mode produces strain-relaxed QDs. Formation of coherently strained SK dots is established by applying a high V/III ratio [46]. It should be noted that both kind of dots show strong PL emission [46–49]. Metalorganic vapor phase epitaxy of GaSb/GaAs showed that Ga-rich conditions also lead to an SK transition yielding relaxed, large islands [50]. Zero-dimensional carrier confinement was still obtained with such growth conditions, if GaSb with a thickness below the critical value for the strain-induced 2D–3D transition was deposited. Localization originated from lateral inhomogeneities in the composition and thickness of the resulting two-dimensional InGaSb layer. True SK quantum dots are obtained by using high V/III ratios [49], similar to the MBE case. In the following, we present the growth dynamics of such dots and their structural properties.

2.4.1 Onset and Dynamics of GaSb/GaAs Quantum-Dot Formation

Growth of coherent GaSb/GaAs SK dots by MOVPE was achieved using triethyl sources for both cations Sb and Ga due to their more complete decomposition with respect to methyl organyls at the low deposition temperatures. Evidence for the SK 2D–3D transition is found for a sample series with gradually increased GaSb layer thickness, deposited at a rate of 0.3 ML/s at 470°C and a high V/III ratio of 5.8; cf. Fig. 2.17. The amount of GaSb is expressed in terms of deposition duration due to an initially very small, nonlinear relation to the layer thickness. In addition, arsenic is introduced into the layer due to the memory effect, yielding about 1 ML $GaSb_{0.4}As_{0.6}$ for a 20 s deposition of GaSb under the given conditions.

The PL spectra show a dominating emission near 1.4 eV related to a 2D wetting layer. For deposition durations longer than 20 s, an additional luminescence

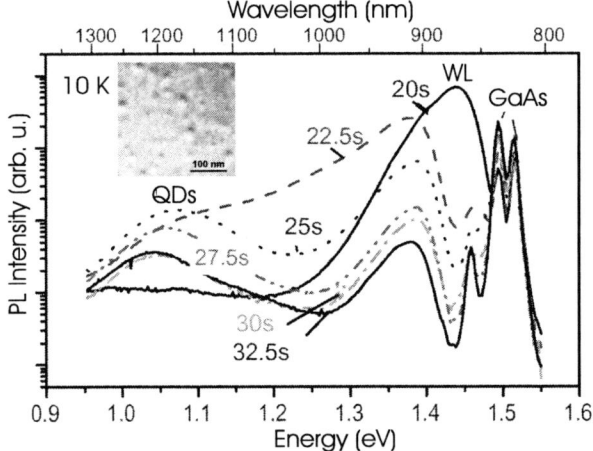

Fig. 2.17. PL spectra of GaSb/GaAs layers with varied thickness. A constant growth interruption of 5 s was applied after deposition. GaAs and WL denote near band-edge emission from the matrix and emission from the wetting layer, respectively. Inset: Plan-view TEM image for 25 s GaSb deposition. From [51]

with a Gaussian shape emerges in the low energy range. The emission appears simultaneously with dots found in transmission-electron micrographs of this series, substantiating the assignment to quantum dots (inset of Fig. 2.17; see also Sect. 4 of Chap. 5). Interestingly, the PL of the wetting layer remains largely unaffected for increased GaSb depositions. The WL emission energy corresponds to a 2 ML thick $GaSb_{0.4}As_{0.6}$ layer, representing some equilibrium value. A similar WL emission energy found in MOVPE samples of other groups (e.g., [52]) and a higher value found in samples grown using molecular beam epitaxy [53] indicates a stronger memory effect for MOVPE conditions. GaSb material in excess of that building the WL is used for QD formation, leading to the gradual red shift of the QD-related PL found in Fig. 2.17. The growth interruption prior to GaAs cap layer deposition has a strong impact on the GaSb-related emissions attributed to QD and WL transitions. The samples characterized in Fig. 2.18 were grown with $\approx$ 4.5 ML GaSb and different durations of the interruption which was performed without Sb flux.

The QDs and WL peaks are present in all spectra. Without growth interruption the peaks are merged with peak energies at 1.13 and 1.33 eV estimated from two Gaussian fits. The interruption leads to a blue shift of the WL and a red shift of the QD peak as shown in the inset. This suggests a gradual increase of the QD size by mass transfer from the 2D wetting layer. This behavior was also observed for InGaAs/GaAs QDs [3, 9], but it is in contrast to that of the InAs QDs presented in Sect. 2.3. From 5 to 10 s interruption the energies of the PL peaks remain unchanged. For longer interruptions the integrated intensity of the PL decreases strongly. This may be related to the sensitivity of the spatially indirect type II exciton to the in-

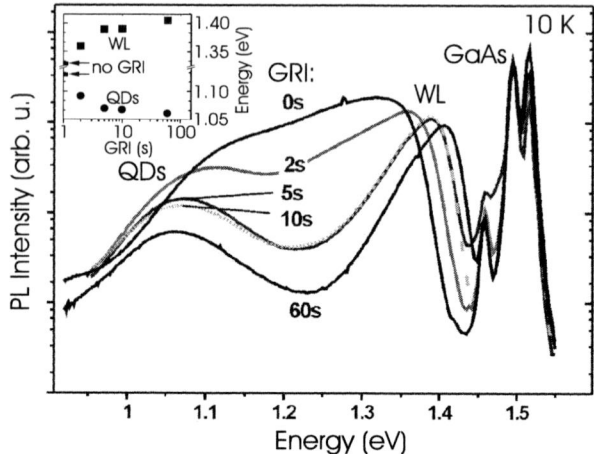

Fig. 2.18. PL spectra of GaSb/GaAs QDs grown with varied growth interruptions. From [49]

terface quality, since the electron is located in the matrix near the interface to the QD.

2.4.2 Structure of GaSb/GaAs Quantum Dots

Structural data from plan-view TEM images provide details of the GaSb QD evolution during the growth interruption. The areal density is found to be largely constant at $\approx 3 \times 10^{10}$ cm^{-2} for the GaSb depositions described earlier. Figure 2.19 shows the lateral size of the QDs from the series given in Fig. 2.18, evaluated from the strain contrast of TEM micrographs [49]. The average lateral size of the QDs increases at up to 5 s interruption, whereas no changes are observed for long interruptions. The mean value of 26 nm base length is close to that reported for MBE-grown samples [54]. A bimodal size distribution of the base length is found for all samples. For the sample grown without GRI and those with GRIs between 2 and 10 s it is indicated by a shoulder.

The constant areal density of the QDs shows that the nucleation occurs during deposition of the GaSb layer. This nonequilibrium stage determines the density of the QDs as previously assumed for InAs/GaAs QDs [55]. Since the size of these QD precursors is far from equilibrium, they subsequently grow by mass transfer from the 2D GaSb WL. The unchanged distribution for samples grown with 5 and 10 s interruptions indicates a saturation of the mass transfer and the formation of a quasi-equilibrium state between the 3D islands and the WL. The presence of both QD precursors and QDs for 10 s GRI indicates a local nature of such equilibrium as predicted by constraint equilibrium theory [55]. The QD size is then determined by the material available from the WL. No significant mass transfer occurs between the QDs, preventing a global equilibrium for the given conditions. PL spectra in Fig. 2.17 showed that no correlation of the QD size and the WL thickness is ob-

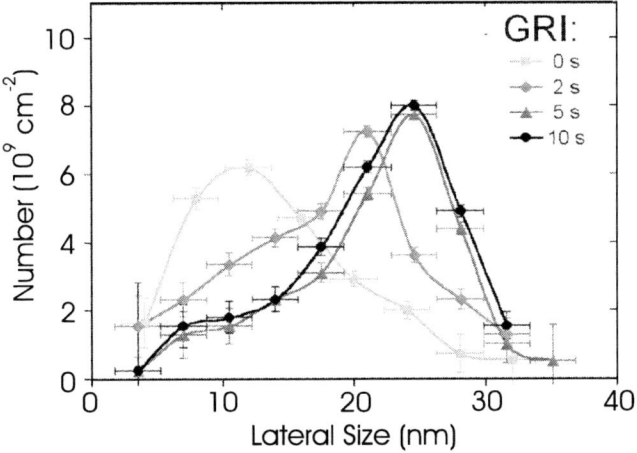

Fig. 2.19. Lateral size distribution of GaSb/GaAs dots grown with different growth interruption durations. From [49]

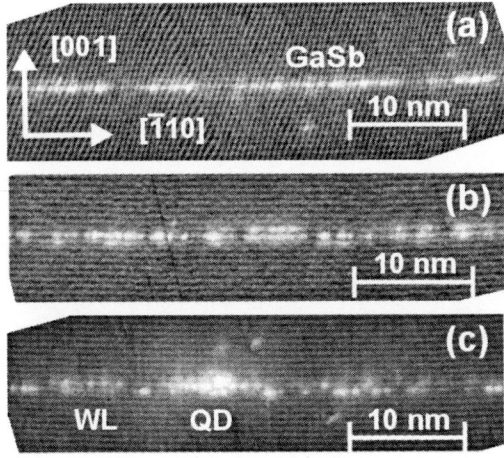

Fig. 2.20. XSTM images of GaSb/GaAs samples with GaSb depositions of (**a**) 21 s, (**b**) 22 s, and (**c**) 25 s. For samples (**b**) and (**c**) a 2 s growth interruption was applied. From [57]

served for the GaSb dots. The WL transition energy remained constant for coverages exceeding the critical value for the SK transition. A constant WL transition energy was also observed in the InAs/GaAs [30] and Ge/Si systems [56]. In the PL spectra (see Fig. 2.18) the bimodal size distribution is masked by the large size inhomogeneity of the QD ensemble. The strong PL signal between the quasi-equilibrium QD and WL emissions is assigned to the large amount of small QDs which are found particularly for short growth interruptions.

A more detailed picture on the local stoichiometry of GaSb dots and wetting layer is provided by cross-sectional STM images. Figure 2.20 shows images of three

GaSb/GaAs samples grown with gradually increased deposition thicknesses [57]. Sample (a) was grown with a GaSb amount expected to be below the critical value for the SK transition and an immediate GaAs cap layer deposition, i.e., without growth interruption. The resulting GaSb quantum well is found to extend over one atomic chain and contains several nanometer-wide gaps. Sample (b) contains a GaSb layer near the critical thickness for the applied growth conditions. It still exhibits a well which extends over two to three atomic chains and is more homogeneous than that of sample (a). In sample (c), with a GaSb layer above the critical value, distinct GaSb dots are found. The areal dot density estimated from the frequency of such dots agrees well with the value 3×10^{10} cm^{-2} determined from plan-view TEM images. The dots have a flat shape with typical base lengths of 4 to 8 nm and heights of 1.5 to 2 nm [58]. The size of the optically active QDs is hence smaller than that estimated for comparable samples from the strain contrast of TEM images, while the shape agrees with published data on MOVPE-grown GaSb dots [49, 52].

The local stoichiometry was obtained from a quantitative analysis of the composition-dependent neighboring atomic chain-distances (pointed out in detail in Chap. 6). For the intermixed GaSb$_x$As$_{1-x}$ dots a typical GaSb content x of 0.6 to 0.7 was derived [58]. In the wetting layer, a smaller Sb content up to $x = 0.5$ was found, being largely in agreement with estimations from PL data.

2.5 Device Applications of InGaAs Quantum Dots

The favorable properties of quantum dots have recently led to a number of applications in novel devices. Incorporation of GaSb dots in charge memories and use of InAs dots in single-photon sources are discussed in Chaps. 11 and 16, respectively. Some achievements in the field of lasers are highlighted in the following. Zero-dimensional charge-carrier localization in the active region of a semiconductor laser was predicted two decades ago to lead to superior device performance, e.g. with respect to a decreased threshold current and high temperature stability [59, 60]. The first realization of a QD injection laser [61] demonstrated the fundamental validity of previous predictions. Subsequently developed concepts for strain engineering and interface control to define QD size and density and to reduce losses have today led to unique device performance for both edge- and surface-emitting QD lasers grown using MOVPE and MBE.

2.5.1 Edge-Emitting Lasers

The first QD lasers with a single layer of self-organized In$_{0.5}$Ga$_{0.5}$As QDs demonstrated the potential of active zero-dimensional nanostructures at cryogenic temperatures, particularly a low lasing threshold of 120 A/cm^2 with a low temperature sensitivity expressed by a characteristic temperature T_0 as high as 350 K, and a very high gain and differential gain [61, 62]. Insufficient carrier confinement, gain saturation and high cavity losses defined avenues for subsequent work. A further demand was

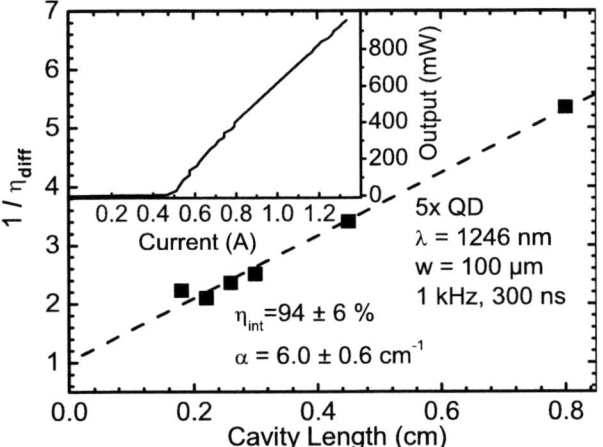

Fig. 2.21. Performance of a ridge waveguide laser with a five-fold InGaAs QD stack measured in pulsed operation. From [27]

a design to tune the emission to the 1.3 µm wavelength range for datacom applications. To increase both confinement factor and gain saturation level, actual QD lasers are composed of a stack of QD layers. The first MOVPE-grown QD lasers reported by us in 1997 used a three-fold InAs QD stack [63] and a ten-fold $In_{0.5}Ga_{0.5}As$ QD stack [64]. Both devices achieved ground-state lasing at room temperature, and a T_0 of 385 K was demonstrated for the ten-fold stack laser up to 50°C. These stacks were grown at the same low temperature for QD layers and GaAs spacers. Substantial progress in laser performance was made particularly by introducing temperature ramping to improve spacer quality and to smooth the interfaces [29]. For three-fold stacked dot layers emitting at 1.16 µm, threshold and transparency current densities of 110 A/cm^2 and 18 A/cm^2, respectively, with internal quantum efficiency exceeding 90% were achieved [15]. An efficient means to extend the emission wavelength toward 1.3 µm is the overgrowth of the In(Ga)As QDs by an InGaAs QW with lower In content to form a dot-in-a-well (DWELL) structure [65, 66]. The QW is also referred to as strain-reducing layer (SRL) due to the strain reduction exerted on the buried QD; see Sect. 14.2.3. Since large, In-rich QDs are required for long wavelength emission the strain may locally be very large, leading to formation of In-rich incoherent clusters which degrade device characteristics. Using the DWELL approach and carefully adjusting the growth parameters for each individual QD layer in the stack to avoid defect formation, lasing at 1250 nm with a very low threshold current density of 66 A/cm^2 and 94% internal quantum efficiency was recently obtained; cf. Fig. 2.21 [27].

2.5.2 Surface-Emitting Lasers

Vertical-cavity surface-emitting lasers (VCSELs) are of particular interest for datacom applications due to advantages with respect to edge emitters, e.g. circular beam

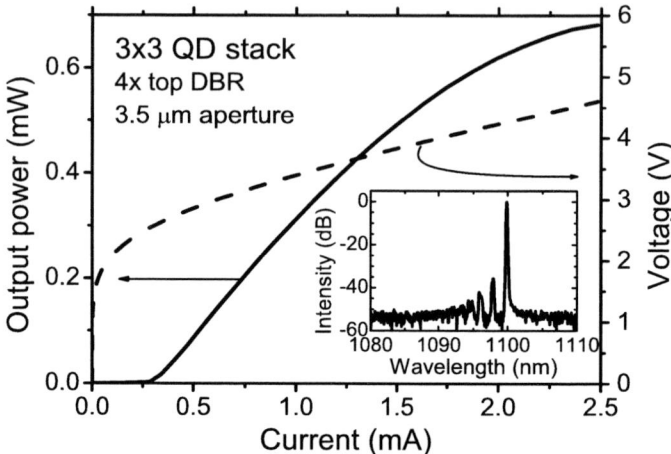

Fig. 2.22. Characteristics of the first electrically driven QD VCSEL grown using MOVPE. Inset: Lasing spectrum at 1 mA injection current. From [66]

profile, temperature-stable operation and planar technology with the possibility of on-wafer testing. Progress on QD-based VCSELs was until recently based solely on material grown using molecular beam epitaxy [67, 68]. The first electrically driven QD VCSEL grown using MOVPE was realized by us using three $In_{0.5}Ga_{0.5}As$ QD layers [66, 69]. The device consists of high-reflection AlO_x/GaAs DBR oxide mirrors cladding a cavity with the stacked QDs, optimized using in situ reflectance [35]. Intra-cavity contacts and one oxide aperture were processed for current injection and constriction, respectively. The VCSEL showed cw lasing near 1 µm up to 240 K, with an output power of 0.7 mW and 35 dB side-mode suppression; cf. Fig. 2.22.

Progress in stacking QD layers enabled the application of an advanced design with a 3×3 grouping of the QD layers in each of the three central field antinodes of a 4λ cavity [69]. This VCSEL achieved ground state lasing at 20°C with 1.45 mW cw (6 mW pulsed) output power at 1.1 µm. Present limitations are basically given by nonradiative recombination in the barrier or wetting layer at high carrier densities.

2.6 Conclusion

Understanding of the self-organized Stranski–Krastanow growth mode for the fabrication of defect-free quantum dots with tailored properties has considerably advanced during the last years. The studied InAs, InGaAs and GaSb dots in GaAs matrix are formed in a regime controlled predominantly by kinetics, though equilibrium-near conditions are found on a local scale. All these dots grow on a two-dimensional wetting layer, but their formation shows distinct differences. A pronounced material transfer from an intermixed WL to the dots is observed for InGaAs and GaSb dots, which both show an inhomogenous composition. In contrast, InAs dot ensembles exhibit material transfer solely between the purely binary dots, leaving the binary WL unaffected. Engineering of QD's properties on the basis of such studies as outlined in this chapter has enabled interesting novel applications in several fields like lasers, charge and spin memories, and quantum information.

References

1. V.A. Shchukin, N.N. Ledentsov, P.S. Kop'ev, D. Bimberg, Phys. Rev. Lett. **75**, 2968 (1995)
2. F. Heinrichsdorff, A. Krost, M. Grundmann, D. Bimberg, A. Kosogov, P. Werner, F. Bertram, J. Christen, in *Proc. 23rd Int. Conf. Physics of Semicond.*, vol. 2 (World Scientific, Singapore, 1996), p. 1321
3. F. Heinrichsdorff, A. Krost, M. Grundmann, D. Bimberg, F. Bertram, J. Christen, A. Kosogov, P. Werner, J. Cryst. Growth **170**, 568 (1997)
4. D. Leonard, M. Krishnamurthy, S. Fafard, J.L. Merz, P.M. Petroff, J. Vac. Sci. Technol. B **12**, 1063 (1994)
5. J. Oshiniowo, M. Nishioka, S. Ishida, Y. Arakawa, Appl. Phys. Lett. **65**, 1421 (1994)
6. H. Lee, R.R. Lowe-Webb, W. Yang, P.C. Sercel, Appl. Phys. Lett. **71**, 2325 (1997)
7. F. Heinrichsdorff, A. Krost, D. Bimberg, A.O. Kosogov, P. Werner, Appl. Surf. Sci. **123/124**, 725 (1998)
8. H. Toyoshima, T. Niwa, J. Yamazaki, A. Okamoto, Appl. Phys. Lett. **63**, 821 (1993)
9. A. Krost, F. Heinrichsdorff, D. Bimberg, A. Darhuber, G. Bauer, Appl. Phys. Lett. **68**, 785 (1996)
10. U. Woggon, W. Langbein, J.M. Hvam, A. Rosenauer, T. Remmele, D. Gerthsen, Appl. Phys. Lett. **71**, 377 (1997)
11. N. Grandjean, J. Massies, O. Totterau, Phys. Rev. B **55**, R10189 (1997)
12. R. Sellin, I. Kaiander, D. Ouyang, T. Kettler, U.W. Pohl, D. Bimberg, Appl. Phys. Lett. **82**, 841 (2003)
13. N. Moll, A. Kley, E. Pehlke, M. Scheffler, Phys. Rev. B **54**, 8844 (1996)
14. E. Pehlke, N. Moll, A. Kley, M. Scheffler, Appl. Phys. A **65**, 525 (1997)
15. R.L. Sellin, C. Ribbat, M. Grundmann, N.N. Ledentsov, D. Bimberg, Appl. Phys. Lett. **78**, 1207 (2001)
16. U.W. Pohl, D. Bimberg, Mat. Res. Soc. Symp. Proc. **794**, 165 (2004)
17. M.V. Maximov, A.F. Tsatsul'nikov, B.V. Volovik, D.S. Sizov, Y.M. Shernyakov, I.N. Kaiander, A.E. Zhukov, A.R. Kovsh, S.S. Mikhrin, V.M. Ustinov, Z.I. Alferov, R. Heitz, V.A. Shchukin, N.N. Ledentsov, D. Bimberg, Y.G. Musikhin, W. Neumann, Phys. Rev. B **62**, 16671 (2000)
18. G. Costantini, A. Rastelli, C. Manzano, P. Acosta-Diaz, R. Songmuang, G. Katsaros, O.G. Schmidt, K. Kern, Phys. Rev. Lett. **96**, 226106 (2006)
19. R. Songmuang, S. Kiravittaya, O.G. Schmidt, J. Cryst. Growth **249**, 416 (2003)
20. N. Liu, J. Tersoff, O. Baklenov, A.L. Holmes, C.K. Shih, Phys. Rev. Lett. **84**, 334 (2000)
21. A. Lenz, R. Timm, H. Eisele, C. Hennig, S.K. Becker, R.L. Sellin, U.W. Pohl, D. Bimberg, M. Dähne, Appl. Phys. Lett. **81**, 5150 (2002)
22. A. Lenz, H. Eisele, R. Timm, C. Henning, S.K. Becker, R.L. Sellin, U.W. Pohl, D. Bimberg, M. Dähne, Appl. Phys. Lett. **85**, 3848 (2004)
23. M. Hanke, D. Grigoriev, M. Schmidbauer, P. Schäfer, R. Köhler, R.L. Sellin, U.W. Pohl, D. Bimberg, Physica E **21**, 684 (2004)

24. F. Heinrichsdorff, A. Krost, D. Bimberg, A.O. Kosogov, P. Werner, Appl. Surf. Sci. **123/124**, 725 (1998)
25. M. Hanke, D. Grigoriev, M. Schmidbauer, P. Schäfer, R. Köhler, R.L. Sellin, U.W. Pohl, D. Bimberg, Appl. Phys. Lett. **85**, 3062 (2004)
26. R. Köhler, D. Grigoriev, M. Hanke, M. Schmidbauer, P. Schäfer, S. Besedin, U.W. Pohl, R.L. Sellin, D. Bimberg, N.D. Zakharov, P. Werner, Mat. Res. Soc. Symp. Proc. **794**, 179 (2004)
27. A. Strittmatter, T.D. Germann, T. Kettler, K. Posilovic, U.W. Pohl, D. Bimberg, Appl. Phys. Lett. **88**, 262104 (2006)
28. W. Seiffert, N. Carlsson, A. Petersson, L.-E. Wernersson, L. Samuelson, Appl. Phys. Lett. **68**, 1684 (1996)
29. R. Sellin, F. Heinrichsdorff, C. Ribbat, M. Grundmann, U.W. Pohl, D. Bimberg, J. Cryst. Growth **221**, 581 (2000)
30. U.W. Pohl, K. Pötschke, A. Schliwa, F. Guffarth, D. Bimberg, N.D. Zakharov, P. Werner, M.B. Lifshits, V.A. Shchukin, D.E. Jesson, Phys. Rev. B **72**, 245332 (2005)
31. J.F. Carlin, R. Houdré, A. Rudra, M. Ilegems, Appl. Phys. Lett. **59**, 3018 (1991)
32. A. Gustafsson, D. Hessmann, L. Samuelson, J.F. Carlin, R. Houdré, A. Rudra, J. Cryst. Growth **147**, 27 (1995)
33. S. Raymond, S. Studenikin, S.-J. Cheng, M. Pioro-Ladrière, M. Ciorga, P.J. Poole, M.D. Robertson, Semicond. Sci. Technol. **18**, 385 (2003)
34. D. Leonard, K. Pond, P.M. Petroff, Phys. Rev. B **50**, 11687 (1994)
35. U.W. Pohl, K. Pötschke, I. Kaiander, J.-T. Zettler, D. Bimberg, J. Cryst. Growth **272**, 143 (2004)
36. U.W. Pohl, K. Pötschke, M.B. Lifshits, V.A. Shchukin, D.E. Jesson, D. Bimberg, N.D. Zakharov, P. Werner, Appl. Surf. Sci. **252**, 5555 (2006)
37. R. Heitz, F. Guffarth, K. Pötschke, A. Schliwa, D. Bimberg, N.D. Zakharov, P. Werner, Phys. Rev. B **71**, 045325 (2005)
38. R. Timm, H. Eisele, A. Lenz, T.-Y. Kim, F. Streicher, K. Pötschke, U.W. Pohl, D. Bimberg, M. Dähne, Physica E **32**, 25 (2006)
39. U.W. Pohl, R. Seguin, S. Rodt, A. Schliwa, K. Pötschke, D. Bimberg, Physica E **35**, 285 (2006)
40. U.W. Pohl, K. Pötschke, A. Schliwa, M.B. Lifshits, V.A. Shchukin, D.E. Jesson, D. Bimberg, Physica E **32**, 9 (2006)
41. K. Pötschke, L. Müller-Kirsch, R. Heitz, R.L. Sellin, U.W. Pohl, D. Bimberg, N. Zakharov, P. Werner, Physica E **21**, 606 (2004)
42. U.W. Pohl, A. Schliwa, R. Seguin, S. Rodt, K. Pötschke, D. Bimberg, Adv. Sol. State Phys. **46** (2006)
43. R.M. Biefeld, S.R. Kurtz, A.A. Allermann, J. Electron. Mater. **26**, 903 (1997)
44. G.B. Stringfellow, *Organometallic Vapor-Phase Epitaxy: Theory and Practice*, 2nd edn. (Academic, London, 1999)
45. M. Peter, N. Herres, F. Fuchs, K. Winkler, K.-H. Bachem, J. Wagner, Appl. Phys. Lett. **74**, 410 (1999)
46. G. Balakrishnan, J. Tatebayashi, A. Khoshakhlagh, S.H. Huang, A. Jallipalli, L.R. Dawson, D.L. Huffaker, Appl. Phys. Lett. **89**, 161104 (2006)
47. F. Hatami, N.N. Ledentsov, M. Grundmann, J. Böhrer, F. Heinrichsdorff, M. Beer, D. Bimberg, S.S. Ruvimov, P. Werner, U. Gösele, J. Heydenreich, U. Richter, S.V. Ivanov, B.Y. Meltser, P.S. Kop'ev, Z.I. Alferov, Appl. Phys. Lett. **67**, 656 (1995)
48. K. Suzuki, R.A. Hogg, Y. Arakawa, J. Appl. Phys. **85**, 8349 (1999)
49. L. Müller-Kirsch, R. Heitz, U.W. Pohl, D. Bimberg, I. Häusler, H. Kirmse, W. Neumann, Appl. Phys. Lett. **79**, 1027 (2001)

50. L. Müller-Kirsch, U.W. Pohl, R. Heitz, H. Kirmse, W. Neumann, D. Bimberg, J. Cryst. Growth **221**, 611 (2000)
51. L. Müller-Kirsch, *Metallorganische Gasphasenepitaxie und Charakterisierung von antimonhaltigen Quantenpunkten*, Berlin Studies in Solid State Physics, vol. 12, ISBN 3-89685-398-8 (Berlin, 1998)
52. Motlan, E.M. Goldys, T.L. Tansley, J. Cryst. Growth **236**, 621 (2002)
53. K. Suzuki, R.A. Hogg, Y. Arakawa, J. Appl. Phys. **85**, 8349 (1999)
54. B.R. Bennett, R. Magno, B.V. Shanabrook, Appl. Phys. Lett. **68**, 505 (1996)
55. L.G. Wang, P. Kratzer, M. Scheffler, N. Moll, Phys. Rev. Lett. **82**, 4042 (1999)
56. H. Sunamura, N. Usami, Y. Shiraki, S. Fukatsu, Appl. Phys. Lett. **66**, 3024 (1995)
57. R. Timm, J. Grabowski, H. Eisele, A. Lenz, S.K. Becker, L. Müller-Kirsch, K. Pötschke, U.W. Pohl, D. Bimberg, M. Dähne, Physica E **26**, 231 (2005)
58. R. Timm, H. Eisele, A. Lenz, S.K. Becker, J. Grabowski, T.-Y. Kim, L. Müller-Kirsch, K. Pötschke, U.W. Pohl, D. Bimberg, M. Dähne, Appl. Phys. Lett. **85**, 5890 (2004)
59. Y. Arakawa, H. Sakaki, Appl. Phys. Lett. **40**, 939 (1982)
60. M. Asada, Y. Miyamoto, Y. Suematsu, IEEE J. Quantum Electron. **QE-22**, 1915 (1986)
61. N. Kirstaedter, N.N. Ledentsov, M. Grundmann, D. Bimberg, V.M. Ustinov, S.S. Ruvimov, M.V. Maximov, P.S. Kop'ev, Z.I. Alferov, U. Richter, P. Werner, U. Gösele, J. Heydenreich, Electron. Lett. **30**, 1416 (1994)
62. N. Kirstaedter, O.G. Schmidt, N.N. Ledentsov, D. Bimberg, V.M. Ustinov, A.Y. Egorov, A.E. Zhukov, M.V. Maximov, P.S. Kop'ev, Z.I. Alferov, Appl. Phys. Lett. **69**, 1226 (1996)
63. F. Heinrichsdorff, M.-H. Mao, N. Kirstaedter, A. Krost, D. Bimberg, A.O. Kosogov, P. Werner, Appl. Phys. Lett. **71**, 22 (1997)
64. M.V. Maximov, I.V. Kochnev, Y.M. Shernyakov, S.V. Zaitsev, N.Y. Gordeev, A.F. Tsatsul'nikov, A.V. Sakharov, I.L. Krestnikov, P.S. Kop'ev, Z.I. Alferov, N.N. Ledentsov, D. Bimberg, A.O. Kosogov, P. Werner, U. Gösele, Jpn. J. Appl. Phys. **36**, 4221 (1997)
65. L.F. Lester, A. Stintz, H. Li, T.C. Newell, E.A. Pease, B.A. Fuchs, K.J. Malloy, IEEE Photonics Technol. Lett. **11**, 931 (1999)
66. I.N. Kaiander, F. Hopfer, T. Kettler, U.W. Pohl, D. Bimberg, J. Cryst. Growth **272**, 154 (2004)
67. D.L. Huffaker, H. Deng, D.G. Deppe, IEEE Photonics Technol. Lett. **10**, 185 (1998)
68. J.A. Lott, N.N. Ledentsov, V.M. Ustinov, N.A. Maleev, A.E. Zhukov, A.R. Kovsh, M.V. Maximov, B.V. Volovik, Z.I. Alferov, D. Bimberg, Electron. Lett. **36**, 1384 (2000)
69. F. Hopfer, I. Kaiander, A. Lochmann, A. Mutig, S. Bognar, M. Kuntz, U.W. Pohl, V. Haisler, D. Bimberg, Appl. Phys. Lett. **89**, 061105 (2006)

3

In-Situ Monitoring for Nano-Structure Growth in MOVPE

Markus Pristovsek and Wolfgang Richter

Abstract. This chapter describes the advances achieved in the last decades for in-situ monitoring of gas phase epitaxial growth (MOVPE) with resolution down to the atomic scale. By spatial resolution electron diffraction would be the tool of choice. However, by mean free path arguments it cannot be applied in the gasphase environment of MOVPE. Optical in-situ techniques on the other hand, easy to setup, have been developed to such a level in the last decades that monolayer resolution is now possible. Even a simple single wavelength reflectance measurement can determine growth rates, composition and temperature. In-situ analysis on the submonolayer scale is routinely possible using Reflectance Anisotropy Spectroscopy (RAS). RAS is best suited to follow online the epitaxial growth evolution of surfaces, revealing surface structure and stochiometry as a function of time (ms). Doping profiling has been achieved as well by combining reflectance and RAS. The chapter closes with an outlook to the ultimate surface tool, the most recent application of a scanning tunneling microscope in MOVPE.

3.1 Introduction

For more than 30 years, Metal Organic Vapor Epitaxy (MOVPE) and Molecular Beam Epitaxy (MBE) have been the main techniques for the epitaxial growth of semiconductors. As far as microscopic knowledge of the growth process and control were concerned, MBE, with its Ultra high Vacuum (UHV) conditions and the standard in-situ analysis tool Reflection High Energy Electron Diffraction (RHEED), was always considered superior to MOVPE. MOVPE on the other hand uses near atmospheric pressure. Because this made upscaling and maintenance easy, MOVPE became the industrial growth method of choice. In other words MBE was the technique with monolayer growth control while MOVPE, with all the electron-based analysis techniques excluded because of the high-pressure environment, had much less control (one or two order of magnitude) and was mainly adjusted and developed empirically by post-growth analysis. Over the last 20 years, however, this situation has changed drastically due to the development of optical surface analysis and other tools. These tools now give monolayer analytical control to MOVPE. Moreover, the optical information usually shows some spectral signature relating to the atomic species and thus gives chemical information (doping, stoichiometry) not directly available by diffraction techniques. Recently, a Scanning Tunneling Microscope for in-situ MOVPE use has been developed and thus one can state that both MOVPE and MBE have equivalent analytical control possibilities.

The oldest optical in-situ tool is optical reflectance, which is quite simple to perform in a nonquantitative manner and still provides information about layer thickness or roughness. The disadvantage is that the signal originates from at least 50 monolayers or more and it is not suitable for the analysis of nanostructures because of the small contribution it carries from the surface. To overcome this problem, other optical in-situ tools were developed in the late 1980s, most importantly Reflectance Anisotropy Spectroscopy (RAS).

Thus, in the early 1990s, the stage was set for atomic-level optical in-situ spectroscopy (epioptics). At that time the standard surface during MOVPE was still widely believed to be either a hydrogen- or a methylene-terminated surface. This was argued in close analogy to the chloride Vapor Phase Epitaxy (VPE), where a chloride-terminated surface was assumed.

In 1990, the first paper by Kisker et al. appeared, using gracing X-ray diffraction (GIXS) which demonstrated unambiguously the presence of a (2×4) reconstruction after annealing GaAs(001) in H_2 [1]. In 1992, the same authors showed that the typical reconstruction under arsenic-rich conditions like in MOVPE is a $c(4 \times 4)$-like structure similar to the one found in UHV/MBE [2]. In the same year the first study using Reflectance Anisotropy Spectroscopy (RAS), which correlates MBE and MOVPE surfaces, appeared by Aspnes and co-workers [3]. They found similarly a (2×4) under hydrogen atmosphere but the main reconstruction was a $c(4 \times 4)$-like under arsenic stabilization in MOVPE. The first study of optical in-situ monitoring of InAs monolayer-by-monolayer deposition dated also from 1992 by Scholz et al. [4]. The study was done in MBE but also on a GaAs(001) $c(4 \times 4)$ surface. Even though at that time the concept of quantum dots was not yet generally known, the spectra recorded and their analysis were very close to the later results. In 1993 monolayer-related oscillations of the RAS signal were demonstrated in MOVPE [5], allowing for monolayer control of thickness and composition for superlattices [6].

Apart from reflectance and RAS, other optical in-situ techniques were tried but not strongly applied to MOVPE growth. Surface Photo Adsorption (SPA), first applied in 1990 to GaAs [7], is a difference technique, i.e. all spectra are measured relative to a reference surface. Reproducibility and quantitative analysis depend very much on the preparation of the reference surface. Moreover, two optical ports are needed. Therefore, this method was applied by only a few groups. Spectral Ellipsometry (SE), developed 100 years ago, also suffers from the disadvantage that two optical ports are needed, that the alignment is not simple and that not many setups are used in growth equipment. However, the possibility of a quantitative analysis compensates for the complicated setup and the relative lack of surface sensitivity. Especially for the growth of noncubic semiconductors (III-nitrides), where RAS cannot be applied, ellipsometry constitutes an essential in-situ analysis tool [8], and even monolayer sensitivity is possible [9].

In recent years, new in-situ techniques were emerging for MOVPE. One of them is in-situ strain monitoring via bending of the substrate. Bending bar-type strain sensors had been used in MBE for basic research and could detect reconstruction changes or the onset of quantum dot formation (e.g. [10]). In MOVPE the development of this technique was strongly driven by the III-Nitrides in order to control the strain during deposition of nitrides on silicon by directly measuring the curvature of a wafer [11]. For the classical semiconductors during nanostructure growth the strain can be converted into a composition via the elastic constants and model simulations [12]. This technique is still very new, but it is already commercialized and can be used with production machines.

All the techniques mentioned so far are averaging the signal over large areas. To get direct insight into nanostructures, there is currently only a single in-situ technique—Scanning Tunneling Microscopy (STM). STM was first applied to GaAs MBE in 1999 by Tsukamoto et al. [13]. Using an innovative setup, the group at TU Berlin also recently succeeded in performing in-situ STM during MOVPE growth [14, 15]. But due to the relative complex setup this technique will not be easily adapted to production machines.

A rather extensive review of in-situ monitoring was given a few years ago [16] and therefore we will describe here only recent developments with a special focus on nanostructures. We will touch in this chapter on reflectance, which has been refined in recent years and is simple to use; on RAS, which constitutes the main MOVPE analysis tool; and on the recently developed STM usable in the high-temperature gas phase environment of MOVPE.

3.2 Reflectance

This method is surely the oldest and simplest optical in-situ technique used especially to determine layer thickness. It relies on the interference of light reflected at the top and the bottom interface of the growing layer. The resulting intensity is modulated with an oscillation corresponding to the optical thickness (layer thickness d multiplied by the refractive index n) which reads for perpendicular incidence

$$I(d) \propto \cos\left(\frac{4\pi}{\lambda}dn\right)\exp(-kd) \qquad (3.1)$$

with d the layer thickness, n the refractive index, k the adsorption coefficient and λ the wavelength of the light.

If the lower layer can be considered as infinite (no light is reflected from the backside) the exact signal amplitude can be also calculated. However, most reflectometers just give relative reflectance signals. In praxis, one observes one maximum and minimum of the reflectance signal relative to a signal from the starting substrate

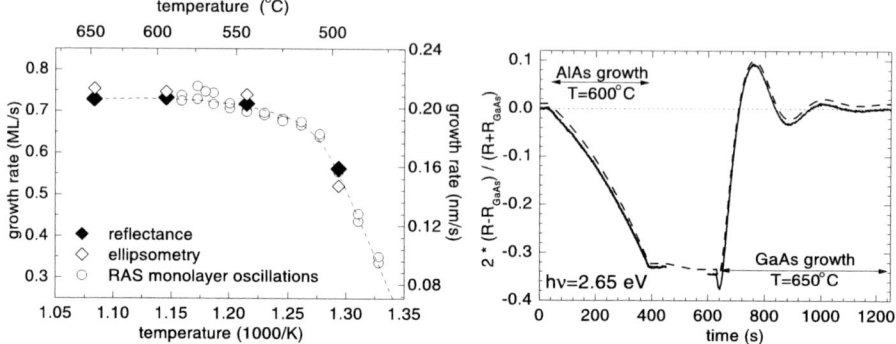

Fig. 3.1. Right side: Evolution of the reflectance signal during growth of AlAs on GaAs (30–400 s) and subsequent GaAs growth on this AlAs layer (600–1200 s). Left side: diagram showing the growth rate obtained from the reflectance signal in comparison to those from RAS monolayer oscillation and from ellipsometric data [18]

for which n and k are usually known (Fig. 3.1). A self-consistent fit via the above formula will yield n, k, and the growth rate. n and k can also give the composition for ternary materials, if a database is available [17].

Therefore, reflection always needs an interface below the surface with a step of the refractive index. Consequently, buffer layer growth of GaAs on GaAs will yield no signal since the GaAs/GaAs interface is not visible to the light and all the light is adsorbed in the 300 μm thick substrate before it reaches the backside (the second hetero-interface GaAs/gas). Thus, no interference will be observed in homoepitaxy. To overcome this problem one can grow a marker layer (from a different material) for the sole purpose of calibration (as has been done in Fig. 3.1). However, for GaN growth on sapphire (or any other heterosubstrate), the resulting sapphire-GaN interface is always visible. Therefore, epitaxy of GaN and other nitrides is the main application for in-situ reflectance. For typical application, like device heterostructures, there is always a heterointerface below the growing layer. Often there are also additional interfaces in the device structure, e.g. the cladding layer or waveguide layer in an edge-emitting laser structure.

The minimum thickness required for a self-consistency is a quarter period of the reflectance oscillation. Therefore, for typical materials with a refractive index of 4 (like GaAs) in the visible range (e.g. 500 nm), a growth rate can be self-consistently determined after $\frac{500}{4 \cdot 2} \approx 60$ nm. Thus, reflectance can be used on waveguides and cladding layers to determine growth rate, composition and even temperature with good accuracy [19]. However, for the nanostructures in the active region, precise data on the refractive index are needed to estimate the thickness (or vice versa).

For ternary materials, if the reflective index of the lower second layer is not known, the composition may also be estimated from the growth rate. The increase of growth rate is then directly related to the additional amount of the third group III material, since in MOVPE the group III supply usually limits the growth rate. This method gives quite good results, as long as the sticking coefficient is not changing significantly. For MOVPE, the sticking coefficient of group III species is nearly always one, thus this method gives excellent values for group-three ternaries like AlGaAs or InGaP at not too low temperatures.

For the analysis of quaternary materials like InGaAsP by reflectance the group III ratio can be determined by the growth rate increase, while the group V ratio is only correlated with to a refractive index. Given a suitable database independent determination of In to Ga and As to P ratio is possible [17]. Even qualitative surface roughness estimates can be performed [20].

The methods described above work with ellipsometry too. Since ellipsometry measures the dielectric function on an absolute scale, its accuracy is much better. Moreover, using spectral information, additional data—e.g. on roughness, ordering and similar data—can be gained even for atomic monolayers. This has recently been exploited intensively in the development of InN epitaxy for MOVPE [8] or the analysis of the nitridation process [21]. The nitridated layer thickness varied only between 0.1 nm and 0.9 nm but could be analysed nevertheless by ellipsometry with excellent accuracy [21].

Finally, for devices like Vertical Cavity Emitting Lasers (VCSEL), where the reflectance of the total layer stack is very crucial, spectral reflectance measurements can help to match upper and lower Bragg mirrors, a task not possible using ex-situ methods [22].

3.3 Reflectance Anisotropy Spectroscopy (RAS)

As mentioned in the above paragraph, reflectance is not suitable for the analysis of nanostructures (i.e., 1–10 nm) because the contribution to the total signal is very small. This problem was solved by Reflectance Anisotropy Spectroscopy (RAS), developed independently by Aspnes and Berkovits [23, 24] which extracts only the anisotropic part of the reflectance. RAS measures the relative change of polarisibility in two perpendicular directions of the surface plane which in optically isotropic materials cannot originate from the bulk but usually originates from anisotropic surface structures (reconstructions). For all classical and technologically important semiconductor surfaces used in materials like Si, GaAs, or InP with zincblende or diamond structure RAS gives a signal related to the surface and allowing for identification of certain surface structures.

The setup of an RAS system in Fig. 3.2 is relatively simple. Polarized light is shone perpendicular on the surface. The polarization direction is 45° to the main eigenaxis of the surface, i.e. along [110] and [$\bar{1}$10] on (001)-surfaces of materials with zincblende or diamond structure. Because of the surface anisotropy the reflected light is then elliptically polarized. This polarization state is transformed into an intensity modulation via a photo-elastic modulator and an analyzer prism (for a more in-depth explanation see [16]). The change of polarization originates in first order from the surface and interfaces, since the bulk is isotropic. Bulk contributions can originate only from symmetry-reducing forces like strain, electric fields (due to bend

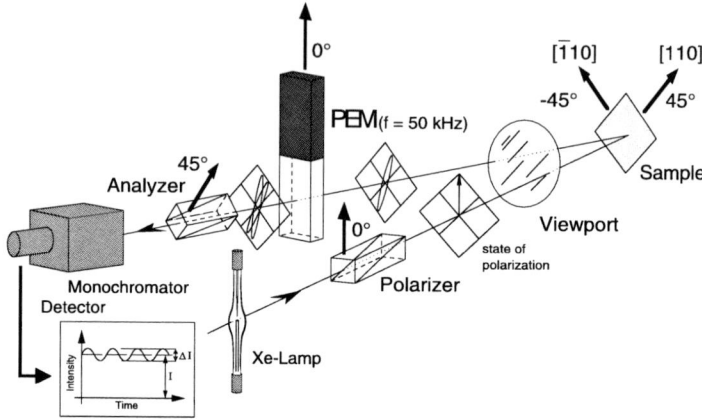

Fig. 3.2. Schematic setup of a typical RAS-system. The diagrams indicate the polarisation of the light at the corresponding positions

bending) or other symmetry-breaking effects. Although nearly all materials experimentally exhibit some RAS response, the understanding of RAS and consequently acceptance of RAS as an analytical tool has essentially benefited from, and would have not been possible without, the success of modern ab-initio theory in describing the spectral dependence shape [25, 26].

3.3.1 RAS Spectra and Surface Reconstruction

RAS has found its foremost application for the analysis of surfaces during epitaxy. Although during the early 1990s, for the most common III–V-semiconductor GaAs the existence of different reconstructions—dependent on the V–III-ratio during growth—was well established in MBE [27] and confirmed by RAS to exist in MOVPE too [3], for the other III–V semiconductor almost no data existed. For this reason the work with RAS was focused in large part on InP as the other most important semiconductor. Based on RAS measurements, STM images and Photoemission Spectroscopy results, together with ab-initio theory calculations, a model for the InP (001) surfaces during growth could be established. An excellent example is the InP (2×4) surface. Several surface models for the (2×4) reconstructions were proposed, and the mixed dimer model showed the best agreement [28]. The most convincing argument was the excellent agreement between RAS measurement at 25 K and ab-initio DFT-GW calculation [29] (Fig. 3.3). For the GaP(001) surface RAS could show that actually two reconstructions with (2×4) symmetry exist. In combination with ab-initio calculations a model for those two surfaces could be derived [30].

Apart from academic interest, the reconstruction of the surface is of utmost importance for switching sequences in order to obtain a proper interface. An arsenic double layer—e.g., as found on the GaAs(001)-c(4×4)—has a higher arsenic carry over into InGaP than a (2 × 4) reconstruction. Thus, a longer purge time or the

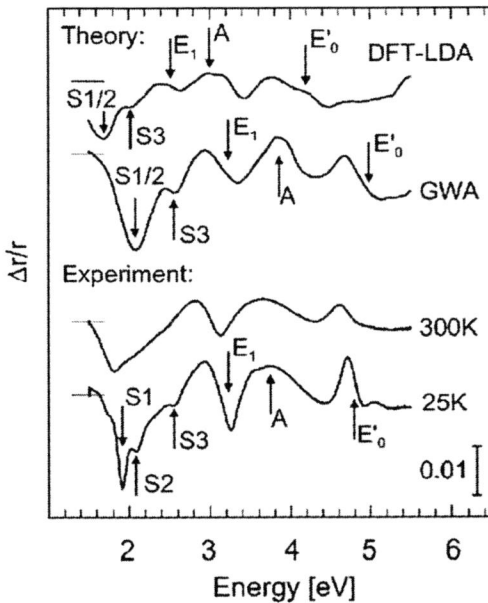

Fig. 3.3. RAS spectrum of the (2 × 4) reconstruction of InP (001) at 300 K and 25 K, together with calculated values from ab-initio theory [29]. The good agreement with the GWA calculation is especially noteworthy

growth of one monolayer of gallium after switching off the arsenic flux can improve this interface. On other interfaces like GaAsSb/InP RAS can also help to control segregation [31].

Furthermore, incorporation is controlled via the adsorption sites of the reconstruction. Phenomena like ordering are directly related to the reconstruction of the growing surface. In case of InGaP on GaAs(001) the strongest ordering was found to occur with a (2 × 1) reconstruction. Forcing a (2 × 4) reconstruction to occur by either high offcut angle, higher temperatures or high doping reduced the ordering and thus the bandgap of the bulk material [32].

Since the main RAS signal originates from the surface-related states which protrude less than 10 monolayers into the bulk, the RAS signal reacts very quickly to changes in the top materials' composition. It can be used to estimate the concentration for very thin layers in the monolayer regime. This was first demonstrated by Zorn et al. for GaAsP [6] (Fig. 3.4). Combining RAS and reflectance, the composition of AlGaInP and other quaternaries could also be obtained [17].

The sensitivity of RAS to the surface chemistry was also exploited to investigate the GaAs overgrowth (capping) of InAs quantum dots on GaAs [33, 34]. The RAS signal indicated clearly when the GaAs surface was completely recovered. Especially for the overgrowth of quantum dots layer stacks, it was found that during a growth interruption after an overgrowth thickness of only 3.5 nm the surfaces became more In-rich again [34]. Together with ex-situ Atomic Force Microscopy images, this

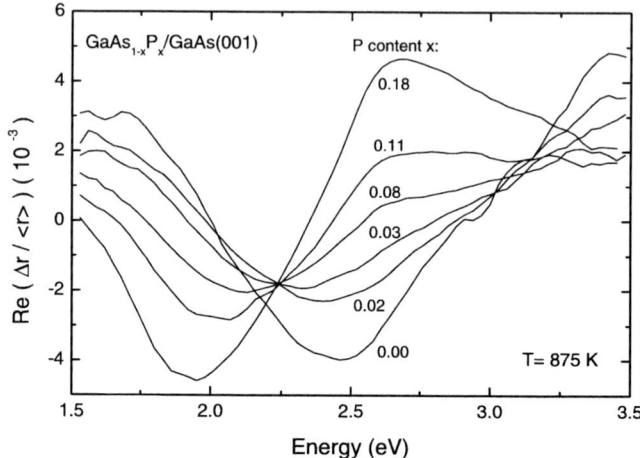

Fig. 3.4. Dependence of the RAS spectra on the atomic composition of GaAsP surfaces

fact gave clear evidence that at this overgrowth thickness larger InAs clusters were not overgrown. As a consequence during growth interruption the material from the clusters diffuses out of the clusters onto the surface, driven by the different surface energies [35]. Further overgrowth then leads to a second wetting layer and to ten times higher photoluminescence signals, since the large clusters with defects had disappeared [34].

3.3.2 Monolayer Oscillations

RAS also allows for a very accurate way to determine the growth rate, even during homoepitaxy, by direct observation of the monolayer-by-monolayer growth oscillations similar as in MBE with RHEED. A real substrate surface always possesses regularly spaced steps, either due to some intentional offcut or by chance. At high enough temperatures, no monolayer oscillations are visible since the energy of the species on the surface (called subsequently monomers) is large enough such that they can diffuse until they reach a preferred site for incorporation—a hole, a kink or a step-edge. This growth mode is called step-flow growth, and the surface appears always the same in RAS as well as in RHEED. Reducing the temperature and/or increasing the growth rate will reduce the diffusion length. When the diffusion length becomes shorter than the spacing between steps (i.e. the terrace width), the growth mode will change to the so-called 2D-island growth. Growth in 2D-island mode occurs via four stages (cf. Fig. 3.5).

I   Nucleation starts from monomers (often atoms) that are unable to reach a step edge. The first islands are small. Monomers density is much higher than island density.

II After island and the monomer densities cross each other, they become inversely proportional ($\propto t^{\frac{1}{3}}$ or $\propto t^{-\frac{1}{3}}$).

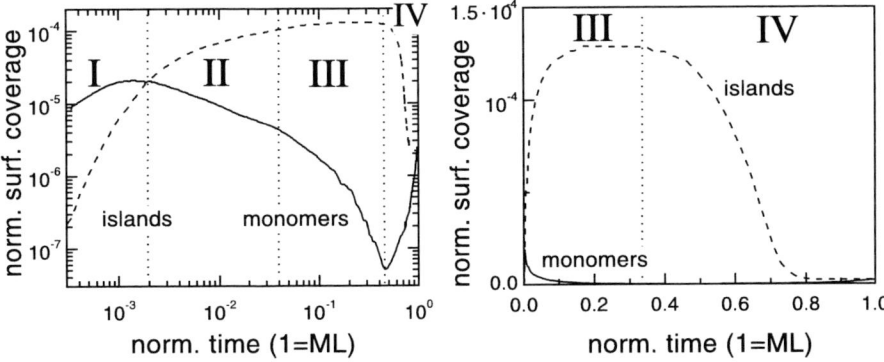

Fig. 3.5. Calculated coverage of monomers (atoms) and islands during the growth of one monolayer of an ideal flat crystal plotted on logarithmic (*left*) and linear scales (*right*) [36]

III When the size of the islands is comparable to their distance, no further islands will nucleate. Nearly all incoming monomers will attach to the borders of the island. Thus, the monomer density is dropping rapidly.

IV Finally, the maximum number of islands is reached. At that moment the number of steps on a terrace is at maximum and thus the surface roughness is maximal. Further monomer incorporation leads to island coalescence and the density of islands decrease.

On top of the islands nucleation is starting again, as soon as the size of an island becomes larger than the diffusion length. Hence the monomer density is increasing again.

These stages are demonstrated in Fig. 3.5, which shows the result of a Monte Carlo simulation for the development of the monomer density and the island density during the growth of a single monolayer [36]. This four regimes can be easily distinguished on the left side of Fig. 3.5. The right-hand side of Fig. 3.5 shows monomer and island densities on a linear time and surface coverage scale. Normally, these scales are relevant for comparison to actual growth experiments. Two observations are remarkable: first, the number of monomers at all times is very small, and therefore monomers are hardly observed experimentally. Second, only the development of the island size in regime III and IV (where unfortunately no good analytical theory exists) is important for comparison of experiment and theory.

This growth mode was first discovered and verified via RHEED intensity oscillations by Joyce and coworker [37]. The RHEED intensity variation during the growth of a monolayer simply correlates with the number of step edges scattering the electrons out of the Bragg directions [38]. At half monolayer coverage (i.e., maximum roughness) the scattering is high and consequently Bragg reflected intensity is low. After completion of a monolayer the surface is again relatively smooth and the reflected intensity increased.

During MOVPE a similar effect was observed during growth of GaAs with GIXS [39, 2], where the intensity of a crystal truncation rod changed due to scattering.

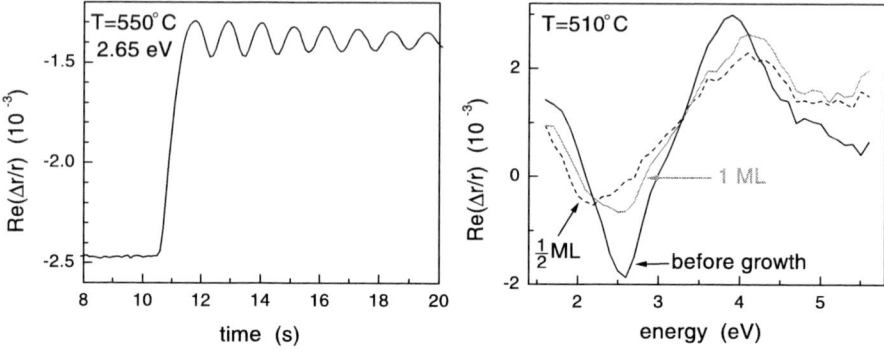

Fig. 3.6. The left side shows an RAS transient during growth of GaAs with 0.5 Pa TMGa at 550°C. The monolayer oscillations are clearly visible even at this temperature. The right side shows spectra composed from 36 transients of the typical surface before, fractional monolayer coverage ($+n\frac{1}{2}$) and full monolayer coverage. With half a monolayer, the spectrum is clearly shifted to lower energies

From the scattering data the minimum distance of islands before nucleation could be estimated as a function of temperature [39]. Typical values were in the range of 30–80 nm [2], resulting in an activation energy for diffusion of around 2.7 eV.

The oscillations of the RAS-signal with monolayer periodicity were observed first during growth of GaAs at 500–550°C [5] (left side of Fig. 3.6). These oscillations were also clearly related to the 2D-island growth mode. Calculations showed, however, that the intensity change due to scattering of light would be too small in order to be observed. Instead, the appearance of the oscillation in the RAS signal is connected to the change of the reconstruction near the step edges. Spectrally resolved measurements of the RAS signal during signal oscillation found that the surfaces oscillate between more gallium covered surfaces at fractional monolayers and more arsenic-rich surfaces near a completed monolayer (Fig. 3.6). Quantitative fits at 510°C gave for the two surfaces a 40% and 60% gallium rich ($n \times 6$) surface and (1×2)-CH$_3$ respective c(4×4) covered surface [40].

The origin of the gallium-rich surface is also related to the number of step edges. This was demonstrated by investigation of vicinal substrates. At higher offcut angle the surfaces became less arsenic rich [41]. Since the number of islands will change the number of steps during 2D island growth mode, oscillation in the RAS signal will be observed. Figure 3.7 demonstrates this schematically. Increasing the temperature and thus increasing arsenic kinetics the recovering of the growth front will become faster. Thus the gallium-rich areas shrink and the oscillation amplitude decreases with increasing temperature.

However, also the diffusion length will increase with increasing temperature and therefore decreases the number of islands on the terraces. This will decrease the oscillation amplitude, too. Thus, the vanishing of monolayer oscillations in the RAS signal with increasing temperature can have two reasons. Either the growth mode is

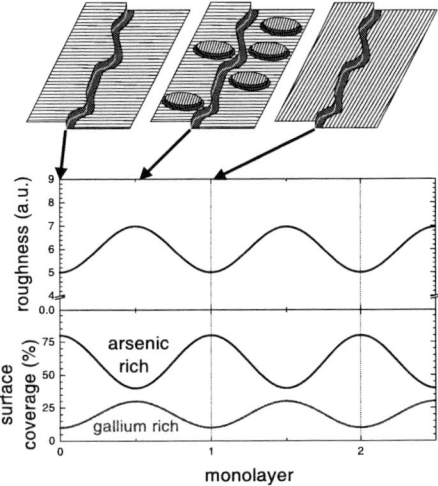

Fig. 3.7. The ideal surface composition and roughness after n, $n + \frac{1}{2}$, and $n + 1$ monolayer growth of GaAs. The top shows schematically the surface around a step-edge, dark areas mark the steps with a gallium-rich reconstruction, the striped areas indicate an arsenic-rich area. Below, the diagrams show the change of roughness (corresponding to the RHEED intensity signal) and surface coverage (corresponding to the RAS signal). In both cases oscillations with monolayer period appear

changing to step-flow growth or the surface kinetics provide an identical surface at all stages.

This can be distinguished by comparison of the transition temperature on different vicinal surfaces. At a higher offcut (i.e. a smaller terrace width) the transition to step-flow growth occurs a lower temperature but the kinetics of AsH$_3$ and TMGa are similar. Comparing growth oscillation on vicinal and nominally flat surfaces, at least up to 590°C the transition to step-flow growth is clearly visible. These data are supported by GIXS truncation rod X-ray scatter data, which observed the transition to step-flow growth between 590°C and 620°C [2].

The RAS oscillations are visible up to 600°C for GaAs and can be used to determine the growth rate (left side of Fig. 3.6). Their maximum amplitude is around 505°C (Fig. 3.8). The change of the surface dielectric function of the different surfaces was even measured quantitatively using spectral ellipsometry, and agreed well with a theoretical calculation [9]. RAS oscillations are also visible for strained InGaAs on GaAs up to 50% Indium [6, 42], and at 400°C one or two oscillations were observed during growth of InAs [4]. Also for growth of GaAsN with up to 7% nitrogen, RAS oscillations were visible.

By measuring the increase of the growth rate of InGaAs relative to GaAs, one can also determine the composition of the InGaAs. A superlattice (SL) grown only via this calibration resulted in 30 periods each consisting of 5 monolayers In$_{0.11}$Ga$_{0.89}$As and 10 monolayers of GaAs, as intended [6]. Figure 3.9 shows the RAS transient of the whole SL and a zoom to a single period. The RAS signal of each SL periods is

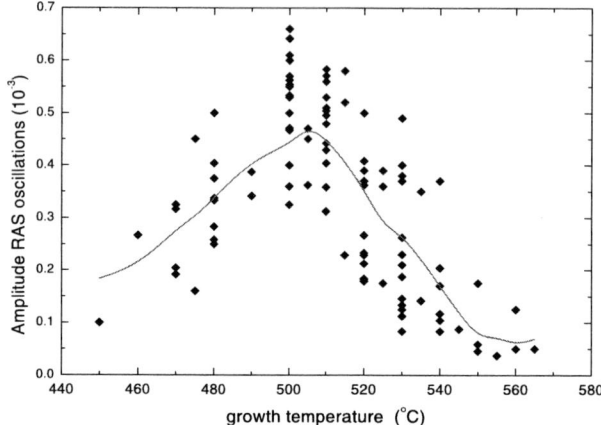

Fig. 3.8. Monolayer oscillation amplitude for growth of GaAs at different temperatures measured between 1994 and 1996 with two different RAS systems at our MOVPE reactor. The thick line is a median of all points

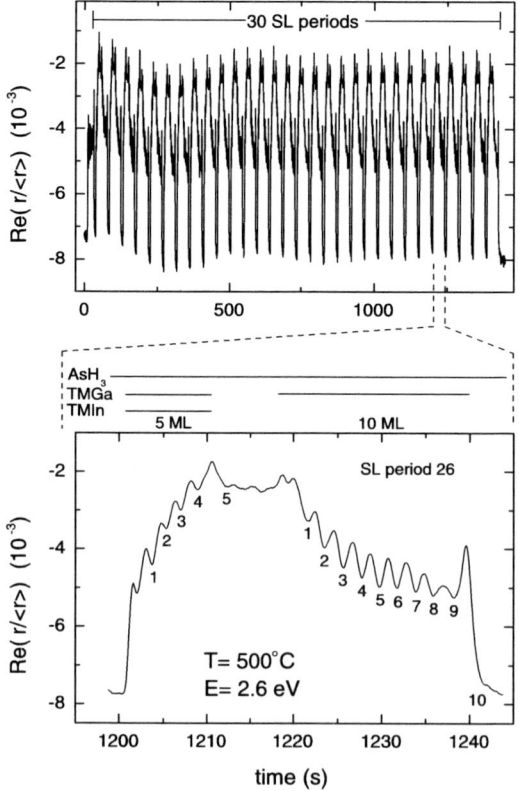

Fig. 3.9. RAS transient of a GaAs/ InGaAs superlattice with 30 periods (*top*) and zoom of a single period (*bottom*)

identical, modulated only by a superimposed interference signal from the thick bulk layer.

There are material systems which apparently lack a different reconstruction near the step-edges, like the phosphides. Therefore, no RAS oscillations are visible for InP, even when the growth proceeds in 2D-island growth mode at low temperatures as found in GIXS experiments [43]. This is, of course, a disadvantage of the RAS method. But for all arsenide-based semiconductors RAS works quite well.

3.3.3 Monitoring of Carrier Concentration

RAS is not only sensitive to the structure of surfaces or interfaces but also to the doping concentration. This was first discovered by Acosta-Ortiz and Lastra-Martinez [44] on oxidized GaAs. The origin lies in the electric field created due to band bending at surfaces or interfaces. This field is always normal to the interface or surface and constitutes a symmetry reduction for cubic materials. It modifies the dielectric properties in lowest order via the linear electro-optic effect. In cubic zincblende materials the third-order electro-optic tensor has just one independent component which causes, with a field normal to the interface, an anisotropic contribution in the plane of the interface [45].

The magnitude of the change, of course, differs at different spectral positions depending on the electronic interband transitions contributing to the dielectric function. Figure 3.10 displays the change of a GaAs RAS spectrum due to doping with 1.8×10^{18} cm^3 silicon. For GaAs the strongest feature occurs at the E_1 critical point

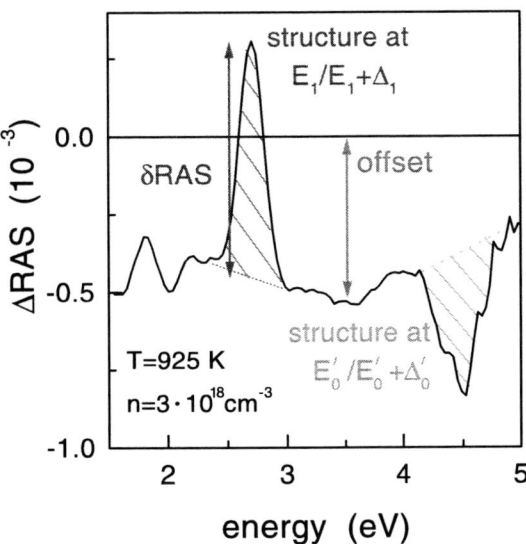

Fig. 3.10. Difference between doped and undoped RAS spectrum (ΔRAS) for a c(4×4)-like reconstructed n-type (1.8×10^{18} cm^3 Si) GaAs (001) surface at 650°C in MOVPE [46]

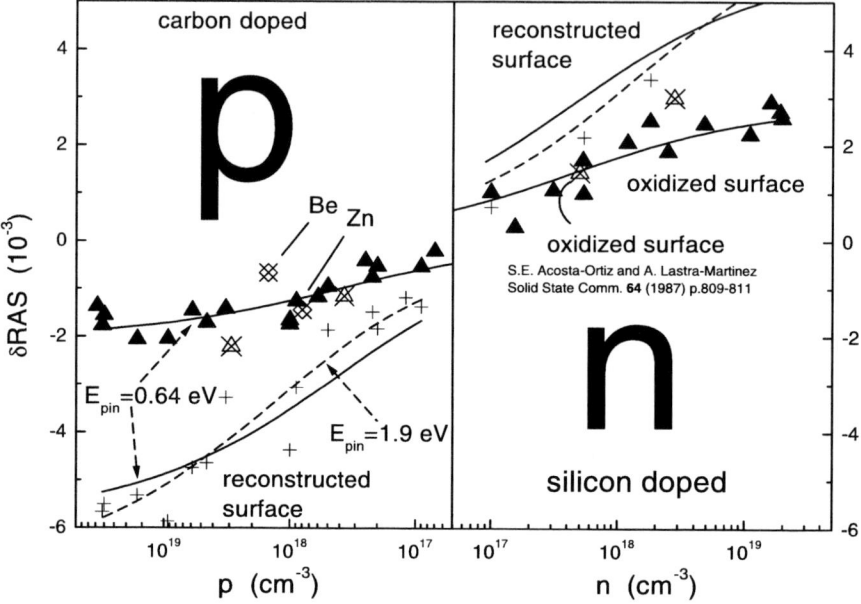

Fig. 3.11. δRAS values obtained on oxidized and (4 × 3) reconstructed surfaces at room temperature. The solid lines are best fits from (3.3) with $E_{\text{pin}} = 0.64\,\text{eV}$, the dashed curves were fitted with $E_{\text{pin}} = 1.9\,\text{eV}$ [47]

together with a smaller one at the E_0'. Furthermore an offset to the spectrum is also visible.

Quantitatively, the effect can be described using the optical indicatrix. For a (001) surface the change of the RAS signal (δRAS) due to an electric field in z-direction is given by

$$\delta\text{RAS} = \frac{\Delta B_{12}}{(n_{10})^2 - 1}(n_{10})^3 \tag{3.2}$$

with ΔB_{12} being the change of the indicatrix along the xy-direction and n is the refractive index ([47]). The amount of ΔB_{12} is linear proportional to the electric field. The electric field at a surface/interface can be estimated via the simple Schottky model. For the optical RAS signal we must also consider the penetration depth of light. The final formula for the effective electric field is

$$\langle E \rangle = \sqrt{\frac{2eN}{\varepsilon_0 \varepsilon}\phi_0} \left\{ 1 + \frac{z_p}{z_d} \left[\exp\left(-\frac{z_d}{z_p}\right) - 1 \right] \right\}, \tag{3.3}$$

where z_p is the penetration depth of light, N the carrier concentration, and z_d the depletion width, which can be calculated from the Schottky model, if the band bending ϕ_0 is known: $z_d = \sqrt{\frac{2\varepsilon_0\varepsilon}{eN}\phi_0}$ [47].

The resulting sensitivity of the change in the RAS signal due to the electric field gives an S-like shape as shown in Fig. 3.11, which shows the change for GaAs as

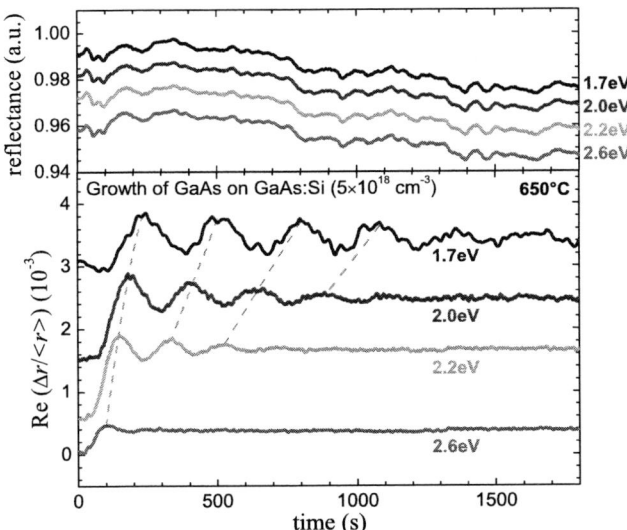

Fig. 3.12. Reflectance and RAS transients recorded simultaneously during growth of GaAs on GaAs:Si (5×10^{18} cm^{-3}). The transients are shifted vertically for better visibility. The gray broken lines indicate the wavelength dependence of the oscillations

a function carrier concentration. The solid lines are fits with (3.3) and resulted in a pinning position in midgap position (0.66 V). From such a calibration the carrier concentration can be directly evaluated [46].

However, this method has some disadvantages. First, the full spectral information is needed; second, an undoped reference surface is needed. (Although the reference can be taken from a database if the alignment and window corrections are done properly.)

We recently discovered another method to determine the doping concentration during layer growth. This method even yields the growth rate during homoepitaxy, and thus provides the possibility for full in-situ doping profiling. During growth of a layer with a doping concentration different than the underlying layer, the RAS signal will oscillate (Fig. 3.12). These oscillations are Fabry–Perot-like interferences due to the change of the refractive index through the electro-optic effect of the electric space charge field at the interface. These anisotropic changes are small (well below 1%), and thus manifest themselves only in the RAS signal.

Figure 3.12 clearly demonstrates that these oscillations are not present in the reflectance but only in the RAS signal. Analytically, the amplitude can be calculated via

$$\delta \text{RAS} = \frac{1}{\hat{n}(\hat{n}^2 - 1)} \exp\left(\frac{i 4\pi d_1 \hat{n}}{\lambda}\right)\left(1 + \exp\left(-\frac{i 4\pi z_d \hat{n}}{\lambda}\right)\right) \Delta \hat{\varepsilon}_2 \quad (3.4)$$

with $\hat{n}$ the complex refractive index of the material at growth temperature, λ the wavelength, z_d the thickness of the space charge layer, and $\Delta \hat{\varepsilon}_2 = \hat{\varepsilon}_{2\alpha} - \hat{\varepsilon}_{2\beta}$ the complex dielectric anisotropy of the space charge layer. As can be expected, the

RAS signal is proportional to the dielectric anisotropy $\Delta\hat{\varepsilon}_2$ [48]. The growth rate can be evaluated as described in the previous section on reflectometry.

3.4 Scanning Tunneling Microscopy (STM)

Scanning Tunneling Microscopy (STM) is a relatively new technique for in-situ analysis. Invented in 1982 by Binning and Rohrer, the technique's ability to measure surfaces in real space with atomic resolution very quickly made it an indispensable tool for surface physics, similar to its derivative, Atomic Force Microscopy (AFM).

The ability to determine the position of each atom on the surface makes STM the ultimate tool for surface growth studies, except for the experimental difficulties of obtaining images at growth conditions. Furthermore, STM is a serial method, i.e. an image is taken by recording the height on each position of the image. Depending on the setup, scan speed, image size, desired spatial resolution, and the number of desired points the time to obtain an image can vary from 10 s up to 10 minutes, which is longer than the usual growth rates of one monolayer per second. For a typical STM a single 256×256 images takes about a minute or longer. Still, within these limits STM can give insight into processes not possible by any other method.

The principle of STM is quite simple (Fig. 3.13). A tip is brought in close proximity to the surface. When a bias voltage is applied between tip and surface a tunnel current will flow. This tunnel current depends exponentially on the distance between sample and tip, and changes more than one order of magnitude for a distance change of 0.2 nm in close proximity. The absolute STM current amplitude is given by the combined density of states integrated up to the current bias voltage for a given tip distance. Thus, from current/voltage measurement at a fixed distance the local band gap or surface states can be determined too. Because of the exponential dependence of the tunnel current, the vibrations between the tip and surface must be well below 0.1 nm. This is a difficult problem in MOVPE because of the vibrations created by the mechanical pumps.

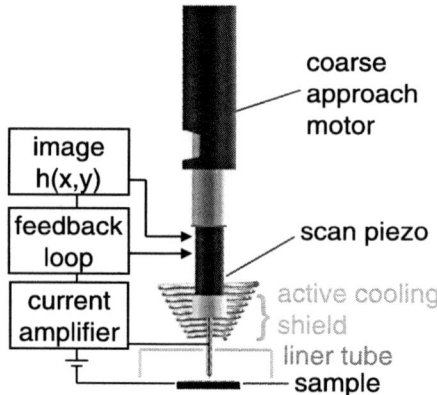

Fig. 3.13. Schematic and principle of the MOVPE-STM, after [14]

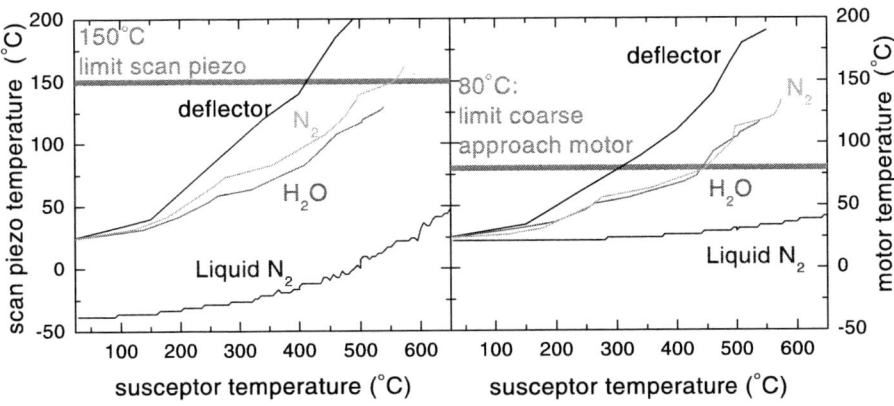

Fig. 3.14. The temperature of two critical parts of the MOVPE in-situ STM with different cooling solutions as mentioned

Apart from the vibration insulation there are other problems that make the application of STM difficult for MOVPE in-situ measurements. One is the thermal movements of the atoms themselves, which will set the natural limit. If the atoms move faster than the average time for scanning a point, the image will be blurred. However, usually at much lower temperatures apparative limits are reached. The piezo-crystals which move the sample with sub-nm precision have a relatively small temperature working range, usually less than 250°C. Comparing this to the typical crystal growth temperatures of III–V semiconductors between 500–1200°C, apparative provisions are needed to prevent heat transfer to the piezo-crystals. In a vacuum (MBE) two solutions are possible: either one uses a very small sample or provides a heat radiation shadow shield. These solutions fail in MOVPE: The gaseous environment transfers heat up to 1000× more effectively than radiation, and the necessary large susceptor areas provide a large hot area. To obtain STM images in MOVPE, an active cooling mechanism is needed. In the published design [14] several heat shields were tested (Fig. 3.14). Only cooling with precooled (80 K) nitrogen gas could maintain the temperature of the piezo-crystals and the piezo coarse approach motor below their respective temperature limits up to a growth temperature of 650°C.

The electric connections (shielded and grounded) constitute another problem with respect to short circuits. Considering the tunnel current to be in the order of nA, even 10 MΩ must be considered a short circuit. This is problematic, since arsenic is semimetallic and is available in abundance during GaAs epitaxy. Thus, condensation of arsenic on all parts of the STM must be avoided. The metallic components (In, Al and Ga) on the other hand have a much smaller vapor pressure and are usually less of a problem. For horizontal MOVPE reactors, condensation can be minimized by placing all sensitive parts outside the inner liner tube and carefully balancing the fluxes inside and outside of the liner tube. Vertical reactors with rotating susceptors at present seem to be not suitable for STM.

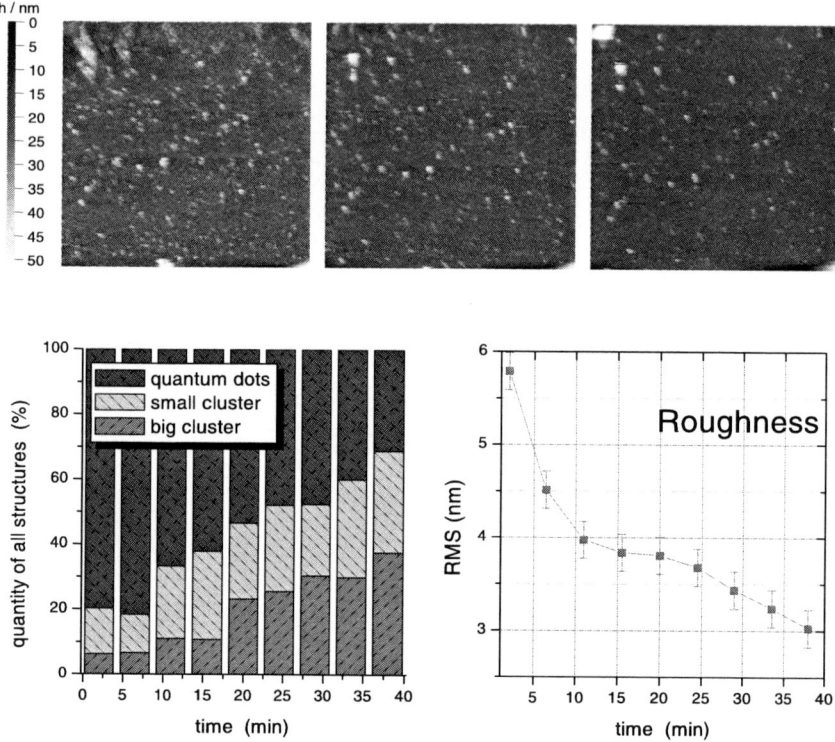

Fig. 3.15. Sequence of three selected images from left to right immediately after InAs dot growth, after 8 min and 16 min annealing at 475°C. The vanishing of small dots due to Ostwald ripening is clearly seen. The lower figure shows the development of the density of dots, small and large clusters, and the lower right figure shows the root means square roughness obtained from these STM images

For in-situ monitoring of III–V semiconductor in MOVPE, currently only a single STM setup at the TU-Berlin exists (for details see [14, 15]). This STM can operate between 0°C and 650°C on AlInGaAsP compounds. A practical example showing the promises of in-situ STM studies is the sequence of three images in Fig. 3.15 during Ostwald ripening of InAs quantum dots on GaAs(001) at 475°C. For all images in this sequence of nine images the densities of quantum dots, small and large clusters were determined and plotted into the lower left part of Fig. 3.15. Clearly the vanishing of the dots by transiting through small clusters into large clusters is visible. Also the root means square roughness decreases accordingly. More studies will be necessary to investigate the exact transfer mechanism, but clearly the sequence of pictures shows the great possibilities of in-situ STM.

3.5 Conclusion

As devices progress to smaller and smaller scales into the realm where even single atoms make important contributions, in-situ monitoring becomes more and more indispensable. Current optical in-situ monitoring using Reflectance Anisotropy Spectroscopy (RAS) can create C/V-profiles during growth with a resolution of 30 nm vertically, can determine the growth rate with an error of 1%, and can even determine the composition within a few percent. Such control allows for processes where a small parameter window must be obeyed or where thickness control on the monolayer regime becomes necessary. This technique may be complemented by optical reflectance for larger or thicker structures or materials like GaN which do not have a cubic structure and RAS cannot be used. The ultimate monitoring of single atoms during growth comes just into view with the development of the first in-situ Scanning Tunneling Microscope for MOVPE. This exciting technique is still in its infancy, but shows great promise.

References

1. D.W. Kisker, P.H. Fuoss, K.L. Tokuda, G. Renaud, S. Brennan, J.L. Kahn, Appl. Phys. Lett. **56**, 2025 (1990)
2. D.W. Kisker, G.B. Stephenson, P.H. Fuoss, F.J. Lamelas, S. Brennan, P. Imperatori, J. Cryst. Growth **124**, 1 (1992)
3. I. Kamiya, D.E. Aspnes, H. Tanaka, L.T. Florez, J.P. Harbison, R. Bhat, Phys. Rev. Lett. **68**, 627 (1992)
4. S.M. Scholz, A.B. Müller, W. Richter, D.R.T. Zahn, D.I. Westwood, D.A. Woolf, R.H. Williams, J. Vac. Sci. Technol. B **10**, 1710 (1992)
5. F. Reinhardt, W. Richter, A. Müller, D. Gutsche, P. Kurpas, K. Ploska, K. Rose, M. Zorn, J. Vac. Sci. Technol. B **11**, 1427 (1993)
6. M. Zorn, J. Jönsson, A. Krost, W. Richter, J.-T. Zettler, K. Ploska, F. Reinhardt, J. Cryst. Growth **145**, 53 (1994)
7. N. Kobayashi, Y. Horikoshi, Jpn. J. Appl. Phys. **29**, L702 (1990)
8. M. Drago, C. Werner, M. Pristovsek, U.W. Pohl, W. Richter, Cryst. Res. Technol. **40**, 993 (2005)
9. J.-T. Zettler, T. Wethkamp, M. Zorn, M. Pristovsek, C. Meyne, K. Ploska, W. Richer, Appl. Phys. Lett. **67**, 3783 (1995)
10. J.P. Silveira, J.M. Garcia, F. Briones, J. Cryst. Growth **227–228**, 995 (2001)
11. A. Krost, A. Dadgar, F. Schulze, J. Bläsing, G. Strassburger, R. Clos, A. Diez, P. Veit, T. Hempel, J. Christen, J. Cryst. Growth **275**, 209 (2005)
12. M. Zorn, F. Bugge, T. Schenk, U. Zeimer, M. Weyers, J.-T. Zettler, Semicond. Sci. Technol. **21**, L45 (2006)
13. S. Tsukamoto, N. Koguchi, J. Cryst. Growth **201/202**, 118 (1999)
14. B. Rähmer, M. Pristovsek, M. Breusing, R. Kremzow, W. Richter, Appl. Phys. Lett. **89**, 063108 (2006)
15. M. Pristovsek, B. Rähmer, M. Breusig, R. Kremzow, W. Richter, J. Cryst. Growth **298**, 8 (2007)
16. M.A. Herman, W. Richter, H. Sitter, *Epitaxy Physical Principles and Technical Implementation*. Springer Series in Materials Science, vol. 62 (Springer, Berlin, 2004). Chap. 10
17. M. Zorn, J.-T. Zettler, Phys. Stat. Sol. (b) **242**, 2587 (2005)
18. J.-T. Zettler, K. Haberland, M. Zorn, M. Pristovsek, W. Richter, P. Kurpas, M. Weyers, J. Cryst. Growth **195**, 151 (1998)

19. K. Haberland, A. Kaluza, M. Zorn, M. Pristovsek, H. Hardtdegen, M. Weyers, J.-T. Zettler, W. Richter, J. Cryst. Growth **240**, 87 (2002)
20. K. Haberland, M. Zorn, A. Klein, A. Bhattacharya, M. Weyers, J.-T. Zettler, W. Richter, J. Cryst. Growth **248**, 194 (2003)
21. M. Drago, C. Werner, M. Pristovsek, U.W. Pohl, W. Richter, Phys. Stat. Sol. (a) **203**, 1622 (2006)
22. M. Zorn, K. Haberland, A. Knigge, A. Bhattacharya, M. Weyers, J.-T. Zettler, W. Richter, J. Cryst. Growth **235**, 25 (2001)
23. V. Berkovits, I. Marenko, T. Minashvili, V. Safarow, Sol. State Commun. **56**, 449 (1985)
24. D.E. Aspnes, J.P. Harbison, A.A. Studna, L.T. Florez, M.K. Kelly, J. Vac. Sci. Technol. A **6**, 1327 (1988)
25. A.I. Shkrebtii, N. Esser, W. Richter, W.G. Schmidt, F. Bechstedt, B.O. Fimland, A. Kley, R. Del Sole, Phys. Rev. Lett. **81**, 721 (1998)
26. O. Pulci, G. Onida, R. Del Sole, L. Reining, Phys. Rev. Lett. **81**, 5374 (1998)
27. L. Däweritz, R. Hey, Surf. Sci. **236**, 15 (1990)
28. W. Schmidt, F. Bechstedt, N. Esser, M. Pristovsek, C. Schultz, W. Richter, Phys. Rev. B **57**, 14596 (1998)
29. W.G. Schmidt, N. Esser, A.M. Frisch, P. Vogt, J. Bernholc, F. Bechstedt, M. Zorn, T. Hannappel, S. Visbeck, F. Willig, W. Richter, Phys. Rev. B **61**, R16335 (2000)
30. A.M. Frisch, W.G. Schmidt, J. Bernholc, M. Pristovsek, N. Esser, W. Richter, Phys. Rev. B **60**, 2488 (1999)
31. S. Weeke, M. Leyer, M. Pristovsek, W. Richter, J. Cryst. Growth **298**, 159 (2007)
32. M. Zorn, P. Kurpas, A.I. Shkrebtii, B. Junno, A. Bhattacharya, K. Knorr, M. Weyers, L. Samuelson, J.T. Zettler, W. Richter, Phys. Rev. B **60**, 8185 (1999)
33. E. Steimetz, W. Richter, F. Schienle, D. Fischer, M. Klein, J.-T. Zettler, Jpn. J. Appl. Phys. **37**, 1483 (1998)
34. E. Steimetz, T. Wehnert, H. Kirmse, F. Poser, J.-T. Zettler, W. Neumann, W. Richter, J. Cryst. Growth **221**, 592 (2000)
35. L.G. Wang, P. Kratzer, M. Scheffler, Q.K.K. Liu, Appl. Phys. A **73**, 161 (2001)
36. J. Amar, F. Family, Phys. Rev. B **50**, 8781 (1994)
37. J.K. Harris, B.A. Joyce, P.J. Dobson, Surf. Sci. Lett. **103**, L90 (1981)
38. J.H. Neave, B.A. Joyce, P.J. Dobson, N. Norton, Appl. Phys. A **31**, 1 (1983)
39. P.H. Fuoss, D.W. Kisker, F.J. Lamelas, G.B. Stephenson, P. Imperatori, S. Brennan, Phys. Rev. Lett. **60**, 2610 (1992)
40. J.-T. Zettler, J. Rumberg, K. Ploska, K. Stahrenberg, M. Pristovsek, W. Richter, M. Wassermeier, P. Schützendüwe, J. Behrend, L. Däweritz, Phys. Stat. Sol. (a) **152**, 35 (1995)
41. K. Ploska, M. Pristovsek, W. Richter, J. Jönsson, I. Kamiya, J.-T. Zettler, Phys. Stat. Sol. (a) **152**, 49 (1995)
42. U.W. Pohl, K. Pötschke, I. Kaiander, J.-T. Zettler, D. Bimberg, J. Cryst. Growth **272**, 143 (2004)
43. T. Kawamura, Y. Watanabe, S. Fujikawa, S. Bhunia, K. Uchida, J. Matsui, Y. Kagoshima, Y. Tsusaka, Surf. Interface Anal. **35**, 72 (2003)
44. S. Acosta-Ortiz, A. Lastra-Martinez, Solid State Commun. **64**, 809 (1987)
45. J.F. Nye, *Physical Properties of Crystals, Their Representation by Tensors and Matrices* (Oxford University Press, Oxford, 1957)
46. M. Pristovsek, S. Tsukamoto, N. Koguchi, B. Han, K. Haberland, J.-T. Zettler, W. Richter, M. Zorn, M. Weyers, Phys. Stat. Sol. (a) **188**, 1423 (2001)
47. M. Pristovsek, S. Tsukamoto, B. Han, J.-T. Zettler, W. Richter, J. Cryst. Growth **248**, 254 (2003)
48. C. Kaspari, M. Pristovsek, W. Richter, J. Cryst. Growth **298**, 46 (2007)

4

Bottom-up Approach to the Nanopatterning of Si(001)

R. Koch

Abstract. The fabrication of single-crystal semiconductor nanostructures which—due to their atomic-like discrete energy levels—are suitable for laser applications, has been a topic of intense research since the early 1990s (Mo et al., *Phys. Rev. Lett.* **65**, 1020, 1990; Eaglesham and Cerullo, *Phys. Rev. Lett.* **64**, 1943, 1990; Lott et al., *Electron. Lett.* **36**, 1384, 2000). A drawback of the common preparation procedure of such quantum dots, namely film growth by the Stranski–Krastanow (SK) mode, is the nonuniform size distribution that broadens the optical spectra compared to quantum-well devices. We follow an alternative approach based on the growth on patterned substrates, which promises dense, uniformly sized, spatially ordered, and wetting-layer-free quantum dot arrays. This article reviews our recent studies on the nanopatterning of Si(001).

4.1 Quantum Dot Growth on Semiconductor Templates

A well-established procedure for the preparation of single-crystal quantum dots on semiconductor templates is film growth by the SK mode [4]. Three-dimensional islands nucleate on top of a continuous wetting layer, in the course of which part of the misfit strain is relieved [5, 6]. In the past years, quantum dot growth has been successfully demonstrated for various semiconductor systems, e.g., Ge/Si(001) [1], SiGe/Si(001) [7], InAs/GaAs(001) [8], PbSe/PbEuTe [9], CdSe/ZnSe [10], etc. Due to the statistical nature of the nucleation process, the obtained quantum dot arrays typically are not very regular with respect to their spatial arrangement and size distribution. For instance, in the case of the SiGe/Si(001) system pyramids, huts and domes coexist in a wide thickness range [11]. Therefore the spectral distribution is broad, although individual dots exhibit sharp resonances [12]. In the case of direct-band-gap emitters, e.g., InAs/GaAs(001), the photoluminescence intensity is high. For the indirect band gap semiconductor Si, for which efficient lasing has been demonstrated in the last year [13, 14], further improvements may be expected from using small, uniformly sized SiGe quantum dots.

Recently, the nanopatterning of substrates with a periodic superstructure has become a feasible alternative to SK growth as it promises a better control of the nucleation process, and thus improved spatial ordering and size uniformity for nanostructure growth. At present, top-down approaches are mostly employed for realizing patterned templates. For instance, a better spatial and size distribution has been obtained for SiGe quantum dots nucleating on the top plane of lithographically fabricated 1D and 2D mesa structures [15–19] due to the change in diffusion from a 2D to a 1D process. Furthermore, patterned templates have been prepared by intentionally generating a dislocation network [20] or by introducing regular arrays of defect centers via focused ion beam lithography [21]. However, whereas spatial and size distribution is improved by these preparation techniques, the achieved quantum dot density is very low (10^8–10^9/cm^2).

In a previous study of the stress evolution of SiGe/Si(001) deposited at 900 K, we found that the growth mode may switch from SK to a kinetic 3D-island mode when the Si(001) surface contains a high density of dimer vacancies [22]. Figure 4.1(a) illustrates the accompanied changes in the stress behavior; the slope of the displayed force vs. thickness curves measured in situ by a cantilever beam technique corresponds to the instantaneous stress. The force curves of the pure Ge film and the SiGe alloy film with a silicon concentration of 14% are characteristic of SK growth. The slope, and thus the stress, is large during growth of the wetting layer (stress regime I), and decreases drastically when the 3D islands nucleate and grow (stress regime II). At Si concentrations above 20%, the stress evolution is reversed with low stress in regime I and larger stress in regime II. A quantitative discussion of the stress results can be found in [22]. The finding that the stress of regime II exceeds that of regime I is in clear contradiction to a SK mode. According to theory [23–25], the 2D/3D transition of the SK mode reduces the elastic energy; it is therefore necessarily accompanied by strain relief and thus a decrease of the instantaneous stress. In situ STM investigations of the $Si_{0.60}Ge_{0.40}$ (Fig. 4.1(b)) and $Si_{0.80}Ge_{0.20}$ films (Fig. 4.1(c)) reveal that both films consist of 3D islands even at the low mean thickness of 1 nm. The average island diameter is about 5 nm and the island height ranges from 1.5 to 2.5 nm with a resulting total island volume that is in good agreement with the deposited film volume. Obviously, at higher Si contents the film growth changes to a 3D island mode without formation of a wetting layer. This is consistent with the low stress values observed in regime I, because the misfit strain is relaxed efficiently by the immediate 3D islanding. These preliminary results therefore indicate that surface defects and, in particular, dimer vacancies, can be used to control the nucleation process on Si(001) and thus to improve the density and size distribution of quantum dot arrays.

4.2 (2 × n) Reconstruction of Si(001)

On Si(001), an intrinsically nanopatterned array of dimer vacancies is provided by the (2 × n) superstructure (Fig. 4.2). It consists of extended (2 × 1) reconstructed stripes separated every few nanometers by a dimer vacancy line (DVL). The

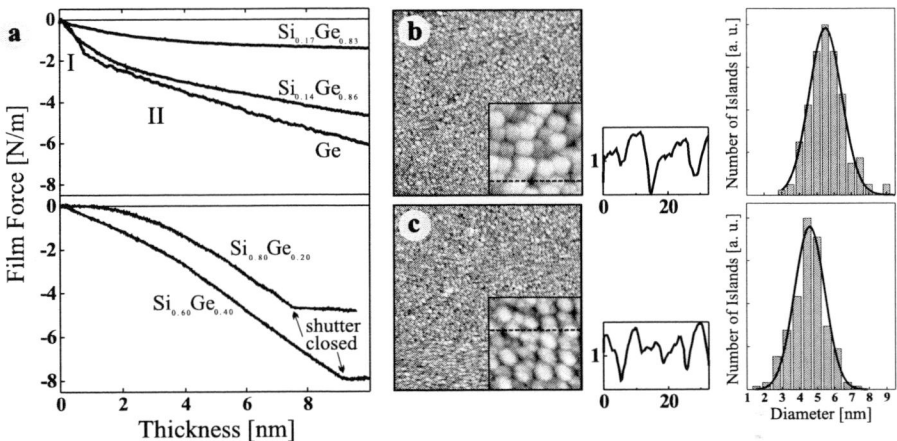

Fig. 4.1. (a) Film forces (i.e. integral forces in films of unit width) measured in real time during the deposition of various Si_xGe_{1-x} films onto Si(001) at 900 K. (b) $250 \times 250\,nm^2$ STM top view of 1 nm $Si_{0.60}Ge_{0.40}$; inset shows an area of $32 \times 32\,nm^2$ with the linescan along the dashed line; units in nm. The size of the quantum dots follows a Gaussian fit. (c) analogous results of a 1 nm $Si_{0.80}Ge_{0.20}$ film

Fig. 4.2. $25 \times 25\,nm^2$ STM top view of low-temperature Si(001) prepared by the Ishizaka–Shiraki etching procedure [37] and heated in UHV for 15 min at 820°C in order to desorb the oxide; $U_{tip} = +2.2\,V$, $I_T = 4\,nA$

Si(001)($2 \times n$) structure has been known for a long time [26] and it was originally attributed to the effect of transition metals, especially Ni contaminants [27, 28]. However, as convincingly shown by Men et al. [29] and others [30–32], the ($2 \times n$) exists also on the uncontaminated Si(001) surface and forms—though only occasionally—after repeated flashing of the sample to 1200°C and subsequent quenching. Furthermore, it has been obtained after ion-bombardment and annealing between 600 and 850°C [33], by etching of Si(001) with reactive gases [34, 35], or by quenching of deposited Si clusters from 1200°C [30].

In a recent study [36, 32] we employed Si(001) substrates that were coated with a very thin oxide by the Ishizaka–Shiraki etching procedure [37]. We found that such "low-temperature" Si(001) surfaces after oxide desorption at 820°C exhibit a relatively high density of intrinsic defects in the topmost surface layer, in particular missing dimers and extended vacancy islands. Occasionally, the surface was patterned by a $(2 \times n)$ configuration, where the missing dimers have self-organized into extended lines running perpendicular to the dimer rows at a distance of about 5 nm. Since the evaporation of the oxide proceeds via the chemical reaction $Si + SiO_2 \rightarrow 2SiO\uparrow$, Si of the Si/SiO$_2$ interface is consumed during desorption of the oxide, thus leaving behind a large number of Si vacancies at the new surface. Our findings therefore suggest that an incompletely filled topmost layer, combined with annealing temperatures and periods that are not sufficient for annihilation of the dimer vacancies at step edges, may be the key ingredients for the preparation of the $Si(001)(2 \times n)$.

4.3 Monte Carlo Simulations on the $(2 \times n)$ Formation

In order to explore the role of the Si deficiency in the surface layer, we developed a realistic Monte Carlo (MC) model that is capable of simulating the diffusion of both monomers and dimers on the Si(001) surface while simultaneously depositing new material. Due to the deposition, the model differs from previous approaches [38–40] where the dimer-vacancies are the only diffusing species. For a detailed discussion of the MC model please refer to [41]. In short, diffusion and deposition proceed on square lattices (80×80 lattice sites) that are stacked vertically as in the diamond crystal lattice. The diffusion barrier is determined by a bond-counting ansatz which takes into account the number and configuration of nearest neighbors at initial and hopping sites. For the diffusion of isolated monomers and dimers the experimental values reported in the literature are used. For diffusion from or to a dimer row along the dimer row axis these diffusion barriers are modified by an expression that depends on the dimer row length and accounts for a repulsive interaction between DVLs. Guided by theoretical studies on the interaction of DVLs [39, 40] a long-range quadratic dependence is used. Again following the theoretical studies we introduce a constant energy term for the diffusion from or to a dimer along the dimer axis which corresponds to a short-range attractive interaction between adjacent dimers and is found to be of minor importance.

Figure 4.3 shows a sequence of MC simulations performed at different coverages of the topmost layer after a total of 10^6 diffusion attempts for each diffusing particle (monomers, dimers). As a starting configuration, the lower layer was filled completely and the top layer at random according to the given coverage. After 25×10^4 cycles typically 95% of the monomers have met another one to form a dimer. The remaining monomers survive for a long time, because their diffusion is strongly constrained by surrounding dimers. From then on, dimer diffusion by far dominates. The

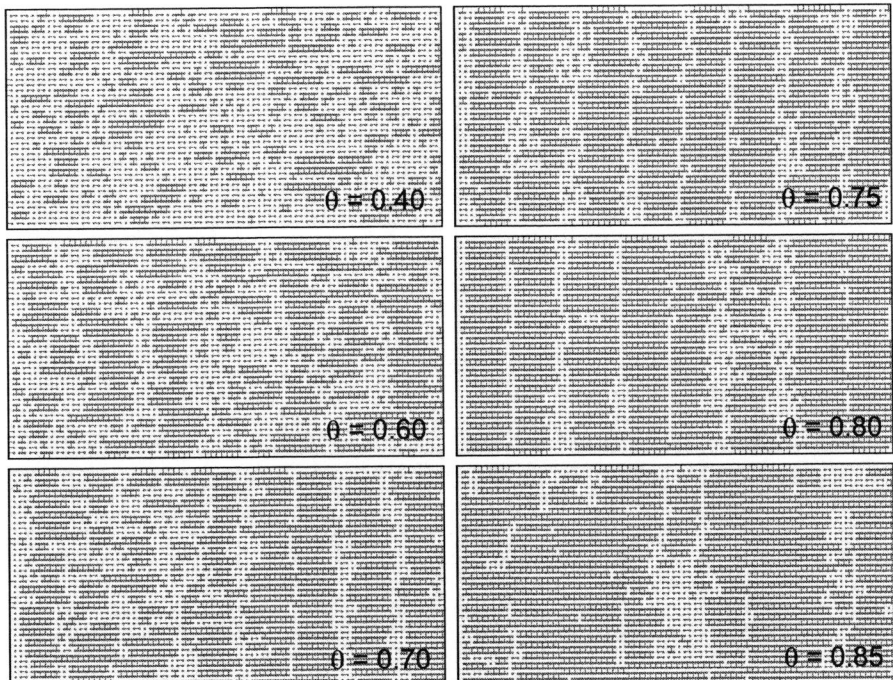

Fig. 4.3. Monte Carlo simulations of Si(001) self-organization performed at different coverages of the topmost layer (*red online*): Each particle (monomer, dimer) at the surface is given 10^6 opportunities in total to hop to a free neighboring site. As a starting configuration, the lower layer (*blue online*) is filled completely and the top layer at random according to the given coverages (compare Fig. 4.5, $\theta = 0$); atoms of a dimer are connected by a solid line

simulations reveal that the tendency to develop a $(2 \times n)$ structure depends strongly on the coverage of the top layer. If it is too low, i.e. <0.5 ML, an irregular array of small islands is formed. On the other hand, if the coverage is too high, i.e. >0.8 ML, extended (2×1) reconstructed areas coexist together with larger vacancy islands. A stripe pattern with DVLs perpendicular to the dimer rows is observed between 0.6 and 0.8 ML, whereby the width of the DVLs decreases with increasing coverage.

Figure 4.4 shows a sequence of MC simulations where 1 ML was deposited atom-by-atom onto a completely filled (2×1) reconstructed surface. After the deposition of an atom, each of the surface monomers and dimers got 1560 chances to hop to a free neighboring site. Simulations with only 700 diffusion cycles yield comparable results. Similar to the coverage-dependent simulations of Fig. 4.3, a (2×10) configuration can be discerned at a coverage of about 0.6 ML. In the case of the deposition series, however, the (2×10) is maintained up to the highest possible coverage of 0.9 ML, where the DVLs are as narrow as one lattice distance. Therefore our MC simulations suggest that deposition of Si may provide an easy means to control and adjust the vacancy concentration at the surface to a value at which the $(2 \times n)$ super-

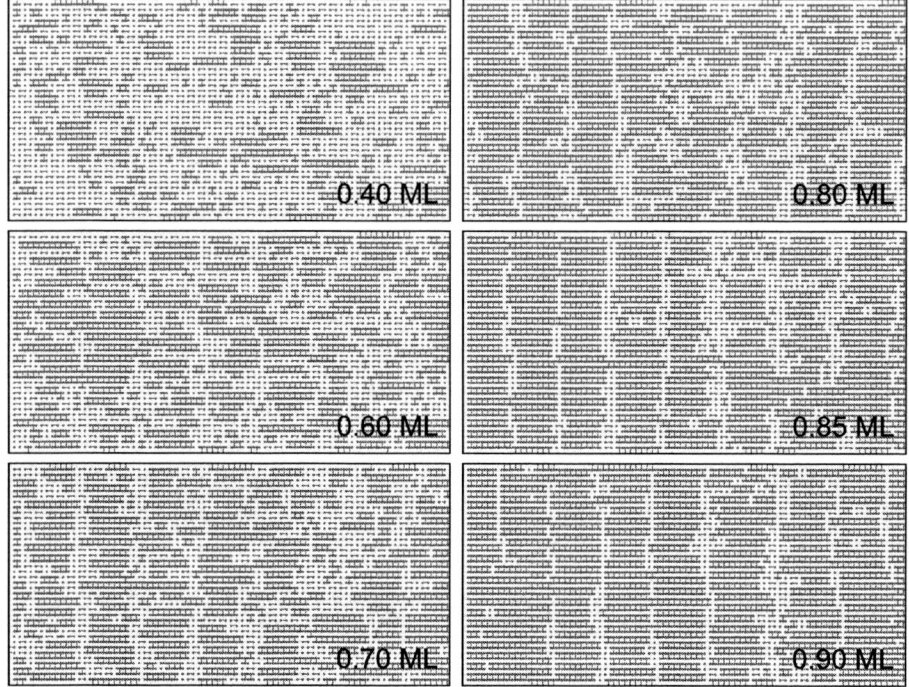

Fig. 4.4. Sequence of MC simulations, where a total of 1 ML was deposited atom-by-atom onto a completely filled (2 × 1) reconstructed surface of Si(001). After an atom was deposited each of the surface monomers and dimers had 1560 opportunities to hop to a free neighboring site

structure forms via a self-organized process. In reality the width of the DVLs may be larger corresponding to 2–4 missing dimers [29], which further narrows the coverage window for the (2 × n) formation.

4.4 Scanning Tunneling Microscopy Results

The experiments were performed in an ultra high vacuum (UHV) system with a base pressure of 1×10^{-10} hPa, equipped with a home-built STM and a four-grid LEED (low-energy electron diffraction) optics. Epi-ready Si(001) samples were mounted onto a Mo sample holder, thereby avoiding any contamination with other metals. In the loading chamber of the UHV system they were baked for four hours at 150°C. After transfer to the main chamber they were further degassed by heating from the back side via the radiation of a Ta foil for one hour at 560°C and two hours at 680°C. Finally the oxide was desorbed by several heating cycles of 5 min at 880–900°C. When the first LEED reflections appeared (i.e., usually after 1–3 heating cycles), the sample was heated for another 5 min at 880–900°C. This procedure routinely

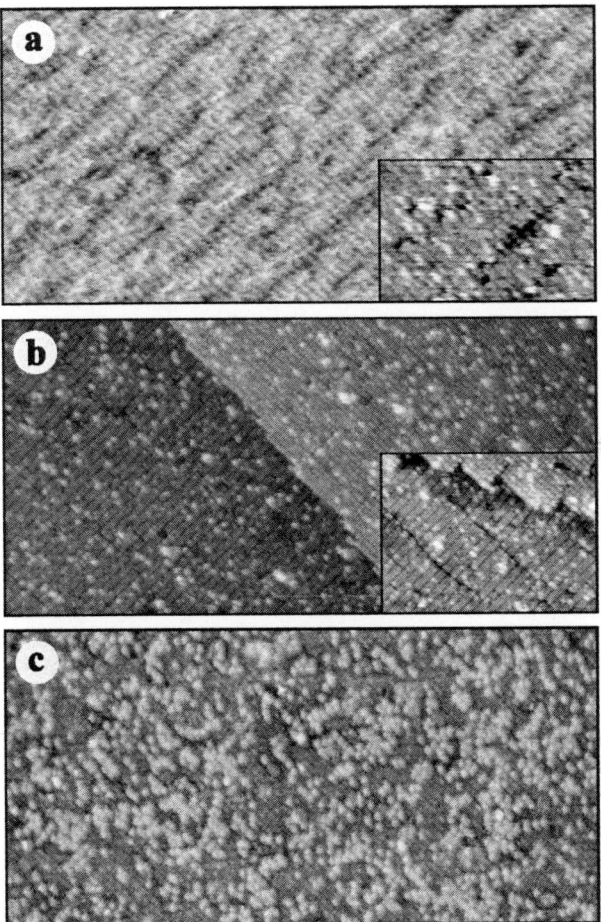

Fig. 4.5. 50×25 nm^2 STM images of Si(001) samples prepared by (**a**) desorption of the oxide at 880–900°C (for details see text), (**b**) annealing for a second time after the first appearance of LEED spots, (**c**) several heating cycles at the higher temperature of 920°C; insets show magnified images with the dimer rows and DVLs clearly resolved

obtained sharp LEED patterns which exhibit the reflections of both the (2×1) and $(2 \times n)$ superstructure. The scanning tunneling microscopy (STM) investigations were performed in the constant current mode with chemically etched tungsten tips.

Figure 4.5(a) shows a typical STM image of a Si(001) sample prepared by the procedure described above. The surface exhibits extended (2×1) reconstructed terraces that contain a high density of DVLs. The DVLs run perpendicular to the dimer rows (see inset). The DVLs form a quite regular stripe pattern with an average stripe distance of about 5 nm. The length of the DVLs ranges between 5 and 20 nm. The long-range order of the stripe pattern is further corroborated by the LEED investigations which reveal pronounced $(2 \times n)$ superstructure reflections in addition to the

(2 × 1) pattern. From the (2 × n) spots a periodicity of 5 nm is determined to be in good agreement with the average stripe distance observed by STM.

The sample displayed in Fig. 4.5(b) has been annealed for a second time after the first appearance of LEED spots. Compared to the preceding experiment the width of the DVLs has shrunk to that of one missing dimer (compare insets of Fig. 4.5(a) and (b)). Because of their narrow width it is difficult to detect the DVLs in the overview of Fig. 4.5(b), although not only the dimer rows but also individual dimers are resolved. Obviously the vacancy concentration decreases during the further annealing step that causes the dimer vacancies to move to and annihilate at the step edges. The STM images of Fig. 4.5(b) also contain a large number of bright protrusions ($\approx$10% of the surface atoms) indicating the presence of point defects. From their height (0.03 nm) and position (either in the left or right half of a dimer, e.g. Fig. 4.5(c)) it seems that the protrusions indicate Si monomers within the dimer rows with the additional dangling bond giving rise to a protrusion.

Sample preparation at slightly higher preparation temperatures further supports the coverage dependence of the (2 × n) formation. The sample shown in Fig. 4.5(c) was prepared by several heating cycles to 920°C. Due to the increased mobility, the Si coverage of the top layer has decreased to 0.3 ± 0.05 ML; the STM height of the 2D Si islands is 0.14–0.16 nm. Under these preparation conditions the Si coverage is obviously too low to favor an ordering according to the (2 × n) superstructure.

4.5 Summary and Outlook

The combined STM and MC study of Si(001) reveals that the formation of the (2 × n) superstructure depends strongly on the Si coverage of the surface layer. Both experiments and simulations show that the (2 × n) formation is restricted to a narrow coverage window. Whereas for coverages lower than 0.5 small randomly distributed 2D Si islands are observed, vacancy islands form at coverages above 0.85 that appear as randomly distributed nanodefects in otherwise (2 × 1) reconstructed terraces. The formation of the (2 × n) takes place only in the coverage range of about 0.6–0.8 ML. The annealing temperature after oxygen desorption plays a decisive role for the development of the (2 × n) reconstruction. If it is too low, the ordering processes are suppressed. If it is too high the dimer vacancies can diffuse to the step edges and are annihilated there. The experiments indicate that a temperature between 900–920°C is required for oxide desorption, whereas the optimum temperature for a subsequent annealing is lower and lies at about 880°C. For annealing at 1100–1200°C—as used for oxygen desorption in the conventional preparation of Si(001)—the MC simulations in agreement with the experiment yield (2 × 1)-reconstructed terraces with only a small number of missing dimers but without the formation of DVLs. The narrow vacancy concentration range explains the occasional appearance of the (2 × n) structure in previous quenching experiments, which obviously are not reproducible enough to establish a defined vacancy concentration in the surface layer. In that respect sputtering and etching methods certainly are more reliable and moreover are easier to control, but a detrimental surface contamination by impurities in the sputtering gas or the etching solution cannot be excluded. As indicated by the MC simulations and presently under investigation, deposition of Si onto the oxygen-desorbed surface seems to be a very promising method for adjusting the dimer vacancy concentration in the surface layer in a well-defined manner, thus providing a feasible procedure for nanopatterning of the Si(001) surface as a template for future quantum dot growth.

Acknowledgements

I thank W. Braun for careful reading of the manuscript. The work was supported by the Deutsche Forschungsgemeinschaft (Sfb 296).

References

1. Y.-W. Mo, D.E. Savage, B.S. Swartzentruber, M.G. Lagally, Phys. Rev. Lett. **65**, 1020 (1990)
2. D.J. Eaglesham, M. Cerullo, Phys. Rev. Lett. **64**, 1943 (1990)
3. J.A. Lott, N.N. Ledentsov, V.M. Ustinov, N.A. Maleev, A.E. Zhukov, A.R. Kovsh, M.V. Maximov, B.V. Volovik, Z.I. Alferov, D. Bimberg, Electron. Lett. **36**, 1384 (2000)
4. I.N. Stranski, L. Krastanow, Sitzungsber. Akad. Wiss. Wien, Abt. IIb **146**, 797 (1937)
5. G. Wedler, J. Walz, T. Hesjedal, E. Chilla, R. Koch, Phys. Rev. Lett. **80**, 2382 (1998)
6. D.M. Schaadt, S. Krauß, R. Koch, K.H. Ploog, Appl. Phys. A **83**, 267 (2006)
7. X. Deng, J.D. Weil, M. Krishnamurthy, Phys. Rev. Lett. **80**, 4721 (1998)
8. D. Leonard, M. Krishnamurthy, C.M. Reaves, S.P. Denbaars, P.M. Petroff, Appl. Phys. Lett **63**, 3203 (1993)
9. G. Springholz, T. Schwarzl, W. Heiss, M. Aigle, H. Pascher, I. Vavra, Appl. Phys. Lett. **79**, 1225 (2001)
10. H. Kirmse, R. Schneider, M. Rabe, W. Neumann, F. Henneberger, Appl. Phys. Lett. **72**, 1329 (1998)
11. J.A. Floro, G.A. Locadomo, E. Chason, L.B. Freund, M. Sinclair, R.D. Twesten, R.Q. Hwang, Phys. Rev. Lett. **80**, 4717 (1998)
12. M. Grundmann, J. Christen, N.N. Ledentsov, J. Böhrer, D. Bimberg, S.S. Ruvimov, P. Werner, U. Richter, U. Gösele, J. Heydenreich, V.M. Ustinov, A.Y. Egorov, A.E. Zhukov, P.S. Kop'ev, Z.I. Alferov, Phys. Rev. Lett. **74**, 4043 (1995)
13. H. Rong, A. Liu, R. Jones, O. Cohen, R. Hak, D. Nicolaescu, A. Fang, M. Paniccia, Nature **433**, 292 (2005)
14. H. Rong, R. Jones, A. Liu, O. Cohen, D. Hak, A. Fang, M. Paniccia, Nature **433**, 725 (2005)
15. T.I. Kamins, R.S. Williams, Appl. Phys. Lett. **71**, 1201 (1997)
16. G. Jin, J.L. Liu, S.G. Thomas, Y.H. Luo, K.L. Wang, B.-Y. Nguyen, Appl. Phys. Lett. **75**, 2752 (1999)
17. T. Kitajima, B. Liu, S.R. Leone, Appl. Phys. Lett. **80**, 497 (2002)
18. Z. Zhong, A. Halilovic, T. Fromherz, F. Schäffler, G. Bauer, Appl. Phys. Lett. **82**, 4779 (2003)
19. B. Yang, F. Liu, M.G. Lagally, Phys. Rev. Lett. **92**, 025502 (2004)
20. Y.H. Xie, S.B. Samavedam, M. Bulsara, T.A. Langdo, E.A. Fitzgerald, Appl. Phys. Lett. **71**, 3567 (1997)
21. M. Kammler, R. Hull, M.C. Reuter, F.M. Ross, Appl. Phys. Lett. **82**, 1093 (2002)

22. R. Koch, G. Wedler, J.J. Schulz, B. Wassermann, Phys. Rev. Lett. **87**, 136104 (2001)
23. J. Tersoff, R.M. Tromp, Phys. Rev. Lett. **70**, 2782 (1993)
24. V.A. Shchukin, N.N. Ledentsov, P.S. Kop'ev, D. Bimberg, Phys. Rev. Lett. **75**, 2968 (1995)
25. E. Pehlke, N. Moll, A. Kley, M. Scheffler, Appl. Phys. A **65**, 525 (1997)
26. K. Müller, E. Lang, L. Hammer, W. Grimm, P. Heilmann, K. Heinz, in *Determination of Surface Structure by LEED*, ed. by P.M. Marcus, F. Jona (Plenum, New York, 1984), p. 483
27. H. Niehus, U.K. Köhler, M. Copel, J.E. Demuth, J. Microsc. **152**, 735 (1988)
28. H.J.W. Zandvliet, H.K. Louwsma, P.E. Hegeman, B. Poelsema, Phys. Rev. Lett. **75**, 3890 (1995)
29. F.-K. Men, A.R. Smith, K.-J. Chao, Z. Zhang, C.-K. Shi, Phys. Rev. B **52**, R8650 (1995)
30. H.Q. Yang, C.X. Zhu, J.N. Gao, Z.Q. Xue, S.J. Pang, Surf. Sci. **412/413**, 236 (1998)
31. M.-H. Tsai, Y.-S. Tsai, C.S. Chang, Y. Wei, I.S.T. Tsong, Phys. Rev. Lett. **56**, 7435 (1997)
32. A. Ney, J.J. Schulz, C. Pampuch, L. Peripelittchenko, R. Koch, Mat. Res. Soc. Symp. Proc. **775**, P9.17.1 (2003)
33. V. Feil, H.J.W. Zandvliet, M.-H. Tsai, J.D. Dow, I.S.T. Tsong, Phys. Rev. Lett. **69**, 3076 (1992)
34. M. Chander, Y.Z. Li, D. Rioux, J.H. Weaver, Phys. Rev. Lett. **71**, 4154 (1993)
35. Y. Wei, L. Li, I.S.T. Tsong, Appl. Phys. Lett. **66**, 1818 (1995)
36. A. Ney, J.J. Schulz, C. Pampuch, L. Peripelittchenko, R. Koch, Surf. Sci. Lett. **520**, L633 (2002)
37. A. Ishizaka, Y. Shiraki, J. Electrochem. Soc. **133**, 666 (1986)
38. P.C. Weakliem, Z. Zhang, H. Metiu, Surf. Sci. **336**, 303 (1995)
39. A. Natori, R. Nishiyama, H. Yasunaga, Surf. Sci. **397**, 71 (1998)
40. C.V. Ciobanu, D.T. Tambe, V.B. Shenoy, Surf. Sci. **556**, 171 (2004)
41. R. Koch, Surf. Sci. **600**, 4694 (2006)

5

Structural Characterisation of Quantum Dots by X-Ray Diffraction and TEM

R. Köhler, W. Neumann, M. Schmidbauer, M. Hanke, D. Grigoriev, P. Schäfer, H. Kirmse, I. Häusler, and R. Schneider

Abstract. X-ray diffraction and transmission electron microscopy (TEM) provide complementary structural data on semiconductor quantum dots. While TEM characterizes single structures with atomic resolution X-ray diffraction yields information on statistical averages of large ensembles. For the work reported here, established methods were refined and some new methods were developed. Materials systems investigated were (In,Ga)As/GaAs, Ga(Sb,As)/GaAs and (Si,Ge)/Si. The composition of wetting layer and quantum dots could be quantitatively determined, showing a good agreement between quite a number of different X-ray and TEM methods. For all systems the depletion of the wetting layer and the enrichment of the strain providing species in the upper part of the quantum dot could be quantitatively analysed. For the model system (Si,Ge)/Si it was found that a ring of deep wetting layer depletion forms around the (Si,Ge)-islands. It was shown that this strain driven depletion prevents further lateral nucleation and thus eventually limits the size of the islands. A specific three-dimensional ordering of multilayered (In,Ga)As quantum dots grown on GaAs high-index surfaces was demonstrated and could be explained in the framework of elasticity theory and surface kinetics.

5.1 Introduction

Structural characterisation plays an essential role for "growth-correlated properties of low-dimensional semiconductor structures"—the topic of this book—because it helps us understand growth and provides the data required for calculating the electronic and optical properties of such low-dimensional structures [1, 2]. As stated in the title, this chapter is dedicated to X-ray diffraction and TEM, whereas structural characterisation on an atomic level by STM and XSTM will be dealt with in the following chapter. In view of understanding self-organised growth of nanostructures, it proves rather helpful to compare several growth methods applied to various material systems. Therefore, the structural characterisation as presented here covers a wide range of different material systems.

Our structural investigations of the system SiGe/Si (Sect. 5.2) clearly prove that growth regime and germanium content are interconnected. In particular there is a self-limitation of growth due to a suppression of nucleation caused by strain in the substrate. For III/V quantum dot (QD) systems (Sects. 5.3, 5.4) our investigations have provided detailed information on QD shape, size, density, chemical composition and three-dimensional positional correlation. Finally, these parameters serve as a basis for a more detailed understanding of the self-organised growth.

In addition to the wide range of materials, we also use a wide range of characterisation techniques. This includes a variety of X-ray diffraction and electron diffraction methods, the latter ranging from high-resolution microscopy to different analytical techniques of transmission electron microscopy. Electron microscopy and X-ray diffraction are complementary as the first one provides local real space information at a microscopic length scale whereas the latter probes reciprocal space with high resolution and thus provides statistical averages over large ensembles. Particularly, X-ray diffraction is highly sensitive to strain. On the other hand, electron microscopy has access to a variety of interaction channels, making the two techniques also complementary in this regard. Therefore our investigations include a comparison of the results acquired with electron microscopy and X-ray diffraction.

The diffraction contrast imaging techniques of conventional transmission electron microscopy (TEM) provides information on the size, shape, and arrangement of quantum dots (QDs). The shape of QDs is quantitatively determined by diffraction contrast simulation for theoretical structure models. The models are calculated by means of molecular dynamics [3]. In special cases, dark-field imaging allows a qualitative imaging of chemical composition using a chemically sensitive reflection. Various methods of quantitative HRTEM (qHRTEM) were applied to measure the local strain and chemical composition on atomic scale [4]. This chemical information was compared with direct analytical measurements by means of energy-dispersive X-ray spectroscopy (EDXS), electron energy loss spectroscopy (EELS), energy-filtered TEM (EFTEM), and the Z-contrast imaging technique.

We applied a variety of different X-ray diffraction methods, mostly by using highly brilliant synchrotron radiation. Within this project we established the use of a CCD-detector for the measurement of X-ray diffuse scattering [5]. This multi-detection technique enables fast data acquisition in all three dimensions of reciprocal space. Eventually, the experimental data are evaluated by comparison with simulations based on the kinematical scattering theory or the distorted-wave Born-approximation (DWBA) [6]. These simulations reside on real space models and numerical strain calculations in the framework of linear elasticity theory using the finite element method (FEM) [7, 8].

5.2 Liquid Phase Epitaxy of SiGe/Si: A Model System for the Stranski–Krastanow Process

5.2.1 Dot Evolution in a Close-to-Equilibrium Regime

It is very interesting to characterise low-dimensional structures grown by MBE or MOCVD with regard to a critical comparison to samples grown by liquid phase epitaxy (LPE). With LPE the growth conditions are much closer to thermodynamic equilibrium, and by changing the Ge content systematically the size of (Si,Ge)/Si structures can be tuned within a wide range while the shape of the objects remains unchanged [9–11]. The structural properties of this system were investigated before by the group of Strunk (see, e.g., [12, 9]). In cooperation with the Institute for Crystal Growth (Berlin) we investigated LPE-grown (Si,Ge)/Si systems. We will show that a detailed investigation of Ge distribution and strain distribution in and around island structures and their relation with self-organised growth gives essential clues regarding the self-limitation of growth.

At comparatively small germanium contents of around 10% (corresponding to a lattice mismatch of 0.42%) different stages of SiGe/Si(001) dots can be easily observed using scanning probe techniques, Fig. 5.1. Elastic strain relaxation initially happens by small and tiny surface undulations (a), which alternate with later morphologies (b) and (c) on the sample. Dots of type (a) appear with a four-fold base and a maximum facet inclination of 15.9°, which corresponds to the presence of {115} side facet. Once those facets have been evolved the dot shape abruptly changes (a → b) at around one third of the final dot height. Truncated pyramids with {111} side and a single (001) top facet appear. However, atomic force micrographs do not provide sufficient information to decide whether morphology change (a, b) performs

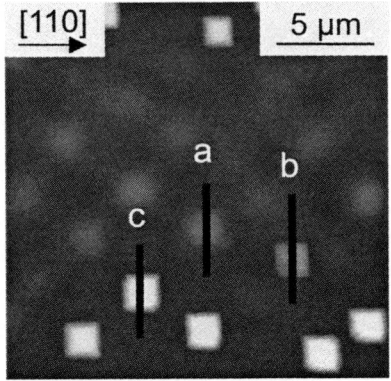
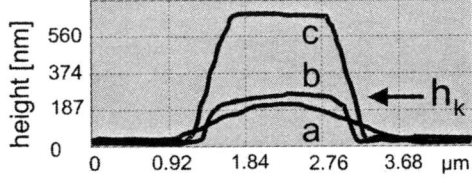

Fig. 5.1. Atomic force micrograph (*left*) and corresponding height profiles (*right*) taken on a sample with a comparatively low germanium content of 10%. A large variety of different island shapes and sizes has been recognised under special growth conditions. At a critical height the shape changes rapidly from a flat lens type (**a**) to truncated pyramids (**b**). Afterwards the growth occurs along [001] (**c**) [13]

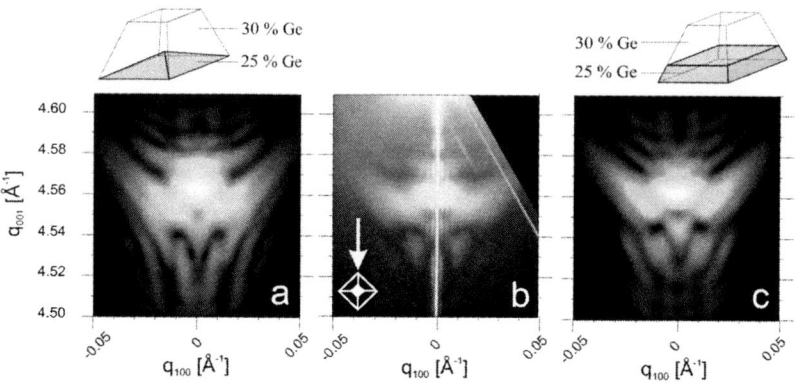

Fig. 5.2. Diffusely scattered intensity around 004 reflection. (**a**) and (**c**) are kinematical simulations for different models and (**b**) for respective measurement. The intense feature on the right-hand side in (**b**) is caused by a detector artifact [13]

by subsequently more steep facets in between {115} and {111}, or by a vertical growth along [001]. Finally the dots grow exclusively along the [001] direction at a nearly constant base.

Is there any relation between this change of growth regime with Ge content? This question was resolved by means of investigations with X-ray diffraction and TEM.

Figure 5.2(b) depicts the diffusely scattered X-ray intensity near the 004 reciprocal lattice point as measured at beamline BW2 (HASYLAB), and two kinematical simulations (Fig. 5.2(a,c)) for individual dots. The numerically derived scattering pattern is based on the three-dimensional strain field as revealed by numerical finite element calculations. Both consider a truncated pyramid with {111} side and a single (001) top facet, whereas the three-dimensional germanium profile differs. Model Fig. 5.2(a) assumes a *low* inner pyramid, while model Fig. 5.2(c) considers a *truncated* $Si_{0.75}Ge_{0.25}$ pyramid. The lack of any lateral mechanical confinement during dot growth enables an effective elastic relaxation, so that a higher germanium content towards the apex becomes energetically favourable. Both models assume a constant germanium content of 30% within the rest of the dot.

A brief comparison reveals an obviously better fit in Fig. 5.2(c). Keeping in mind a constant germanium content within the lower third of the dot we can conclude that the particular growth transition (a → b) in Fig. 5.1 happens very similarly to the evolution of surface undulations. Namely the dot growth performs by subsequently more steep facets which eventually leads to {111} facets. The step-like change in the germanium content is obviously related to switching from lateral growth leading to more steep side facets to growth in the (001) direction.

In addition to X-ray measurements the size, shape and arrangement of the SiGe islands were investigated by conventional TEM in the diffraction contrast imaging mode [14]. In order to get information about the structural properties at atomic scale, HRTEM imaging was done. A detailed analysis of the strain state of (Si,Ge) islands near the silicon interface in different viewing directions was carried out [15]. An

5 Structural Characterisation of Quantum Dots by X-Ray Diffraction and TEM 101

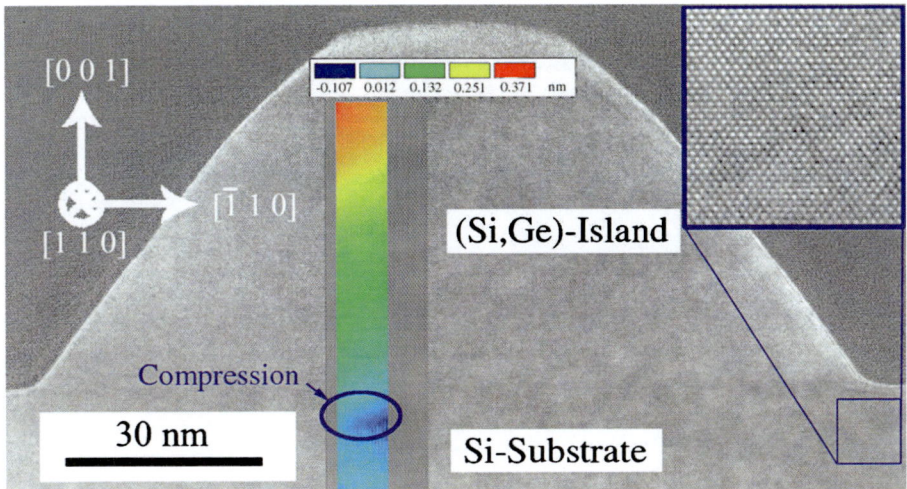

Fig. 5.3. C_S-corrector HRTEM image of a (Si,Ge) island analysed by DALI (REF, reference lattice)

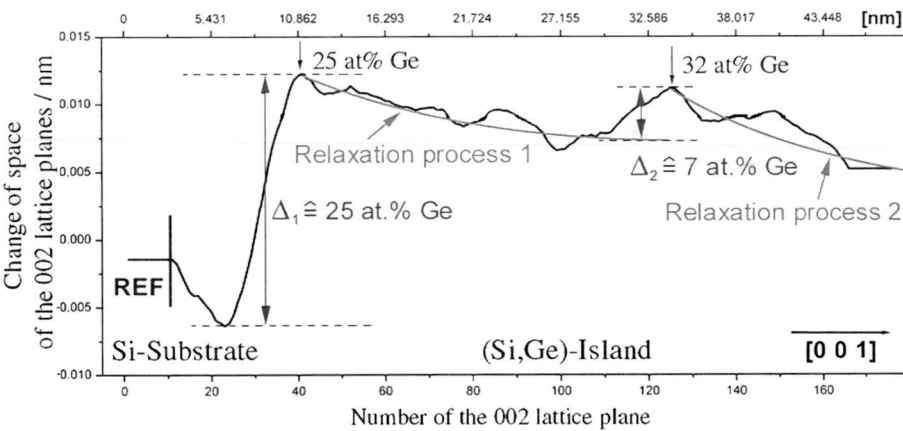

Fig. 5.4. First derivation of displacements of the (002) lattice planes

HRTEM image of a typical (Si,Ge) island, which was analysed by means of the program package DALI, is shown in Fig. 5.3. The island contains no crystal defects in the interface region or in the whole island area. The viewing direction of the island is [110], i.e. along the edge of the basal plane. The colour-coded map of displacement $u_{[001]}$ of the (002) lattice planes within a selected area is superimposed on the image. The reference lattice (REF) is defined in the Si substrate (blue, in colour online). The colour-coded map represents the displacement of the atom columns in the [001] direction compared with the reference lattice. In order to get information on the chemical composition via strain one has to calculate the first derivative with respect to the displacement plot's position. Figure 5.4 shows the derivative of dis-

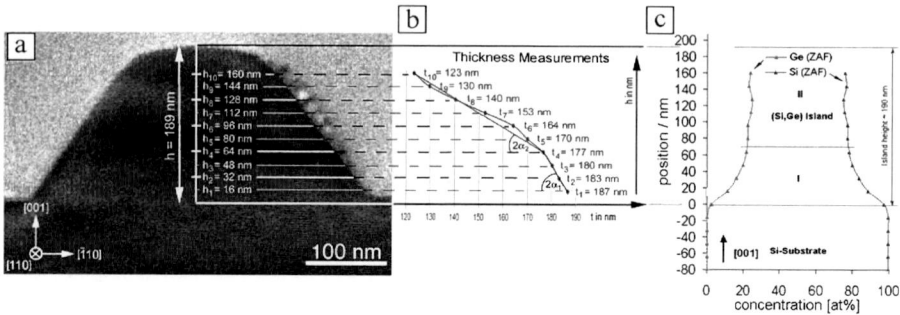

Fig. 5.5. Determination of the germanium concentration of a $Si_{1-x}Ge_x$ island grown on (001)-orientated Si substrate. (**a**) Cross-sectional bright-field TEM image of a $Si_{1-x}Ge_x$ island. (**b**) Correlation between the island height (see (**a**)) and the sample thickness. (**c**) Germanium and silicon concentration determined by EDX analysis

placements of the (002) lattice planes. The first maximum then corresponds to a Ge content of about 25 at.%. At the island's midheight the concentration of germanium is approximately 32 at.%. The results hint at a jump in the Ge concentration within the island.

Furthermore, the microchemistry of the layer/island system was investigated by EDXS, EELS and EFTEM analysis [16–19]. Figure 5.5 shows a cross-sectional bright-field image of a SiGe island and the corresponding thickness from bottom up to the top of the island. The Ge composition was determined using a modified ZAF correction method [19]. The analysis clearly exhibits an increase of Ge until 25 at.% is reached (Fig. 5.5(c)). The germanium content of the upper part of the islands remains constant. The results are in good agreement with X-ray measurements [8, 13, 20].

It is well known that the final base size w of a heteroepitaxially grown, dislocation-free dot inversely scales with the applied lattice mismatch x: $w \propto x^{-2.03}$ [9]. This universal scaling law is frequently discussed in terms of an inverse quadratic strain dependence. Regions of comparatively high strain energy tend to be dissolved during subsequent growth and can serve as a monitor of strain distribution. Figure 5.6 shows a typical cloverleaf-like wetting layer depletion in the vicinity of a $Si_{0.9}Ge_{0.1}$ dot, whose particular shape reflects the four-fold symmetry of the strain energy distribution (not shown here). Strain induced wetting layer depletion has also been observed for MBE and other material systems [21, 22].

Figure 5.7 depicts an intermediate stage (a) and a fully developed $Si_{0.91}Ge_{0.09}$ dot (b) with a final aspect ratio island base/height of nearly two. The scanning electron micrographs clearly reveal the microstructure of the $\langle 101 \rangle$-edges, which are assembled of small {111}-type facets. Obviously the lateral dot dimension increases after formation of a truncated pyramid due to additional {111}-slabs. Since there is no two-dimensional nucleation at {111}, subsequent slabs (see feature (F) in Fig. 5.7) exclusively nucleate at the dot bottom. However, they do not grow into spatially complete facets since they stop in the dot's corners (K) at the bottom. Finite element

5 Structural Characterisation of Quantum Dots by X-Ray Diffraction and TEM

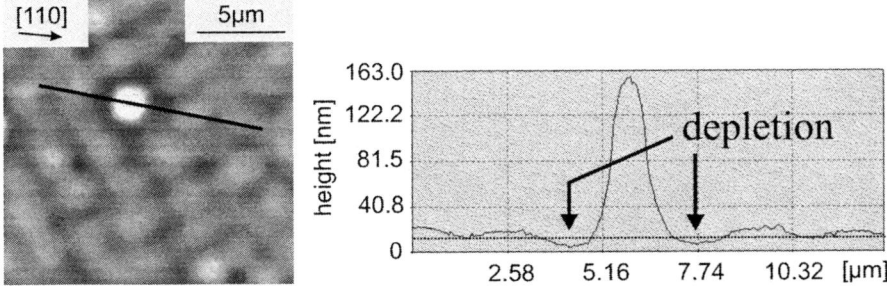

Fig. 5.6. Atomic force micrograph of a sample containing 10% germanium (*left*). Due to high strain energy in the vicinity of a dot the wetting layer was depleted (*right*). Lateral dimension of this area is in the order of the dot size itself [13]

Fig. 5.7. Scanning electron micrograph of a $Si_{0.91}Ge_{0.09}/Si(001)$ island. The $\langle 101 \rangle$-edges show a pronounced dissection indicating to a suspended nucleation at the island bottom [13]

calculations predict a strongly enhanced strain energy near these corners [13]. Therefore the growth process initially suspends around these points. Later on, nucleation will be completely suppressed after a strain-induced wetting layer depletion and finally the dot growth stops. Thus, the final dot size can be related to a kinetically restricted nucleation problem rather than a transport phenomenon. Consequently we presume that the absence of further nucleation even holds as a key argument to a finite dot size in case of other growth techniques (cf. the discussion in Chap. 1).

Besides, this strain distribution in the substrate—due to the fully developed island—is also responsible for the formation of island chains in early stages of growth with low island densities [23].

5.3 (In,Ga)As Quantum Dots on GaAs

5.3.1 Shape, Size, Strain and Composition Gradient in InGaAs QD Arrays

Self-organised (In,Ga)As/GaAs quantum dots (QDs) as formed in the Stranski–Krastanov growth mode are zero-dimensional electronic systems [1, 2, 24] which are

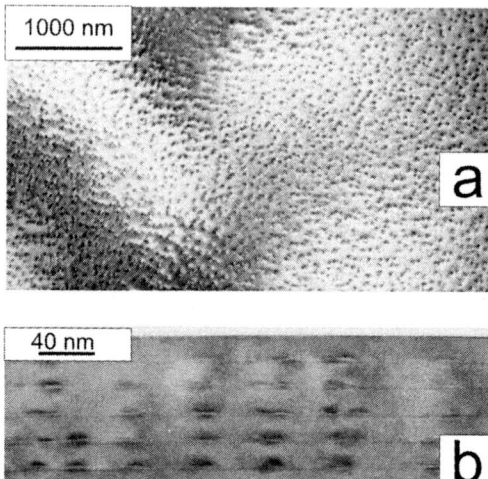

Fig. 5.8. Transmission electron micrograph in plan-view (**a**) and cross-section (**b**). A pronounced vertical ordering of the QDs is visible (**b**), while no indication of a lateral ordering is found (**a**) [29]

highly interesting for both the study of new physical phenomena and device applications. The In distribution within (In,Ga)As QDs and the exact shape of such QDs are among the most important issues to be solved owing to their direct impact on the alignment of confined electron and hole wave functions [25, 26].

The sample investigated here was grown by MOCVD on a GaAs(001) substrate— see for details of the growth procedure (see Sect. 2.2.2). The sample consists of a five-fold stack of self-organised $In_{0.60}Ga_{0.40}As$ quantum dots, separated by nominally 20 nm GaAs spacer layers. In order to stack subsequent layers closely with a vertical period around 20 nm, a surface-flattening technique during growth of the respective GaAs spacer layers was applied [27]. TEM micrographs (Fig. 5.8) depict that the QDs reside on a very thin and continuous wetting layer (WL). Enforced by the long-range strain field of the QDs and the comparatively thin GaAs spacers between the dot planes, the QDs are vertically arranged into columns, whereas they do not exhibit a significant lateral ordering (cf. Fig. 5.8(a)) [28].

Figure 5.9 shows an experimental reciprocal space map around the GaAs 224 reflection (Fig. 5.9(d)) and simulated intensity distributions (Fig. 5.9(a–c)) for different QD shapes and In compositions as described later. The X-ray diffuse scattering in Fig. 5.9(d) exhibits a complicated intensity pattern which is caused by the complex interplay between geometry, strain and chemical composition profile within the QDs. The signal also contains contributions from the (In,Ga)As WLs and the GaAs matrix. The equally spaced numbered satellites on the crystal truncation rod (CTR) are caused by the vertical periodicity of the sample structure. Their vertical spacing of $0.0327\,\text{Å}^{-1}$ reveals a QD layer-to-layer period of 19.2 nm, in good agreement with the nominal value of 20 nm. The weaker ancillary maxima between these satellites reflect the overall thickness of the five-fold QD stack.

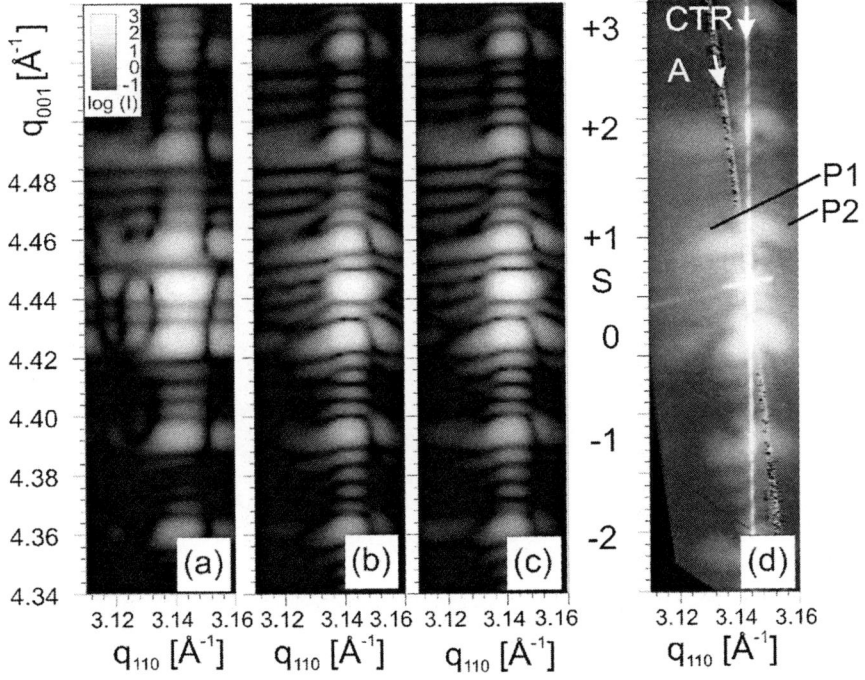

Fig. 5.9. Measured X-ray intensity distribution in the vicinity of the asymmetrical 224 reciprocal lattice point (**d**) compared to simulations of the scattered intensity calculated using different QD shapes and composition profiles (**a–c**) as described in Fig. 5.10. Feature (A) is caused by a detector artifact [28]

The vertical position of the 0th order satellite on the CTR at $q_{001} = 4.427\,\text{Å}^{-1}$ enables a measurement of the mean vertical strain inside the superlattice. For given thicknesses of the GaAs spacer (18.1 nm) and the WL (two GaAs lattice constants $a = 5.6532\,\text{Å}$) the In content of the WL can be deduced to about 37%. It must be noted that variations of the WL thickness in the simulations essentially alter the intensities of higher-order satellites only, provided that the product of WL thickness and In concentration is kept constant. The assumed thickness corresponds to values previously reported for similar (In,Ga)As/GaAs QD stacks [28] as revealed by transmission electron microscopy. The In concentration of 37% is considerably smaller than the nominal value of 60%. This difference can be explained by In migration from the WL towards the QDs during island formation [30, 31] as substantiated later.

Finite element calculations (FEM) of the elastic strain field [29] show that only the dot stack comprises regions of nonzero lateral strain. Therefore, only these areas give rise to the diffuse scattering on both sides of the CTR. This makes the diffuse signal an appropriate means to probe the In distribution within the QDs and the transition regions to the adjacent GaAs matrix. However, the complexity of the intensity distribution prevents straightforward conclusions on the QD morphology. To

 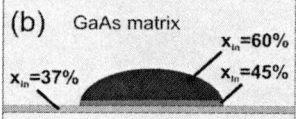 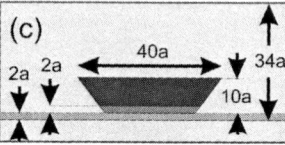

Fig. 5.10. Model QD shapes and compositions used for dynamical scattering simulations. The wetting layer contains 37% Indium in all three structures. (**a**) Lens-shaped QD with homogeneous QD apex, containing 60% In. (**b**) Lens-like QD with a $2 \times a$ GaAs thick transition layer containing 45% In. (**c**) Inverted-cone shaped QD with a $2 \times a$ GaAs transition layer containing 45% In. Dot dimensions are given in units of the GaAs lattice parameter a [28]

extract detailed quantitative information, we performed dynamical scattering simulations probing different QD shapes, namely cuboids [28], pyramids, flat lenses and inverted cones. In turn, different In concentration profiles were assumed for the respective shapes. Figure 5.10 shows cross-sectional schemes of three selected model QDs. A lens-like QD with 60% In content on top of a WL is shown in Fig. 5.10(a). The second model QD shown in Fig. 5.10(b) has the same shape, size and In concentration as the previous one but comprises a $2 \times a$ thick $In_{0.45}Ga_{0.55}As$ transition layer between WL and the QD apex. The third model QD depicted in Fig. 5.10(c) exhibits the same vertical chemical composition profile as that of Fig. 5.10(b) but has an inverted-cone shape.

The calculated intensity distributions of the three QDs in Figs. 5.10(a–c) are shown in Figs. 5.9(a–c), respectively. All simulations reproduce the periodicity of the five-fold QD stack. However, Fig. 5.9(a), referring to a lens-like QD without transition layer, does not describe any detail of the diffuse scattering from the dots. This is particularly obvious if the measured features P1 and P2 are compared to the simulation. In contrast to Fig. 5.9(a), Figs. 5.9(b) and (c), referring to the lens-like and inverted-cone QD shape with transition layer, respectively, reflect the particular shapes of P1 and P2 as well as other dot-related scattering intensity distributions very well. The rough approximation of a gradually increasing concentration profile by just a single step may be justified by the fact that the exact characteristic of the vertical profile essentially affects the intensities of high-order satellites only. Such high-order satellites are not shown here because variations of their intensities cannot be resolved experimentally due to their low signal strengths.

The remarkable similarity of Figs. 5.9(b) and (c) shows that the diffusely scattered X-ray signal of the measured low-order satellites is largely insensitive to variations of the QD shape. Therefore, the QD shape could not be determined further using the available data. In spite of such experimental limitations, our findings clearly indicate a nonhomogeneous vertical In concentration which increases towards the QD apex. During formation, the upper parts of uncovered QDs can elastically relax well since no lateral constraints prevent the lattice to expand. Therefore, the QD strain decreases from the bottom to the top of the QD. As long as the QDs are uncovered, both lateral In migration from the WL to the QDs as well as vertical In segregation from the QD bottom to the apex [32] are favourable in terms of strain energy.

5.3.2 Chemical Composition of (In,Ga)As QDs Determined by TEM

The chemical composition of QDs can also be determined by analysing diffraction contrast dark-field images, when chemically sensitive reflections exist for the structure to be investigated. The amplitude of 002 reflection of (In,Ga)As (sphalerite structure) predominantly depends on the chemical composition [33] and can therefore be used for a quantitative analysis of the various elements of (In,Ga)As. Strictly speaking, the 002 amplitude is a function of the difference between the mean atomic number of the cations and that of the anions in this compound. The amplitude of the 002 beam of GaAs ($\Delta Z = 2$) is low compared to that of InAs ($\Delta Z = 16$). Hence, in an 002 dark-field image InAs will appear brighter than GaAs. For a reliable quantification of the In content it is necessary to have either standards for comparison or fix points of known composition. The 002 intensity of $In_xGa_{1-x}As$ ($0 \leq x \leq 1$) exhibits a quadratic curve with two fixed values. The first one is the 002 intensity of GaAs usually chosen as the substrate and cap. The minimum of the curve is correlated to an In content of 18% [34]. Thus, the composition analysis can be applied to every cross-sectional 002 dark-field image with GaAs and $In_xGa_{1-x}As$ with $x > 0.18$.

An example of the extraction of chemical information of a single layer of (In,Ga)As QDs embedded in GaAs (001) is shown in Fig. 5.11. This sample was grown by MOCVD under similar conditions compared to the sample discussed in Sect. 5.3.1. The 002 dark-field image (Fig. 5.11(a)) exhibits a single layer of (In,Ga)As QDs embedded in GaAs, where the self-organised QD growth was performed by MOCVD at 793 K. The nominal thickness of InAs was 2.6 monolayers.

The noise visible within the GaAs area (see Fig. 5.11(a)) is caused by the amorphous layers covering the TEM specimen on both sides due to the ion milling preparation. Nevertheless, the (In,Ga)As layer containing the QDs can clearly be recognised. The cross-section of the QDs exhibits a lens-like shape realised by high-index

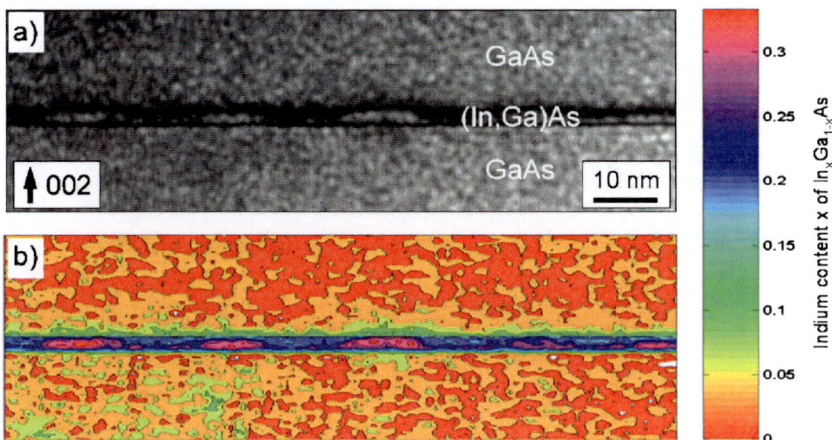

Fig. 5.11. Analysis of chemical composition of (In,Ga)As QDs embedded in GaAs matrix. (**a**) Chemically sensitive diffraction contrast dark-field TEM image. (**b**) Map of In content

facets, which is also expected from scanning tunnelling microscopy investigations [35]. Obviously, the growth of GaAs cap does not alter the shape strongly. The bright-appearing QDs are accompanied by a dark rim where the intensity minimum corresponds to $x_{In} = 0.18$.

The map of the In content given in Fig. 5.11(b) was derived by a complex digital image analysis. As the TEM specimen may exhibit a wedge shape, a correction of the thickness contribution to the 002 intensity was applied. After subtraction of this contribution, the modulation of In content within the GaAs region defines the error bar of the method. In the present case the error bar amounts to about $\Delta x = \pm 0.05$. The residual intensity is used for the determination of the In content. Due to the quadratic behaviour of the 002 intensity with respect to the In content there are always two solutions for one and the same intensity value. The correct solution is found by plausibility considerations.

Approaching the (In,Ga)As QD layer from the underlying GaAs a steep increase of the In content is found (cf. Fig. 5.11(b)). This is also evident from the line scans given in Fig. 5.12 extracted for the individual QDs 1 to 3 starting from the left in Fig. 5.11(a). The maximum of the In content amounts to about $x_{In} = 0.35 \pm 0.05$ and is similar for every QD under investigation. The wetting layer exhibits a reduced In content of $x_{In} = 0.27 \pm 0.05$.

This result of higher indium content in the QD as compared to the WL is in qualitative agreement with the X-ray results as discussed in Sect. 5.3.1. However, the X-ray results propose an additional transition layer between the WL and the QD while the island apex exhibits a constant In content. On the other hand, the TEM results suggest an indium composition gradient inside the QDs. At the QD apex this gradient is smaller as compared to the QD bottom. Two reasons could be

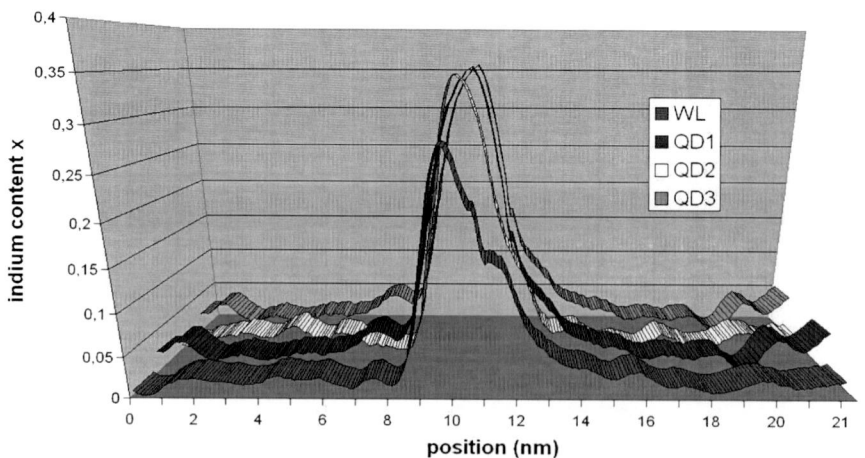

Fig. 5.12. Line profiles of In content of $In_x Ga_{1-x} As$ QDs as determined by digital analysis of chemically sensitive 002 dark-field images. QD 1 to 3 refer to the QDs visible in Fig. 5.11 starting from the left. The In profile of the wetting (WL) is given for comparison

responsible for this behaviour. On the one hand, the lens-like shape of the QD leads to an increase of the GaAs fraction within the volume transmitted by the electron beam close to the apex of the QD. On the other hand, segregation of In causes a delayed incorporation of In into the GaAs cap layer. Beside the chemical composition, the size of the (In,Ga)As QDs can be determined to be about 3 nm in height and about 15 nm in width. There is a significant difference for the determination of QD size basing on either a merely composition-sensitive image or a strain-sensitive image. Regarding the length scale of the surrounding strain field as a measure of the QD size usually leads to an overestimation.

5.3.3 Controlling 3D Ordering in (In,Ga)As QD Arrays through GaAs Surface Orientation

Three-dimensional (3D) arrays of QDs [36, 37] have attracted increasing interest because they hold promise for possible applications based on the coupling of the electronic wave functions of adjacent QDs. Although both planar and vertical correlation of the QD positions have been predicted [38, 39] and experimentally observed in many materials systems [40–43] the control of the QDs' positions through sophisticated growth techniques remains the ultimate goal. We will demonstrate that, e.g., the symmetry and the planar/vertical tilt angles of 3D (In,Ga)As QD lattices can be tuned by growing the QDs on GaAs substrates with different crystallographic orientations.

In the Stranski–Krastanow growth mode the growth conditions can be optimised to produce nanostructures of near identical size and shape. However, often only a random spatial distribution of the QD is observed for a single layer of QDs [44, 28]. Nevertheless, for multiple layers a range of different results—from near-perfect QD chains to three-dimensional (3D) lattices—has been reported [45, 46, 36, 37]. We report here on experiments that use natural surface steps on high index substrates to clarify the role of both surface diffusion and strain in producing 3D ordering of QDs. The samples were grown by solid-source molecular beam epitaxy (MBE) on singular (100), and GaAs $(n11)$B substrates where n is equal to 9, 7, 5, 4 and 3. The high index surfaces are tilted toward (111)B surface. Consequently, B-type steps are formed running along [011]-direction, while the direction perpendicular to the step edges is $[2n\bar{n}]$. After a half micrometer thick GaAs buffer layer grown at 580°C the substrate temperature was lowered to 540°C for the deposition of the (In,Ga)As/GaAs multi-layered structure. The multilayer structure consists of 16.5 periods of 10 monolayers (ML) In$_{0.40}$Ga$_{0.60}$As QDs and 120 ML GaAs spacers. The last layer of QDs was left exposed for topographic atomic force microscopy (AFM) imaging under ambient conditions.

The 3D self-organised arrays on a (100) GaAs substrate show highly aligned QD chains oriented along the $[0\bar{1}1]$ direction, Fig. 5.13(a), while vertically aligned perpendicular to the (100) surface. This situation is dramatically changed, however, when the substrate surface orientation is different from (100). For example, for a (511)B substrate orientation the 3D array is laterally aligned in a two-dimensional lattice, Fig. 5.13(b). This behaviour can be understood because, for the high index

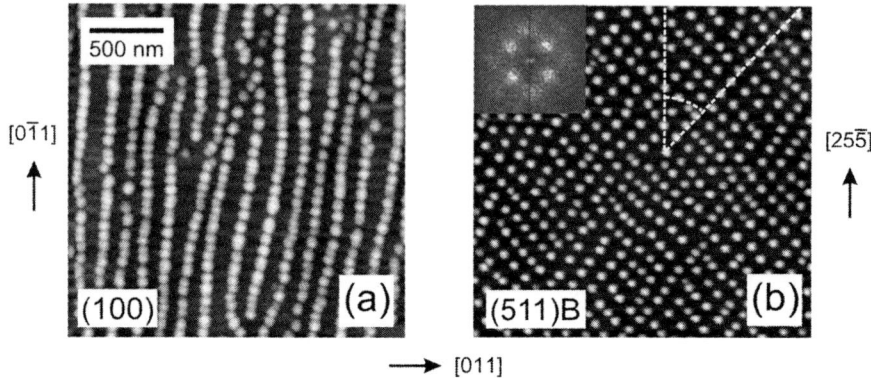

Fig. 5.13. AFM images of the surface morphology of multilayered $In_{0.40}Ga_{0.60}As$ QDs grown on GaAs (100) (**a**) and (511)B (**b**). The inset is a fast Fourier transform taken from the AFM image. The in-plane deflection angle θ between the nearest-neighbour direction and $[25\bar{5}]$-direction is indicated [47]

(n11)B surfaces under study, B-type surface steps with variable separations are introduced along the [011] direction. By adjusting the index of the substrate the separation S between steps can be varied systematically from $S = 0.60$ nm for (311)B to $S = 1.80$ nm for (911)B creating the opportunity to change the surface diffusion pattern for adatom migration. For growth on a (100) substrate, the adatom diffusion is large along the $[0\bar{1}1]$-direction, while it is comparatively low along the [011]-direction. Therefore, strain can be more efficiently relaxed along the $[0\bar{1}1]$-direction than along the [011]-direction. As a consequence, the strain field that is transferred from one QD layer to the next forces the nearest neighbour QD to be along the $[0\bar{1}1]$-direction. Eventually, a 3D array of QD chains oriented along the $[0\bar{1}1]$-direction is formed. The same mechanism is responsible for pattern formation on high index substrates. Here, the diffusion along the $[0\bar{1}1]$-direction can be gradually tuned to be comparable to that along the [011]-direction as is apparently the case for substrates ranging from (911)B to (311)B. As a quantitative measure of this effect GISAXS data have been evaluated in order to determine the in-plane deflection angle θ (Fig. 5.14(a)). It is experimentally seen to vary linearly with $1/S$, where S is the step separation, Fig. 5.14(b). Therefore, the natural surface steps on high index templates can be used to tune the 2D QD lateral ordering on the surface.

The lateral ordering in our 3D QD lattices can be understood to depend on the starting substrate. We will further demonstrate that the vertical ordering is also dependent but for a totally different reason. Our observations show that the vertical ordering in 3D lattices is not strictly along the growth direction as might be naively expected. Rather, the vertical alignment deviates from the surface normal by a significant inclination angle α. This is demonstrated in Fig. 5.15 which shows two 2D sections of the X-ray diffuse scattered intensity from the (511)B-sample in the vicinity of the GaAs $5\bar{1}1$ reciprocal lattice point. Figure 5.15(a) displays the X-ray diffuse scattering in a plane containing the $[25\bar{5}]$-direction while Fig. 5.15(b) shows the

5 Structural Characterisation of Quantum Dots by X-Ray Diffraction and TEM   111

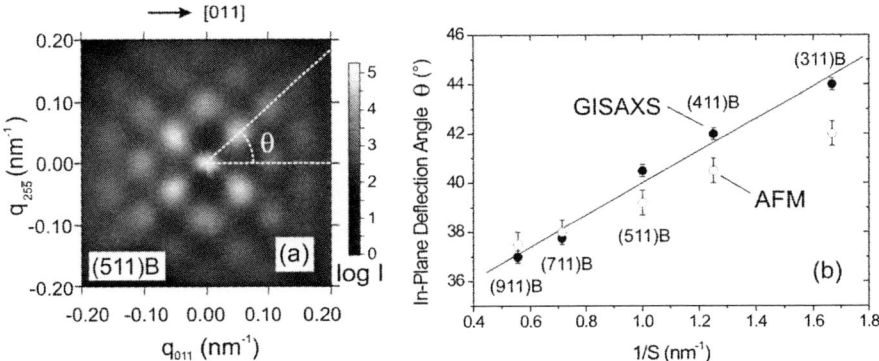

Fig. 5.14. (a) GISAXS from sample with (511)B substrate orientation. (b) In-plane deflection angle θ of the nearest-neighbour direction from $[2n\bar{n}]$ as a function of $1/S$ (S: nominal surface step separation) of the high index surfaces under study; ○, AFM-FFT; •, GISAXS; the solid line represents a linear fit to the GISAXS data [47]

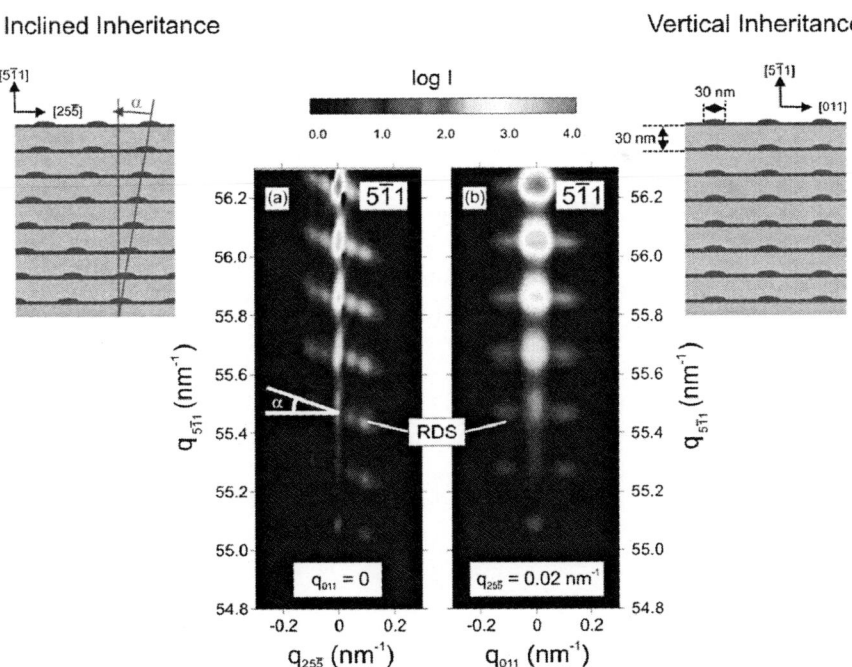

Fig. 5.15. X-ray diffuse scattering (*bottom*) from the (511)B-sample in the vicinity of the GaAs $5\bar{1}1$ reciprocal lattice point. In the direction along $[25\bar{5}]$ (a) the X-ray diffuse scattering (which is concentrated in RDS-sheets) is inclined by the angle α while in the direction along $[01\bar{1}]$. (b) The RDS-sheets are oriented horizontally. This proves (c) inclined vertical inheritance of the horizontal QD positions towards the $[25\bar{5}]$-direction (*top*) [47]

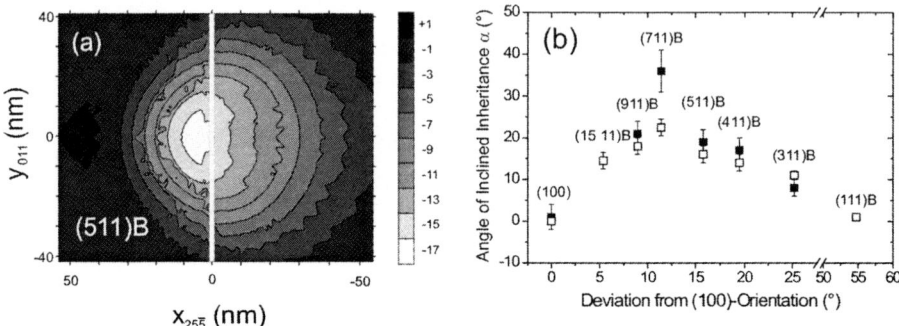

Fig. 5.16. (a) Calculated elastic strain energy density (*linear grey scale* in arb. units) on the (511)B-surface above a strained $In_{0.40}Ga_{0.60}As$ QD which is embedded into a GaAs matrix at 30 nm below the surface. (b) Angle of inclined inheritance α of the lateral QD positions to the surface normal; (*open squares*) calculations; (*filled squares*) experimental [47]

X-ray diffuse scattering in a plane containing the [011]-direction. The diffuse intensity is vertically bunched into resonant diffuse scattering (RDS) sheets, indicating that the horizontal QD positions are correlated between adjacent QD-layers. However, in the direction along [25$\bar{5}$], Fig. 5.15(a), the RDS sheets are tilted from the exact horizontal orientation. It is well accepted that the angle of inclination α exactly corresponds to the deviation of the vertical arrangement of the QD positions from the surface normal towards the [25$\bar{5}$]-direction [48]. By contrast, the RDS sheets are not inclined in the direction along [011], Fig. 5.15(b). The inclination angles α were determined experimentally for all samples under study, Fig. 5.16(b). From these evaluations we can conclude that the observed inclination always points towards the [$2n\bar{n}$]-direction, i.e. the direction perpendicular to the nominal step edge direction. On the other hand the inclination toward [011] always vanishes.

In order to understand this behaviour, numerical model calculations using linear elasticity theory were performed. Our calculations are based on the finite-element method (FEM) and take into account the QD size, shape and the full elastic anisotropy of the involved materials. For the samples under study here, the horizontal extension of the QDs is of the same magnitude as the thickness of the spacer layer ($t = 30$ nm). Therefore, results of exact theoretical treatment are expected to still depend on the actual shape and size of the QDs. For a lens shaped $In_{0.40}Ga_{0.60}As$ dot (base width $w = 30$ nm, height $h = 5$ nm) located 30 nm below the (511)B surface of a GaAs matrix numerical results of the elastic energy density are presented in Fig. 5.16(a). These calculations show a shallow energetic minimum, however, shifted towards the [25$\bar{5}$]-direction. This result confirms the experimental result of inclined inheritance toward this direction, Fig. 5.16(a).

We performed a systematic series of FEM calculations for different GaAs surface orientations of the type (n11)B. From the shift of the minimum in the elastic energy density and the spacer layer thickness angles α of inclined inheritance are deduced and compared with corresponding experimental results, Fig. 5.16(b).

For (111)B and (100) surface orientations the energy density shows the minima just above the underlying QD in the previous layer, thus favouring exact vertical correlation [49]. However, for surface orientations on the path between (100) and (111)B, i.e. (911)B, (711)B, ..., (311)B, remarkably high values for α are observed, with a maximum for the (711)B surface orientation exactly matching the experimental observation.

Although overall agreement between the FEM calculations and the experimental values for α is observed, the experimentally derived values are slightly larger than the predicted ones. This deviation cannot be explained by the finite size of the QDs. Our calculations show that increased QD sizes—as compared to the spacer thickness—lead to a reduction of the inclination angle α, while a decrease of the QD size does not further significantly affect our results. Obviously, the present geometrical parameters ($w = 30$ nm, $h = 5$ nm, $t = 30$ nm) are close to the "far-field limit" considered by Holý et al. [49] for which the inclined inheritance angle is insensitive to the finite QD size. This can be also inspected by comparing $\alpha_{511} = 10°$ as predicted by Holý et al. with $\alpha_{511} = 11°$ as calculated in our study.

5.4 Ga(Sb,As) Quantum Dots on GaAs

The materials system GaSb/GaAs shows a type II band alignment, with the holes trapped in the GaSb quantum dots (QDs) and the electrons localised in the adjacent GaAs barriers. These properties, especially the strong hole confinement, makes the quantum dot structures favourable for storage devices [50]. The strain-induced self-organisation of GaSb/GaAs QDs was mainly performed successfully by means of molecular beam epitaxy [51–53]. The GaSb/GaAs QDs described here were grown by MOCVD [54], see a detailed discussion in Sect. 2.4.1.

The electronic and optical properties of the QDs structure strongly depend on the structural and chemical peculiarities. The growth experiments aim at their optimisation, i.e., narrow size distribution and homogeneous area density, lateral arrangement and even chemical composition among the QDs.

Structural properties of the QDs were investigated as a function of both the nominal thickness of the GaSb layer as well as of the growth interruption time after GaSb deposition. All samples were grown at a temperature of about 743 K. Two different series of samples were generated, where for the first one the nominal GaSb layer thickness was 3.6, 4.0, 4.5, 5.0, and 5.8 monolayers (ML) at a growth interruption time of 5 s. As to the second type, a growth interruption time of 0, 2, 5, and 10 s was applied to a nominal 4.5 ML thick GaSb layer.

Plan-view TEM investigations give information about the size distribution, the area density, the lateral arrangement of the QDs as well as about defect formation. A sequence of 220 dark-field images is shown in Fig. 5.17. Here, the image contrast is presumably caused by the strain field in the surrounding of the QDs. The images were taken from samples grown with different GaSb layer thicknesses but equal growth interruption time of 5 s. The increase of the area density of the QDs with increasing layer thickness is clearly visible and is emphasised by the graphs given in

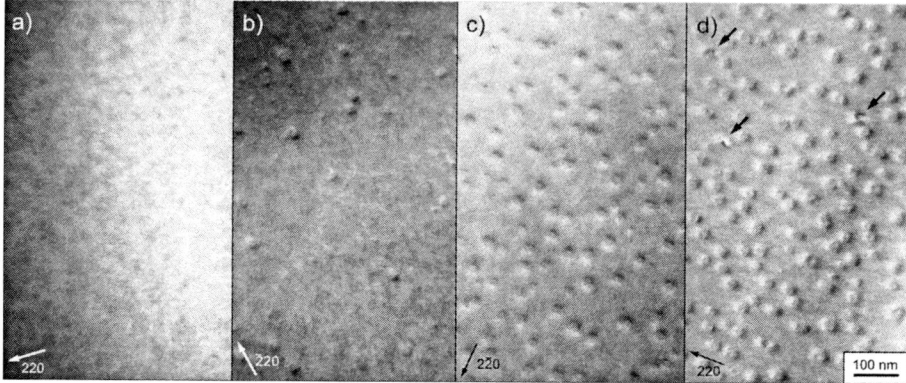

Fig. 5.17. Plan-view TEM dark-field images of a series of samples grown at different GaSb layer thicknesses. (**a**) 3.6 ML, (**b**) 4.0 ML, (**c**) 5.0 ML and (**d**) 5.8 ML always followed by a growth interruption of 5 s. Deposition of 5.8 ML GaSb causes defect formation (cf. arrows in Fig. 5.17(d))

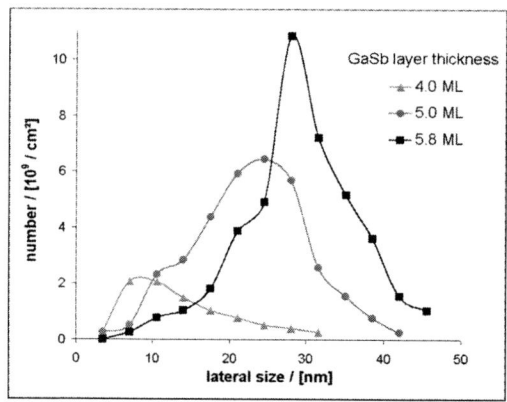

Fig. 5.18. Distribution of the lateral size of the GaSb quantum dots shown in Fig. 5.17(b–d)

Fig. 5.18. The area density of the quantum dots corresponds to the integral value of the graphs and amounts to about $1 \cdot 10^{10}$ cm^{-2} for a GaSb layer thickness of 4.0 ML, $3 \cdot 10^{10}$ cm^{-2} for 5.0 ML and $4 \cdot 10^{10}$ cm^{-2} for 5.8 ML. For the latter case the largest QDs exceed the critical size for plastic relaxation resulting in a formation of defects (cf. arrows in Fig. 5.17(d)). This is also verified by photoluminescence (PL) investigations where the intensity of the peak related to the QDs decreases with increasing GaSb layer thickness [55]. The rising size of the QDs with increasing layer thickness (see Fig. 5.18) is also confirmed by a red shift of this PL peak. Additionally, the results are in very good agreement with ex-situ atomic force microscopy measurements of uncapped samples. On the sample shown in Fig. 5.17(a) only atomic steps were detected proving the 2D growth regime. QDs having a lateral size of about 30 nm and a height of 3–4 nm were found on a second sample ($d_{GaSb} = 4.5$ ML and $t_{GRI} = 5$ s,

5 Structural Characterisation of Quantum Dots by X-Ray Diffraction and TEM

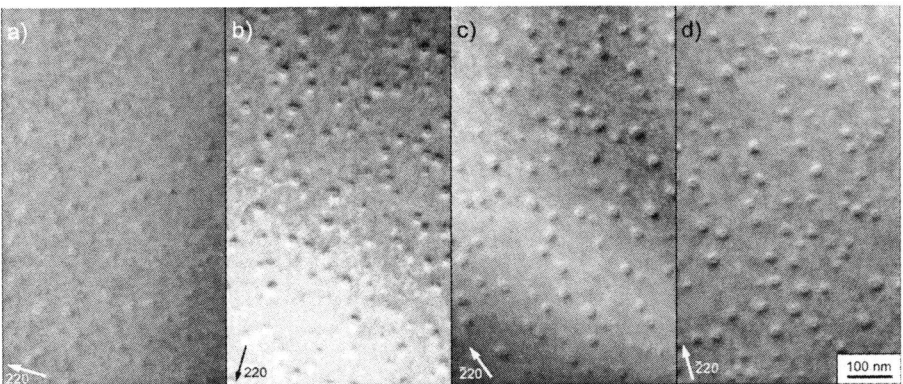

Fig. 5.19. Plan-view TEM dark-field images of GaSb quantum dots for different growth interruption times at a nominal 4.5 ML thick GaSb layer. (**a**) 0 s, (**b**) 2 s, (**c**) 5 s and (**d**) 10 s

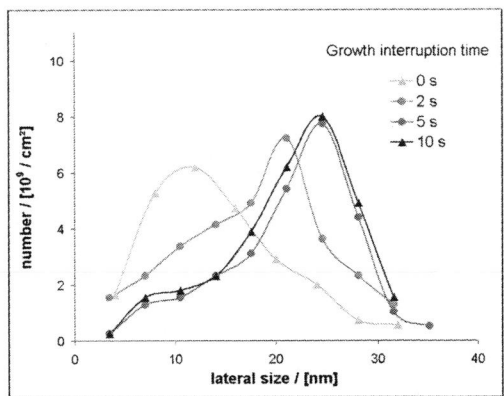

Fig. 5.20. Distribution of the lateral size of the GaSb quantum dots given in Fig. 5.19

[55, 54, 56]. The plan-view dark-field image of this sample is given in Fig. 5.19(c). In general, a lateral arrangement is not observed due to the smaller area density in comparison to, e.g., lateral ordered InP/(In,Ga)P QDs [57].

Another series of 220 dark-field plan-view images is shown in Fig. 5.19. The images were received from samples of a series where different growth interruption times (0 s, 2 s, 5 s, and 10 s) were applied after the deposition of GaSb. The nominal thickness of the GaSb layer was always 4.5 ML. The area density of the QDs amounts to about $3 \cdot 10^{10}$ cm^{-2} which remains the same for all samples of this series. In Fig. 5.20 the development of the size distribution of the QDs can be observed. The size distribution of the QDs remains stable for a time of growth interruption exceeding 2 s. Hence the formation of the QDs reaches an equilibrium state after about 2 s.

Among others, two conclusions can be drawn from the plan-view investigations. First, the final lateral size of the QDs is about 25 nm, and second, a regular lateral

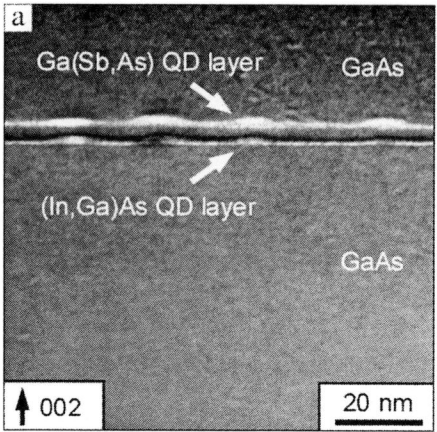 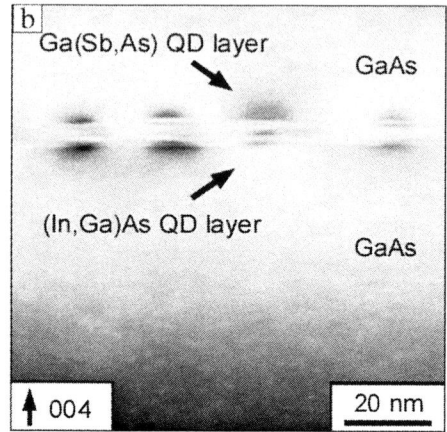

Fig. 5.21. Correlation of strain field and vertical arrangement of QDs. (**a**) Predominantly composition-sensitive 002 dark-field image. (**b**) Strain-sensitive TEM 004 dark-field image

arrangement of the QDs is missing. For tailoring of both parameters the GaSb/GaAs QDs were grown on an (In,Ga)As seed layer capped by a thin spacer layer of GaAs. Formation of (In,Ga)As QDs grown on GaAs substrate by self-organisation, lead to a strain field in the intermediate GaAs spacer layer. The position of the GaSb/GaAs QDs on top of the spacer is coupled with this strain field. Finally, the GaSb/GaAs are capped with GaAs.

The arrangement of the QDs in vertical direction can be visualised by cross-sectional TEM imaging. Moreover, diffraction-contrast TEM investigations in cross-sectional view were performed in order to receive information about the vertical and lateral size of the Ga(Sb,As) QDs as well as of the extension of the strain field caused by the QDs.

In Fig. 5.21(a) the GaSb QD layer is visualised using the 002 reflection, which is predominantly sensitive to the composition for materials having sphalerite structure [33]. Mainly the difference of the atomic scattering factors contributes to the contrast of an 002 dark-field image. The narrow bright lines represent the (In,Ga)As and the Ga(Sb,As) wetting layer, respectively, which is broadened at points where the QDs are located. For (In,Ga)As the bright regions are accompanied by two dark lines above and underneath indicating regions where the composition of $In_xGa_{1-x}As$ is between $0 \leq x \leq 0.4$ [33]. For these values the intensity of the 002 beam is smaller than that of GaAs presuming constant thickness of the sample. Compared to GaAs the higher brightness of the inner line of (In,Ga)As wetting layer and seed QDs denotes that the maximum In content is significantly higher than 0.4. Contrary to that, the Ga(Sb,As) QD layer shows a gradual change of the brightness without any intermediate minimum. This gives a monotonic dependence of the amplitude of the 002 beam on the Sb content in the system Ga(Sb,As).

The strain field surrounding the QDs is imaged in Fig. 5.21(b) using the 004 reflection for dark-field imaging. The image was taken from exactly the same region as shown in Fig. 5.21(a). The extensions of the strain field down into the substrate as well as into the cap layer above are evident from the contrast behaviour adjacent to QDs. The direct vertical correlation of the QDs as seen in Fig. 5.21(a) indicates that the strain field of the seed QDs has reached up to the surface onto which the Ga(Sb,As) was deposited. Hence, the thickness of the GaAs spacer layer of 4.5 nm is appropriate for the growth of vertically correlated QDs.

The lateral size of the Ga(Sb,As) QDs derived from their cross-sections amounts to 10 to 15 nm. The height is about 2 to 3 nm giving an aspect ratio of width over height of 5. The QDs were found free of defects. Hence, the lattice mismatch is purely elastically relaxed by the formation of QDs on top of a wetting layer.

5.4.1 Structural Characterisation of Ga(Sb,As) QDs by High-Resolution TEM Imaging

Methods of quantitative HRTEM (qHRTEM) were used to measure the local strain field and chemical composition of In and Sb on atomic scale [15, 4, 58]. HRTEM images were recorded at a C_s-corrected microscope (installed at the Ernst Ruska-Centre at Research Centre Juelich) where delocalisation phenomena are eliminated ensuring the imaging of interfaces without broadening. The position as well as the size of the stacked QDs are recognisable by visual inspection of the contrast pattern (Fig. 5.22(a)). The wetting layers are invisible due to their small thickness. The results of the analysis of strain and composition are illustrated in Fig. 5.22(b), where the left part shows a colour-coded plot of the local displacement relative to a reference lattice. The indicator of the strain is the displacement $u_{[001]}$ along the [001] direction between positions of atom columns in the inspected region compared to an unstrained reference lattice. The reference lattice (RL) was defined in an area located well underneath the (In,Ga)As QD. Approaching the (In,Ga)As QD a compression of the (002) lattice planes is detected whereas on top of the QD the lattice planes are

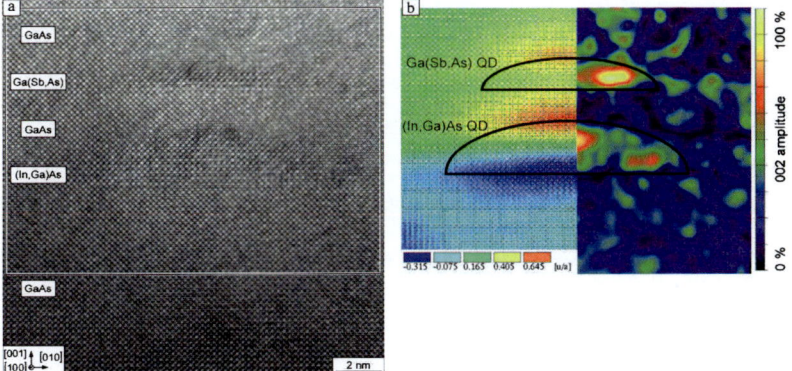

Fig. 5.22. Quantitative HRTEM of QDs. (**a**) HRTEM image in [100] projection. (**b**) Displacement of the 002 lattice planes as derived by DALI (*left*) and amplitude of the 002 beam determined by JCMP (*right*)

relaxed in consistence with the higher In content. The GaAs spacer layer causes a decrease of the vertical lattice distance. Near the lower interface of the Ga(Sb,As) QD the (002) lattice plane distance increases again due to the larger lattice parameters of Ga(Sb,As) compared to GaAs up to a maximum value representing the apex of the QD. In the region above the two QDs and beside the GaAs lattice is undeformed.

The right part of Fig. 5.22 gives the intensity of the predominantly chemically sensitive 002 reflection gained by Bragg filtering of the HRTEM image. The QDs can be identified easily by their bright contrast. Around the (In,Ga)As QD a dark rim can be recognised, which originates from the node in the amplitude of the 002 reflection at $x_{In} \approx 0.18$. This effect is not observed in $GaSb_yAs_{1-y}$.

5.4.2 Chemical Characterisation of Ga(Sb,As) QDs by HAADF STEM Imaging

In order to image the QD structure without contributions due to strain and oscillating intensity of diffracted beams STEM Z-contrast imaging has to be applied. Figure 5.23(a) shows an atomically resolved high-angle annular dark field (HAADF)-STEM image. Presuming a constant thickness of the specimen within the field of view the signal registered at the annular dark-field detector is exclusively a function of the mean atomic number. The vertically stacked QDs of (In,Ga)As as well as of Ga(Sb,As) are clearly visible with bright contrast owing to the higher mean atomic

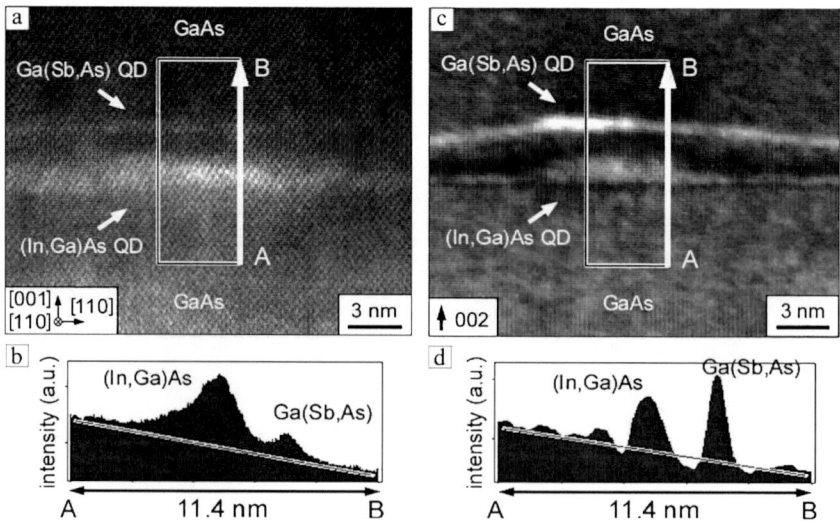

Fig. 5.23. Composition-sensitive imaging of QDs. (a) High-angle annular dark-field STEM image. (b) HAADF STEM intensity profile across the two QDs from A to B. The values were averaged parallel to the interfaces within the framed area of (a). The linearly descending background is due to the thickness wedge of the TEM specimen. (c) 002 dark-field TEM image. (d) 002 intensity profile across the two QDs

number Z in comparison to GaAs. The spacer thickness between the two QD layers is only 3.5 nm. Owing to this, incomplete flattening of the surface of the GaAs spacer is seen. The Ga(Sb,As) QDs form on the bumps directly above the (In,Ga)As seed QDs. Hence, the cross-section of the Ga(Sb,As) QD is crescent-like. In order to better visualise the intensity distribution a line scan from A to B is drawn (see Fig. 5.23(b)). Perpendicular to the line from A to B the intensity is averaged within the area marked by the rectangle. Both QDs are clearly indicated in the line scan as intensity maxima. Additionally, a background signal is detected showing a decrease approaching position B. This is due to the decrease of the sample thickness.

For comparison a 002 dark-field diffraction contrast image (Fig. 5.23(c)) is shown of the same specimen area. As described earlier, the intensity of the 002 beam exhibits a minimum for (In,Ga)As at an In content of about 18% instead of Ga. Consequently, there is no direct correlation between the intensity in the image and the content. The intensity profile given in Fig. 5.23(d) reveals local minima underneath and above the Ga(Sb,As) QD as well.

The comparison of the results of both methods gives a more direct visualisation of the composition by using HAADF STEM imaging technique, although one must consider an influence of thickness variation. After subtraction of the background intensity due to thickness wedge the signal is merely depending on the mean atomic number. For quantification of the signal a standard is necessary.

References

1. D. Bimberg, M. Grundmann, N.N. Ledentsov, *Quantum Dot Heterostructures* (Wiley, Chichester, 1999)
2. D. Bimberg, N.N. Ledentsov, J. Phys.: Condens. Matter **15**, R1 (2003)
3. K. Scheerschmidt, D. Conrad, H. Kirmse, R. Schneider, W. Neumann, Ultramicroscopy **81**, 289 (2000)
4. W. Neumann, H. Kirmse, I. Häusler, R. Otto, J. Microsc. **223**, 200 (2006)
5. R. Köhler, P. Schäfer, T. Panzner, M. Schmidbauer, R. Feidenhans'l, HASYLAB Annual Report. p. 60, 2001
6. M. Schmidbauer, D. Grigoriev, M. Hanke, P. Schäfer, T. Wiebach, R. Köhler, Phys. Rev. B **71**, 115324 (2005)
7. Th. Wiebach, M. Schmidbauer, M. Hanke, H. Raidt, R. Köhler, H. Wawra, Phys. Rev. B **61**, 5571 (2000)
8. M. Schmidbauer, *X-Ray Diffuse Scattering from Self-Organized Mesoscopic Semi-Conductor Structures*. Springer Tracts in Modern Physics, vol. 199 (Springer, Berlin, 2004)
9. W. Dorsch, H.P. Strunk, H. Wawra, G. Wagner, J. Groenen, R. Carles, Appl. Phys. Lett. **72**, 179 (1998)
10. A.-K. Gerlitzke, *Heteroepitaktisches Wachstum von $Si_{1-x}Ge_x$-Nanostrukturen auf Si-Substraten mittels Flüssigphasenepitaxie* (Mensch & Buch, Berlin, 2006)
11. Th. Teubner, T. Boeck, J. Cryst. Growth **289**, 366 (2006)
12. S. Christiansen, M. Albrecht, H.P. Strunk, H.J. Maier, Appl. Phys. Lett. **64**, 3617 (1994)
13. M. Hanke, M. Schmidbauer, D. Grigoriev, P. Schäfer, R. Köhler, A.-K. Gerlitzke, H. Wawra, Phys. Rev. B **69**, 075317 (2004)

14. T. Tham, Ph.D. thesis, Humboldt-Universität zu Berlin, 2002
15. W. Neumann, H. Kirmse, I. Häusler, R. Otto, I. Hähnert, J. Alloys Comp. **382**, 2 (2004)
16. R. Schneider, I. Häusler, A.-K. Gerlitzke, W. Neumann, in *Microscopy Conference 2005*, Davos, 2005
17. I. Häusler, H. Schwabe, R. Schneider, W. Neumann, M. Hanke, R. Köhler, A. Gerlitzke, in *Microscopy Conference 2005*, Davos, 2005
18. I. Häusler, H. Schwabe, H. Kirmse, W. Neumann, in *14. Jahrestagung der DGK 2006*, Freiburg, 2006
19. I. Häusler, H. Kirmse, W. Neumann, in *Proc. 16. Intern. Microscopy Congress*, Sapporo, 2006
20. M. Hanke, Ph.D. thesis, Humboldt-Universität zu Berlin, 2002
21. S.A. Chaparro, Y. Zhang, J. Drucker, Appl. Phys. Lett. **76**, 3534 (2000)
22. U. Denker, O.G. Schmidt, N.Y. Jin-Phillip, K. Eberl, Appl. Phys. Lett. **78**, 3723 (2001)
23. M. Hanke, H. Raidt, R. Köhler, H. Wawra, Appl. Phys. Lett. **83**, 4927 (2003)
24. V.A. Shchukin, N.N. Ledentsov, D. Bimberg, *Epitaxy of Nanostructures* (Springer, New York, 2004)
25. P.W. Fry, I.E. Itskevich, D.J. Mowbray, M.S. Skolnick, J.J. Finley, J.A. Barker, E.P. O'Reilly, L.R. Wilson, I.A. Larkin, P.A. Maksym, M. Hopkinson, M. Al-Khafaji, J.P.R. David, A.G. Cullis, G. Hill, J.C. Clark, Phys. Rev. Lett. **74**, 733 (2000)
26. P. Jayavel, H. Tanaka, T. Kita, O. Wada, H. Ebe, M. Sugawara, J. Tatebayashi, Y. Arakawa, Y. Nakata, T. Akiyama, Appl. Phys. Lett. **84**, 1820 (2004)
27. R.L. Sellin, F. Heinrichsdorff, C. Ribbat, M. Grundmann, U.W. Pohl, D. Bimberg, J. Cryst. Growth **221**, 581 (2000)
28. M. Hanke, D. Grigoriev, M. Schmidbauer, P. Schäfer, R. Köhler, R.L. Sellin, U.W. Pohl, D. Bimberg, Appl. Phys. Lett. **85**, 3062 (2004)
29. M. Hanke, D. Grigoriev, M. Schmidbauer, P. Schäfer, R. Köhler, U.W. Pohl, R.L. Sellin, D. Bimberg, N.D. Zakharov, P. Werner, Physica E **21**, 684 (2004)
30. A. Krost, F. Heinrichsdorff, D. Bimberg, A. Darhuber, G. Bauer, Appl. Phys. Lett. **68**, 785 (1996)
31. I. Kegel, T.H. Metzger, A. Lorke, J. Peisl, J. Stangl, G. Bauer, K. Nordlund, W.V. Schoenfeld, P.M. Petroff, Phys. Rev. B **63**, 035318 (2001)
32. L.G. Wang, P. Kratzer, N. Moll, M. Scheffler, Phys. Rev. B **62**, 1897 (2000)
33. A. Rosenauer, U. Fischer, D. Gerthsen, A. Förster, Ultramicroscopy **72**, 121 (1998)
34. A. Lemaître, G. Patriarche, F. Glas, Appl. Phys. Lett. **85**, 3717 (2004)
35. J. Márquez, L. Geelhaar, K. Jacobi, Appl. Phys. Lett. **78**, 2309 (2001)
36. G. Springholz, V. Holý, M. Pinczolits, G. Bauer, Science **282**, 734 (1998)
37. G. Springholz, M. Pinczolits, P. Mayer, V. Holý, G. Bauer, H.H. Kang, L. Salamanca-Riba, Phys. Rev. Lett. **84**, 4669 (2000)
38. J. Tersoff, C. Teichert, M.G. Lagally, Phys. Rev. Lett. **76**, 1675 (1996)
39. Q. Xie, A. Madhukar, P. Chen, N. Kobayashi, Phys. Rev. Lett. **75**, 2542 (1995)
40. M. Schmidbauer, T. Wiebach, H. Raidt, M. Hanke, R. Köhler, H. Wawra, Phys. Rev. B **58**, 10523 (1998)
41. M. Schmidbauer, T. Wiebach, H. Raidt, M. Hanke, R. Köhler, H. Wawra, J. Phys. D: Appl. Phys. **32**, 230 (1999)
42. M. Schmidbauer, F. Hatami, M. Hanke, P. Schäfer, K. Braune, W.T. Masselink, R. Köhler, M. Ramsteiner, Phys. Rev. B **65**, 125320 (2002)
43. M. Schmidbauer, F. Hatami, P. Schäfer, M. Hanke, T. Wiebach, H. Niehus, W.T. Masselink, R. Köhler, Mat. Res. Soc. Symp. Proc. **642**, J6.8 (2001)
44. D. Leonard, M. Krishnamurthy, C.M. Reaves, S.P. Denbaars, P.M. Petroff, Appl. Phys. Lett. **63**, 3203 (1993)

45. T. Mano, R. Nötzel, G.J. Hamhuis, T.J. Eijkemans, J.H. Wolter, Appl. Phys. Lett. **81**, 1705 (2002)
46. Y.I. Mazur, W.G. Ma, X. Wang, Z.M. Wang, G.J. Salamo, M. Xiaoa, T.D. Mishima, M.B. Johnson, Appl. Phys. Lett. **83**, 987 (2003)
47. M. Schmidbauer, S. Seydmohamadi, D. Grigoriev, Z.M. Wang, Y.I. Mazur, P. Schäfer, M. Hanke, R. Köhler, G.J. Salamo, Phys. Rev. Lett. **96**, 066108 (2006)
48. E.A. Kondrashkina, S.A. Stepanov, R. Opitz, M. Schmidbauer, R. Köhler, R. Hey, M. Wassermeier, D.V. Novikov, Phys. Rev. B **56**, 10469 (1997)
49. V. Holý, G. Springholz, M. Pinczolits, G. Bauer, Phys. Rev. Lett. **83**, 356 (1999)
50. Y. Sugiyama, Y. Nakata, Y. Muto, S. Futatsugi, T. Yokoyama, N. Fujitsu Labs. Ltd., Atsugi, J. Sel. Top. Quantum Electron. **4**, 880 (1998)
51. F. Hatami, N.N. Ledentsov, M. Grundmann, J. Böhrer, F. Heinrichsdorff, M. Beer, D. Bimberg, S.S. Ruvimov, P. Werner, U. Gösele, J. Heydenreich, U. Richter, S.V. Ivanov, B.Y. Meltser, P.S. Kop'ev, Z.I. Alferov, Appl. Phys. Lett. **67**, 656 (1995)
52. J.M. Gérard, J.B. Génin, J. Lefebvre, J.M. Moison, N. Lebouché, F. Barthe, J. Cryst. Growth **150**, 351 (1995)
53. K. Suzuki, R.A. Hogg, K. Tachibana, Y. Arakawa, Jpn. J. Appl. Phys. **37**, L203 (1998)
54. L. Müller-Kirsch, Ph.D. thesis, Technische Universität Berlin, 2002
55. L. Müller-Kirsch, R. Heitz, U.W. Pohl, D. Bimberg, I. Häusler, H. Kirmse, W. Neumann, Appl. Phys. Lett. **79**, 1027 (2001)
56. I. Häusler, Master's thesis, Humboldt-Universität zu Berlin, 2001
57. F. Hatami, U. Müller, H. Kissel, K. Braune, R.-P. Blum, S. Rogaschewski, H. Niehus, H. Kirmse, W. Neumann, M. Schmidbauer, R. Köhler, W.T. Masselink, J. Cryst. Growth **216**, 26 (2000)
58. R. Otto, H. Kirmse, I. Häusler, A. Rosenauer, D. Bimberg, L. Müller-Kirsch, Appl. Phys. Lett. **85**(21), 4908 (2004)

6

The Atomic Structure of Quantum Dots

Mario Dähne, Holger Eisele, and Karl Jacobi

Abstract. In this chapter, the atomic structure of both uncapped and buried quantum dots is described as derived from scanning tunneling microscopy in both top-view and cross-sectional geometry. Important conclusions are drawn also on the growth processes during quantum dot formation as well as during overgrowth. It is demonstrated that uncapped InAs quantum dots on GaAs(001) have a pyramidal shape with dominating {137} side facets and—in the case of larger dots—also {101} and {111} side facets. Buried InAs and InGaAs quantum dots, in contrast, are characterized by a truncated pyramidal shape with a (001) top facet and rather steep side facets. In addition, segregation processes during capping lead to strong intermixing and—under special overgrowth conditions—even to concave top facets or to the formation of nanovoids. Buried GaSb quantum dots are found to be much smaller, but also show a truncated pyramidal shape and strong intermixing effects. The experimental results will be discussed in the framework of the strain-induced segregation processes occurring during the different stages of quantum dot formation and overgrowth.

6.1 Introduction

Strained quantum dots are currently of high interest because of their self-organized preparation by Stranski–Krastanow growth and their localized δ-like electronic states correlated with their spatial structure [1, 2]. In order to understand the physical properties of quantum dots as well as their growth mechanisms, the knowledge of the exact atomic structure, mainly characterized by size, shape and stoichiometry, is essential.

The atomic structure of quantum dots is often studied in reciprocal space, e.g. by reflection high-energy electron diffraction (RHEED) or X-ray diffraction (XRD) as well as in real space by transmission electron microscopy (TEM), atomic force microscopy (AFM), (top-view) scanning tunneling microscopy (STM) or cross-sectional scanning tunneling microscopy (XSTM). While XRD and TEM are discussed in detail in the previous chapter, here we give an overview of the determination of the quantum dot atomic structure using STM and XSTM.

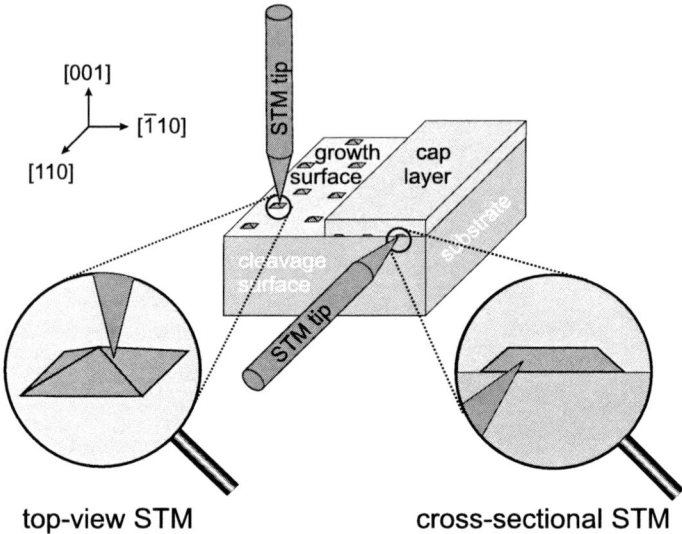

Fig. 6.1. Comparison of STM (*left*) and XSTM experiments (*right*)

The difference between STM and XSTM experiments is shown in Fig. 6.1. In STM experiments (left) the tunneling tip probes the growth surface, allowing us to study the atomic structure of uncapped quantum dots. In contrast, XSTM experiments (right) are performed with the tip probing a cross-sectional surface prepared by cleavage, enabling us to determine the atomic structure of buried quantum dots. Combining these techniques reveals a rather clear picture of the atomic structure of uncovered and covered quantum dots. This also allows us to study the mechanisms at work during growth and capping.

6.2 Experimental Details

The experiments were performed in two different ultra-high-vacuum systems. One setup combines a home-built molecular beam epitaxy (MBE) chamber for InAs-GaAs growth with a Park Instruments STM chamber for the in-situ analysis of the growth surface. The other setup consists of a home-built microscope specially designed for XSTM experiments including a cleavage stage for sample preparation. The samples for the XSTM experiments were grown by MBE and metal-organic chemical vapor deposition (MOCVD).

6.3 STM Studies of InAs Quantum Dots on the Growth Surface

In the following, we present data on the atomic structure of uncovered InAs quantum dots. Figure 6.2 shows a top-view STM image of a rather small InAs quantum

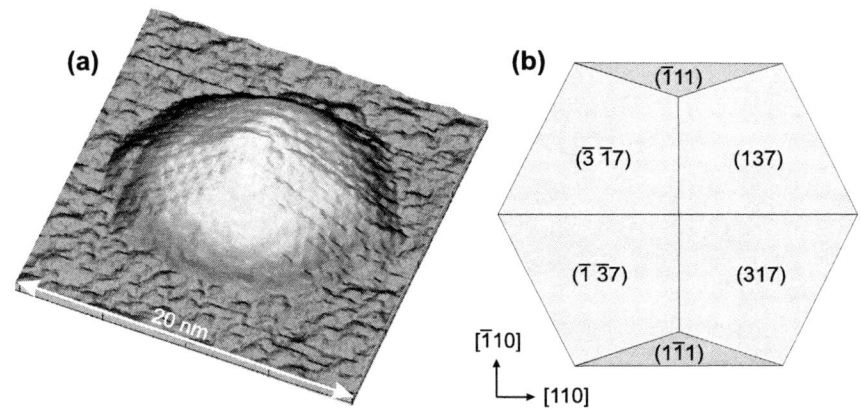

Fig. 6.2. (a) Top-view STM image of a small InAs quantum dot and (b) corresponding structure model

dot prepared by MBE growth of 1.8 monolayers (ML) InAs with a growth rate of 0.027 ML/s at a substrate temperature of 450°C, together with the corresponding structure model. The dot is characterized by a pyramidal shape and has the same spatial C_{2v} symmetry as the substrate surface. The four main side facets are atomically resolved and show a periodic corrugation, allowing us to determine their exact crystalline orientation as {137}. This was concluded from a comparison with STM images from several differently oriented GaAs surfaces [3, 4]. The smaller facets along {$\bar{1}$10} directions appear rounded and are less resolved, but can be assigned to {$\bar{1}$11} surfaces.

It should be noted here that the previously unexpected assignment to the high-indexed {137} surfaces was only possible due to the exact knowledge of the surface structure and the discovery of the new low-energy GaAs(2 5 11) surface at almost the same time [5–7], which comprises the (137) surface as a subunit. A facet determination based only on its spatial orientation using STM images without atomic resolution, in contrast, has proved to be insufficient in many cases.

Upon further InAs deposition, the shape of the quantum dots undergoes considerable change [8]. Figure 6.3 shows the structural evolution with dot size increasing from (a) to (j). While the small dots are dominated by high-indexed {137} facets, larger dots are characterized by additional {101} and {111} facets, which finally dominate. At intermediate stages, small {$\bar{1}$35} and {$\bar{1}$12} facets were also observed. However, the quantum dots keep their pyramidal shape, and (001) top facets were never found. Similar results were obtained by Costantini et al. [9, 10]. The observed shapes are in contrast to results from earlier energy calculations of the strained surfaces, where only low-index facets were considered [11], but agree well with recent theoretical studies [12, 13] (see also Chap. 1).

The observed behavior is related to the strain that increases considerably with the size of the interface area between dot and substrate, limiting lateral dot growth and favoring an increase of dot height in order to accommodate more dot material. Such

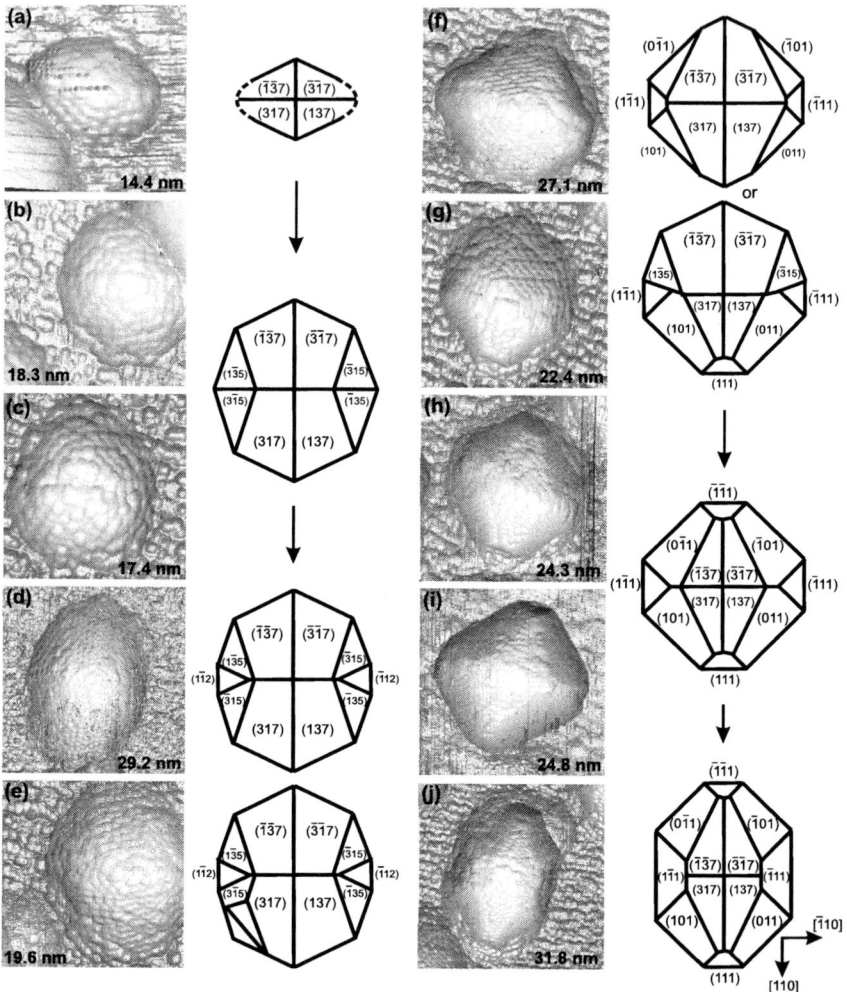

Fig. 6.3. Evolution of the dot shape with dot size increasing from (**a**) to (**j**). At the bottom of each image, its size is given

a shape can only be realized with flanks considerably steeper than the {137} facets, which—for the case of unstrained material—have a gradient angle of only about 24°. {101} and {111} facets, in contrast, are characterized by much steeper gradient angles of about 45° and 55°. This view is further supported by the appearance of {$\bar{1}11$} facets already at the smallest dots, considerably limiting the length of an otherwise purely {137}-faceted dot. The resulting lateral aspect ratio close to one leads to a strong reduction of the strain energy. A schematic model for the quantum dot growth on the GaAs(001) surface is shown in Fig. 6.4, showing the shell-like growth starting with {137} facets, which are successively replaced by {101} facets at the lower dot flanks when the dot size increases.

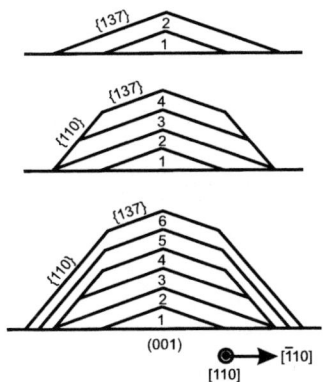

Fig. 6.4. Schematic model for the shell-like growth of InAs quantum dots

In addition to these results, the growth of InAs quantum dots on GaAs substrates with orientations different from (001) has also been studied intensively using top-view STM [4, 14–17]. Here a dominance of low-index facets like {101} and {111} was observed, and the symmetry of the substrate surface was again found to determine the symmetry of the quantum dots.

6.4 XSTM Studies of Buried Nanostructures

In the following we present XSTM studies on the atomic structure of buried quantum dots. For this purpose, the (110) and/or ($\bar{1}$10) cleavage surfaces through InAs, InGaAs, and GaSb quantum dot heterostructures in a GaAs matrix were studied using XSTM.

6.4.1 InAs Quantum Dots

Section 6.3 presented top-view STM data on the atomic structure of uncovered InAs quantum dots on GaAs(001). In order to study the possible structural changes upon capping using XSTM [18], quantum dot samples were prepared in the same MBE growth chamber as used for the top-view investigations described above, followed by rapid capping of the dots with a thick GaAs layer at relatively low temperatures in order to minimize segregation effects.

Figure 6.5 shows representative XSTM images of such quantum dots for both possible orientations of the {110} cleavage surface. These dots were prepared by deposition of 1.8 ML InAs with a growth rate of 0.027 ML/s at a substrate temperature of 450°C. Thus, the preparation conditions correspond to those of the top-view image of the uncapped quantum dot shown in Fig. 6.2(a). After deposition of the quantum dot material, growth was interrupted 10 s for dot formation. Subsequently, the dots were capped with about 10 nm GaAs at 0.15 ML/s and 450°C and further overgrown by 800 nm GaAs while the substrate temperature was gradually raised to 520°C.

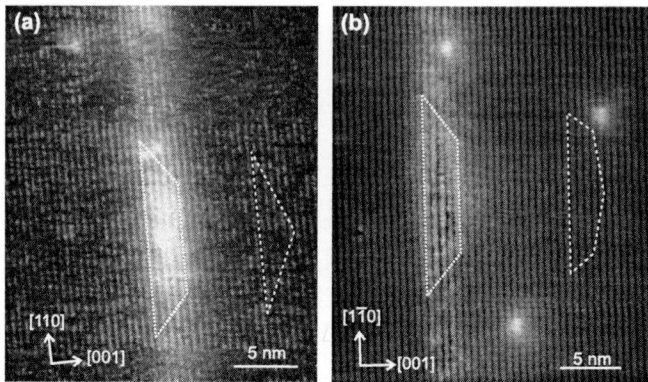

Fig. 6.5. Filled state XSTM images of small MBE-grown InAs quantum dots, taken (**a**) at the ($\bar{1}10$) and (**b**) at the (110) cleavage surface. The dotted lines mark the observed contours, while the dashed lines show the cross-sections expected for the uncapped dots such as those in Fig. 6.2

The vertical lines in Fig. 6.5 represent the atomic chains at the {110} cleavage surfaces. Due to the negative sample bias, the filled states are imaged, which are mostly related to the dangling bonds of the group-V atoms. It should be noted that only every second monolayer along the growth direction is imaged, since the atoms from the monolayers in between are located underneath the {110} cleavage surfaces.

In the center of each image, a quantum dot is visible due to its brighter appearance. This bright contrast results from a combination of structural and electronic effects. After cleavage the compressively strained InAs material within the GaAs host relaxes out of the cleavage surface, leading to a bright contrast [19]. This structural contrast is very pronounced at quantum dots, because of the large strain of zero-dimensional objects, but it is much weaker at the wetting layer, which is considerably less strained [19–21]. The structural contrast is further enhanced by electronic effects, since electronically different materials show different tunneling probabilities. In a simple picture, the electronic states of InAs quantum dots, the wetting layer, or even single indium atoms lead to a higher tunneling probability than at the GaAs host, resulting in a retraction of the tip in order to keep the tunneling current constant and consequently gives a brighter appearance.

The observed contours of the quantum dots in Fig. 6.5 consequently in a brighter are marked by dotted lines. The trapezoidal shape of both cross-sections indicates a truncated pyramidal shape of the quantum dots with a large (001) top face. This is in contrast to the untruncated pyramidal shape of the uncapped quantum dots, which would lead to the triangular or pentagonal cross-sections marked by the dashed lines in Fig. 6.5. In addition, the side flanks show an angle around 35°, which disagrees strongly with the small angles of 11° and 22° expected for the ($\bar{1}10$) and (110) cross-sections of the {137} facets, respectively. On the other hand, the observed angles correspond with cross-sections of {101} facets. Thus it is found that the quantum dots have undergone remarkable changes in shape upon capping.

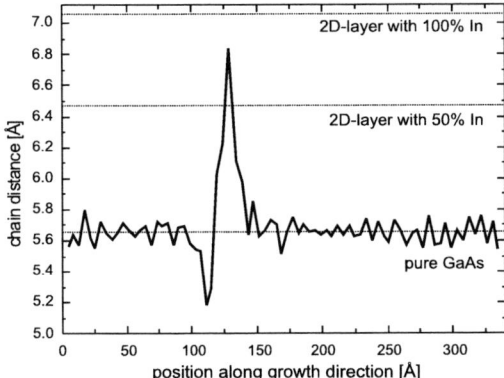

Fig. 6.6. Determination of the local stoichiometry of the center region of the small MBE-grown quantum dot shown in Fig. 6.5(a)

Furthermore, the stoichiometry is also different from pure InAs. A rather smooth appearance of the atomic chains would be expected in XSTM images only for the case of pure InAs quantum dots [22]. On the other hand, considerable contrast variations are observed in Fig. 6.5, in particular at the dot sides, indicating the formation of ternary InGaAs. In order to determine the stoichiometry quantitatively, we can take advantage of the lattice mismatch of 7.2% between InAs and GaAs, leading to a correlation between lattice constant and stoichiometry. This effect is shown in Fig. 6.6, where the measured distance between neighboring atomic chains is plotted as a function of position along growth direction for the quantum dot shown in Fig. 6.5(a).

While the typical lattice constant of GaAs is measured deep in the substrate and the overlayer, considerable changes are observed in the dot region, with a strong increase at the dot itself. For a quantitative analysis, the anisotropic strain in the system has to be considered. Using strain relaxation simulations on an atomic scale [19–21], the lattice constants of two-dimensional InGaAs layers with different indium content embedded in a GaAs host were determined, as indicated by the dotted lines in Fig. 6.6. However, the strain in zero-dimensional quantum dots is even larger, so that corrections of the lattice constant have to be performed. This effect is manifested, e.g., by a local lattice constant even smaller than that of GaAs just underneath the dot. This is due to the compressive strain also within the underlying GaAs that would be absent in the two-dimensional case. With these corrections, an indium content of about 80% at the dot center is evaluated, decreasing considerably at the dot sides.

Like the quantum dot, the wetting layer also shows strong structural modifications upon capping. As observed in Fig. 6.5, the wetting layer appears almost as thick as the quantum dot height. From a detailed investigation of the local lattice constant, a rather low indium content of about 25% is derived at the bottom of the wetting layer, further decreasing exponentially in growth direction. This behavior indicates a strong indium segregation into the growing capping layer.

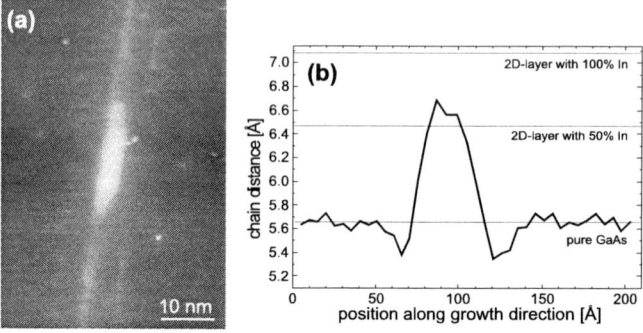

Fig. 6.7. (a) Filled state XSTM image of a large MBE-grown quantum dot and (b) the local stoichiometry analysis of its center region

Quantum dots grown with more material were found to be much larger, while the other structural parameters are similar. This is shown in Fig. 6.7 for the case of a quantum dot grown using slightly more InAs at the same substrate temperature of 450°C. Again, a truncated pyramidal shape is found, and the analysis of the local stoichiometry yields a maximum indium content of around 80% at the dot center. Furthermore, the outer parts and in particular the side flanks are characterized by a much lower indium content. The high compressive strain in the GaAs material directly above and below the dot is again manifested by a local lattice constant below that of GaAs. The wetting layer of this sample is also characterized by an indium content reaching only about 20% and decaying exponentially in growth direction.

The strong structural modifications upon capping can mostly be related to strain energy reduction. Before capping, the maximum strain is found at the lower side flanks of the dot, while strain is almost relaxed at the top of the dot. Capping induces a strong strain energy increase both at the top and at the side flanks, in particular close to the dot base. The strain is relaxed by lateral segregation at the growth surface resulting in a truncation of the pyramid, a mixed stoichiometry at the dot sides, and steeper side facets: The GaAs material deposited on the dot surface segregates to the sides of the dots, taking along the InAs apex material. This results in a truncation of the pyramid and a deposition of intermixed InGaAs material at the sides of the dot. Furthermore, the side facets become steeper in order to accommodate the additional material while limiting the area of the bottom interface of the dot. In this way, the strain energy is considerably reduced.

It should be noted that the amount of such overgrowth-induced modifications critically depends on how much time the system has to react and the special conditions at the growth surface, in particular during growth interruptions. In XSTM experiments at MOCVD-grown InAs dots, a pure InAs stoichiometry of dots and wetting layer was observed [21, 23]. These dots were grown and overgrown rather fast at a growth rate of 0.3 ML/s and a substrate temperature of 485°C. Furthermore, switching off the arsine dose during a 4 s growth interruption after InAs deposition reduced the mobility of the species at the growth surface, resulting in a freezing of indium and gallium segregation. This view is supported by a recent photoluminescence and TEM study of the ripening behavior during a growth interruption without arsine, demonstrating that the flat shape of the quantum dots and their high indium content are maintained [24]. Both the fast overgrowth and the lack of arsine pressure during dot ripening are thus assumed to result in the stoichiometrically pure InAs quantum dots.

6.4.2 InGaAs Quantum Dots

Here we discuss the structural properties of intermixed InGaAs quantum dots [25, 26]. The aim of the sample grown by MOCVD was to tune the transition wavelength to the technologically important 1.3 μm, which generally requires rather large and thus more strained dots. For this purpose, an $In_{0.80}Ga_{0.20}As$ quantum-dot layer was deposited, followed by a 60 s long growth interruption to enable dot formation. Then the dots were covered by a thin $In_{0.10}Ga_{0.90}As$ layer to reduce overgrowth-induced strain and to reduce the carrier confinement. This part of the growth was performed at a rate of 1 ML/s and a substrate temperature of 500°C. After a subsequent 5 nm thin GaAs overlayer, a 600 s long growth interruption was introduced at an elevated substrate temperature of 600°C in order to flatten the surface and to anneal possible point defects.

The overview XSTM image in Fig. 6.8 shows that three different dot types have formed. The shapes range from truncated pyramids (type-1) over dots with a concave top facet (type-2) to structures characterized by nanoholes in the center (type-3).

A typical XSTM image of the type-1 structure is shown in Fig. 6.9(a). While the overall shape can be described by a truncated pyramid with rather steep side flanks, as indicated by the dotted lines, the brightest contrast and therefore the highest indium concentration is observed at the top center, resulting in a reversed truncated

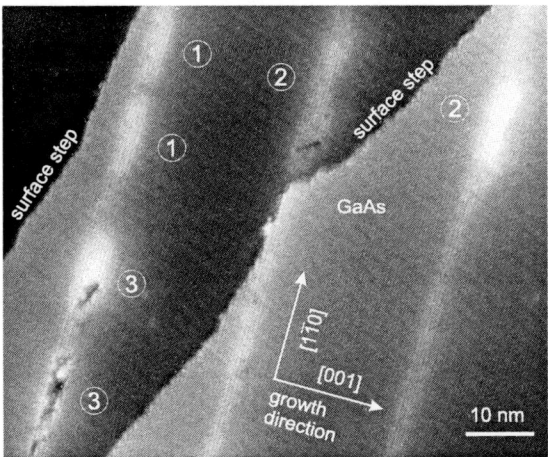

Fig. 6.8. Filled state overview image of the different types of InGaAs nanostructures observed by XSTM

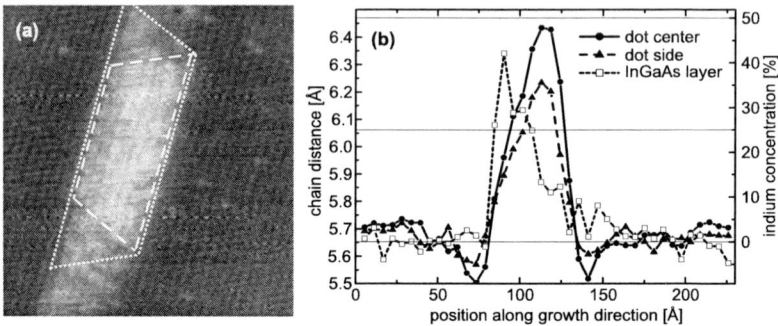

Fig. 6.9. (a) Empty state XSTM image of a type-1 InGaAs quantum dot, and (b) investigation of the local stoichiometry

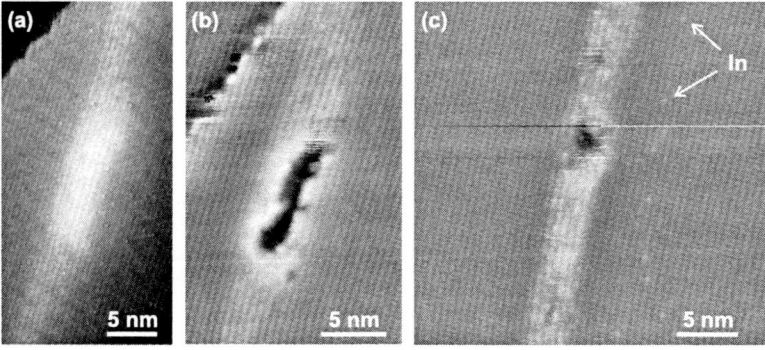

Fig. 6.10. XSTM data on (a) a type-2 InGaAs dot and (b, c) type-3 InGaAs structures with nanovoids

cone shape of the indium-rich core of the quantum dot, as indicated by the dashed lines [25]. This view is further supported by the evaluation of the local lattice constant, shown in Fig. 6.9(b), resulting in a maximum indium content of about 60% in the top center of the dot and again considerably less at the sides of the dot. The wetting layer is found to be intermixed with an indium content reaching 40% at its base and decreasing almost exponentially along growth direction, again indicating strong indium segregation. It should be noted that a stoichiometry distribution with an indium-rich center and gallium-rich side flanks has been observed frequently for highly strained dots—e.g., in Fig. 6.7 or in the case of MBE-grown InGaAs dots, e.g. in Fig. 6.7 [27].

Figure 6.10(a) shows a detailed image of a type-2 InGaAs quantum dot. Its top facet is clearly characterized by a concave shape with a depression amounting to about 1 nm. The dot is again characterized by a reversed cone shape of its indium-rich center. Type-2 dots have a slightly larger base than the type-1 dots.

Representative XSTM images of type-3 InGaAs structures are shown in Fig. 6.10(b) and (c). They are characterized by a nanovoid, meaning that material

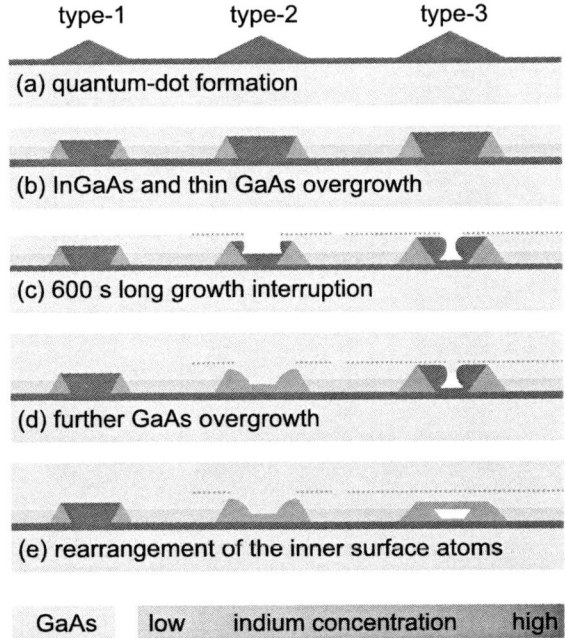

Fig. 6.11. Model for the formation of InGaAs dots and nanovoids during the different growth stages (**a**) to (**e**)

is completely missing in the dot center. Such nanovoids were mainly observed for the largest structures found in this sample.

Figure 6.11 shows the model for the growth of the different structure types within the same sample [26], which is also supported by theoretical calculations [28]. After formation (a), the different quantum dot structures have a pyramidal shape and are mainly distinguished by their size. Subsequent overgrowth (b) leads to a truncation, as discussed in Sect. 6.4.1, and a gallium enrichment at the flanks. The long growth interruption of 600 s (c) affects the three types in a different way: While the small type-1 dots are not modified, strong indium segregation occurs from the center of the larger and thus more strained type-2 and type-3 dots toward the present growth surface (c). This view is supported by the observation of an indium submonolayer just at the position where the growth was interrupted for 600 s, as visible in Fig. 6.10(c). Upon subsequent fast overgrowth (d), the large crater of the type-3 structure cannot be filled completely and a nanovoid remains, which may later reorganize its shape (e). The smaller crater of the less-strained type-2 dot, in contrast, can be filled completely during overgrowth, leaving a depression in the top of the dot. Such ring-like structures with craters have also been observed at InAs and InGaAs dots covered by thin GaAs overlayers using top-view AFM [29, 30].

A different strategy for growing larger InAs or InGaAs dots involves the use of antimony surfactants during growth. In such MOCVD-grown samples, where antimony was provided only during dot growth, the formation of stoichiometrically rather pure InAs dots was observed [31]. In the case of an additional antimony supply prior to dot deposition, again rather pure InAs dots were found, while an antimony content of about 25% was found in the wetting layer. This can be explained by the slightly tensile strain along growth direction at the wetting layer, preferring the incorporation of larger antimony atoms, which are already present at the growth surface. In the case of the InAs quantum dots, in contrast, the strong compressive strain prevents an incorporation of antimony atoms.

6.4.3 GaSb Quantum Dots

GaSb quantum dots embedded in GaAs are characterized by a type-II band alignment with a large hole binding energy and a spatial separation of the electrons and holes, resulting in long carrier lifetimes [32, 33]. This makes them interesting for applications like nanoscale charge-storage devices [1, 34], as discussed in detail in Chap. 11. However, the growth of GaSb quantum dots is not well understood, and data on the critical thickness and the structural properties of the dots vary considerably in the literature [35–38].

Here we present XSTM results on MOCVD-grown GaSb quantum dots grown with a rate of about 0.1 ML/s at a substrate temperature of 470°C [39–42]. The early stages of the growth of GaSb quantum dots are shown in Fig. 6.12, with the amount of deposited GaSb increasing from (a) to (c). The wetting layers have an inhomogeneous appearance and are characterized by small gaps of pure GaAs. An analysis of the local stoichiometry of all three wetting layers yields maximum antimony contents below 50%, as shown in Fig. 6.13(b). Furthermore, a total incorporation of about 1 ML GaSb is evaluated, indicating a considerably smaller critical thickness for dot formation than in the case of the InAs–GaAs system [42].

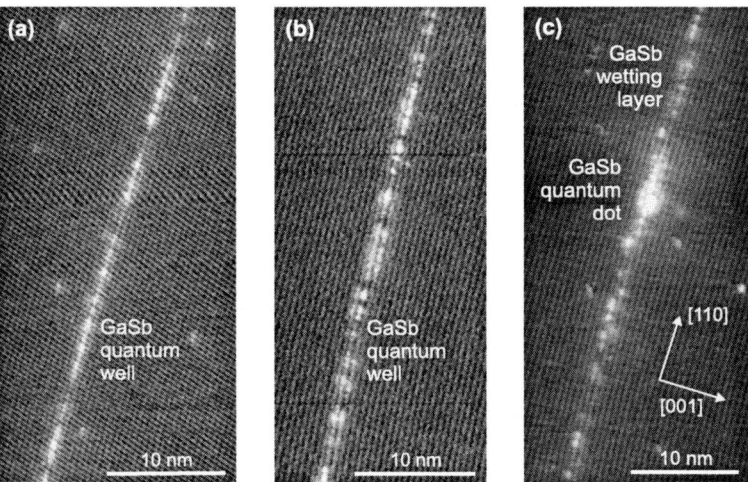

Fig. 6.12. Filled state XSTM images with GaSb exposure increasing from (**a**) to (**c**), showing the onset of GaSb quantum dot growth

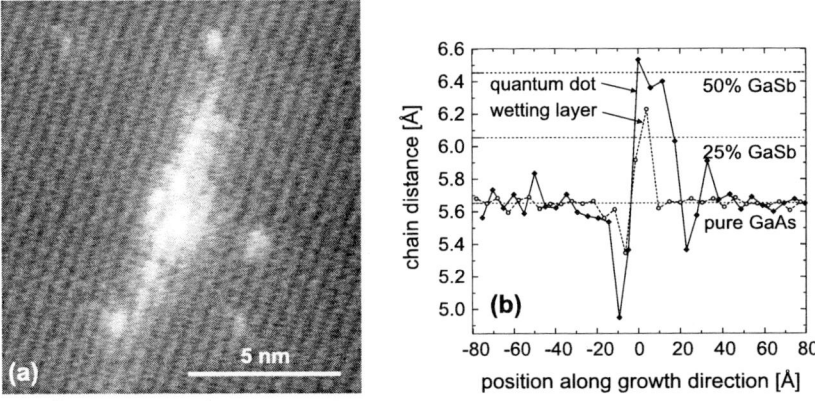

Fig. 6.13. (a) Filled state XSTM image of a MOCVD-grown GaSb quantum dot and (b) determination of the local stoichiometry of the center of the quantum dot and a representative section of the wetting layer

With increasing GaSb deposition, flat two-dimensional and very small three-dimensional islands appear, which finally develop to quantum dots, as shown in Fig. 6.12(c). These optically active quantum dots, however, are considerably smaller than typical InAs and InGaAs dots. In Fig. 6.13, XSTM data on the atomic structure of a representative GaSb quantum dot are shown. This quantum dot is characterized by an inhomogeneous stoichiometry with a maximum antimony content reaching 60%, but even smaller dots with almost pure GaSb stoichiometry were also observed [40, 41].

Finally it should be noted that in recent XSTM experiments on MBE-grown GaSb samples much larger ring-like quantum dots were observed. These quantum rings are characterized by a truncated pyramidal shape, a large hole in the center, and 25–100% antimony content [43].

6.5 Conclusion

In this chapter, we demonstrated the potential of atomically resolved top-view and cross-sectional scanning tunneling microscopy experiments to resolve the atomic structure of quantum dots. In particular the combination of both experiments allows us to obtain detailed information on the processes occurring during quantum-dot growth and subsequent capping. We found that energetic aspects manifested by local strain and surface energies as well as kinetic effects tuned by the growth parameters play an important role during formation and capping of the quantum dots.

Acknowledgements

We thank F. Heinrichsdorff, A. Krost, R.L. Sellin, L. Müller-Kirsch, K. Pötschke, U.W. Pohl, and D. Bimberg for the MOCVD samples, G. Balakrishnan and D.L. Huffaker for the GaSb MBE samples, J. Márquez, L. Geelhaar, Y. Temko, M.C. Xu, and T. Suzuki for the experiments comprising MBE growth and STM analysis, and O. Flebbe, A. Lenz, R. Timm, J. Grabowski, and L. Ivanova for performing the XSTM experiments. Financial support by the Deutsche Forschungsgemeinschaft, Sonderforschungsbereich 296, projects A02 and A04 is gratefully acknowledged.

References

1. M. Grundmann (ed.), *Nano-Optoelectronics* (Springer, New York, 2002)
2. V.A. Shchukin, N.N. Ledentsov, D. Bimberg, *Epitaxy of Nanostructures* (Springer, New York, 2004)
3. J. Márquez, L. Geelhaar, K. Jacobi, Appl. Phys. Lett. **78**, 2309 (2001)
4. K. Jacobi, Prog. Surf. Sci. **71**, 185 (2003)
5. L. Geelhaar, J. Márquez, P. Kratzer, K. Jacobi, Phys. Rev. Lett. **86**, 3815 (2001)
6. K. Jacobi, L. Geelhaar, J. Márquez, Appl. Phys. A **75**, 113 (2002)
7. L. Geelhaar, J. Márquez, P. Kratzer, K. Jacobi, Phys. Rev. B **60**, 155308 (2002)
8. M.C. Xu, Y. Temko, T. Suzuki, K. Jacobi, J. Appl. Phys. **98**, 083525 (2005)
9. G. Costantini, C. Manzano, R. Songmuang, O.G. Schmidt, K. Kern, Appl. Phys. Lett. **82**, 3194 (2003)
10. G. Costantini, A. Rastelli, C. Manzano, R. Songmuang, O.G. Schmidt, K. Kern, H. von Känel, Appl. Phys. Lett. **85**, 5673 (2004)
11. N. Moll, M. Scheffler, E. Pehlke, Phys. Rev. B **58**, 4566 (1998)
12. P. Kratzer, Q.K.K. Liu, P. Acosta-Diaz, C. Manzano, G. Costantini, R. Songmuang, A. Rastelli, O.G. Schmidt, K. Kern, Phys. Rev. B **73**, 205347 (2006)
13. T. Hammerschmidt, Ph.D. thesis, Technische Universität Berlin, http://opus.kobv.de/tuberlin/volltexte/2006/1358/, 2006
14. T. Suzuki, Y. Temko, M.C. Xu, K. Jacobi, J. Appl. Phys. **96**, 6398 (2004)
15. T. Suzuki, Y. Temko, M.C. Xu, K. Jacobi, Phys. Rev. B **69**, 235302 (2004)
16. M.C. Xu, Y. Temko, T. Suzuki, K. Jacobi, Appl. Phys. Lett. **84**, 2283 (2004)
17. Y. Temko, T. Suzuki, M.C. Xu, K. Jacobi, Phys. Rev. B **71**, 045336 (2005)
18. H. Eisele, A. Lenz, R. Heitz, R. Timm, M. Dähne, Y. Temko, T. Suzuki, K. Jacobi, Unpublished
19. H. Eisele, Ph.D. thesis, Technische Universität Berlin (Wissenschaft und Technik Verlag, Berlin, 2002)
20. H. Eisele, O. Flebbe, T. Kalka, M. Dähne-Prietsch, Surf. Interface Anal. **27**, 537 (1999)
21. O. Flebbe, H. Eisele, T. Kalka, F. Heinrichsdorff, A. Krost, D. Bimberg, M. Dähne-Prietsch, J. Vac. Sci. Technol. B **17**, 1639 (1999)
22. H. Eisele, O. Flebbe, T. Kalka, F. Heinrichsdorff, A. Krost, D. Bimberg, M. Dähne-Prietsch, Phys. Stat. Sol. (b) **215**, 865 (1999)
23. H. Eisele, O. Flebbe, T. Kalka, C. Preinesberger, F. Heinrichsdorff, A. Krost, D. Bimberg, M. Dähne-Prietsch, Appl. Phys. Lett. **75**, 106 (1999)
24. U.W. Pohl, K. Pötschke, A. Schliwa, F. Guffarth, D. Bimberg, H.D. Zakharov, P. Werner, M.B. Lifshits, V.A. Shchukin, D.E. Jesson, Phys. Rev. B **72**, 245332 (2005)
25. A. Lenz, R. Timm, H. Eisele, C. Hennig, S.K. Becker, R.L. Sellin, U.W. Pohl, D. Bimberg, M. Dähne, Appl. Phys. Lett. **81**, 5150 (2002)
26. A. Lenz, H. Eisele, R. Timm, S.K. Becker, R.L. Sellin, U.W. Pohl, D. Bimberg, M. Dähne, Appl. Phys. Lett. **85**, 3848 (2004)
27. N. Liu, J. Tersoff, O. Baklenov, A.L. Holmes, C.K. Shih, Phys. Rev. Lett. **84**, 334 (2000)

28. L.G. Wang, P. Kratzer, M. Scheffler, Q.K.K. Liu, Appl. Phys. A **73**, 161 (2001)
29. J.M. García, G. Medeiros-Ribeiro, K. Schmidt, T. Ngo, J.L. Feng, A. Lorke, J. Kotthaus, P.M. Petroff, Appl. Phys. Lett. **71**, 2014 (1997)
30. J.-S. Lee, H.-W. Ren, S. Sugou, Y. Masumoto, J. Appl. Phys. **84**, 6686 (1998)
31. R. Timm, H. Eisele, A. Lenz, T.-Y. Kim, F. Streicher, K. Pötschke, U.W. Pohl, D. Bimberg, M. Dähne, Physica E **32**, 25 (2006)
32. F. Hatami, N.N. Ledentsov, M. Grundmann, J. Böhrer, F. Heinrichsdorff, M. Beer, D. Bimberg, S.S. Ruvimov, P. Werner, U. Gösele, J. Heydenreich, U. Richter, S.V. Ivanov, B.Y. Meltser, P.S. Kop'ev, Z.I. Alferov, Appl. Phys. Lett. **67**, 656 (1995)
33. M. Hayne, J. Maes, S. Bersier, V.V. Moshchalkov, A. Schliwa, L. Müller-Kirsch, C. Kapteyn, R. Heitz, D. Bimberg, Appl. Phys. Lett. **82**, 4355 (2003)
34. M. Geller, C. Kapteyn, L. Müller-Kirsch, R. Heitz, D. Bimberg, Appl. Phys. Lett. **82**, 2706 (2003)
35. P.M. Thibado, B.R. Bennett, M.E. Twigg, B.V. Shanabrook, L.J. Whitman, J. Vac. Sci. Technol. A **14**, 885 (1996)
36. K. Suzuki, R.A. Hogg, Y. Arakawa, J. Appl. Phys. **85**, 8349 (1999)
37. Motlan, E.M. Goldys, L.V. Dao, J. Vac. Sci. Technol. B **20**, 291 (2002)
38. I. Farrer, M.J. Murphy, D.A. Ritchie, A.J. Shields, J. Cryst. Growth **251**, 771 (2003)
39. L. Müller-Kirsch, R. Heitz, U.W. Pohl, D. Bimberg, I. Häusler, H. Kirmse, W. Neumann, Appl. Phys. Lett. **79**, 1027 (2001)
40. R. Timm, H. Eisele, A. Lenz, S.K. Becker, J. Grabowski, T.-Y. Kim, L. Müller-Kirsch, K. Pötschke, U.W. Pohl, D. Bimberg, M. Dähne, Appl. Phys. Lett. **85**, 5890 (2004)
41. R. Timm, J. Grabowski, H. Eisele, A. Lenz, S.K. Becker, L. Müller-Kirsch, K. Pötschke, U.W. Pohl, D. Bimberg, M. Dähne, Physica E **26**, 231 (2005)
42. R. Timm, A. Lenz, H. Eisele, L. Ivanova, K. Pötschke, U.W. Pohl, D. Bimberg, G. Balakrishnan, D.L. Huffaker, M. Dähne, Phys. Stat. Sol. (c) **3**, 3971 (2006)
43. R. Timm, A. Lenz, H. Eisele, L. Ivanova, G. Balakrishnan, D.L. Huffaker, M. Dähne, Unpublished

7
Theory of Excitons in InGaAs/GaAs Quantum Dots

Andrei Schliwa and Momme Winkelnkemper

Abstract. We employ the configuration interaction (CI) method in order to discuss many-particle properties of quantum dots as functions of the dots' size, shape and composition. Single-particle states, necessary for the CI-basis expansion, are calculated by eight-band $k \cdot p$ theory. Special emphasis is put on the role of strain and piezoelectricity, where the latter is treated up to second order. Finally, we address the inverse problem of fitting spectroscopic data to our detailed theoretical model leading to the determination of size, shape and composition as adjustable parameters.

7.1 Introduction

A great deal of interest in quantum dots persists due to their numerous actual and potential applications [1] and, of course, because of the paradigm-violating nature of many of their properties. These properties include the exciton fine-structure [2], the electron p-state splitting [3] or the shifts of the few-particle state recombinations of the biexciton (XX) [4–6], trions ($X^{\pm}$) [7–10] or charged biexcitons ($XX^{\pm}$) relative to the exciton [11–13]. These physical quantities are sensitively affected by the QDs' size, shape and composition via the inhomogeneous strain, the piezoelectric field and via the atomistic symmetry anisotropy (ASA).

The interrelation of the latter quantities and the Coulomb interaction will be the article's main subject. As soon as more than one carrier is confined in a QD the different manifestations of the Coulomb interaction—like the direct terms, exchange, and correlation—lead to the formation of distinct many-body states with different energies and transition characteristics.

Despite the tremendous advances in structural characterization techniques, the real shape and composition of capped quantum dots (QD), which present the decisive input parameters for all modeling [14, 15, 3], are often only poorly known. Despite their atomic resolution, even the most sophisticated STM techniques either provide only cross-sections of capped [16] or surface images of uncapped [17] QDs. Here, we show that specific spectroscopic properties—namely the binding energies of $X^{\pm}$, and XX relative to the exciton ground state—are very sensitive to the geometry and composition of QDs. Therefore, these quantities are fingerprints for a specific QD structure. This knowledge can be used to address the inverse problem of fitting calculated to measured data by using the QD structure as adjustable parameter.

Eight-band $\mathbf{k} \cdot \mathbf{p}$ theory enables us to obtain the electronic structure, thus taking into account arbitrary QD-shapes, as well as strain, piezoelectricity and band mixing effects [3]. The model provides, at reasonable computational cost, a fast and transparent relation between the electronic structure of QDs and bulk properties of the constituent materials.

Single-particle orbitals obtained by the eight-band $\mathbf{k} \cdot \mathbf{p}$ method are used to create a set of Slater determinants, which in turn serve as basis for the configuration interaction method. This method is employed for the calculation of few-particle energies, and allows us to incorporate direct Coulomb interaction, exchange and correlation effects.

7.2 Interrelation of QD-Structure, Strain and Piezoelectricity, and Coulomb Interaction

7.2.1 The Binding Energies of the Few Particle Complexes

In experiments, only energy differences have been observed. Relevant for our discussion are the exciton (X^0) recombination energy $E(X^0)$

$$E(X^0) = [\mathcal{E}_0^{(e)} - \mathcal{E}_0^{(h)}] + J_{00}^{(eh)} + \delta_{\text{Corr}}(X^0),$$

and the binding energies of biexciton (XX^0) and trion ($X^{\pm}$). The latter two are defined with reference to the exciton recombination

$$\begin{aligned}
\Delta(X^+) &= [-\mathcal{E}_0^{(h)} + E^{(0)}(X^0)] - E^{(0)}(X^+) \\
&= -J_{00}^{(eh)} - J_{00}^{(hh)} + \delta_{\text{Corr}}(X^0) - \delta_{\text{Corr}}(X^+), \\
\Delta(X^-) &= [\mathcal{E}_0^{(e)} + E^{(0)}(X^0)] - E^{(0)}(X^-) \\
&= -J_{00}^{(eh)} - J_{00}^{(ee)} + \delta_{\text{Corr}}(X^0) - \delta_{\text{Corr}}(X^-), \\
\Delta(XX^0) &= 2E^{(0)}(X^0) - E^{(0)}(XX^0) \\
&= -J_{00}^{(eh)} - J_{00}^{(ee)} - J_{00}^{(hh)} + 2\delta_{\text{Corr}}(X^0) - \delta_{\text{Corr}}(XX^0).
\end{aligned} \quad (7.1)$$

$J_{00}^{(ij)}$ is the direct Coulomb integral between states Ψ_0^i and Ψ_0^j being either electron or hole ground state, $\mathcal{E}_0^{(e/h)}$ the respective single-particle energies, and $\delta_{\text{Corr}}(\chi^q)$ the energy correction due to self-consistency and correlation effects. $E^{(0)}(\chi^q)$ is the total energy of particle χ^q including all Coulomb effects. Figure 7.1 depicts the relative importance of the direct Coulomb terms and correlation energies with respect to the binding energies $\Delta(\chi^q)$. Both contributions and their relation to the QD structure properties will be generally discussed in the following.

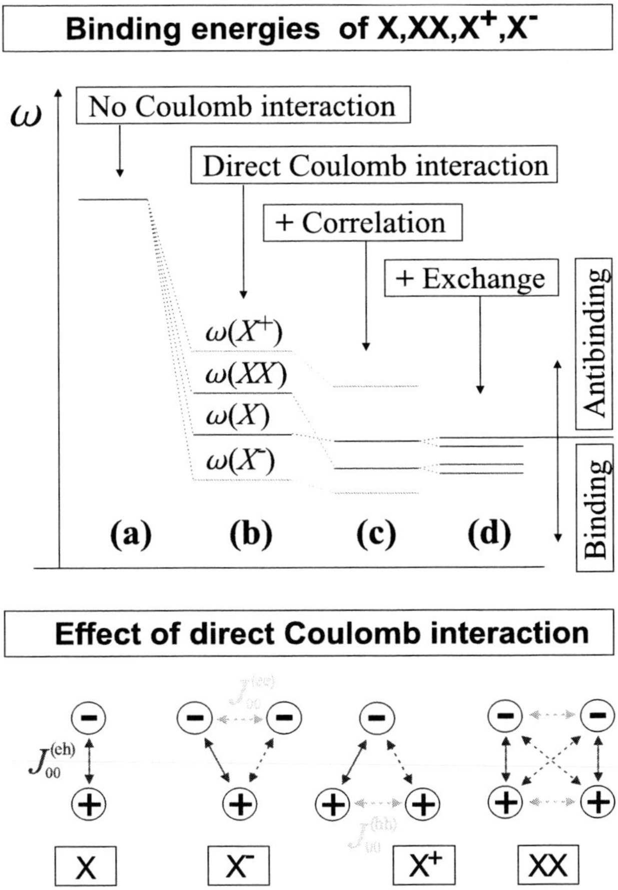

7 Theory of Excitons in InGaAs/GaAs Quantum Dots   141

Binding energies of X, XX, X$^+$, X$^-$

Effect of direct Coulomb interaction

Fig. 7.1. Upper panel: The evolution of the multiparticle transition energies ω: (**a**) no Coulomb interaction is taken into account, the transition energies $\omega(\chi^q)$ are degenerate. (**b**) Coulomb interaction lifts the degeneracy of the transition energies. Four different combinations in the order of $\omega(\chi^q)$ can occur (see Fig. 7.2). (**c**) Adding correlation can produce a binding biexciton. In principle 24, different sequences of $\omega(\chi^q)$ are now possible. (**d**) Exchange splits the exciton ground state into two nondegenerate dark and two nondegenerate bright states. Since both bright states act as final states for the biexciton decay, $\omega(XX)$ is split as well. Lower panel: Energetics of the direct Coulomb interaction of the four many-particle complexes. The appearance of the negative trion X^--decay on the higher or lower energy side of the exciton line depends on the relative strength of the additional terms $J_{00}^{(ee)}$ and $J_{00}^{(eh)}$ (additional forces are marked as dashed arrows). If $J_{00}^{(ee)} > |J_{00}^{(eh)}|$, the X^- decay line appears on the high-energy side of X, otherwise it appears on the low-energy side. The same rationale applies to the X^+. For the biexciton XX decay the additional forces are not sufficient to create a binding biexciton, since always $J_{00}^{(ee)} + J_{00}^{(hh)} \geq 2|J_{00}^{(eh)}|$ holds. From [18]

Direct Coulomb Interaction

The direct Coulomb integrals $J_{00}^{(ij)}$ are calculated employing the Poisson approach:

$$J_{00}^{(ij)} = q_i \int d\mathbf{r} |\Psi_0^i|^2 V_0^j,$$
$$q_j |\Psi_0^j|^2 = \epsilon_0 \nabla \cdot (\epsilon_s \nabla V_0^j). \tag{7.2}$$

Image charge effects arising from the material-dependent static dielectric constant are taken into account. The magnitude of J depends on the particle types, being either repulsive as for $J_{00}^{(ee)}$ and $J_{00}^{(hh)}$ or attractive like $J_{00}^{(eh)}$, on the spatial extent of the wave functions and, for $J_{00}^{(eh)}$, on the relative position of electron and hole orbitals. Four different cases with respect to size and position of electron and hole wave function can arise (Fig. 7.2(a–d)).

(a) Electron and hole ground state wave function share the same barycenter, but the extent of the electron wave function is larger than that of the hole. This is generally considered the archetype situation for InAs/GaAs QDs, since the resulting order of our four excitonic complexes is thought to be the most typical case encountered in experiments, apart from the antibinding position of the biexciton XX, which results from the thus-far neglected correlation effect.

(b) A vertical electric field can pull apart the electron and hole, leading to the configuration displayed in Fig. 7.2(b).

(c) Reversing the size of electron and hole wave function leads to the case considered in Fig. 7.2(c). The resulting order $|J_{00}^{(eh)}| < J_{00}^{(hh)} < J_{00}^{(ee)}$ can also be a consequence of a large piezoelectric field inside the QD, as can be seen from Fig. 7.9 for large pyramidal QDs. The electron ground state wave function remains in the

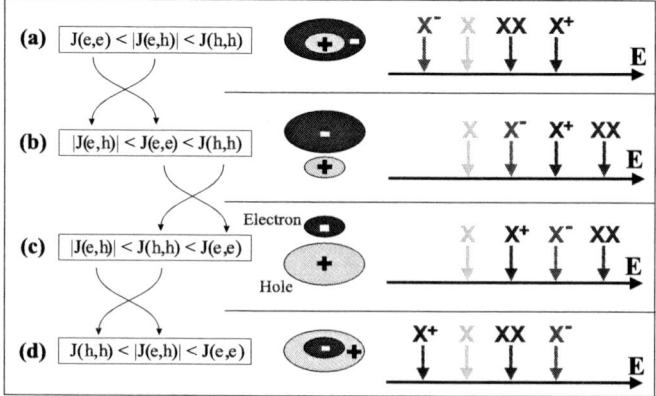

Fig. 7.2. The recombination energy $\omega(\chi^q)$ of X, X^+ and X^- relative to the exciton X (spectroscopic shift), depends first on the Coulomb energies $J_{00}^{(eh)}$, $J_{00}^{(ee)}$ and $J_{00}^{(hh)}$. These quantities in turn depend on the electron and hole wave function size and their mutual position creating four different possibilities, labeled with **(a)** to **(d)**. From [18]

QD-center whereas the hole wave function is driven into the two opposite QD-edges where the piezoelectric potential has its minimum. The resulting electron hole overlap decreases and, hence, $|J_{00}^{(eh)}|$ becomes smaller than $J_{00}^{(hh)}$ or $J_{00}^{(ee)}$.

(d) The last possible configuration is an electron and hole ground state sharing the same barycenter, but other than in case (a), the electron wave function extent is smaller than that of the hole state. Typically, QDs with a large vertical aspect ratio feature this type of configuration.

Neglecting correlation, these four cases lead to four different arrangements of the recombination energies $\omega(X)$, $\omega(XX)$, $\omega(X^+)$ and $\omega(X^-)$ as can be seen in Fig. 7.2 (right panel). $\omega(X)$ is defined as the recombination energy of the first exciton bright state and $\omega(XX)$ as the recombination into the first exciton bright state.

Correlation

The effect of correlation is more subtle. So far the Coulomb integrals are evaluated only for fixed single-particle states. By using self-consistent mean field theory, the single-particle orbitals are allowed to alter their shape and location in response to Coulomb attraction and repulsion. The Coulomb integrals are solved self-consistently. Energy corrections beyond that limit due to the use of correlated methods like the configuration interaction method or the quantum Monte Carlo method. In this case, the probability distribution of one carrier is allowed to become dependent on the positions of the other carriers. This is, e.g., impossible if the excitonic wave function is represented as a product of electron and hole wave function only, like for the Hartree-method. In the Hartree–Fock method, the ground state is expressed as single Slater-determinant. Therefore electron and hole can become correlated by their spin. This is called Fermi correlation. Coulomb correlation is accessible only if linear combinations of multiple Slater-determinants are allowed as in the CI method. Because we omit the self-consistency step in this article, the correlation energies in (7.1) cover both the correlation energy and the self-consistency corrections.

As we will see in this work, the correlation energy is specific for each particle type and its magnitude can be very different for the various complexes χ^q. Hence, correlation may alter the order of the four transition energies, leading to 24 theoretically possible sequences. An important consequence is the appearance of binding biexcitons, which cannot be achieved by considering direct Coulomb effects alone. As we will see, the magnitude of the correlation is larger the more dissimilar electron and hole wave functions are, the larger their distance or the smaller their spatial overlap is. Further on, the correlation energies of the four complexes relative to each other depend on the spectral density of electron and hole states.

7.3 Method of Calculation

The electronic and optical properties are calculated for self-organized QDs based on a three-dimensional implementation of the eight-band $k \cdot p$ model for the single-particle states and a configuration interaction scheme for the multiparticle states

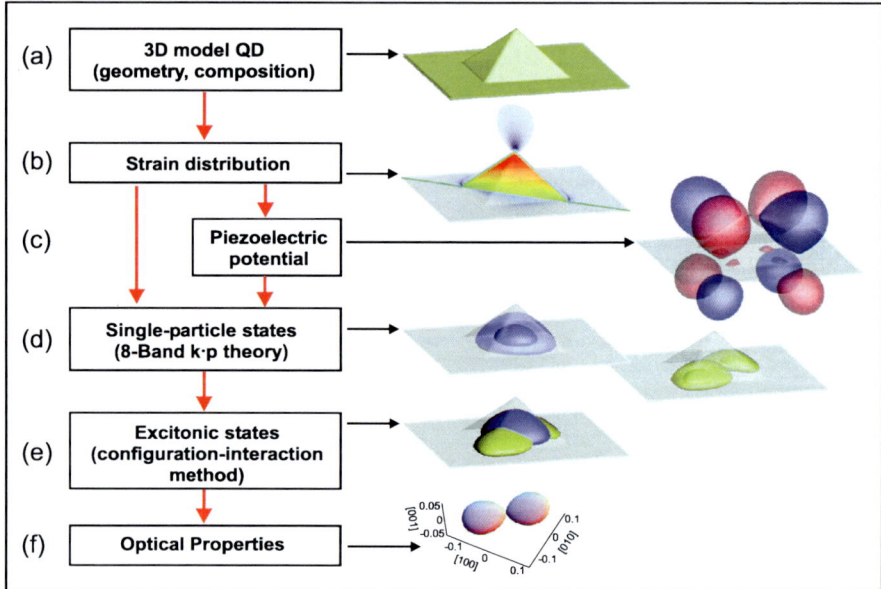

Fig. 7.3. The steps involved in modeling the electronic and optical QD-properties (see text), shown for a pyramidal InAs/GaAs QD as an example

[14, 3, 19]. The calculations account for the inhomogeneous strain distribution, the first- and second-order piezoelectric effects [14, 20], interband mixing and Coulomb interactions. A schematic presentation of the six steps involved is shown in Fig. 7.3(a–f) and works in the following way: (a) The modeling process starts with assumptions of shape, size and composition guided either by structural investigations of a QD sample or—as in this work—suggested only by peculiarities of the optical spectra. Next, the strain distribution (b) and the piezoelectric potential (first- and second-order) (c) are calculated, which enter (d) the strain-dependent eight-band $\boldsymbol{k}\cdot\boldsymbol{p}$ Hamiltonian. By solving the Schrödinger equation we obtain single-particle wave functions. The parameters entering this Hamiltonian are based on experimental values for the required bulk material Γ-point band structure parameters. Free, adjustable parameters are *not* present in this model. (e) The single-particle states provide a basis for the configuration interaction model which is applied to calculate excitonic properties, including correlation and exchange. (f) Finally, the optical absorption spectra are computed.

7.3.1 Calculation of Strain

Since the impact of strain on the confinement is comparable to that of the band offsets at the heterojunctions, the wave functions and energies are very sensitive to the underlying strain distribution.

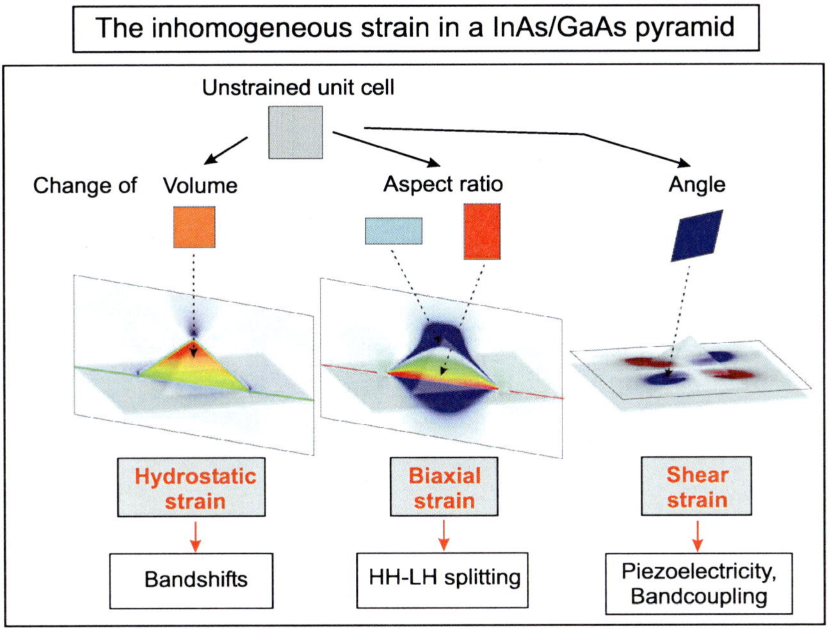

7 Theory of Excitons in InGaAs/GaAs Quantum Dots

The inhomogeneous strain in a InAs/GaAs pyramid

Fig. 7.4. Different types of strain shown as vertical or horizontal scans through a pyramidal InAs/GaAs quantum dot. The hydrostatic strain, $\epsilon_H = \epsilon_{xx} + \epsilon_{yy} + \epsilon_{zz}$, shifts the local conduction and valence bands. The biaxial strain, $\epsilon_B = \epsilon_{xx} + \epsilon_{yy} - 2\epsilon_{zz}$, splits the local HH and LH band edges. The shear strain components are responsible for piezoelectricity and band coupling

The continuum elastic model (CM) is chosen for the current implementation of the eight-band $\mathbf{k} \cdot \mathbf{p}$ model. The total strain energy of the CM model is given by [3]

$$U_{cm} = \frac{1}{2} \sum_{i,j,k,l} C_{ijkl} \epsilon_{ij} \epsilon_{kl}.$$

U is minimized for a given structure, using finite differences for the strains $\epsilon_{ij} = \partial u_i / \partial x_j$, where $\mathbf{u}$ is the displacement vector field. The compliances C_{ijkl} are represented by the parameters C_{11}, C_{12} and C_{44} for cubic crystals.

Results for the strain distribution in capped InAs pyramides were shown by Grundmann et al. in [14] and are depicted in Fig. 7.4.

7.3.2 Piezoelectricity and the Reduction of Lateral Symmetry

Piezoelectricity is defined as the generation of electric polarization by application of stress to a crystal lacking a center of symmetry [21]. The zincblende structure is one of the simplest examples of such a lattice and the strength of the resulting polarization is described by one parameter alone, e_{14}, for the linear case, resulting in a polarization P_1, and three parameters, B_{114}, B_{124} and B_{156}, for the quadratic case [20] resulting in a polarization P_2. Their relation to the strain tensor field is given by

$$P_1 = 2e_{14} \begin{pmatrix} \epsilon_{yz} \\ \epsilon_{xz} \\ \epsilon_{xy} \end{pmatrix},$$

$$P_2 = 2B_{114} \begin{pmatrix} \epsilon_{xx}\epsilon_{yz} \\ \epsilon_{yy}\epsilon_{xz} \\ \epsilon_{zz}\epsilon_{xy} \end{pmatrix}$$

$$+ 2B_{124} \begin{pmatrix} \epsilon_{yz}(\epsilon_{yy} + \epsilon_{zz}) \\ \epsilon_{xz}(\epsilon_{zz} + \epsilon_{xx}) \\ \epsilon_{xy}(\epsilon_{xx} + \epsilon_{yy}) \end{pmatrix} \quad (7.3)$$

$$+ 4B_{156} \begin{pmatrix} \epsilon_{xz}\epsilon_{xy} \\ \epsilon_{yz}\epsilon_{xy} \\ \epsilon_{yz}\epsilon_{xz} \end{pmatrix}.$$

Piezoelectric charges, ρ_{piezo}, arise from the polarizations

$$\rho_{\text{piezo}}(r) = -\nabla \cdot P,$$
$$P = P_1 + P_2.$$

The resulting piezoelectric potential is obtained by solving Poisson's equation taking into account the material dependent static dielectric constants, $\epsilon_s(r)$

$$\rho_p(r) = \epsilon_0 \nabla \cdot [\epsilon_s(r) \nabla V_p(r)], \quad (7.4)$$

$$\Leftrightarrow$$

$$\Delta V_p(r) = \frac{\rho_p(r)}{\epsilon_0 \epsilon_s(r)} - \frac{1}{\epsilon_s(r)} \nabla V_p(r) \cdot \nabla \epsilon_s(r). \quad (7.5)$$

The first term on the right-hand side of (7.5) refers to the true three-dimensional charge density while the second is the contribution of polarization interface charge densities due to a discontinous $\epsilon_s(r)$ across heterointerfaces.

The importance of the second-order term, P_2, for InGaAs/GaAs(111) quantum wells (QWs) and QDs was pointed out by Bester and coworkers [20, 22]. They found that in QWs the linear and quadratic contributions have opposite effects on the field, and for large strain the quadratic term even dominates. For InAs/GaAs QDs, however, the situation is more complex. In addition to the large strain their three-dimensional structure comes into play: The linear term generates a quadrupole-like potential, which reduces a structural C_{4v}- or $C_{\infty v}$-symmetry of a QD to C_{2v} [14, 23]. The effect of the quadratic term has been evaluated for lens-shaped QDs [22]. It was found to cancel the first-order potential inside the QD, leading to a field-free QD. This investigation was extended to a variety of more realistic structures in [24]. For a pyramidal QD with a base length of 17 nm and {101} side facets, Fig. 7.5 shows the strength and distribution of the piezoelectric potential resulting from the two orders of the piezoelectric tensor. Apart from the different orientation and sign of the two

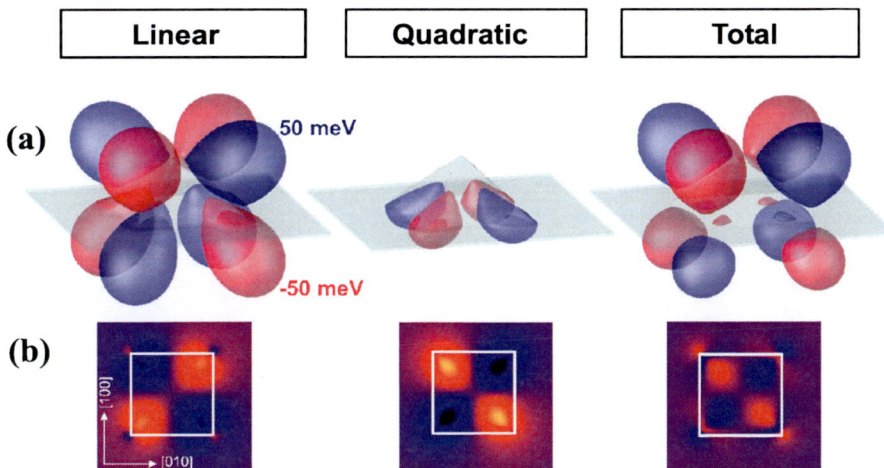

Fig. 7.5. (a) The piezoelectric potential isosurfaces at ±50 meV of a pyramidal InAs quantum dot with 17 nm base length shown for the linear, the quadratic part and for sum of both. (b) Contour plots of the piezoelectric potential 2 nm above the wetting layer

contributions, an important peculiarity of the second-order potential is its restriction to the interior of the QD, which is in apparent contrast to the widespread first-order field. This difference is linked to the origin of the polarization P: P_1 is a function of the shear-strain components alone, whereas P_2 results mainly from the product of the diagonal and the shear-strain. However, in contrast to the shear-strain components, the diagonal elements ϵ_{ii} are large only inside the QD and its close vicinity. Therefore, P_2-charges can only be created in this region.

7.3.3 Single Particle States

The energy levels and wave functions of bound electron and hole states are calculated using the eight-band $k \cdot p$ model. It was originally developed for the description of electronic states in bulk material [25–28]. For use in heterostructures, the envelope function version of the model has been applied to quantum wells [29], quantum wires [15] and quantum dots [30, 31, 3, 32]. Details of the principles of our implementation are outlined in [15].

This model enables us to treat QDs of arbitrary shape and material composition, including the effect of strain, piezoelectricity, VB mixing and CB–VB interaction. The strain enters our model via the use of deformation potentials as outlined by Bahder [33]. Its impact on the local band edges as a function of the QD geometry will be discussed in the next section.

The $k \cdot p$ model, when applied to small quantum structures, has in principle a few well-known drawbacks which have been examined in detail [34, 35]. They are basically related to the fixed number of Bloch functions used for the wave function expansion, the restriction to the close vicinity of the Brillouin zone center and to the limited ability to account for the symmetry of the underlying lattice. These problems do not arise in microscopic theories like the empirical pseudopotential method [34] (EPM) or the empirical tight-binding method (ETB) [36, 37], which a priori have greater potential accuracy. This potential, however, can only be exploited if the corresponding input parameter—the form factors in the EPM or the tight-binding parameter and their strain dependence in the ETB—are known with sufficient accuracy. Reliable generation of these parameters, however, is highly nontrivial and is, for now, controversial. One of the most appealing features of the $\bm{k} \cdot \bm{p}$ model, in contrast, is the direct availability of all parameters entering the calculations and the corresponding transparency of the method. Additionally, the required computational expense of the method is comparatively small. The material parameters used in this work are taken from [3].

7.3.4 Many-Particle States

As soon as more than one charge carrier is confined in the QD, the influence of direct Coulomb interaction, exchange effects and correlation lead to the formation of distinct multiparticle states. These states are calculated here using the configuration interaction method. This method rests on a basis expansion of the excitonic Hamiltonians into Slater determinants. These consist of antisymmetrized products of single-particle wave functions being obtained from eight-band $\bm{k} \cdot \bm{p}$ theory in our case. The method is applicable in the strong confinement regime, since the obtained basis functions are already similar to the weakly correlated many-body states.

7.3.5 The Configuration Interaction Model

Configuration interaction (CI) is a linear variational method for solving the few-particle Schrödinger equation. Two meanings are connected to the term configuration interaction in this context. Mathematically, configuration simply describes the linear combination of Slater determinants (SD) used for the wave functions. In terms of a specification of orbital occupation, interaction means the mixing (interaction) of different electronic configurations (states).

In order to account for correlation, CI uses a wave function $|\Psi_N^\alpha\rangle$ (N is the number of particles, and α is an index to label the few-particle states), which is a linear combination of Slater determinants $|\Phi_{a,b,c..}\rangle$ built up from single-particle orbitals:

$$|\Psi_N^\alpha\rangle = \sum_{a,b,c..} C_{a,b,c..}^\alpha |\Phi_{a,b,c..}\rangle. \tag{7.6}$$

As an example, the X^+ ground state configuration, (2, 2), consists of the four Slater determinants $|\Phi_{e_i,h_j,h_k}\rangle$ with i, j, k being the index of electron and hole ground state, $i, j, k \in 1, 2$. The wave function results as a linear combination of these four Slater determinants:

$$|\Psi_2^{X+}\rangle = \sum_{i,j,k=1}^{2} C_{i,j,k}^{X+} |\Phi_{e_i,h_j,h_k}\rangle. \tag{7.7}$$

If the expansion includes all possible CSFs, it is called a full configuration interaction (FCI) procedure which exactly solves the Schrödinger equation within the space spanned by the one-particle basis set. This is not feasible in our case. Therefore we restrict ourselves to the inclusion of all bound orbitals, where i runs over all confined electron and j, k over all confined hole states. Still we refer to this approach to as FCI.

Other CI methods use an even more restricted basis set: The allowed Slater determinants can be characterized by the number of excited state orbitals. If only one orbital is excited, it is referred to as a single excitation determinant (single CI). If one or two excited state orbitals exist, it is a single-double excitation determinant (single-double CI [SDCI]) and so on. These derivatives are used to limit the number of determinants in the expansion.

Role of the Basis Size

Since we restrict ourselves to a basis built up from bound orbitals only, part of the correlation energy is not included in our approach. Shumway and coworkers [38] estimated this defect by comparing their FCI results for spherical QDs having the same restricted expansion basis to a quantum Monte Carlo treatment. They found that their CI calculations cover about 80% of the total correlation energy.

As an example, in Fig. 7.6 we compare the biexciton correlation energies, $\delta_{\text{Corr}}(XX)$, for a FCI and a SDCI calculation for an InAs pyramid (17.2 nm base length) as a function of the number of configurations that are taken into account. For the biexciton-configurations $(2, 2)$, $(2, 10)$ and $(10, 2)$, the FCI and SDCI are equivalent. Two prominent features are highlighted:

First, the correlation energy is much more sensitive to the number of hole states than to the number of electron states, as can be seen by comparing the $(2, 2)$ to $(2, 10)$ and the $(2, 2)$ to $(10, 2)$ configuration.

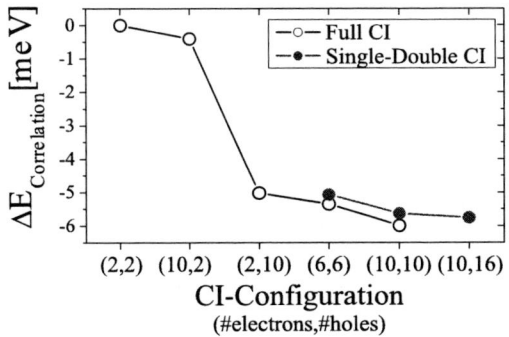

Fig. 7.6. The biexciton correlation energy as a function of the used CI-configuration (i, j), where i is the number of electron states and j the number of hole states being included to built up the CI-basis. Results for Full CI are compared to that of single-double CI. For the configurations $(2, 2)$, $(10, 2)$ and $(2, 10)$, FCI and SDCI are identical. From [18]

Table 7.1. Relation between particle type, basis size, number of matrix elements (ME) and number of *nonzero* ME if 6 electron and 10 hole states contribute to the full configuration

Particle type	Basis size	No. of ME	No. of *nonzero* ME
X	60	3 600	1 830
X^-	150	22 500	7 275
X^+	270	72 900	17 685
XX	675	455 625	63 450

Second, the results for the FCI deviate very little, if at all, from SDCI. The appealing property of the SDCI in this context is the largely reduced number of required matrix elements: For the (10, 10) FCI, e.g., eight times more elements are to be evaluated than for the (10, 16) SDCI.

Table 7.1 shows the evolution of the basis size with increasing number of carriers for the full CI method. It highlights the factorial growth in the number of matrix elements, which inhibits the usage of the FCI method for larger number of carriers in an excitonic complex.

7.3.6 Interband Spectra

The interband absorption spectra are calculated by Fermi's golden rule applied to excitonic states obtained from the configuration interaction method [19].

There exist no strict selection rules for the decay of excitons. As a rule of thumb one can say that those transitions have a large oscillator strength where electron and hole state share the same symmetry properties *and* have a sizable spatial overlap. However, since the hole states consist of HH and LH parts, each with its own symmetry, they have finite recombination probabilities with several different electron states of different symmetry character.

7.4 The Investigated Structures: Variation of Size, Shape and Composition

Our selection of model QDs is guided by the reported broad variation of structures observed through experiments (see, e.g., [39, 40] and references therein). The following series are considered:

Series A: The pyramidal InAs/GaAs structures similar to [3] with base lengths 10.2 nm (A1), 13.6 nm (A2), 17.0 nm (A3), and 20.4 nm (A4).

Series B: Starting with the 17 nm base length pyramid of series A, the vertical aspect ratio, ar_V, is varied between 0.5 (full pyramid) and 0.04 (very flat QD).

Series C: The QDs have a circular base and their vertical aspect ratio varies between 0.5 (half-sphere) and 0.17.

Series D: Starting again with the 17 nm base length pyramid of series A an elongation in [110] and [1$\bar{1}$0] direction is explored. The lateral aspect ratio, ar_L, (length in [110] direction divided by length in [1$\bar{1}$0] direction) varies between 2 and 0.5 (a value of 1 corresponds to the square base).

It is important to note that the QD volume has been kept constant for series B, C and D.

Series E: A homogeneous variation of the In-content for $In_xGa_{1-x}As$/GaAs is considered. The starting point again is the 17 nm base length pyramid of series A. The In content decreases in steps of 10% from 100% to 70%.

Series F: The QDs of this series have a circular base together with a trumpet-shaped like InGaAs composition profile. The integral In amount of the QDs is equal to QD A3.

Series G: By applying a smoothing algorithm on structure A3 with variable smoothing steps (N) the process of Fickian diffusion as a result of an annealing procedure is simulated.

Series H: In addition to the previous list of series we add a special series that is closely related to experiments carried out by Rodt et al. [7]. Due to the multimodal distribution of the PL peaks of the investigated samples it was possible to derive the structure of the participating QDs to unprecedented detail. It was found that the QDs responsible for each of the nine well-separated peaks differ by one monolayer in height and base length. The smallest QD starts with a height of 3 monolayers and a base length of 9.1 nm and the largest one ends with a height of 11 monolayers and a base length of 13.6 nm, respectively. This series will be the basis for comparison in the article by U.W. Pohl and S. Rodt in this book.

7.5 The Impact of QD Size

To study the impact of the QD size we focus on series A and H. The pyramidal shape of the former series represents a model QD structure introduced by Grundmann et al. [14]. The other series is closely modeled to reproduce the spectroscopic peculiarities observed by Heitz et al. [41] describing an onion-like size distribution. Both series together encompass an X^0-energy range of 300 meV starting at 1.3 eV for the smallest QD of series H to as little as 1 eV for the largest full pyramid of series A.

Figure 7.7 shows the single particle energies of both series, and Figs. 7.8–7.9 show the shape of the ground state wave functions. Their position and spatial extent determine the direct Coulomb energies.

Series H: For the smallest QDs of series H we observe a larger spread of the electron orbital into the surrounding matrix as compared to that of the hole orbital, which is always strongly confined inside the QD. Consequently, the Coulomb repulsion between two electrons is much smaller than for two holes, occupying the ground state. Therefore, the negative trion is binding and the positive trion is antibinding (see Fig. 7.10). The biexciton is antibinding too, since

$$\left|J_{00}^{(eh)}\right| < \left(J_{00}^{(ee)} + J_{00}^{(hh)}\right)/2$$

holds (in case of an equal sign, the biexciton transition would be degenerate with the exciton transition). The smaller the electron–hole size disparity is, the smaller

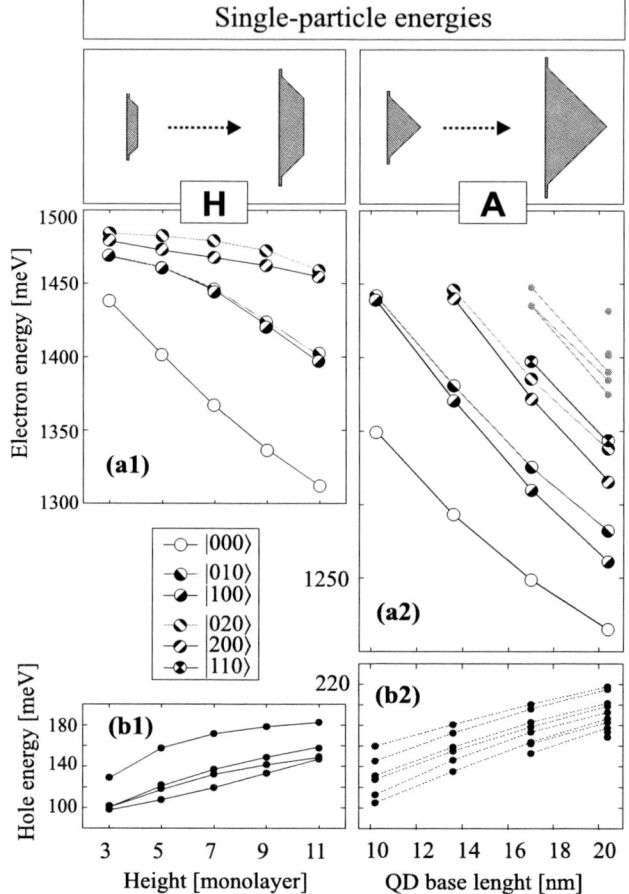

Fig. 7.7. Single-particle electron and hole energies shown for series H and A

are the binding energies. At a height of 11 ML, the correlation energy $\delta_{\text{Corr}}(XX)$ is large enough to create a binding biexciton, although $|\delta_{\text{Corr}}(XX)|$ is decreasing with increasing QD height. For the absolute values of the correlation energies we observe the order

$$|\delta_{\text{Corr}}(X)| < |\delta_{\text{Corr}}(X^-)| < |\delta_{\text{Corr}}(XX)| < |\delta_{\text{Corr}}(X^+)|.$$

The former two values are increasing and the latter two are decreasing upon size increase, however, without changing their order.

Series A: The trend with respect of the relative electron-hole extent continues with series A. For the smallest pyramid of this series the direct Coulomb energies are equal with respect to their absolute value. Hence, the order of the recombination energies is determined by the correlation energies alone. Now we find a binding positive trion and an antibinding negative trion and the biexciton is also in a binding state.

The larger the pyramids of series A become, the larger the hole wave function becomes relative to the electron wave function. This happens for two reasons. First, the biaxial strain and its sign change at the QD center enforces a hole position at the QD bottom. Second, since the lateral QD-extent is largest at the pyramid base the hole orbital can cover a comparatively large space, larger than the electron orbital can take. This results in a larger Coulomb energy between two electrons than for two holes occupying the ground state level. The absolute value of the electron hole Coulomb attraction $|J_{00}^{(eh)}|$ is even smaller than $J_{00}^{(ee)}$ and $J_{00}^{(hh)}$, which results from the piezoelectric field, as will be shown in the next section.

For the correlation energies we observe

$$|\delta_{Corr}(X)| < |\delta_{Corr}(X^-)| < |\delta_{Corr}(XX)| \approx |\delta_{Corr}(X^+)|.$$

The last two quantities exhibit an enormous increase upon increasing pyramid size.

7.5.1 The Role of the Piezoelectric Field

Section 7.2.1 demonstrated that the relative size and position of electron and hole wave functions is decisive for the XX, X^+ and X^- binding energies. From our previous work [14, 24] we know that the piezoelectric field strongly affects the order and the orientation of the single particle orbitals and it leads to a spatial separation of electron and hole wave function (see Fig. 7.9) for series A. As a result, the electron-hole overlap and, hence, their Coulomb attraction decreases and $|J_{00}^{(eh)}|$ can become smaller than $J_{00}^{(ee)}$ and $J_{00}^{(hh)}$ (see, e.g., Table 7.2). Consequently, according to Fig. 7.2, the XX, X^+ and X^- recombination peaks are blue-shifted relative to the exciton line X, as a result of the piezoelectric effect. By taking correlation into account, the picture changes and in most cases we encounter at least a binding positive trion X^+ and sometimes also a binding biexciton XX.

A strong piezoelectric field has a large impact on the shape and orientation of the hole wave functions, especially for the ground state $|h_0\rangle$. In contrast to the electron ground state $|e_0\rangle$, $|h_0\rangle$ is strongly elongated and distorted in the direction of

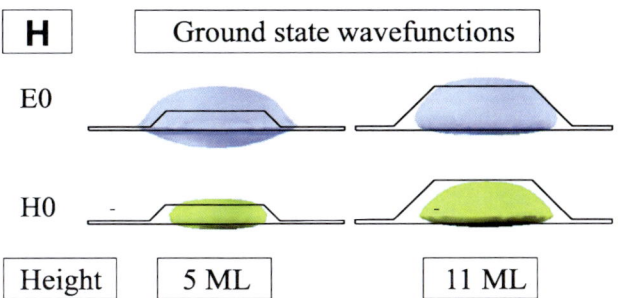

Fig. 7.8. Probability density at 65% of electron and hole ground state wave functions for QDs of series H having a height of 5 ML and 11 ML, respectively

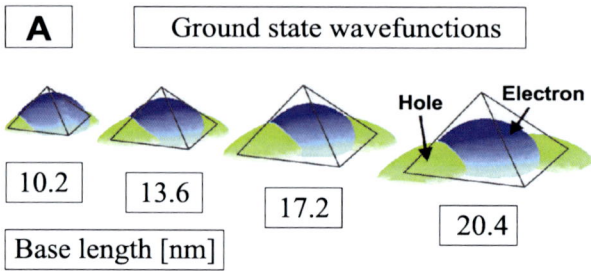

Fig. 7.9. Probability density at 65% of electron and hole ground state wave functions for pyramidal QDs of series A. Due to the increase of the piezoelectric field with larger QD size, the hole wave function (*yellow*) tends to elongate along $[1\bar{1}0]$. As a result the electron hole overlap decreases and $|J_{00}^{(eh)}|$ becomes smaller than $J_{00}^{(hh)}$ and $J_{00}^{(ee)}$

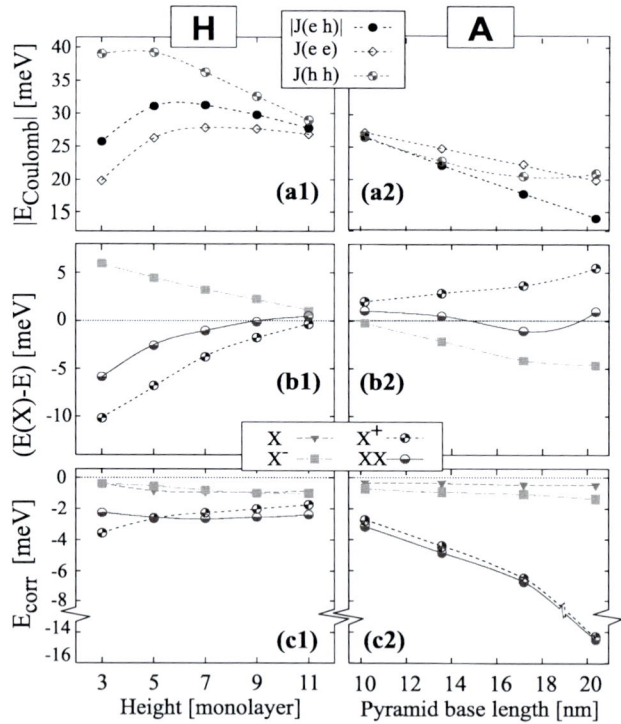

Fig. 7.10. (a) The direct Coulomb energies $J_{00}^{(ee)}$, $J_{00}^{(hh)}$ and $J_{00}^{(eh)}$; (b) the difference of the binding energies with respect to the exciton energy, calculated for a configuration (10e, 10h); and (c) the correlation energies $\delta_{\text{Corr}}(\chi^q)$ are shown for series A and H. First- and second-order piezoelectric effects are included. From [18]

Table 7.2. Direct Coulomb energies, Correlation and Binding energies for a (full) pyramidal QD (base length 17.2 nm) in absence (a) and presence (b) of the piezoelectric field. For the latter, first and second order components are taken into account

	Coulomb energies [meV]			Correlation energies [meV]		Binding energies [meV]		
	$J_{00}^{(eh)}$	$J_{00}^{(ee)}$	$J_{00}^{(hh)}$	$\delta_{\text{Corr},XX}$	δ_{Corr,X^+}	ΔE_{XX}	ΔE_{X^+}	ΔE_{X^-}
(a) No piezo	−19.0	22.8	17.7	−3.4	−3.4	−1.3	4.0	−3.9
(b) With piezo	−17.5	22.5	20.5	−6.4	−6.6	−0.4	3.5	−3.1

the piezoelectric potential minima. The maximum of the probability density of $|h_0\rangle$ resides not in the dot center anymore, but is shifted to the corners where the piezoelectric potential has its minimum. Since the probability density is rising thereby, the Coulomb repulsion between two holes $J_{00}^{(hh)}$ that occupy the ground state is increased. This result is displayed in Table 7.2, where we assess the change of Coulomb and correlation energies for the full pyramid with 17.2 nm base length upon introduction of a piezoelectric field: $|J_{00}^{(eh)}|$ decreases by 1.5 meV and $J_{00}^{(hh)}$ increases by 2.8 meV.

Still the repulsion between two holes remains smaller than for two electrons, $J_{00}^{(hh)} < J_{00}^{(ee)}$, which is a peculiarity of the pyramidal shape, where the special strain conditions force the hole ground state to be located at the QD bottom. Therefore the resulting sequence of Coulomb energies changes from case (d) in Fig. 7.2 to case (c), along with a crossing of the X^+ from binding to antibinding. However, if correlation is added to the calculations, X^+ becomes binding again, due to the comparatively large $\delta_{\text{Corr}}(X^+)$.

7.6 The Aspect Ratio

7.6.1 Vertical Aspect Ratio

Different Types of Charge Separation Effects

In Fig. 7.11 we can identify two regimes for series B in terms of the relative Coulomb binding energies [Fig. 7.11(a)] and the resulting relative binding energies [Fig. 7.11(b)]: The first regime holds for $\text{ar}_V < 0.2$ and can be described as the unequal vertical wave function spread out of electron and hole ground state. Since both share the same barycenter and, apart from their size, have a similar shape, we encounter case (a) of Fig. 7.2 with the same order of X^+, XX and X^-. The only difference to Fig. 7.2(a) is their relative position compared to the exciton, which is changed due to correlation. In this range of ar_V, the very strong z-confinement leads to a wave function spill-over of electron and hole state, visible through the decreasing Coulomb energies upon smaller aspect ratio.

The other regime—Stier et al. [42] coined the term piezoelectric regime—holds for $\text{ar}_V > 0.2$. It is characterized by a larger hole wave function extent, due to the

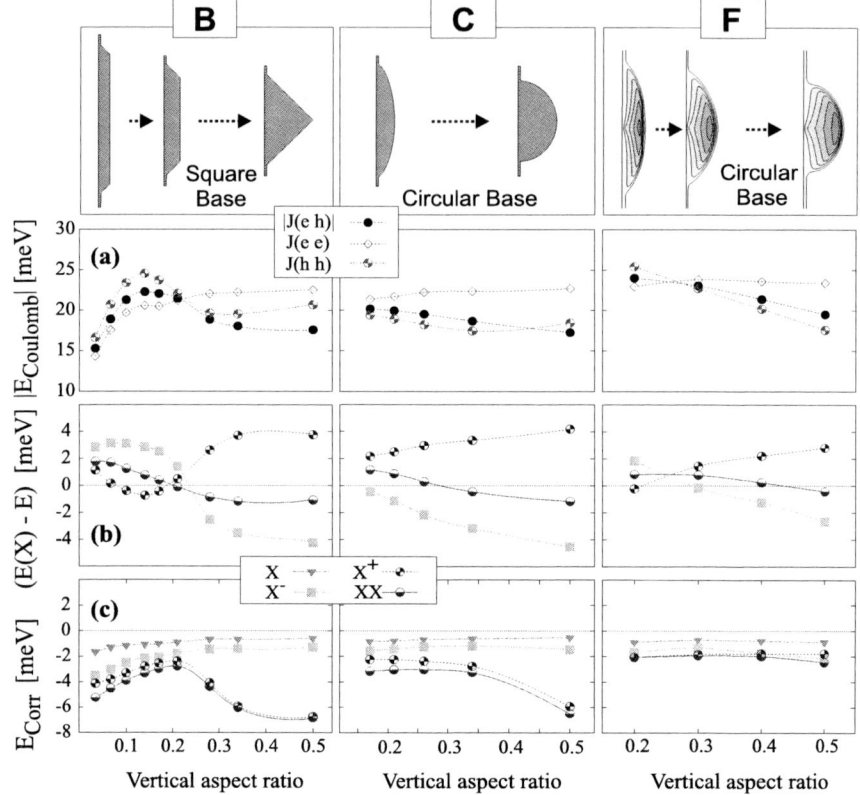

Fig. 7.11. (a) The direct Coulomb energies $J_{00}^{(ee)}$, $J_{00}^{(hh)}$ and $J_{00}^{(eh)}$; (b) the spectroscopic shifts, calculated for a configuration (10e, 10h); and (c) the degree of correlation shown for series B, C and F. First- and second-order piezoelectric effects are included. From [18]

pyramidal shape (see Sect. 7.5), and a reduced electron hole overlap as a result of a larger piezoelectric field. Hence, the Coulomb attraction of electron and hole $|J_{00}^{(eh)}|$ becomes smaller than $J_{00}^{(ee)}$ or $J_{00}^{(hh)}$.

Series C and F exhibit similar behavior in terms of the peak order, although the piezoelectricity plays only a minor role in these structures (see [22, 24]), which becomes visible through the relatively larger electron hole attraction, $|J_{00}^{(eh)}| > J_{00}^{(hh)}$ (except for the half-sphere in series C).

The charge separation in series F is induced by the trumpet-shaped composition profile forcing the hole ground state to be located above the electron ground state. Since the In core extends toward the tip, the hole state can expand in a lateral direction more efficiently than the electron. Hence, as in the case of the full pyramid, we find for $ar_V > 0.2$: $J_{00}^{(ee)} > J_{00}^{(hh)}$.

The Biexciton Binding Energy

For all three series B, C and F, where the vertical aspect ratio is the variation parameter, the biexciton changes from antibinding (ar$_V$ = 0.5) to binding for smaller aspect ratios. The crossover point is different in each series, but the spectroscopic shift is monotonically decreasing with increasing ar$_V$ in all three series.

Correlation

Within our three series B, C and F, we again observe that the correlation energies of biexciton and positive trion increase drastically when the attractive Coulomb force $|J_{00}^{(eh)}|$ becomes smaller than the repulsive terms $J_{00}^{(ee)}$ and $J_{00}^{(hh)}$, respectively.

7.6.2 Lateral Aspect Ratio

A QD elongation away from the square basis, as shown in series D, is often discussed as a possible source of the fine-structure splitting, since it introduces a symmetry reduction from C_{4v} to C_{2v} already on the level of the QD structure. However, as long as no piezoelectricity is included (and/or the ASA in the case of atomistic models), there is no distinction possible for the single particle energies or the peak energies of the excitonic spectra between the two possible elongations [110] and [1$\bar{1}$0]. In this case only the polarization delivers the information on the QD orientation. Even more, there is no change of the direct Coulomb energies J and the corresponding few-particle binding energies visible throughout series D, as can be seen in the left panels of Fig. 7.12.

The situation changes when piezoelectricity is taken into account. First, the electron-hole attraction $J_{00}^{(eh)}$ becomes smaller than the repulsive terms $J_{00}^{(ee)}$ and $J_{00}^{(hh)}$. Second, the direct Coulomb energies and the resulting binding energies of XX, X^+ and X^- are different for both directions (see Fig. 7.12(a) and (b)). The order of the excitonic complexes, however, remains unchanged. Third, the degree of correlation for the biexciton and the positive trion becomes largest for a large elongation along [110], where the attractive electron–hole Coulomb forces reach their minimum and the repulsive Coulomb forces their maximum. There is no experimental evidence from HRTEM etc. for such asymmetry. Fourth, without correlation we observe a completely different order of the excitonic complexes namely (X, X^+, X^-, XX) from lower to higher transition energies.

7.7 Different Composition Profiles

7.7.1 Inverted Cone-Like Composition Profile

In order to identify the consequences of an inhomogeneous composition profile like in series F, we compare the flattest QD of this series (ar$_V$ = 0.2) (further referred to as QD$_{F\text{-inhom}}^{0.2}$) to the pure InAs lens-shaped QD from series C having the same

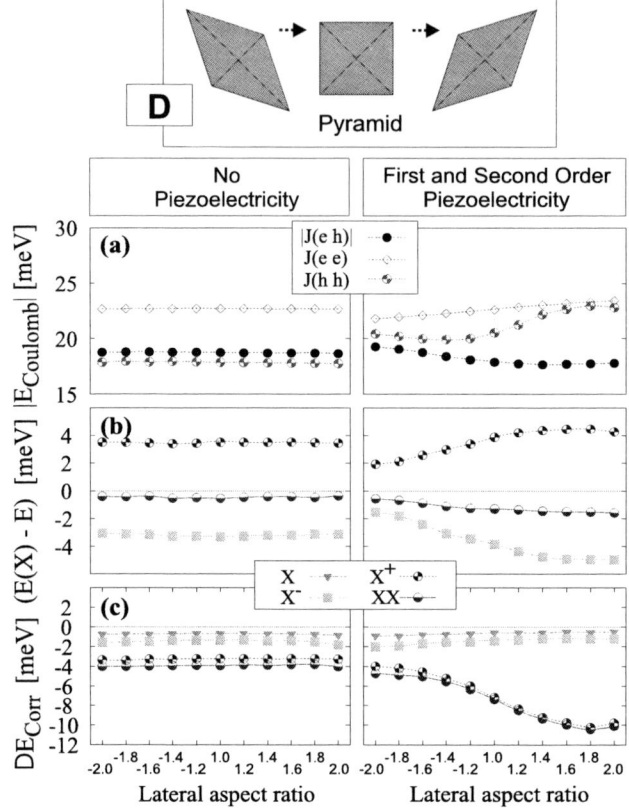

Fig. 7.12. (a) The Coulomb energies $J_{00}^{(ee)}$, $J_{00}^{(hh)}$ and $J_{00}^{(eh)}$; (b) the spectroscopic shifts, calculated for a configuration (10e, 10h); and (c) the degree of correlation shown for series of elongated QDs (D) in the cases of excluded piezoelectric field (*left panel*), and if first- and second-order effects are taken into account (*right panel*). From [18]

vertical aspect ratio (further referred to as $QD_{C-hom}^{0.2}$). Both QDs are designed to contain the same integral amount of InAs. Compared to the archetype pyramidal QD the electron hole alignment of $QD_{F-inhom}^{0.2}$ is reversed and their barycenters are 0.2 nm apart.

Prominent differences and similarities between $QD_{C-hom}^{0.2}$ and $QD_{F-inhom}^{0.2}$ are:

(i) Both ground state wave functions (not shown here) of $QD_{F-inhom}^{0.2}$ are stronger localized than their $QD_{C-hom}^{0.2}$ counterparts, resulting in significantly larger Coulomb energies $J_{00}^{(eh)}$, $J_{00}^{(ee)}$ and $J_{00}^{(hh)}$.

(ii) In contrast to $QD_{C-hom}^{0.2}$, $J_{00}^{(hh)}$ is larger than $J_{00}^{(ee)}$ for $QD_{F-inhom}^{0.2}$, since the hole ground state is stronger localized than the electron ground state. Consequently we find a different pattern for the binding energies of XX, X^+ and X^- as can be seen from Fig. 7.11.

7.7.2 Annealed QDs

In series G we simulate the effect of annealing on the electronic properties for a pyramidal QD, originally having a base length of 17.2 nm. Prominent features are:

(i) Both ground state wave functions increase their localization resulting in larger Coulomb energies $J_{00}^{(eh)}$, $J_{00}^{(ee)}$ and $J_{00}^{(hh)}$.

(ii) Due to the decreasing piezoelectric field upon annealing, $|J_{00}^{(eh)}|$ becomes larger than $J_{00}^{(hh)}$ with the first annealing step. Therefore the biexciton can change its sign and become binding.

(iii) We again observe a dramatic drop of the XX and X^+ correlation energies when $|J_{00}^{(eh)}|$ becomes larger than $J_{00}^{(hh)}$.

7.7.3 InGaAs QDs with Uniform Composition

A variation of the average $In_xGa_{1-x}As$ composition of QDs is typically employed to tailor the emission wavelength. The impact of the Ga content $(1 - x)$ on the multiparticle electronic properties is investigated using series E. From our earlier work [24] we know that with increasing Ga content inside the QD, the second-order piezoelectric effect, which dominates for 100% InAs, is reduced. Since the first-order terms remain unaffected, we encounter a sign change of the total piezoelectric field.

The presence of piezoelectricity constrains the wave function to a smaller volume leading to an increase of $J_{00}^{(hh)}$. Therefore, we find the smallest value of $J_{00}^{(hh)}$ for a Ga content of 15% (see Fig. 7.13) where the QD interior is virtually piezoelectric field-free. As a result, we observe a nonmonotonic behavior of biexciton binding energy.

The correlation energies of XX and X^+ exhibit this nonmonotonic behavior too, which is related to the characteristics of $|J_{00}^{(eh)}| - J_{00}^{(hh)}$ turning from negative to positive values first and then to almost zero.

7.8 Correlation vs. QD Size, Shape and Particle Type

Correlation is an effect that seeks to minimize the total energy in the tradeoff between the kinetic energy of the involved single particle states and their mutual Coulomb attraction or repulsion. Figures 7.10, 7.11, 7.12 and 7.13(c) show that the magnitude of the energy correction due to correlation is specific for each particle type and foremost dependent on the QDs size and shape. This is especially obvious in series A (see Fig. 7.10), where $|\delta_{Corr}(XX)| \approx |\delta_{Corr}(X^+)| > |\delta_{Corr}(X^-)| > |\delta_{Corr}(X)|$ and all the correlation energies increase with increasing dot size.

In general, there are three main factors, which determine the degree of correlation:

(i) First, the absolute and relative size of electron and hole wave function and their mutual position; in other words, how much Coulomb energy can be gained (J_{eh}) or saved (J_{hh}, J_{ee}) by relocating to a more favorable place, changing shape or size? We observe, for instance, strong correlation effects for the positive trion and biexciton in the following cases:

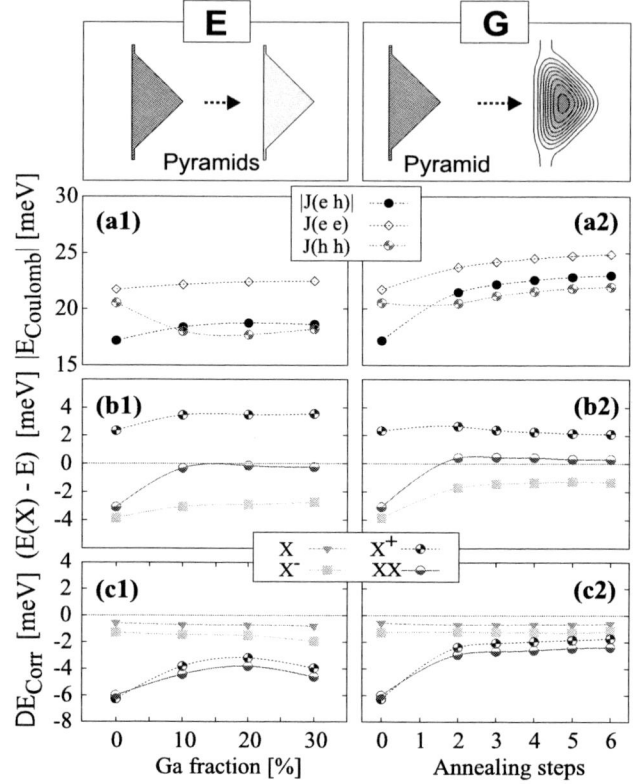

Fig. 7.13. (a) The direct Coulomb energies $J_{00}^{(ee)}$, $J_{00}^{(hh)}$ and $J_{00}^{(eh)}$; (b) the spectroscopic shifts, calculated for a configuration (10e, 10h); and (c) and the degree of correlation shown for series E and G. First- and second-order piezoelectric effects are included. From [18]

(1) Electron and hole orbital size are very dissimilar in series H, due to the asymmetric spread of electron and hole wave function. (2) The barycenter of both carriers are at different vertical positions as in the case of the half sphere. We find a dipole of 1.5 nm resulting in a smaller Coulomb attraction between electron and hole $J_{00}^{(eh)}$ compared to $J_{00}^{(ee)}$ and $J_{00}^{(hh)}$ (Fig. 7.11(c)). (3) The largest correlation energies are observed for the largest full pyramid of series A. Electron and hole orbital are pulled in different directions due to strong piezoelectricity. Their spatial overlap reduces and $J_{00}^{(eh)}$ becomes smaller than $J_{00}^{(ee)}$ and $J_{00}^{(hh)}$. For this case we observe correlation energies of the order of 15 meV. This large value, however, is also related to the large wave function extent and the accompanying small kinetic energy. This will further detailed below in case (iii).

(ii) Second, the particle type. Here, the ratio of the number of Coulomb interactions within a particle type to the number of carriers within this complex is important. Since the CI method is a variational method, this ratio weights the importance of the Coulomb interaction relative to the kinetic energies of the single carriers. For the exciton this ratio is 1/2, one Coulomb integral compared to two carriers. For the biexciton we have the ratio 6/4, six Coulomb integrals and four particles. Therefore the relative importance of the Coulomb interaction is larger for the biexciton than for the exciton and, hence, more emphasis is put on minimizing the total Coulomb energy in the case of the biexciton, which is done by mixing higher excited state configurations into the CI ground state.

For both trions we encounter a ratio 3/3, three Coulomb integrals and three particles. Hence, this point cannot explain why the correlation energies are mostly very different for the positive and the negative trion, which will be explained by the next point.

(iii) Third, the sensitivity of the kinetic energy to small wave function size variations or equivalently the unequal density of the electron and hole spectrum. Both quantities are a function of the effective mass and the spatial wave function extent. A larger effective mass translates into a smaller kinetic energy and a smaller energetic separation of ground and excited states. A larger wave function extent in real space transforms in a smaller extent in k-space, hence, the kinetic-energy integral

$$E_{\text{kin}} \approx \int_{V_k} \Psi^*(\boldsymbol{k}) E(\boldsymbol{k}) \Psi(\boldsymbol{k}) \, d\boldsymbol{k},$$

becomes smaller too. Consequently, a variation δE_{kin} in response to a small wave function size variation $\delta \eta$ is larger, the steeper the dispersion $E(\boldsymbol{k})$ (as result of a smaller effective mass) and the smaller the wave function extent in real space is. In other words, δE_{kin} is large for small electrons and small for large hole orbitals. If δE_{kin} is small, the wave function can more easily reshape to save Coulomb energy. The "reshaping" in our case is performed by mixing higher excited state configurations into the ground state configuration.

The importance of this point becomes visible by comparing $\delta_{\text{Corr}}(X^+)$ and $\delta_{\text{Corr}}(X^-)$: both complexes $X^+(h_0^2 e_0^1)$ and $X^-(h_0^1 e_0^2)$ share the same number of particles but in different configurations. Since the density of the hole spectrum is larger than that of the electron spectrum, the energy of those complexes, containing a larger number of positive carriers, is more affected by correlation. Therefore, $\delta_{\text{Corr}}(X^+) > \delta_{\text{Corr}}(X^-)$ holds. The more dense the spectrum becomes, the larger can be the correlation, as can be seen from the rising correlation energies of series A with increasing size in Fig. 7.10 (the corresponding spectral density can be derived from Fig. 7.7).

Another striking example is series B: here the absolute *and* the relative spectral densities change with the aspect ratio. For the pyramid (ar$_V = 0.5$) again we observe a large spectral hole density and small electron density. The electron density (see [24]) becomes larger with smaller aspect ratio resulting in a rising correlation energy of X and X^-. The hole density, in contrast, becomes smaller until ar$_V = 0.1$. Below that value the spectral hole density is rising again. This translates into a nonmonotonic δ_{Corr} behavior of the particles containing two holes, the XX and the X^+ (see Fig. 7.11(c1)).

7.9 Conclusions

For a large number of QD structures of varying size, shape and composition we investigated the relationship between the structural properties and the energies of a selected set of few-particle states (X, XX, $X^{\pm}$), which can be easily traced in experiments. The resulting binding energies turn out to be very sensitive to the various morphological peculiarities. We analyzed in detail the relationship between the QD geometry, the resulting shape and position of electron and hole wave functions, the direct Coulomb energies and changes introduced by correlation effects. The correlation effects are larger for biexciton and positive trion states, which we attribute to the larger spectral density of the hole subsystem. This in turn is a result of the larger hole effective mass. Correlation is very sensitive to the relative size and position of the electron and hole ground state orbitals, influencing equally the direct Coulomb energies $J_{00}^{(eh)}$, $J_{00}^{(ee)}$ and $J_{00}^{(hh)}$. Large correlation energies $\delta_{Corr}(XX)$ and $\delta_{Corr}(X^+f)$ are observed in those cases, where the absolute value of the attractive Coulomb term $J_{00}^{(eh)}$ falls below the values of the repulsive terms $J_{00}^{(ee)}$ and $J_{00}^{(hh)}$.

In Fig. 7.14 we compiled some of our data to visualize the wealth in the variation of the spectroscopic properties for different QD structures. We found the difference between the exciton s- and p-channel transitions as a function of the exciton energy (an easily observable quantity in experiments) an impressive fingerprint in order to distinguish between different types of InGaAs/GaAs QD structures. If, in addition, data on the spectroscopic shift of XX, $X^{\pm}$ relative to X are available, we are quite confident that, by the spectroscopic signature alone, large parts of the QD

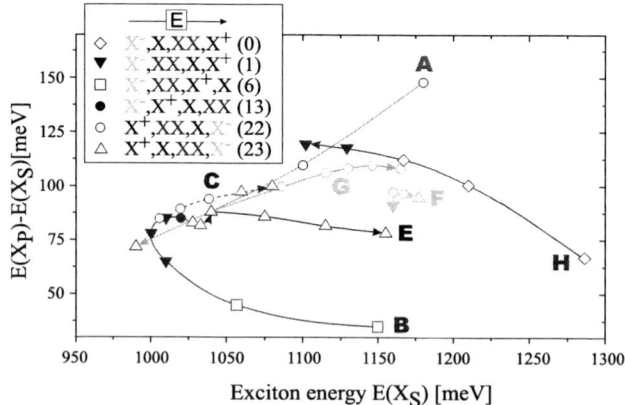

Fig. 7.14. The difference $(E(X_P) - E(X_S))$ plotted versus $E(X_S)$. With $E(X_P)$ we denote the average value of the P-channel transitions, taken form the excitonic absorption spectra. The data are shown for the series A, B, C, E, F, G and H. The symbols are used to denote the order of the calculated few-particle energies X, XX and $X^{\pm}$. The small arrows indicate the order of structures within each series as shown in the structure figures attached to the electronic data (Figs. 7.7, 7.11 and 7.13). From [18]

morphology can be deduced. One example of this approach will be detailed in the next chapter.

References

1. D. Bimberg, M. Grundmann, N.N. Ledentsov, *Quantum Dot Heterostructures* (Wiley, New York, 1999)
2. R. Seguin, A. Schliwa, S. Rodt, K. Pötschke, U.W. Pohl, D. Bimberg, Phys. Rev. Lett. **95**, 257402 (2005)
3. O. Stier, M. Grundmann, D. Bimberg, Phys. Rev. B **59**, 5688 (1999)
4. M. Bayer, T. Gutbrod, A. Forchel, V.D. Kulakovskii, A. Gorbunov, M. Michel, R. Steffen, K.H. Wang, Phys. Rev. B **58**, 4740 (1998)
5. S. Rodt, R. Heitz, A. Schliwa, R.L. Sellin, F. Guffarth, D. Bimberg, Phys. Rev. B **68**, 035331 (2003)
6. L. Landin, M.S. Miller, M. Pistol, C.E. Pryor, L. Samuelson, Science **280**, 282 (1998)
7. S. Rodt, A. Schliwa, K. Pötschke, F. Guffarth, D. Bimberg, Phys. Rev. B **71**, 155325 (2005)
8. M.E. Ware, A.S. Bracker, E. Stinaff, D. Gammon, D. Gershoni, V.L. Korenev, Physica E **13**, 55 (2005)
9. B. Urbaszek, R.J. Warburton, K. Karrai, B.D. Gerardot, P. Petroff, J. Garcia, Phys. Rev. Lett. **90**, 247403 (2003)
10. G.A. Narvaez, G. Bester, A. Zunger, Phys. Rev. B **72**, 245318 (2005)
11. I.A. Akimov, A. Hundt, T. Flissikowski, F. Henneberger, Appl. Phys. Lett. **81**, 4730 (2002)
12. R. Seguin, S. Rodt, A. Schliwa, K. Pötschke, U.W. Pohl, D. Bimberg, Phys. Stat. Sol. (b) **243**, 3937 (2006)
13. M. Ediger, G. Bester, B.D. Gerardot, A. Badolato, P.M. Petroff, K. Karrai, A. Zunger, R.J. Warburton, Phys. Rev. Lett. **98**, 036808 (2007)
14. M. Grundmann, O. Stier, D. Bimberg, Phys. Rev. B **52**, 11969 (1995)
15. O. Stier, D. Bimberg, Phys. Rev. B **55**, 7726 (1997)
16. H. Eisele, O. Flebbe, T. Kalka, C. Preinesberger, F. Heinrichsdorff, A. Krost, D. Bimberg, M. Dähne Pietsch, Appl. Phys. Lett. **75**, 106 (1999)
17. J. Marquez, L. Geelhaar, K. Jacobi, Appl. Phys. Lett. **78**, 2309 (2001)
18. A. Schliwa, M. Winkelnkemper, D. Bimberg, Phys. Rev. B (2007, submitted)
19. O. Stier, *Electronic and Optical Properties of Quantum Dots and Wires*. Berlin Studies in Solid State Physics, vol. 7 (Wissenschaft und Technik Verlag, Berlin, 2001)
20. G. Bester, X. Wu, D. Vanderbilt, A. Zunger, Phys. Rev. Lett. **96**, 187602 (2006)
21. W.F. Cady, *Piezoelectricity* (McGraw-Hill, New York, 1946)
22. G. Bester, A. Zunger, X. Wu, D. Vanderbilt, Phys. Rev. B **74**, 081305 (2006)
23. G. Bester, A. Zunger, Phys. Rev. B **71**, 045318 (2005)
24. A. Schliwa, M. Winkelnkemper, D. Bimberg, Phys. Rev. B **76**, 205324 (2007)
25. P. Enders, A. Bärwolf, M.W.D. Suisky, Phys. Rev. B **51**, 16695 (1995)
26. E.O. Kane, in *Band Theory and Transport Properties*, ed. by W. Paul. Handbook on Semiconductors, vol. 1 (North-Holland, Amsterdam, 1982), p. 194
27. F.H. Pollak, Semicond. Semimet. **32**, 17 (1990)
28. P. Enders, Phys. Stat. Sol. (b) **187**, 541 (1995)
29. D. Gershoni, C.H. Henry, G.A. Baraff, IEEE J. Quantum Electron. **29**, 2433 (1993)
30. H. Jiang, J. Singh, Phys. Rev. B **56**, 4696 (1997)

31. C. Pryor, Phys. Rev. B **57**, 7190 (1998)
32. J.A. Majewski, S. Birner, A. Trellakis, M. Sabathil, P. Vogl, Phys. Stat. Sol. (c) **8**, 2003 (2004)
33. T.B. Bahder, Phys. Rev. B **41**, 11992 (1990)
34. J. Kim, L.W. Wang, A. Zunger, Phys. Rev. B **57**, R9408 (1998)
35. H. Fu, L.-W. Wang, A. Zunger, Phys. Rev. B **57**, 9971 (1998)
36. R. Santoprete, B. Koiller, B. Capaz, P. Kratzer, Q.K.K. Liu, M. Scheffler, Phys. Rev. B **68**, 235311 (2003)
37. S. Lee, F. Oyafuso, P. von Allmen, G. Klimeck, Phys. Rev. B **69**, 045316 (2004)
38. J. Shumway, A. Franceschetti, A. Zunger, Phys. Rev. B **63**, 155316 (2001)
39. A. Lenz, R. Timm, H. Eisele, C. Hennig, S.K. Becker, R.L. Sellin, U.W. Pohl, D. Bimberg, M. Dähne, Appl. Phys. Lett. **81**, 5150 (2002)
40. S. Ruvimov, P. Werner, K. Scheerschmidt, U. Gösele, J. Heydenreich, U.R.N.N. Ledentsov, M. Grundmann, D. Bimberg, V.M. Ustinov, A.Y. Egorov et al., Phys. Rev. B **51**, 14766 (1995)
41. R. Heitz, F. Guffarth, K. Pötschke, A. Schliwa, D. Bimberg, Phys. Rev. B **71**, 045325 (2005)
42. O. Stier, A. Schliwa, R. Heitz, M. Grundmann, D. Bimberg, Phys. Stat. Sol. (b) **224**, 115 (2001)

8

Phonons in Quantum Dots and Their Role in Exciton Dephasing

F. Grosse, E.A. Muljarov, and R. Zimmermann

Abstract. The acoustic phonon spectrum is significantly modified for embedded quantum dots by inhomogeneous change of material properties and intrinsic strain. The change of the local elastic properties due to strain is calculated employing density functional theory and used as input for phonon calculations within continuum elasticity model. It is demonstrated that overall the exciton–phonon coupling strength is reduced, characteristic oscillations appear in the excitonic polarization, and the spectral broadband is modified compared to a bulk phonon assumption. The zero phonon line broadening is discussed in terms of real and virtual phonon-assisted transitions between different exciton levels in a quantum dot. A microscopic theory of the excitonic multilevel system coupled to acoustic phonons is developed, and the full time-dependent polarization and absorption are calculated using the cumulant expansion. Examples are given for dephasing of optical excitations in single and vertically coupled quantum dots.

8.1 Introduction

Significant progress has been achieved over the last years in tailoring properties of semiconductor nanostructures, in particular quantum dots (QDs). The state of the art is reported in detail in this book. A large application-driven interest stems from the fact that QDs are possible candidates (and are used already to some extent) in promising applications like QD lasers, single photon sources, and quantum information devices. In this respect, the dephasing of the optical excitations (excitons) is an essential issue. A significant loss of coherence would, e.g., forbid application of semiconductor nanostructures in devices for quantum computation. A central role in the dephasing is played by the interaction with lattice vibrations (phonons), in particular with the acoustic phonons [1, 2]. The exciton-acoustic phonon coupling is usually described by deformation potential coupling. Progress has been made in the description of the charged carrier properties specifically associated with the nanostructure (confinement). The phonons, however, are usually taken as bulk phonons—

although it is known that they can be significantly modified due to inhomogeneous changes of elastic properties present in nanostructures [3].

Here we discuss recent theoretical progress in the description of phonons in semiconductor nanostructures and their relevance to dephasing of optical excitations. After a short introduction to structural specifics of semiconductor nanostructures in Sect. 8.2, the theory of acoustic phonons in QDs is explained in Sect. 8.3. Special emphasis is given to continuum elasticity, but we go beyond a linear description. The basic theory of the exciton-acoustic phonon coupling is developed in Sect. 8.4. Different approaches to optical dephasing, starting with the independent boson model, are discussed in Sect. 8.5. A quadratic coupling model is derived and solved exactly summing up the cumulant expansion. Finally, the excitonic multilevel system is treated with applications to single and coupled QDs.

8.2 Structural Properties of Semiconductor Nanostructures

First we review some important facts that are needed to understand the specifics of the coupling between excitons and acoustic phonons in semiconductor nanostructures. The localization (confinement) of carriers or excitons can be achieved by embedding a small band gap material like InAs into a material with a larger band gap like GaAs. Since the band gap decreases with increasing of the equilibrium lattice constant, it means that the material with the larger lattice constant has to be embedded and forms the region where the exciton will be confined (quantum dot). This lattice constant relation is also favorable for the formation of QDs in the Stranski–Krastanow growth mode [4]. It also implies that the dot material is compressed if the structure is grown pseudomorphically, i.e. without defects, which is necessary for usability in optical setups.

In continuum elasticity theory [8] the acoustic phonons are solely described by elastic constants and the corresponding mass density (see Sect. 8.3). The numerical values of the zincblende elastic constants of various semiconductors [5–7] are to a good approximation a unique function of the equilibrium lattice constant and decrease with increasing volume of the unit cell, see Fig. 8.1. Thus, together with the dependence of the band gap on the lattice constant, one finds: The dot material, which forms the confinement for the charge carrier, is softer (smaller elastic constant) than the barrier material. However, due to intrinsic strain in pseudomorphic nanostructures, the elastic constants are effectively changed. It is intuitively clear that the compression of the dot material leads to an effective increase of elastic constants, i.e. the material becomes harder. We will discuss in the following how to overcome the restriction to bulk phonons by including the different and strain-dependent elastic properties of dot and barrier material.

8.3 Theory of Acoustic Phonons in Quantum Dots

Acoustic phonons can be described either by continuum elasticity or by atomistic methods. The large number of atoms involved in nanostructures makes the descrip-

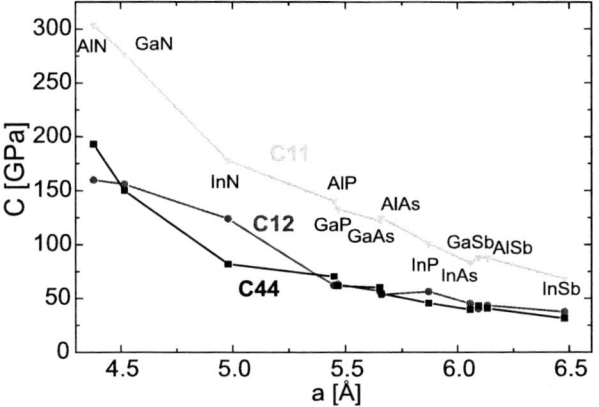

Fig. 8.1. Dependence of the elastic constants C_{11}, C_{12}, and C_{44} on the lattice constant, given for various III–V compounds including the group III-nitrides. Values are taken from [5]. The values for the nitride compounds are taken from density functional calculations [6, 7]

tion by the latter computationally very demanding although not impossible. Continuum elasticity naturally cannot describe certain details exactly, which are related to atomistic length scales, e.g. alloying, ordering, and specific interface formations, for instance in noncommon anion systems. These effects play only minor roles for the spherical quantum dot model we will investigate to gain a qualitative understanding. Therefore, continuum elasticity, which allows an analytic treatment, is used to determine the phonon properties of the nanostructure.

8.3.1 Continuum Elasticity Model of Phonons

Continuum elasticity considers small volume elements and their displacement $\boldsymbol{u}(\boldsymbol{r})$ under deformations. The components of the Lagrangian strain tensor $\overset{\leftrightarrow}{\epsilon}$ are given with respect to the displacement field $\boldsymbol{u}(\boldsymbol{r})$ for a homogeneously deformed system by [9]

$$\epsilon_{kl} = \frac{1}{2}\left(\frac{\partial u_k}{\partial x_l} + \frac{\partial u_l}{\partial x_k} + 2\frac{\partial u_k}{\partial x_l}\frac{\partial u_l}{\partial x_k}\right). \tag{8.1}$$

The relation between stress $\overset{\leftrightarrow}{\sigma}$ and strain $\overset{\leftrightarrow}{\epsilon}$ is given using the equilibrium mass density ρ_0 and the free energy density ϕ by

$$\sigma_{ij} = \sum_{kl} \frac{\partial \rho_0 \phi}{\partial \epsilon_{kl}}\left(\delta_{ik} - \frac{\partial u_i}{\partial x_k}\right)\left(\delta_{jl} - \frac{\partial u_j}{\partial x_l}\right). \tag{8.2}$$

No linear approximation has been made in either of the above equations. The free energy density ϕ can be expanded around the equilibrium in terms of the strain tensor

$$\rho_0 \phi = \frac{1}{2}\sum_{I,J=1}^{6} C_{IJ}\epsilon_I\epsilon_J + \frac{1}{3!}\sum_{I,J,K=1}^{6} C_{IJK}\epsilon_I\epsilon_J\epsilon_K + \cdots, \tag{8.3}$$

here given up to third-order using the Voigt notation [8] indicated by capital indices. The expansion coefficients C_{IJ} are the second-order and C_{IJK} are the third-order elastic constants. Due to symmetry, the number of independent elastic constants is reduced. For zincblende crystals there are only three independent second-order elastic constants: C_{11}, C_{12}, and C_{44}.

The equation of motion for the displacement is given by

$$-\omega^2 \rho(\mathbf{r}) u_i(\mathbf{r}) = \sum_j \frac{\partial}{\partial x_j} \sigma_{ij}(\mathbf{r}). \tag{8.4}$$

For nanostructures the mass density $\rho(\mathbf{r})$ and the stress tensor $\overleftrightarrow{\sigma}(\mathbf{r})$ are spatially dependent due to local variations of the material (concentration).

In linear elasticity, i.e. restricting to small displacements only, which corresponds to phonons, the strain definition (8.1) and the stress–strain relation (8.2), using the free energy density expansion (8.3), reduce to

$$\epsilon_{ij} = \frac{1}{2}\left(\frac{\partial u_i}{\partial x_j} + \frac{\partial u_j}{\partial x_i}\right) \quad \text{and Hooke's law:} \quad \sigma_I = \sum_J C_{IJ} \epsilon_J. \tag{8.5}$$

In the description of the phonons we will restrict ourselves to isotropic media where only two elastic constants are independent, $C_{12} = C_{11} - 2C_{44}$ (Lamé coefficients $C_{12} = \lambda$ and $C_{44} = \mu$). Transforming to convenient spherical coordinates, since we consider a spherical inclusion in the next section, and focusing on the most relevant components for the electron-acoustic phonon coupling with angular momentum zero $\mathbf{u}(\mathbf{r}) = \mathbf{e}_r u(r)$, the wave equation in isotropic media becomes

$$-\omega^2 \rho(r) u(r) = -4 \frac{dC_{44}(r)}{dr} \frac{u(r)}{r} + \frac{d}{dr}\left[C_{11}(r)\left(\frac{du(r)}{dr} + \frac{2u(r)}{r}\right)\right]. \tag{8.6}$$

This equation naturally contains the usual boundary conditions

$$u \quad \text{and} \quad C_{11}\left(\frac{du}{dr} + \frac{2u}{r}\right) - 4C_{44}\frac{u}{r} \quad \text{continuous} \tag{8.7}$$

and is also applicable to situations where elastic constants or mass density change gradually without sharp boundaries. The displacement eigenmodes have to fulfill the normalization condition [10]

$$\int_\Omega d^3r\, \rho(\mathbf{r}) \mathbf{u}_\nu(\mathbf{r}) \cdot \mathbf{u}_{\nu'}(\mathbf{r}) = \delta_{\nu,\nu'} \frac{\hbar}{2\omega_\nu}. \tag{8.8}$$

The question remains which elastic constants have to be used in (8.6). The pseudomorphic growth, i.e. matching the crystal structure of different materials on the atomic scale without defects, leads to intrinsic strain if both materials have different lattice constants. The starting point of the phonon calculation in nanostructures is therefore an inhomogeneously strained crystal. Since a typical lattice mismatch is more than 2% in nanostructures (for InAs and GaAs e.g., around 7%) higher orders in the free energy expansion (8.3) contribute significantly. Under compression this leads to an effective hardening of the material. In the following we will restrict ourselves to hydrostatic strain. This ensures that the crystal symmetry is preserved. The numerical values of the effective strain-dependent elastic constants are calculated in the following. They are used as input for (8.6), where their spatial dependence is due to the change of the material *and* the spatially dependent strain, i.e. $C_{IJ} = C_{IJ}(\overleftrightarrow{\sigma}(r))$. Therefore, we will calculate the phonons in the model of linear elasticity but the effect of higher expansion terms in the free energy is included via effective elastic constants, which differ from the bare ones.

The calculation of elastic properties of small periodic cells is possible directly by employing *ab initio* density functional theory. Here the focus is on the calculation of effective elastic constants for hydrostatically deformed InAs and GaAs unit cells by density functional perturbation theory. These calculations are carried out with the ABINIT computer code [11–13]. The local density approximation (LDA) is applied for the exchange-correlation. Soft norm-conserving pseudopotentials are taken from the code of the Fritz-Haber Institute, Berlin [14]. Wave functions are expanded into plane waves with converged Monkhorst–Pack meshes of $k = 8 \times 8 \times 8$ per 1×1 unit cell and a cutoff energy of $E_{\text{cut}} = 36\,\text{Ha} \equiv 979\,\text{eV}$.

The result of the calculation is shown in Fig. 8.2, where the dependence of the effective elastic constants on the local lattice constant is given for InAs and GaAs. As expected, the numerical values decrease with increasing lattice constant for C_{11} and C_{12}. From linear elasticity one would expect a merely linear dependence. The

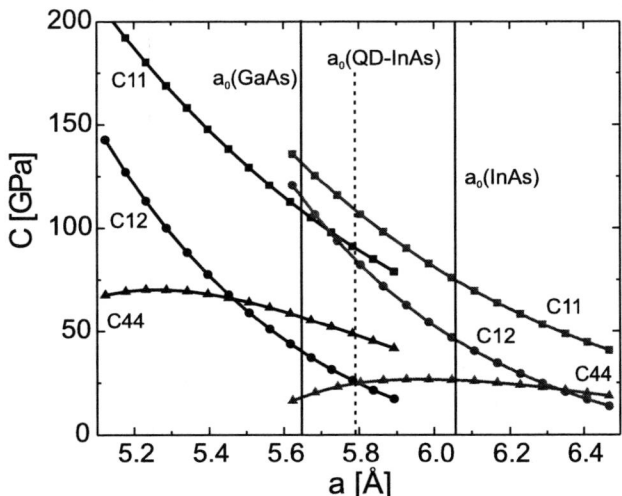

Fig. 8.2. Effective elastic constants for GaAs (*black*) and InAs (*red*) calculated by employing density functional perturbation theory for hydrostatic deformations. The experimental equilibrium lattice constants are given by vertical lines together with the relaxed lattice constant (*dashed line*) of the embedded InAs QD in a GaAs matrix

strong bowing, especially for C_{44}, which has a maximum close to the equilibrium lattice constant is due to nonlinear dependencies. Knowing the local lattice constant in the relaxed situation allows us to determine the spatially dependent effective elastic constants to be used in the phonon calculation.

8.3.2 Phonons in Quantum Dots

We will present calculations of elastic properties and acoustic phonons for a spherical inclusion with a radius of $R = 5$ nm and $R = 10$ nm in an infinitely extended host crystal [15, 16]. The system is chosen to achieve the most results analytically at a minimal numerical cost, but still demonstrate the influence of specific features of the nanostructure phonons in the exciton-acoustic phonon coupling in the next section. Numerical values for the parameters used in the calculation are given in Table 8.1.

A hydrostatic deformation is characterized by a simple scaling of all coordinates $x'_j = (1 - \alpha)x_j$. The coefficient α can be deduced from the static solution ($\omega = 0$) of (8.6) including the boundary conditions from (8.7) accordingly

$$\alpha = \alpha_0 \frac{C_{44,\text{GaAs}}}{C_{44,\text{InAs}} - C_{44,\text{GaAs}} + \frac{3}{4}C_{11,\text{InAs}}} \quad (8.9)$$

(lattice mismatch equal to $\alpha_0 = (a_{0,\text{InAs}} - a_{0,\text{GaAs}})/a_{0,\text{InAs}} = 6.8\%$ is found using the experimental values given in Table 8.1). The relaxed lattice constant a of the embedded InAs QD, which is compressed hydrostatically, is therefore given by $a_{\text{InAs}} = a_{0,\text{InAs}}(1-\alpha)$ (last row of Table 8.1, and the dashed line in Fig. 8.2). The volume of the host material GaAs is unchanged with a compensating radial compression and a lateral dilation. Considering only the hydrostatic part, the elastic properties of the host material will be treated unchanged.

Solving (8.6) for bulk situations, i.e. with no spatial dependence of either elastic constants or mass density, the solution is simply

$$u^{(\text{B})}(r) = j_1(k_\text{B}r) \quad (8.10)$$

with $j_1(x)$ being the spherical Bessel function of the first kind, and the wave number being related to the phonon energy via $E = \hbar v_\text{lB} k_\text{B}$ ($v_\text{l} = \sqrt{C_{11}/\rho}$ is the longitudinal sound velocity).

Table 8.1. Effective masses for electron m_e and hole $m_{\text{hh,av}}$ (averaged isotropic), deformation potentials for conduction D_C and valence band D_V, mass density ρ, and lattice constant a_0 for unstrained GaAs and InAs used in the calculations [5, 17]. Last row: Resulting values for the embedded strained InAs QD (s-InAs)

	m_e $[m_0]$	$m_{\text{hh,av}}$ $[m_0]$	D_C [eV]	D_V [eV]	ρ [g/cm^3]	a_0(Exp) [Å]
GaAs	0.067	0.501	−7.17	1.16	5.33	5.65
InAs	0.026	0.388	−5.08	1.00	5.66	6.06
s-InAs	0.026	0.388	−5.08	1.00	6.50	5.79

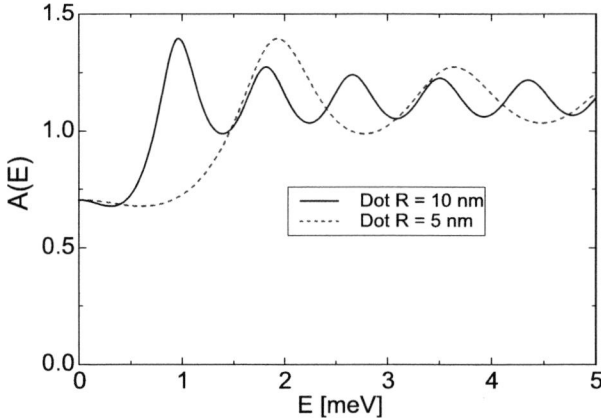

Fig. 8.3. Energy dependence of the phonon amplitudes for two spherical InAs inclusions in a GaAs host crystal with radii $R = 5$ nm (*red dashed line*) and $R = 10$ nm (*black solid line*)

Treating the spatial dependence in (8.6) leads to a modification of the phonon modes. The displacement around the origin inside the dot (index D) is scaled by an energy-dependent prefactor,

$$u(r) = A'(E) j_1(k_D r), \quad (8.11)$$

which has to be determined by solving (8.6) numerically. This prefactor is the main source of changes in the electron–phonon coupling. For later comparison with the coupling function we define a phonon amplitude as

$$A(E) = A'(E) \frac{v_{lB}}{v_{lD}}. \quad (8.12)$$

By definition, for bulk barrier (index B) phonons, $A(E) \equiv 1$ which is the normalization used in Fig. 8.3. Due to embedding the sphere in an *infinite* elastic medium there is a continuous manifold of phonon modes in contrast to a situation where the sphere is free or rigidly clamped. In the latter case one gets discrete eigenmodes [18]. The oscillating behavior in Fig. 8.3 resembles some properties of these discrete spectra.

8.4 Exciton-Acoustic Phonon Coupling in Quantum Dots

The electron-acoustic phonon coupling has two major contributions: deformation potential coupling and piezoelectric coupling. Since the latter was shown to be small in QDs [19], we will focus on the first in the following.

Both electron and hole levels couple to volume deformations (div $u = \text{tr}\, \overleftrightarrow{\epsilon}$) whereas only the hole additionally couples to uniaxial and shear contributions. The relaxed static strain distribution for the embedded sphere contains only volume changes, which are incorporated in the changes of the band offsets (see inset in Fig. 8.4). The electron-acoustic phonon interaction Hamiltonian is [10]

$$H_{\text{def}}(r_e, r_h) = \sum_\nu D_C(r_e) a_\nu^\dagger \text{div}\, \boldsymbol{u}_\nu(r_e) - D_V(r_h) a_\nu^\dagger \text{div}\, \boldsymbol{u}_\nu(r_h) + \text{h.c.} \quad (8.13)$$

with the (material-dependent) deformation potential constant for the conduction D_C and valence band D_V. Here the deformation $\boldsymbol{u}(r)$ contains only the phonon displacement, which is measured with respect to the relaxed position. Therefore, one has to use the solution of (8.6) where effective elastic constants determined by the local static strain are used. The phonon creation is described by the operator $a_\nu^\dagger$. In a QD where the exciton (electron-hole pair) is described by the confinement function $\Psi_n(r_e, r_h)$ the relevant exciton–phonon matrix elements are

$$M_\nu^{nm} = \int dr_e \int dr_h$$
$$\times \Psi_n^*(r_e, r_h) \{ D_C(r_e) \text{div}\, \boldsymbol{u}_\nu(r_e) - D_V(r_h) \text{div}\, \boldsymbol{u}_\nu(r_h) \} \Psi_m(r_e, r_h). \quad (8.14)$$

For strong confinement the wave function can be approximated by a product of one-particle wave functions. Restricting to the lowest electron and hole level in a spherical QD, we have $\Psi_1(r_e, r_h) = \varphi_{e,1}(r_e) \varphi_{h,1}(r_h)$, which couples only to the $l = 0$ acoustic phonon modes ν. For the polaronic modification of a single excitonic transition within the independent Boson model [20], the diagonal element in (8.14) is of central importance. The exciton–phonon coupling function for the lowest sublevels reads [21]

$$f(E) = \sum_\nu \delta(E - \hbar \omega_\nu) |M_\nu^{11}|^2. \quad (8.15)$$

Insertion of the bulk expression (8.10) into (8.14) leads to

$$f^{(B)}(E) = \frac{E^3}{4\pi^2 \hbar^3 \rho_B v_{\text{lB}}^5} \left| \int_0^\infty dr\, r^2 [D_{CB} \varphi_{e,1}^2(r) - D_{VB} \varphi_{h,1}^2(r)] j_0(k_B r) \right|^2. \quad (8.16)$$

In this approximation, the coupling function (divided by E^3) follows the (squared) Fourier transform of the electron and hole charge density, which decays on a scale of $E \approx \hbar v_{\text{lB}}/R$ (R is the QD radius).

The coupling function $f(E)$ calculated from the numerical solution of (8.6) is plotted in Fig. 8.4 for the dot radius of $R = 10$ nm. It is compared to situations where phonons are taken either for barrier or dot bulk material. Qualitatively, the full solution contains structures, which are due to the resonance-like features in the phonon amplitude shown in Fig. 8.3. Additionally, we find an overall reduction of the coupling strength, especially for low energies. Figure 8.5 compares calculations of different inclusion sizes. Decreasing the size of the QD from $R = 10$ nm to 5 nm leads to a stronger coupling. Increasing the energetic distance of the resonances in the phonon amplitude for smaller dots the structure in the coupling function reduces.

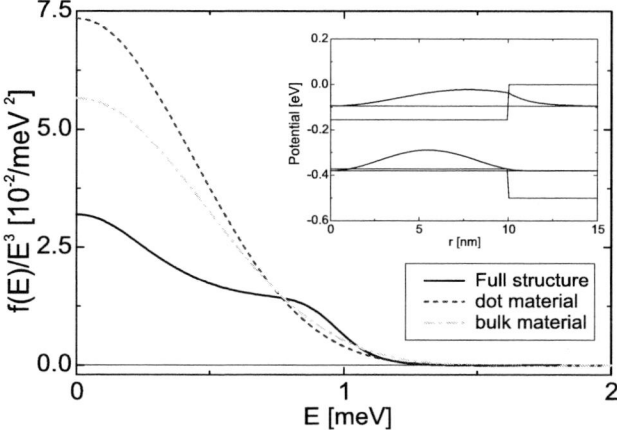

Fig. 8.4. Exciton–phonon coupling function compared for different treatments of acoustic phonons for a spherical InAs inclusion with a radius of $R = 10$ nm in a GaAs host. Black solid line: full solution; red dashed: bulk dot material (strained); green dot-dashed: bulk barrier material. The inset shows the band alignment, the corresponding one-particle eigenenergies and charge-densities for electron and hole ground state

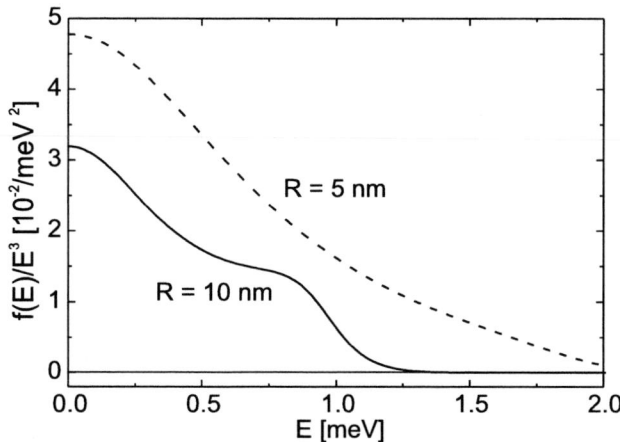

Fig. 8.5. Exciton–phonon coupling function compared for radii $R = 10$ nm (*solid line*) and $R = 5$ nm of the InAs spherical inclusion

8.5 Dephasing of the Exciton Polarization in Quantum Dots

Now we are ready to address the dephasing of the optical polarization due to the coupling between excitonic states and acoustic phonons. In InAs/GaAs QDs and QD molecules the dephasing has been studied by time-integrated four-wave mixing (FWM) measurements [22, 23]. These experiments revealed the following general trends in the optical polarization: The FWM polarization as a function of the de-

174   F. Grosse et al.

lay time between two excitation pulses first experiences a quick initial decay on a few picosecond time scale; at later times the FWM polarization shows a much slower exponential or even multiexponential [23, 24] decay which strongly depends on temperature. In optical spectra these basic features manifest themselves, respectively, as a broadband and a much narrower Lorentzian zero-phonon line (ZPL) with a temperature-dependent width. They have been observed by microphotoluminescence studies in InGaAs QDs [25–29], and also in CdZnTe [30, 31], CuNaCl [32], CdZnTe QDs [33, 34], and in strain-free GaAlAs QDs [35–37].

There have been many attempts in the literature to describe theoretically these properties of the QD dephasing, see e.g. [1, 31, 38–46]. Most of these works focused mainly on the independent boson model (IBM) [20], in the limit of a single exciton state. This well-known model can describe satisfactorily the measured initial decay of the FWM polarization [47]. However, it is unable to describe any decay at long times. Therefore, the IBM cannot be used for a dephasing calculation since the width of the ZPL is exactly zero within this approximation.

Recently, we developed a new approach to the exciton–phonon problem in QDs [2, 48, 49]. The full temporal decay of the linear and FWM polarizations could be explained in terms of real and virtual phonon-assisted transitions. In the following sections we present the main results of this theory, demonstrating the basic sources of the dephasing on some exactly solvable models (such as IBM and quadratic coupling model). We also discuss the general solution of the exciton–phonon problem which is obtained using the cumulant expansion. Then, we present some numerical results for the dephasing of the ground exciton state in single QDs and of a few lowest exciton states in asymmetric coupled QDs.

8.5.1 Single Exciton Level: Independent Boson Model

Restricting to a single (ground) exciton level in a QD, the independent boson model [20] was used, e.g., in [1, 31, 38, 40] to calculate optical polarization, absorption, and PL, assuming bulk acoustic phonons. In a similar manner, we apply the IBM to the ground exciton state in a QD, now taking into account the modification of the acoustic phonon modes due to the QD structure which was studied in Sects. 8.3 and 8.4.

The IBM Hamiltonian has the form:

$$H = \sum_\nu \hbar\omega_\nu a_\nu^\dagger a_\nu + \left[\hbar\omega_{\text{ex}} + \sum_\nu M_\nu^{11}\left(a_\nu + a_\nu^\dagger\right)\right]|1\rangle\langle 1|, \qquad (8.17)$$

where the phonon index ν takes all possible quantum numbers of acoustic phonons coupled to the ground exciton (or electron–hole pair) state $|1\rangle$, $\hbar\omega_{\text{ex}}$ is the bare exciton transition energy, and M_ν^{11} is the coupling matrix element defined in (8.14).

In this model, any optical response function can be written explicitly by introducing a phonon propagator:

$$D(t) = \hbar^{-2} \sum_\nu |M_\nu^{11}|^2 \left[(N_\nu + 1)e^{-i\omega_\nu|t|} + N_\nu e^{i\omega_\nu|t|}\right], \qquad (8.18)$$

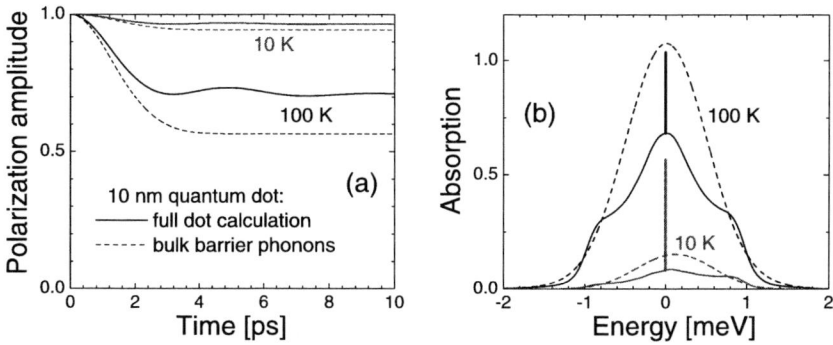

Fig. 8.6. Polarization amplitude (**a**) and absorption (**b**) of a 10 nm InGaAs QD, calculated with structurally modified phonons and with bulk barrier (GaAs) phonons, for $T = 10$ K and 100 K. Lengths of vertical lines are proportional to the ZPL weights. All parameters are the same as used in Sect. 8.4

where $N_\nu = 1/[\exp(\hbar\omega_\nu/k_B T) - 1]$ is the Bose-distributed phonon occupation number. The linear polarization (a response on a delta-pulse excitation) takes the form

$$P(t) = \exp[-i\omega_{ex} t + K(t)], \quad (8.19)$$

where the cumulant

$$K(t) = -\int_0^t d\tau \, (t-\tau) D(\tau) \quad (8.20)$$

provides the exact solution already in first-order. In fact, $K(t)$ is proportional to $|M_\nu^{11}|^2$ and thus accounts for only a single-phonon (first-order) process.

In the limit of long times the cumulant behaves as $K(t) \to -i\Delta t - S$, where

$$\Delta = -i \int_0^\infty d\tau \, D(\tau) = \frac{1}{\hbar} \int_0^\infty \frac{dE}{E} \frac{f(E)}{e^{E/k_B T} - 1} \quad (8.21)$$

is the polaron shift and

$$S = -\int_0^\infty \tau \, d\tau \, D(\tau) = \int_0^\infty \frac{dE}{E^2} f(E) \coth(E/2k_B T) \quad (8.22)$$

is the Huang–Rhys factor [50]. Note that the coupling function $f(E)$ defined by (8.15) has contribution from both electron and hole densities.

The calculated linear polarization, Fig. 8.6(a), has a quick (3–5 ps) decay which reflects the formation of a steady-state polaron cloud around the QD. Compared to the model with bulk phonons of the barrier material (dashed curves), the full calculation with QD renormalized phonons (solid curves) shows typical oscillations that actually originate from the same features in the coupling function, Fig. 8.5, and can be even better seen in the absorption,

$$\alpha(\omega) = \text{Re} \int_0^\infty dt \, P(t) e^{i\omega t}, \quad (8.23)$$

plotted in Fig. 8.6(b).

As the IBM provides no broadening mechanism of the ZPL, the latter has zero width. However, its spectral weight, proportional to the height of the delta-like line in Fig. 8.6(b), is given by $\exp(-S)$ and depends on temperature. The broadband is strongly asymmetric at low temperatures and gets more symmetric with respect to the ZPL position (here taken as zero of energy) as the temperature grows.

8.5.2 Multilevel System: Real and Virtual Phonon-Assisted Transitions

The virtual process taken into account by the IBM is displayed schematically in Fig. 8.7 (left diagram). The ground exciton state in a QD (here labeled by 1) interacts with a single acoustic phonon shown by the dashed line which starts and ends at the same exciton level. In other words, the interaction is diagonal with respect to the excitonic state. As we saw in Fig. 8.6(a), such a process results only in a partial decoherence, with no further decay of the steady-state exciton polaron.

In order to describe the long-time decay of the optical polarization which is observed experimentally, one necessarily has to take into account a nondiagonal coupling between different excitonic states in the QD. This coupling results in two types of processes: virtual transitions (middle) and real transitions (right diagram). These virtual transitions also start and end at the same exciton level. However, at least two phonons participate in them, and the nondiagonal matrix element (between state 1 and the exciton excited state 2) is now involved. They also result in the pure dephasing but mainly produce the long-time decay of the excitonic polarization. Real transitions, in turn, change the exciton occupation number, thus leading to a population decay. Both of these two processes, virtual and real transitions, originating from the nondiagonal coupling, are responsible for the microscopic broadening of the ZPL.

To describe the full time dependence of the optical polarization we have to include in the model this nondiagonal coupling as well. In a single-exciton representation, the Hamiltonian of the multilevel system linearly coupled to acoustic phonons has the form [48]:

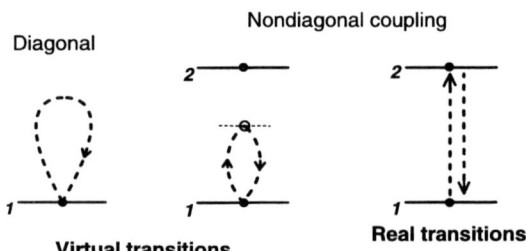

Fig. 8.7. Schematic diagrams describing virtual and real phonon-assisted transition due to diagonal and nondiagonal exciton–phonon coupling in a QD with two exciton levels

8 Phonons in Quantum Dots and Their Role in Exciton Dephasing

$$H = \sum_n \hbar\omega_n |n\rangle\langle n| + \sum_q \hbar\omega_q^{ac} a_q^\dagger a_q + \hbar V, \tag{8.24}$$

$$V = \sum_{n,m} |n\rangle\langle m| \sum_q M_{nm}(q)(a_q + a_{-q}^\dagger), \tag{8.25}$$

where we now deal with bulk phonons for simplicity. The matrix elements of the exciton–phonon coupling due to the deformation potentials are given by

$$M_{nm}(q) = \sqrt{\frac{\omega_q}{2\rho v_l^2 V}} \int dr_e\, dr_h\, \Psi_n^*(r_e, r_h) \Psi_m(r_e, r_h) \left[D_C e^{iq\, r_e} - D_V e^{iq\, r_h} \right], \tag{8.26}$$

where $\Psi_n(r_e, r_h)$ and $\hbar\omega_n$ are, respectively the wave function and the transition energy of the QD exciton state $|n\rangle$. Note that the IBM accounts for only diagonal $n = m$ terms in (8.25). We are now interested in a proper treatment of both diagonal and nondiagonal terms.

Dephasing of the Ground Exciton State: Quadratic Coupling Model

The effect of virtual transitions shown in Fig. 8.7 (middle diagram) can be taken into account by the quadratic coupling model which allows an exact solution [2]. In order to bring the initial Hamiltonian (8.24)–(8.25) into this form, we apply a unitary transformation which eliminates the nondiagonal terms in first-order. If we concentrate on the exciton ground state transition, the resulting effective Hamiltonian then has the form

$$H = \sum_q \hbar\omega_q^{ac} a_q^\dagger a_q + (V_L + V_Q) |1\rangle\langle 1|, \tag{8.27}$$

where

$$V_L = \sum_q M_{11}(q)(a_q + a_{-q}^\dagger) \tag{8.28}$$

is the linear diagonal coupling similar to that in (8.17) and treated exactly in the IBM, while

$$V_Q = \frac{1}{\hbar} \sum_{pq} \sum_{n \neq 1} \frac{M_{1n}(p) M_{n1}(q)}{\omega_1 - \omega_n} (a_p + a_{-p}^\dagger)(a_q + a_{-q}^\dagger) \tag{8.29}$$

is a new term found by mapping the nondiagonal coupling into a diagonal one which, however, is quadratic in the phonon displacement operators.

With this quadratic term, the Hamiltonian can be solved exactly by summing up an infinite series of diagrams in the cumulant [2]. In case of only one excited level (the sum in (8.29) is reduced to $n = 2$) the polarization can be written explicitly as

$$P(t) = \exp[-i\omega_1 t + K_L(t) + K_Q(t)] \tag{8.30}$$

with the linear cumulant $K_L(t)$ defined by (8.20), where the phonon quantum number ν and the matrix element M_ν^{11} have to be replaced in (8.18) by, respectively, the bulk-phonon momentum q and $M_{11}(q)$. For the quadratic cumulant we have an infinite series

$$K_Q(t) = \frac{1}{2} \sum_{k=1}^{\infty} \frac{1}{k} \int_0^t dt_1 \ldots dt_k \, D_Q(t_1 - t_2) D_Q(t_2 - t_3) \ldots D_Q(t_k - t_1), \quad (8.31)$$

where another phonon propagator,

$$D_Q(t) = \frac{2i}{\hbar^2} \frac{1}{\omega_2 - \omega_1} \sum_q |M_{12}(\boldsymbol{q})|^2 [(N_q + 1)e^{-i\omega_q |t|} + N_q e^{i\omega_q |t|}], \quad (8.32)$$

is introduced. The quadratic cumulant (8.31) is calculated by solving for each time t the Fredholm eigenvalue problem

$$\int_0^t dt_2 \, D_Q(t_1 - t_2) u_j(t_2) = \Lambda_j u_j(t_1) \quad (8.33)$$

and then converting a k-fold convolution of $D_Q(t)$ in (8.31) into a power Λ_j^k. The subsequent infinite summation over k yields

$$K_Q(t) = -\frac{1}{2} \sum_j \ln(1 - \Lambda_j). \quad (8.34)$$

In the long-time limit, the result can even be given analytically. The asymptotic behavior of the quadratic cumulant is $K_Q(t) \to -S_Q - \Gamma t - i\Delta_Q t$, where the microscopic ZPL width Γ can be expressed via the Fourier transform $D_Q(\omega)$ of the phonon propagator (8.32),

$$\Gamma = \text{Re} \int_0^\infty \frac{d\omega}{2\pi} \ln[1 - D_Q(\omega)] = \int_0^\infty \frac{d\omega}{4\pi} |D_Q(\omega)|^2 + \cdots . \quad (8.35)$$

In lowest order, Γ is proportional to $|D_Q|^2$, since one has to take the real part of the full expression. Thus we immediately see that the virtual process taken into account by this term is of second-order and contains the off-diagonal matrix element $M_{12}(\boldsymbol{q})$ in fourth power. Moreover, in accordance with the definition (8.32), Γ is proportional to $N_p(1 + N_q)$, which naturally follows from the fact that this virtual process involves one-phonon absorption and one-phonon emission (see the middle diagram in Fig. 8.7). The quadratic addition S_Q to the full Huang–Rhys factor is in general complex, which leads to a (small) dispersive contribution to the ZPL Lorentzian.

Analytical expressions for the ZPL width, similar to (8.35), have been derived earlier in different models of the quadratic electron–phonon coupling [51–55]. In the present approach, however, we are interested in the full temporal dynamics of the optical response.

The calculated polarization amplitude shown in Fig. 8.8 (solid curve) has both characteristic features discussed above: A quick initial decoherence and a long-time decay. Remarkably, this full time dependence cannot be reproduced by adding simply by hand the exponential decay to the IBM result (see dotted and dashed curves, respectively). The calculation is done for an InAs spherical QD with parabolic potentials, using only one adjustable parameter, namely a Gauss decay length of the

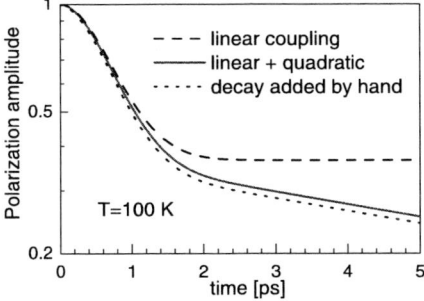

Fig. 8.8. Polarization amplitude calculated for an InAs QD with at $T = 100$ K with linear and quadratic coupling to acoustic phonons. For the dotted curve, a finite decay $1/\Gamma = 12.3$ ps has been added by hand to the linear result

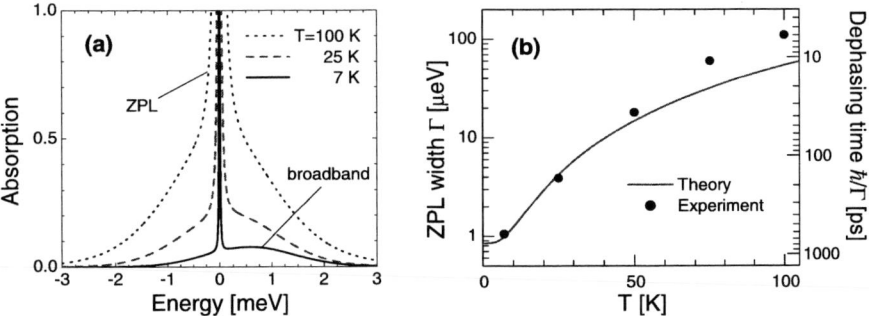

Fig. 8.9. (a) Absorption spectra of an InAs QD at different temperatures. The ZPL transition energy is taken as zero of energy. (b) Calculated ZPL width/dephasing time compared with the experimental results by Borri et al. [22] (*circles*)

confinement functions. Here, it was taken equal to 3.3 nm (the same for electron and hole). In the quadratic coupling (8.29), we account for three electron and three hole excited states having excitation energies of 59 meV and 6 meV, respectively. The parameters of acoustic phonons are given in [2].

The absorption spectrum of the ground exciton state is shown in Fig. 8.9(a) for different temperatures. A nearly Lorentzian ZPL is superimposed to the phonon broadband. At elevated temperatures, the spectrum gets more symmetric and the broadband fraction becomes larger. The width of the broadband does not change much, but the ZPL width grows very quickly. Its temperature dependence shown in Fig. 8.9(b) is in good agreement with the experimental data (circles) extracted from the FWM polarization decay [22]. Note that for the ground exciton state, the calculated phonon-induced linewidth Γ strictly vanishes at zero temperature. Thus, for an adequate comparison with the experiment, a radiative decay of 0.85 µeV is added to the Γ.

Cumulant Expansion

In typical QDs, virtual processes are the major mechanisms of the dephasing of the ground exciton state, since the energy distances to higher confined states are much larger than the energy range of acoustic phonons coupled to the QD. In contrast, the energy distance between excited states in QDs and QD molecules can be small thus allowing real phonon-assisted transitions between the levels to come into play, too [48, 49, 56].

The cumulant expansion [20, 57] which provides the exact solution of both the independent boson and the quadratic coupling model is now used to solve the general Hamiltonian (8.24)–(8.25). Historically, the cumulant expansion is not a very popular diagram technique compared to, e.g., the self-energy approach. However, it turns out to be the best method to calculate the phonon-induced exciton dephasing in nanostructures and even in bulk semiconductors [58, 59].

In case of the multilevel problem (8.24)–(8.25), the full linear polarization takes the form

$$P(t) = \sum_n |\mu_n|^2 e^{-i\omega_n t} P_n(t), \tag{8.36}$$

where $\mu_n = \mu_{cv} \int d\mathbf{r}\, \Psi_n(\mathbf{r}, \mathbf{r})$ are the exciton matrix elements of the interband dipole moment operator $\hat{\mu}$. The polarization $P_n(t)$ due to exciton state $|n\rangle$ can be written as an infinite perturbation series

$$P_n(t) = \left\langle \langle n|\mathcal{T} \exp\left\{-i \int_0^t d\tau\, V(\tau)\right\}|n\rangle \right\rangle = 1 + P_n^{(1)} + P_n^{(2)} + \cdots, \tag{8.37}$$

where the expectation value (external brackets) is taken over the phonon system in thermal equilibrium, and $V(t)$ is written in the interaction representation.

The perturbation series in (8.37) can be visualized by drawing diagrams which are standard in the electron–phonon problem. They are shown in Fig. 8.10 up to second-order. Note that the phonon Green's function (dashed lines) depends on four exciton state indices. Some of them are obviously different if phonon-assisted processes between different exciton states are involved. The first-order diagram (1) is just what is accounted for in Fermi's golden rule to describe real transitions [60]. The second-order diagrams (2b) and (2c) give rise to virtual transitions which have been treated by the quadratic coupling model.

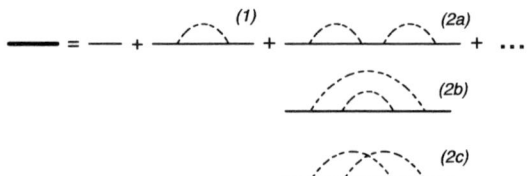

Fig. 8.10. Diagram representation of the perturbation series for the full exciton polarization $P_n(t)$, up to second-order

To sum up the perturbation series, we introduce for each state the cumulant function $K_n(t)$ defined as

$$P_n(t) = e^{K_n(t)} \tag{8.38}$$

and calculate it using the series

$$K_n(t) = P_n^{(1)}(t) + P_n^{(2)}(t) - \frac{1}{2} P_n^{(1)}(t) P_n^{(1)}(t) + \cdots \tag{8.39}$$

reconstructed from the plain expansion for $P_n(t)$, (8.37).

If all off-diagonal elements of the exciton–phonon interaction are neglected, the cumulant expansion ends in first-order: The contribution of all higher terms of the polarization is exactly canceled in the cumulant, (8.39), by lower order products. This result permits, in particular, the exact solution of the IBM. The inclusion of the nondiagonal interaction leads to nonvanishing terms in the cumulant in any order. Still, there is a partial cancellation of diagrams which provides a large-time asymptotics of the cumulant, $K_n(t) \to -S_n - i\Delta_n t - \Gamma_n t$, where S_n, Δ_n, and Γ_n, are, respectively, Huang–Rhys factor, polaron shift, and ZPL width of state $|n\rangle$. For example, diagrams (1), (2b), and (2c) in Fig. 8.10 behave linear in time at $t \gg R/v_l$ (R is the QD radius, v_l is the sound velocity), while diagram (2a) has a leading t^2 behavior. In the cumulant, however, this quadratic term is canceled exactly by the square of diagram (1), $P_n^{(1)} P_n^{(1)}/2$. Thus, taking into account higher orders in the cumulant expansion leads only to higher-order corrections in the ZPL width (as well as in S_n and Δ_n).

The broadening of the ZPL is exclusively due to the nondiagonal interaction. Real transitions already contribute in first-order; see the right diagram in Fig. 8.7 where only one phonon participates. The cumulant expansion taking them into account in first-order reproduces exactly Fermi's golden rule. The calculated ZPL width is quadratic in the off-diagonal element $M_{12}(q)$; thus it is generally stronger than that due to virtual transitions. However, it exhibits a sharp maximum at the energy of phonons coupled to the QD (which is of order of $\hbar v_l/R$) and then drops quickly [together with $M_{12}(q)$] as the level distance $\omega_2 - \omega_1$ increases; see the dashed curve in Fig. 8.11. Such a sharp dependence on the level distance in case of real transitions is a manifestation of the so-called phonon bottleneck effect in QDs.

Virtual transitions, in turn, have a relatively weak dependence on the level distance; see the dash-dotted curve in Fig. 8.11. In fact, in the lowest (second-) order, the ZPL width is proportional to $1/(\omega_2 - \omega_1)^2$. Also, the contribution of virtual transitions is generally smaller as it is determined by the fourth power of $M_{12}(q)$. However, virtual transitions are always present in QDs and lead to a nonvanishing broadening of the ZPL everywhere.

In the full solution of the multilevel problem (Fig. 8.11, full curve) we account for both real and virtual transitions on an equal footing and proceed up to second-order in the cumulant, thus covering both limiting cases where real or virtual transitions dominate. Note that the interplay between real and virtual transitions is not simply additive in this case, as there is also a second-order correction due to real transitions which slightly reduces the ZPL width.

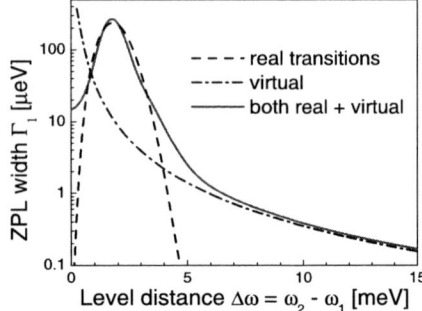

Fig. 8.11. Broadening of the exciton ground state ZPL as a function of the exciton level distance in a spherical InAs QD, calculated at $T = 100$ K with account for only real transitions (Fermi's golden rule), only virtual transitions, and for both real and virtual transitions up to second-order in the cumulant

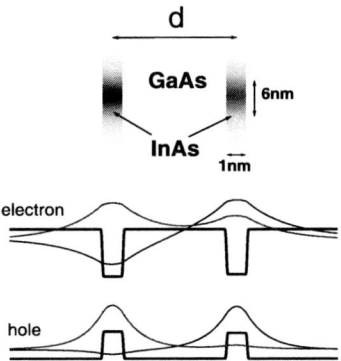

Fig. 8.12. Schematic view of an InGaAs QDM and the wave functions of the lowest two states for electron and hole with heterostructure potentials along the growth direction

8.5.3 Application to Coupled Quantum Dots

Now that we have the full solution of the multilevel problem at hand, which accounts for both real and virtual phonon-assisted transitions, we are able to study the dephasing in coupled quantum dots, or QD molecules (QDMs) [48].

We consider vertically coupled QDs, schematically shown in Fig. 8.12. To describe excitonic states in such a molecule, we restrict ourselves to a four-level model, taking into account the two lowest electron and hole states, also shown in Fig. 8.12. Without Coulomb interaction the electron-hole pair state is a direct product of the one-particle states. Thus we have a basis of four states. Then we include the Coulomb interaction and diagonalize a 4×4 Hamiltonian. To be closer to the experimental situation we take a slightly asymmetric QDM, where the potential of the left dot is 2% deeper than those of the right one.

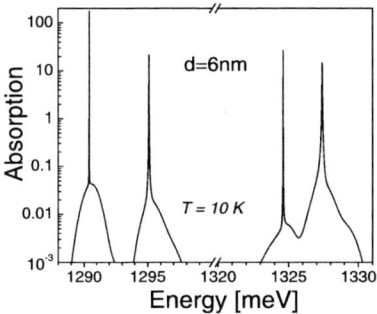

Fig. 8.13. Absorption spectrum (log scale) of a $d = 6$ nm QDM calculated in the model of four s-shell excitonic states

We assume both dots to have the same cylindrical form with height $L_z = 1$ nm (in the growth direction) adjusted from the comparison with experimentally measured transition energies [23, 61, 62] (taking 92% of In concentration). Given that s-, p-, d-, and f-shells in the luminescence spectra of QDMs have nearly equidistant positions [63], the in-plane confining potentials are taken parabolic with Gaussian localization lengths of carriers adjusted to $l_e = 6.0$ nm and $l_h = 6.5$ nm. The electronic band parameters are taken from [64] and the acoustic phonon parameters are the same as used previously [2].

In the absorption spectrum of a QDM having the (center-to-center) interdot distance $d = 6$ nm, we see four finite-width Lorentzian lines on the top of broadbands, Fig. 8.13. The width of the broadband is of the order of the typical energy of phonons participating in the transitions, which is about 2 meV. If two levels come close to each other and the broadbands start to overlap, the ZPLs get considerably wider, due to real phonon-assisted transitions between neighboring levels (see the upper two levels). Such a mechanism of dephasing is very efficient if the level spacing is small; the latter, in turn, is controlled in QDMs by the tunneling which induces a level repulsion at short interdot distances.

The dependence of the absorption on the interdot distance d, Fig. 8.14, clearly demonstrates this effect. However, there is another effect that leads to a quite unexpected result: Coulomb anticrossing. As d increases, the two higher levels come close to each other and should exchange phonons more efficiently. Nevertheless, when the anticrossing is reached at around $d = 8.5$ nm, the ZPL width of the highest level suddenly drops and never restores at larger d.

This can be seen even more clearly from the separate plots in Fig. 8.15 of exciton energies (with oscillator strengths given by the circle area) and ZPL widths as functions of d. We can see that, due to the Coulomb anticrossing, there is a sudden switch of the symmetry of states $|3\rangle$ and $|4\rangle$, so that the exciton–phonon matrix element between them drops quickly and their ZPL widths become very small. Real transitions between these two states are suppressed by symmetry, and the dephasing is due only to virtual transitions into rather distant lower states $|1\rangle$ and $|2\rangle$.

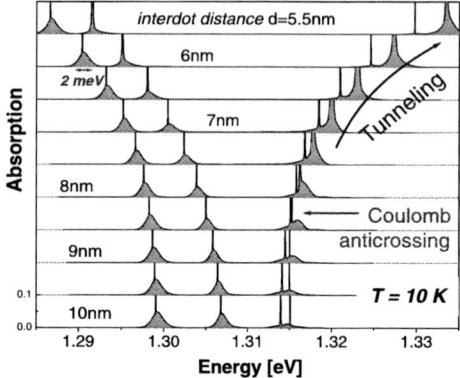

Fig. 8.14. Absorption spectrum (linear scale) of an asymmetric InGaAs QDM calculated at $T = 10\,\text{K}$ for different interdot distances d. The peaks of the ZPLs are truncated

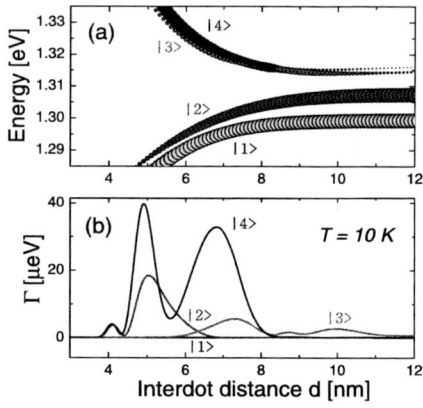

Fig. 8.15. Exciton energies (**a**) and dephasing rates (**b**) of the four s-shell excitonic states in an InGaAs QDM as functions of the interdot distance d, at $T = 10\,\text{K}$. The oscillator strength is given in panel (**a**) by the circle area

There are also oscillations in Γ clearly seen in Fig. 8.15(b) for levels $|2\rangle$ and $|4\rangle$ in the region between $d = 4\,\text{nm}$ and $8\,\text{nm}$. For an explanation, note that the matrix elements $M_{nm}(\mathbf{q})$, (8.26), are Fourier transforms of the electron (hole) probabilities. When located in different QDs, they carry a factor of $\exp(iqd)$. Since the typical phonon momentum participating in real transitions is given by $\hbar/L_z$, one could expect that $\Gamma(d)$ has maxima spaced by a length of order L_z.

8.6 Summary

We developed a microscopic theory of the dephasing of optical excitations in QDs and QDMs due to acoustic phonons. The phonon spectrum is strongly affected by the change in material properties as well as intrinsic strain in nanostructures. The inclusion of this improved description compared to assuming bulk phonons leads to characteristic oscillations in the polarization, and to a modulation of the phonon-induced broadband in the absorption. The broadening of the zero phonon line (the long-time decay of the polarization) is due to real and virtual phonon-assisted transitions. They can be treated separately by, respectively, Fermi's golden rule and the exactly solvable quadratic coupling model. Going beyond these two approaches we calculate the full time-dependent polarization and absorption using the cumulant expansion which turns out to be a very powerful tool in the dephasing problem. For excited states in single and coupled QDs, a complicated interplay between level spacing and phonon coupling leads to a rich dephasing scenario.

References

1. B. Krummheuer, V.M. Axt, T. Kuhn, Phys. Rev. B **65**, 195313 (2002)
2. E.A. Muljarov, R. Zimmermann, Phys. Rev. Lett. **93**, 237401 (2004)
3. M. Stroscio, M. Dutta, *Phonons in Nanostructures* (Cambridge University Press, Cambridge, 2001)
4. V.A. Schukin, N.N. Ledentsov, D. Bimberg, *Epitaxy of Nanostructures* (Springer, New York, 2003)
5. I. Vurgaftman, J.R. Meyer, L.R. Ram-Mohan, J. Appl. Phys. **89**, 5815 (2001)
6. A. Wright, J. Appl. Phys. **82**, 2833 (1997)
7. F. Grosse, J. Neugebauer, Phys. Rev. B **63**, 085207 (2001)
8. C. Kittel, *Introduction to Solid State Physics*, 8th edn. (Wiley, New York, 2005)
9. F. Birch, Phys. Rev. **71**, 809 (1947)
10. C. Trallero-Giner, F. Comas, F. García-Moliner, Phys. Rev. B **50**, 1755 (1994)
11. X. Gonze, J.-M. Beuken, R. Caracas, F. Detraux, M. Fuchs, G.-M. Rignanese, L. Sindic, M. Verstraete, G. Zerah, F. Jollet, M. Torrent, A. Roy, M. Mikami, P. Ghosez, J.-Y. Raty, D. Allan, Comput. Mater. Sci. **25**, 478 (2002)
12. X. Gonze, G.-M. Rignanese, M. Verstraete, J.-M. Beuken, Y. Pouillon, R. Caracas, F. Jollet, M. Torrent, G. Zerah, M. Mikami, P. Ghosez, M. Veithen, J.-Y. Raty, V. Olevano, F. Bruneval, L. Reining, R. Godby, G. Onida, D. Hamann, D. Allan, Z. Kristallogr. **220**, 558 (2005)
13. The ABINIT code is a common project of the Université Catholique de Louvain, Corning Incorporated, and other contributors, v 4.6.5. http://www.abinit.org
14. M. Fuchs, M. Scheffler, Comput. Phys. Commun. **116**, 1 (1999)
15. F. Grosse, R. Zimmermann, *Physics of Semiconductors*, ed. by W. Jantsch, F. Schäffler, AIP Conf. Proc. **893**, 1007 (2007)
16. F. Grosse, R. Zimmermann, Phys. Rev. B **75**, 235320 (2007)
17. C.G. Van de Walle, Phys. Rev. B **39**, 1871 (1989)
18. H. Lamb, Proc. Lond. Math. Soc. **13**, 189 (1882)
19. T. Takagahara, Phys. Rev. B **60**, 2638 (1999)
20. G. Mahan, *Many-Particle Physics* (Plenum, New York, 1990)
21. G. Mannarini, R. Zimmermann, Phys. Rev. B **73**, 115325 (2006)
22. P. Borri, W. Langbein, S. Schneider, U. Woggon, R.L. Sellin, D. Ouyang, D. Bimberg, Phys. Rev. Lett. **87**, 157401 (2001)

23. P. Borri, W. Langbein, U. Woggon, M. Schwab, M. Bayer, S. Fafard, Z. Wasilewski, P. Hawrylak, Phys. Rev. Lett. **91**, 267401 (2003)
24. P. Borri, W. Langbein, E.A. Muljarov, R. Zimmermann, Phys. Status Solidi (b) **243**, 3890 (2006)
25. C. Kammerer, G. Cassabois, C. Voisin, C. Delalande, P. Roussignol, A. Lemaître, J.M. Gérard, Phys. Rev. B **65**, 033313 (2001)
26. C. Kammerer, C. Voisin, G. Cassabois, C. Delalande, P. Roussignol, F. Klopf, J.P. Reithmaier, A. Forchel, J.M. Gérard, Phys. Rev. B **66**, 041306 (2002)
27. M. Bayer, A. Forchel, Phys. Rev. B **65**, 041308 (2002)
28. K. Leosson, D. Birkedal, I. Magnusdottir, W. Langbein, J. Hvam, Physica E **17**, 1 (2003)
29. I. Favero, G. Cassabois, R. Ferreira, D. Darson, C. Voisin, J. Tignon, C. Delalande, G. Bastard, P. Roussignol, J.M. Gérard, Phys. Rev. B **68**, 233301 (2003)
30. L. Besombes, K. Kheng, L. Marsal, H. Mariette, Phys. Rev. B **63**, 155307 (2001)
31. S. Moehl, F. Tinjod, K. Kheng, H. Mariette, Phys. Rev. B **69**, 245318 (2004)
32. K. Edamatsu, T. Itoh, K. Matsuda, S. Saikan, Phys. Rev. B **64**, 195317 (2001)
33. K. Sebald, P. Michler, T. Passow, D. Hommel, G. Bacher, A. Forchel, Appl. Phys. Lett. **81**, 2920 (2002)
34. K. Ikeda, Y. Ogawa, F. Minami, S. Kuroda, K. Takita, Phys. Status Solidi (c) **3**, 874 (2006)
35. E. Peter, J. Hours, P. Senellart, A. Vasanelli, A. Cavanna, J. Bloch, J.M. Gérard, Phys. Rev. B **69**, 041307 (2004)
36. N.K.K. Ikeda, F. Minami, Microelectron. J. **36**, 247 (2005)
37. S. Sanguinetti, E. Poliani, M. Bonfanti, M. Guzzi, E. Grilli, M. Gurioli, N. Koguchi, Phys. Rev. B **73**, 125342 (2006)
38. R. Zimmermann, E. Runge, in *Proc. 26th ICPS Edinburgh, IOP Conf. Series 171*, ed. by A.R. Long, J.H. Davies (IOP Publishing, Bristol, 2002). Paper M 3.1
39. J. Förstner, C. Weber, J. Danckwerts, A. Knorr, Phys. Rev. Lett. **91**, 127401 (2003)
40. P. Palinginis, H. Wang, S.V. Goupalov, D.S. Citrin, M. Dobrowolska, J.K. Furdyna, Phys. Rev. B **70**, 073302 (2004)
41. K.J. Ahn, J. Förstner, A. Knorr, Phys. Rev. B **71**, 153309 (2005)
42. L. Jacak, J. Krasnyj, W. Jacak, R. Gonczarek, P. Machnikowski, Phys. Rev. B **72**, 245309 (2005)
43. V.N. Stavrou, X. Hu, Phys. Rev. B **72**, 075362 (2005)
44. V.N. Stavrou, X. Hu, Phys. Rev. B **73**, 205313 (2006)
45. G. Mannarini, R. Zimmermann, Phys. Rev. B **73**, 115325 (2006)
46. P. Machnikowski, Phys. Rev. Lett. **96**, 140405 (2006)
47. A. Vagov, V.M. Axt, T. Kuhn, W. Langbein, P. Borri, U. Woggon, Phys. Rev. B **70**, 201305 (2004)
48. E.A. Muljarov, T. Takagahara, R. Zimmermann, Phys. Rev. Lett. **95**, 177405 (2005)
49. E.A. Muljarov, R. Zimmermann, Phys. Status Solidi (b) **243**, 2252 (2006)
50. K. Huang, A. Rhys, Proc. R. Soc. Lond. A **204**, 406 (1950)
51. G.F. Levenson, Phys. Status Solidi (b) **43**, 739 (1971)
52. I. Osad'ko, Sov. Phys. Solid State **14**, 2522 (1973)
53. D. Hsu, J. Skinner, J. Chem. Phys. **81**, 1604 (1984)
54. S.V. Goupalov, R.A. Suris, P. Lavallard, D. Citrin, IEEE J. Sel. Topics Quantum Electron. **8**, 1009 (2002)
55. V. Hizhnyakov, H. Kaasik, I. Sildos, Phys. Status Solidi B **234**, 644 (2002)
56. E.A. Muljarov, R. Zimmermann, *Physics of Semiconductors*, ed. by W. Jantsch, F. Schäffler. AIP Conf. Proc. **893**, 915 (2007)

57. R. Kubo, J. Phys. Soc. Jpn. **17**, 1100 (1962)
58. D. Dunn, Can. J. Phys. **53**, 321 (1975)
59. B. Gumhalter, Phys. Rev. B **72**, 165406 (2005)
60. U. Bockelmann, Phys. Rev. B **50**, 17271 (1994)
61. M. Bayer, P. Hawrylak, K. Hinzer, S. Fafard, M. Korkusinski, Z. Wasilewski, O. Stern, A. Forchel, Science **291**, 451 (2001)
62. G. Ortner, M. Bayer, Y. Lyanda-Geller, T.L. Reinecke, A. Kress, J.P. Reithmaier, A. Forchel, Phys. Rev. Lett. **94**, 157401 (2005)
63. S. Fafard, M. Spanner, J.P. McCaffrey, Z.R. Wasilewski, Appl. Phys. Lett. **76**, 2268 (2000)
64. *Semiconductors*, Landolt-Börnstein New Series, ed. by O. Madelung, vols. 17b, 22a (Springer, Berlin, 1986)

9

Theory of the Optical Response of Single and Coupled Semiconductor Quantum Dots

C. Weber, M. Richter, S. Ritter, and A. Knorr

Abstract. Due to their quasi-zero-dimensional structure, quantum dots show optical properties which are different from those of nanostructures with spatial confinement in less than three dimensions. In this chapter, the theory of both the linear optical properties and nonlinear dynamics of semiconductor quantum dots is discussed. The main focus is on the experimentally accessible quantities such as absorption/luminescence and pump-probe spectra. The results are calculated for single and coupled quantum dots (Förster coupling) as well as quantum dot ensembles. The focus is on obtaining a microscopic understanding of the interactions of optically excited quantum dot electrons with the surrounding crystal vibrations (electron–phonon coupling). The discussed interactions are important for applications in, e.g., quantum information processing and laser devices.

9.1 Introduction

This chapter discusses the linear and nonlinear optical properties of semiconductor quantum dots (QDs). QD electrons experience different interactions due to their environment: coupling to phonons and photons. Due to the discrete energy states caused by the three-dimensional spatial confinement, the optical dephasing witnessed in the linear spectra as well as the nonlinear dynamics is quite different compared to that in bulk semiconductors [1].

We start by considering the optical properties of a single QD, both in the linear regime with absorption and resonance fluorescence spectra as well as in the nonlinear regime with Rabi oscillations and differential transmission spectra. The concept of THz acoustoluminescence in QDs is discussed. In a next step, we generalize our approach to two coupled QDs and discuss their dynamics with respect to Coulomb and radiative coupling. Finally, we present the nonlinear response of an ensemble of QDs as well as the optical properties of a set of coupled QDs.

9.2 Theory

This section focuses on the theoretical foundations. After a discussion of the QD model, we present the interaction Hamiltonians. Then we introduce the formalisms to describe the optical response.

9.2.1 Quantum Dot Model

For the QDs treated in this chapter, we use the envelope function ansatz for the wave functions: a product ansatz of the Bloch part (to take into account the periodic potential) $u_{k\approx 0}(x)$ at the band edge and an envelope function $\xi(x)$ (to describe the additional confinement of the QD) [2]:

$$\varphi(x) = u_{k\approx 0}(x)\,\xi(x), \tag{9.1}$$

where in effective mass approximation $\xi(x)$ (x is the three-dimensional spatial coordinate) is a solution of the eigenvalue equation

$$\left[-\frac{\hbar^2}{2m^*}\Delta + U(x)\right]\xi(x) = (E - E_{k\approx 0})\xi(x) \tag{9.2}$$

with the effective mass m^* and the mesoscopic confinement potential $U(x)$ (for models going beyond the above ansatz; see, e.g., [3, 4]). In this work, we apply different models for $\xi(x)$.

Spherical Harmonic Oscillator Model [1]: The advantage of the spherically symmetric three-dimensional harmonic oscillator model, described by the potential $U(x) = 1/2\,m^*\omega^2|x|^2$, is its simplicity. The ground state envelope functions are given by

$$\xi(x) = \left(\frac{m^*\omega}{\pi\hbar}\right)^{\frac{3}{4}} \exp\left(-\frac{m^*\omega}{2\hbar}|x|^2\right), \tag{9.3}$$

where the frequency ω is defined via the quantization energy E_{qt} [see Fig. 9.1(a)] by $E_{qt} = 3/2\,\hbar\omega$.

Harmonic Disc Model [1]: A more realistic model is the disc-like two-dimensional harmonic oscillator model with infinite confinement in the growth direction, described by $U(x) = 1/2\,m^*\omega^2(x^2 + y^2) + V_0[\theta(z - L/2) + \theta(-z - L/2)]$, $V_0 \to \infty$, where L is the extension in the growth direction. This confinement potential yields the ground state envelope functions

$$\xi(x) = \left(\frac{m^*\omega}{\pi\hbar}\right)^{\frac{1}{2}} \exp\left[-\frac{m^*\omega}{2\hbar}(x^2 + y^2)\right]\sqrt{\frac{2}{L}}\cos\left(\frac{\pi}{L}z\right), \tag{9.4}$$

with the quantization energy $E_{qt} = \hbar\omega + (\hbar^2\pi^2/2m^*L^2)$.

Eight-band $k\cdot p$ Wave Functions: More realistic QD envelope functions and eigenenergies taking into account strain effects and band mixing can be calculated within the eight-band $k\cdot p$ theory [5]. For this approach, the product ansatz (9.1) is generalized to include the two energetically lowest conduction band and the six energetically highest valence band sublevels.

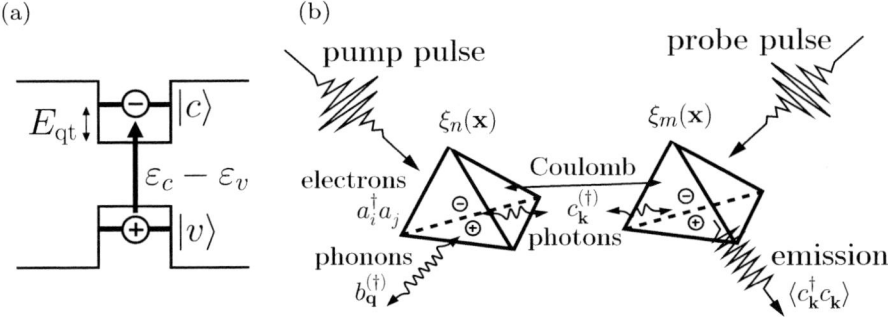

Fig. 9.1. (a) A two-level QD with one conduction band $|c\rangle$ and one valence band $|v\rangle$ level (transition energy $\varepsilon_c - \varepsilon_v$, quantization energy E_{qt}). (b) General experimental and theoretical setup of single and coupled QD optical response

9.2.2 Hamiltonian

In the calculations, the electron picture is applied. We consider parameters suitable for two general types of QD models: (i) self-assembled InAs/GaAs QDs in the strong confinement regime considered, e.g., in [6, 7] and (ii) AlAs/GaAs interface QDs in the weak confinement regime considered, e.g., in [8]. In the first case, excitons can be considered as a small perturbation and neglected [9]; in the second case, the electron picture must be viewed as a theoretical simplification, especially in nonlinear optics [10].

Figure 9.1(b) shows the general setup for considering the optical response of single and coupled QDs. Via an external excitation, electrons are excited from the valence into the conduction band [see Fig. 9.1(a) for a two-level system] or between conduction band levels. Here, $a_{i\lambda}^\dagger$ ($a_{i\lambda}$) denotes the fermionic creation (annihilation) operator of a confined electron in the state $|i\lambda\rangle$ with energy $\varepsilon_{i\lambda}$, where i denotes the level and λ the sublevel index of the QD. The electrons interact with phonons [with the bosonic creation (annihilation) operator $b_q^\dagger$ (b_q), three-dimensional quasi-momentum q, and energy dispersion ω_q] and photons [with the bosonic creation (annihilation) operator $c_{k\sigma}^\dagger$ ($c_{k\sigma}$), three-dimensional momentum k, polarization index σ, and energy dispersion ω_k]. The response is considered either by means of a probe pulse or the photon emission spectrum.

Electron–Phonon Coupling

For the interaction of the electronic system with phonons, we restrict our considerations mainly to longitudinal acoustic (LA) phonons coupling via the deformation potential [11]. It has been shown that for interband transitions, this coupling dominates over the polar coupling to optical as well as the piezoelectric coupling to acoustic phonons [7]. We consider only bulk phonon modes.[1] The interaction Hamiltonian is

---
[1] Note that this approximation cannot be applied to colloidal QDs where the confined phonon modes must be considered.

given by [11, 12]

$$H_{\text{el-ph}} = \sum_q \sum_{i\lambda, j\mu} g^q_{i\lambda, j\mu} a^\dagger_{i\lambda} a_{j\mu} (b_q + b^\dagger_{-q}), \qquad (9.5)$$

with the deformation potential coupling element

$$g^{q(\text{LA})}_{i\lambda, j\mu} = \sqrt{\frac{\hbar |q|}{2\rho_m c_L V}} D_i F^q_{i\lambda, i\mu} \delta_{ij}, \qquad (9.6)$$

where ρ_m is the mass density, c_L the longitudinal velocity of sound in media, D_i the band deformation potential, V the quantization volume, and $F^q_{i\lambda, j\mu} = \int d^3 x \, \xi^*_{i\lambda}(x) \times e^{i q \cdot x} \xi_{j\mu}(x)$ the form factor. For the discussion in Sect. 9.3.1, the polar Fröhlich coupling element to longitudinal optical (LO) phonons is given as well:

$$g^{q(\text{LO})}_{i\lambda, j\mu} = \sqrt{\frac{e^2 \hbar \omega_{\text{LO}}}{2\varepsilon_0 V} \left(\frac{1}{\varepsilon_\infty} - \frac{1}{\varepsilon_s}\right)} \frac{1}{|q|} F^q_{i\lambda, i\mu} \delta_{ij} \qquad (9.7)$$

with the constant LO phonon energy $\hbar \omega_{\text{LO}}$ and the static and high-frequency dielectric constants ε_s and ε_∞, respectively.

Interaction with the Quantized Electromagnetic Field

The interaction of the electronic system with the quantized electromagnetic field is given in dipole approximation by [13, 14]

$$H_{\text{el-light}} = -\sum_{k\sigma} \sum_{i\lambda, j\mu} M^{k\sigma}_{i\lambda, j\mu} a^\dagger_{i\lambda} a_{j\mu} (c_{k\sigma} - c^\dagger_{k\sigma}) \qquad (9.8)$$

with the coupling element

$$M^{k\sigma}_{i\lambda, j\mu} = i \sqrt{\frac{\hbar \omega_k}{2\varepsilon_0 V}} e_{k\sigma} \cdot (d_{ij} X^{i\lambda}_{j\mu} + D^i_{\lambda\mu} \delta_{ij}), \qquad (9.9)$$

where $d_{ij} = -e \Omega^{-1}_{\text{ec}} \int_{\text{ec}} d^3 x \, u^*_i(x) x u_j(x)$ ($e > 0$) is the microscopic interband (integration over elementary cell Ω_{ec}) and $D^i_{\lambda\mu} = -e \sum_n \xi^*_{i\lambda}(X_n) X_n \xi_{i\mu}(X_n)$ (X_n: lattice vector) the mesoscopic intraband dipole moment. The factor $X^{i\lambda}_{j\mu} = \sum_n \xi^*_{i\lambda}(X_n) \xi_{j\mu}(X_n)$ describes the overlap of the envelope functions between different sublevel states, while $e_{k\sigma}$ denotes the polarization vector of the light field.

Interaction with an External Optical Field

The external optical excitation of the system is described by a coherent (classical) light source in dipole approximation [15, 16]:

9 Theory of the Optical Response of Single and Coupled Semiconductor   193

$$H_{\text{el-light}} = -\sum_{i\lambda,j\mu} \boldsymbol{E}(t) \cdot (\boldsymbol{d}_{ij} X^{i\lambda}_{j\mu} + \boldsymbol{D}^{i}_{\lambda\mu}\delta_{ij}) a^{\dagger}_{i\lambda} a_{j\mu}. \tag{9.10}$$

As a source, a modulated Gaussian or cw excitation is considered, $\boldsymbol{E}(t) = \tilde{\boldsymbol{E}}(t)\cos(\omega_L t)$, where for a Gaussian pulse $\tilde{\boldsymbol{E}}(t) = \boldsymbol{E}_0 \exp[-(t-t_0)^2/\tau^2]$ with the temporal offset t_0, the length τ, and the frequency ω_L is used. The pulse area is defined via the envelope, $\Theta = \int dt' \tilde{\Omega}(t')$, with the Rabi frequency $\tilde{\Omega}(t) = \boldsymbol{d}_{ij} X^{i\lambda}_{j\mu} \cdot \tilde{\boldsymbol{E}}(t)/\hbar$ for an interband transition $|i\lambda\rangle \rightarrow |j\mu\rangle$.

Coulomb Interaction

The electrons interact via the Coulomb interaction, given by the Hamiltonian [15, 17]

$$H_c = \frac{1}{2}\sum_{i,j,k,l} V_{ijkl} a^{\dagger}_i a^{\dagger}_j a_l a_k, \tag{9.11}$$

where the Coulomb matrix element is given by

$$V_{ijkl} = \frac{e^2}{4\pi\varepsilon_0\varepsilon_r} \int d^3x \int d^3x' \varphi^*_i(\boldsymbol{x}) \varphi^*_j(\boldsymbol{x}') \frac{1}{|\boldsymbol{x}-\boldsymbol{x}'|} \varphi_k(\boldsymbol{x}) \varphi_l(\boldsymbol{x}'). \tag{9.12}$$

Here, ε_r is the dielectric constant of the medium, and the sublevel index λ is included in $|i\rangle \equiv |i\lambda\rangle$.

To calculate the dynamics of two coupled QDs (each modeled as an interband two-level system with one conduction and one valence level $|c\rangle$ and $|v\rangle$, respectively) which have no electronic overlap, an expansion of the potential is performed [17]: (i) a long-range expansion about a reference point of each QD, varying on a mesoscopic scale and neglecting the variation on the scale of the elementary cell—this yields level diagonal contributions in the Hamiltonian $H_{cc} = \sum_{i>j} V^{ij}_{cc} a^{\dagger}_{c_i} a^{\dagger}_{c_j} a_{c_j} a_{c_i}$ and $H_{cv} = \sum_{i \neq j} V^{ij}_{cv} a^{\dagger}_{c_i} a^{\dagger}_{v_j} a_{v_j} a_{c_i}$ [$a^{(\dagger)}_{v_i}$ ($a^{(\dagger)}_{c_i}$) denotes the operator of an electron in QD i in the state $|v\rangle$ ($|c\rangle$)]; and (ii) a short-range expansion about an arbitrary lattice vector, taking into account the microscopic variation of the QD—this yields nondiagonal contributions $H_F = \sum_{i \neq j} V^{ij}_F a^{\dagger}_{c_i} a^{\dagger}_{v_j} a_{c_j} a_{v_i}$. On the dipole–dipole level, the level diagonal elements correspond to an electrostatic energetic shift of the system (biexcitonic shift $V_{bs} = V_{cv} - V_{cc}$), while the nondiagonal elements, the so-called Förster coupling elements V_F [18, 19], correspond to an excitation transfer between the different QDs. We restrict the considerations above to one electron in each QD.

9.2.3 Mathematical Formalisms

In order to investigate the optical properties of QDs, the microscopic polarizations $\langle a^{\dagger}_i a_j \rangle$ ($i \neq j$) and densities $\langle a^{\dagger}_i a_i \rangle$, where $\langle a^{\dagger}_i a_j \rangle = \text{tr}(a^{\dagger}_i a_j \rho)$ (density matrix ρ), must be calculated. The absorption spectrum, for example, is calculated via the complex susceptibility of the quantum system, $\alpha(\omega) \propto \omega \,\text{Im}(P(\omega)/E(\omega))$, where

$P(t) = n_D \sum_{i,j} d_{ij} \langle a_i^\dagger a_j \rangle(t)$ (constant QD number density n_D) is the macroscopic polarization for an interband system. To describe the dynamics of a quantum mechanical system in the Schrödinger picture, the dynamics of the density matrix $\rho(t)$ has to be determined. For complex systems including many electrons, phonons, and/or photons, the complete dynamics of the density matrix can be treated only in rare cases (see, e.g., [20] for a numerically exact treatment). In the other cases, one has to apply perturbational schemes in order to describe the dynamics appropriate for the experimentally relevant observables. Analogous to the Schrödinger picture, the dynamics of the system in the Heisenberg picture can be calculated for the operators $a_i^\dagger(t)$ and $a_j(t)$.

Correlation Expansion

One of the approximation schemes is the so-called correlation expansion employed in the density matrix formalism [9, 15, 21]. The main idea of this scheme is that a distinct set of observables determines the complete dynamics of the system, thus truncating the infinite many-body hierarchy. The correlation expansion is especially appropriate in the case of arbitrarily strong laser fields and the weak coupling regime for the electron–phonon (and Coulomb) interaction.

As an example, for a single-particle observable, we have to calculate the equation of motion

$$\partial_t \langle a_i^\dagger a_j \rangle \equiv \partial_t \mathrm{tr}\left(a_i^\dagger a_j \rho\right) = \frac{i}{\hbar} \mathrm{tr}\left\{\rho\left[H, a_i^\dagger a_j\right]_-\right\}. \tag{9.13}$$

If the Hamiltonian contains many-particle interactions, the temporal derivative of the single-particle observable couples to multiple-particle observables; e.g., in the case of Coulomb interaction, it couples to $\langle a_i^\dagger a_j^\dagger a_k a_l \rangle$, in the case of electron–phonon interaction, it couples to phonon-assisted observables $\langle a_i^\dagger b_q^{(\dagger)} a_j \rangle$, etc. This usually ends up in an infinite hierarchy of equations of motion. To break this hierarchy, it is necessary to focus on only a small set of observables $\mathcal{O}$ and assume that the density matrix takes the canonical form $\rho(t) = Z^{-1} \exp[-\sum_{\hat{O} \in \mathcal{O}} \lambda_{\hat{O}}(t) \hat{O}]$ with $Z = \mathrm{tr}\{\exp[-\sum_{\hat{O} \in \mathcal{O}} \lambda_{\hat{O}}(t) \hat{O}]\}$ [22]. Depending on the chosen set of observables, the hierarchy is broken at different levels, e.g. if the expansion is limited to single-electron observables, one obtains the Hartree–Fock factorization rule

$$\langle a_i^\dagger a_j^\dagger a_k a_l \rangle \approx \langle a_i^\dagger a_l \rangle \langle a_j^\dagger a_k \rangle - \langle a_i^\dagger a_k \rangle \langle a_j^\dagger a_l \rangle. \tag{9.14}$$

If two-particle observables are also considered in the expansion, it is sometimes common (but not necessary) to calculate the difference to the Hartree–Fock factorization, the so-called two-particle correlation,

$$\langle a_i^\dagger a_j^\dagger a_k a_l \rangle^c = \langle a_i^\dagger a_j^\dagger a_k a_l \rangle - \langle a_i^\dagger a_l \rangle \langle a_j^\dagger a_k \rangle + \langle a_i^\dagger a_k \rangle \langle a_j^\dagger a_l \rangle, \tag{9.15}$$

in order to estimate the difference of the full two-particle observable to the two-particle observable approximated by single-particle observables. Similar schemes are applied for the correlations between phonons and electrons and between photons and electrons [22–24]. In the case of the electron–phonon interaction, again treating only single-particle observables, this leads, e.g., in second Born approximation to the factorization rule

9 Theory of the Optical Response of Single and Coupled Semiconductor

$$\langle a_i^\dagger a_j b_q^\dagger b_{q'}\rangle \approx \langle a_i^\dagger a_j\rangle\langle b_q^\dagger b_{q'}\rangle, \tag{9.16}$$

and for the calculation of the higher order particle correlation,

$$\langle a_i^\dagger a_j b_q^\dagger b_{q'}\rangle^c = \langle a_i^\dagger a_j b_q^\dagger b_{q'}\rangle - \langle a_i^\dagger a_j\rangle\langle b_q^\dagger b_{q'}\rangle. \tag{9.17}$$

As an example of the resulting equations of motion, the electron–phonon interaction for an interband two-level system in a non-Markovian second Born approximation (in bath approximation) leads to [16]

$$\partial_t\langle a_v^\dagger a_c\rangle = -\frac{i}{\hbar}(\varepsilon_c - \varepsilon_v)\langle a_v^\dagger a_c\rangle + i\Omega(t)\left(1 - 2\langle a_c^\dagger a_c\rangle\right),$$
$$+ \frac{i}{\hbar}\sum_q \left(g_{vv}^q - g_{cc}^q\right)\left(\langle a_v^\dagger b_q a_c\rangle + \langle a_v^\dagger b_{-q}^\dagger a_c\rangle\right) \tag{9.18}$$

$$\partial_t\langle a_v^\dagger b_q a_c\rangle = -\frac{i}{\hbar}(\varepsilon_c - \varepsilon_v + \hbar\omega_q)\langle a_v^\dagger b_q a_c\rangle - 2i\Omega(t)\langle a_c^\dagger b_q a_c\rangle$$
$$+ \frac{i}{\hbar}\langle a_v^\dagger a_c\rangle\left(g_{vv}^{-q} - g_{cc}^{-q}\right)(n_q + 1) \tag{9.19}$$

as well as analogous equations for the other correlations.

Related to the correlation expansion is the Nakajima–Zwanzig projection operator technique [25] which yields similar or identical results, especially for electron–phonon coupling. A further related technique is the dynamics-controlled truncation scheme [26], which is especially suitable for a treatment where the calculation can be restricted to a given order in the external electric field.

TCL Formalism

In contrast to the correlation expansion, if we focus on strong electron–phonon coupling and weak electromagnetic fields (susceptibility expansion possible), the so-called time convolutionless (TCL) expansion scheme is appropriate [25]. For example, sometimes a certain part of all infinitely many phonon processes can be considered [27]. The density matrix is divided into a system and a bath part, where only the observables of the system are relevant. Therefore, a projector $\mathcal{P}\rho = \text{tr}_B(\rho) \otimes \rho_B$ is introduced which traces out the degrees of freedom of the bath B and replaces them with the bath density matrix in thermal equilibrium ρ_B. Thus, the density matrix consists of a relevant $\mathcal{P}\rho$ and an irrelevant part $\mathcal{Q}\rho$ with $\mathcal{Q} = 1 - \mathcal{P}$. The dynamics is then calculated in the interaction picture, chosen so that the dynamics of the density matrix is solely determined by the system–bath interaction, e.g. the electron–phonon interaction. The basic idea for deriving this expansion scheme is that it is assumed that the equations of motion of the relevant density matrix have the form [25]

$$\partial_t \mathcal{P}\rho(t) = \mathcal{K}(t)\mathcal{P}\rho(t), \tag{9.20}$$

i.e. that the equations of motion are time local in the relevant density matrix $\mathcal{P}\rho(t)$ and no convolution is present as in the Nakajima–Zwanzig formalism. The concrete derivation and the expansion scheme for $\mathcal{K}(t)$ can be found in [25]. Here, only the term in second order in the system–bath interaction is given, which is exact for a harmonic bath and a system–bath coupling diagonal in the system energy eigenstates:

$$\mathcal{K}_2(t) = \int_0^t dt_1 \mathcal{P}\mathcal{L}(t)\mathcal{L}(t_1)\mathcal{P}, \qquad (9.21)$$

where the Liouvillian $\mathcal{L}(s)$ is defined by the system–bath interaction according to $\mathcal{L}(s)\rho(t) = \frac{i}{\hbar}[H_{S-B}(s), \rho(t)]_-$. A technique related to the TCL formalism is the cumulant expansion technique [28], which was originally developed in statistical physics [29]. In the case of diagonal electron–phonon coupling, the TCL theory yields the same results as the exact cumulant expansion.

Independent Boson Model

For the case of a two-level QD interacting with a phonon bath via a coupling diagonal in the electronic states under δ-pulse excitation, the dynamics can be solved exactly. The solution of the so-called independent Boson model (IBM) for $p(t) = \langle a_v^\dagger a_c \rangle$ is given by [7, 30, 31]

$$p(t) = p_0 \exp\left\{\left(-i\omega_g + i\sum_q \frac{|\tilde{g}_{vc}^q|^2}{\hbar^2 \omega_q}\right)t - \sum_q \frac{|\tilde{g}_{vc}^q|^2}{\hbar^2 \omega_q^2}[n_q e^{i\omega_q t} + (n_q + 1)e^{-i\omega_q t}]\right\}, \qquad (9.22)$$

with the interband gap energy $\omega_g = \omega_c - \omega_v$, the Bose–Einstein distribution $n_q = [\exp(\hbar\omega_q/k_B T) - 1]^{-1}$, and $\tilde{g}_{vc}^q = g_{vv}^q - g_{cc}^q$ [cf. (9.6) and (9.7)]. For the derivation of (9.22), it is important that the Hamiltonian is defined so that the basic assumption of no interaction between the bosons and the electrons at the equilibrium positions of the Born–Oppenheimer approximation holds, i.e. that the Hamiltonian of the electron–phonon interaction does not act on the ground state, $[H_{el-ph}, a_v^\dagger a_v] = 0$. Using the TCL scheme in the interaction picture with the density matrix dynamics solely governed by H_{el-ph}, (9.22) is reproduced.

9.3 Single Quantum Dot Response

With the foundations of the previous section, we turn to the optical properties of a single QD. For the calculations, we use parameters suitable for self-assembled InAs/GaAs QDs in the strong confinement regime as a model system (cf. Sect. 9.2.2). Furthermore, we restrict to an interband or intraband two-level system.

9.3.1 Linear Absorption Spectra and Quantum Optics

This section discusses the absorption [7, 30] and resonance fluorescence (RF) spectra [14] of QDs. For the absorption, the exactly solvable model for arbitrarily strong electron–phonon coupling is compared to a perturbative approach. Furthermore, linear as well as nonlinear RF spectra [32, 33] including the electron–phonon interaction and microscopically calculated radiative effects are presented.

Absorption: Arbitrarily Strong Electron–Phonon Coupling

Electron–phonon coupling for an interband two-level system (valence level $|v\rangle$, conduction level $|c\rangle$) results in the polarization $p(t)$ given by the independent Boson model (9.22). Figure 9.2(a,b) shows the polarization decay and the linear absorption spectra for the interaction with LA phonons at different temperatures. An elaborate discussion of the solution including electron–phonon coupling mechanisms beyond the LA phonon deformation potential interaction can be found in [7]. Characteristic for the interaction of the electronic system with LA phonons are (i) two distinct time scales in the time domain due to the non-Markovian character of the pure dephasing and (ii) broad phonon sidebands in the frequency domain which form to both sides of the zero-phonon line (ZPL) describing the absorption and emission of phonons [30]. It should be noted that in our model, the ZPL is caused by a phenomenological dephasing describing a purely radiative broadening, calculated microscopically for instance in [13, 14]. Further ZPL broadening mechanisms are discussed in the literature [34–36]. An overview of the ZPL treatment can be found in the article by Grosse et al. in this book (Chap. 8). Experimental results on phonon sidebands can be found in [37–39]. The polar coupling to LO phonons with a constant dispersion does not lead to a polarization decay, but rather to quantum beats in the temporal evolution of the polarization and discrete phonon peaks in the absorption spectrum [broadened in Fig. 9.2(c) due to the LA phonons] [7]. For a polarization dephasing via coupling to LO phonons, considerations beyond the model considered here must be applied [40–42].

Absorption: Perturbative Theory

Applying the correlation expansion (Sect. 9.2.3), which resembles the IBM correlations only to a finite order, we can derive equations of motion up to an arbitrary order in the electron–phonon coupling strength. In practice, it is necessary to restrict the evaluation to the fourth or sixth order in the electron–phonon coupling strength g [12, 16]. The advantage compared to the exact solution is that the results can be

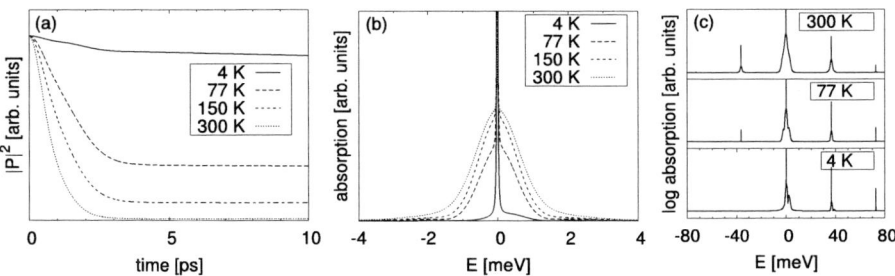

Fig. 9.2. (a) Polarization decay and (b) linear absorption spectra of a QD interacting with LA phonons at different temperatures for the independent Boson model. (c) Linear absorption spectra at different temperatures for the independent Boson model including both LA and LO phonons

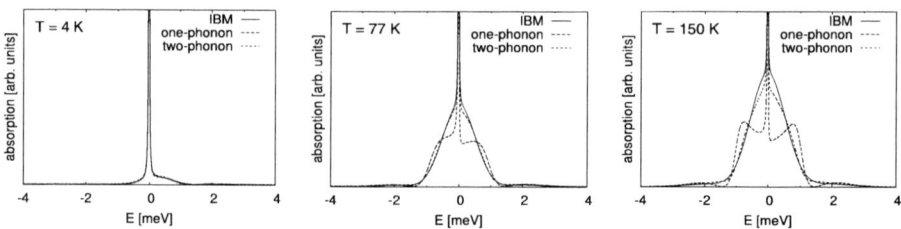

Fig. 9.3. Linear absorption spectra of a QD interacting with LA phonons at different temperatures, comparing the independent Boson model with the correlation expansion results for up to two-phonon processes

extended to nonlinear optics up to infinite order in the optical field. Figure 9.3 shows a comparison of the exact model with the results of the correlation expansion of one- and up to two-phonon processes for a QD interacting with LA phonons (see also [7]). It can be seen that, for this system, the correlation expansion to the applied order is a good approximation up to ca. 120 K; after this, the multiphonon processes (second order in $g \,\widehat{=}\,$ one-phonon process, etc.) due to the increased phonon population become more important.

Linear Resonance Fluorescence Spectra

As complementary information to the absorption, it is interesting to investigate the signatures of the electron–phonon interaction in resonance fluorescence spectra as well. For this, the quantization of the radiation field is necessary (see Sect. 9.2.2). The stationary power spectrum for a two-level system is determined by the photon number density [13, 21],

$$S(\omega) \propto \frac{\langle c_k^\dagger c_k \rangle(t)}{|M_{vc}^k|^2 t} \quad (t \gg T^*), \tag{9.23}$$

where T^* is a typical relaxation time. The power spectrum captures both the coherent ($\propto \langle c_k^\dagger \rangle, \langle c_k \rangle$) and the incoherent ($\propto \langle c_k^\dagger c_k \rangle^c = \langle c_k^\dagger c_k \rangle - |\langle c_k \rangle|^2$) light emission of a QD.

Figure 9.4(a) shows the RF spectra of a QD interacting with LA phonons and the radiation field for different temperatures. It is found that, compared to the absorption spectra which are independent of the pulse shape, the RF spectra depend strongly on the pulse length [14]. For a short pulse (left), the influence of the electron–phonon interaction is very weak since the non-Markovian interaction leads to pure dephasing which is witnessed only in the decrease of the polarization, but not in an electronic density relaxation (which determines the dynamics of the photon number density). For a long pulse (right), the electron–phonon coupling acts indirectly via the polarization back on the electronic density and thus affects the RF spectra which show phonon sidebands.

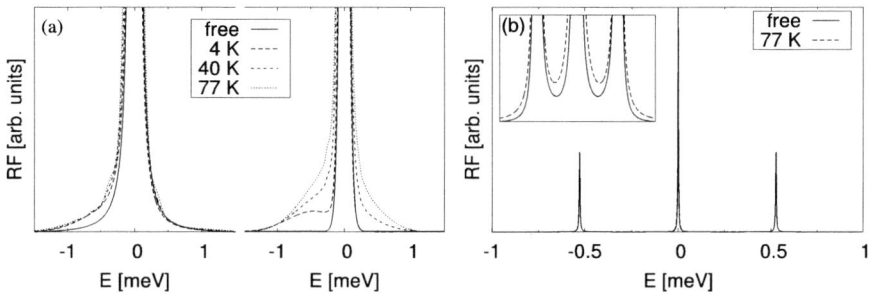

Fig. 9.4. (a) Linear RF spectra of a QD interacting with LA phonons and the radiation field at different temperatures under (*left*) δ-pulse excitation and (*right*) Gaussian excitation with a pulse length of 8 ps [calculated by K.J. Ahn]. (b) Nonlinear RF spectra showing Mollow splitting and the signatures of the electron–phonon interaction (magnification of the phonon sidebands in the inset) [calculated by K.J. Ahn]

Mollow Splitting

Under excitation with a strong field, the RF spectrum shows Mollow splitting signatures [43], i.e. optical Stark effect line splitting. Without scattering, the spectrum can be calculated analytically within the Markovian approximation, showing a linewidth ratio of 3:2:3 and an intensity ratio of 1:3:1 [13]. Adding the electron–LA phonon interaction, the RF spectrum shows weak phonon sidebands (see Fig. 9.4(b)). Due to the phonon emission/absorption asymmetry (Fig. 9.3), the emission intensity peak ratio changes and becomes asymmetric.

9.3.2 Semiclassical Nonlinear Dynamics

In this section, we discuss the semiclassical dynamics where the nonlinear excitation is caused by a coherent classical field.

Rabi Oscillations

The non-Markovian nature of the dephasing due to the electron–LA phonon interaction can also be studied in the decoherence of Rabi oscillations, a famous nonlinear optical effect in which a large pulse area leads to coherent carrier density oscillations [44, 45]. Figure 9.5(a) shows Rabi oscillations for two distinct time scales [given by the pulse length τ larger than ($\tau = 20$ ps) and comparable to ($\tau = 5$ ps) the electron–phonon scattering time] for (i) a QD interacting with LA phonons (solid line), (ii) an artificial two-level system with exponential pure dephasing ($T_2 = 4$ ps) and no energy relaxation (dashed line), and (iii) an undamped two-level system (dotted line). Compared to the exponential dephasing [system (ii)], where the damping of the Rabi oscillations increases monotonously for longer Gaussian pulse lengths until the incoherent limit is reached, the electron–phonon interaction in the QD [system (i)] leads to a maximal dephasing for a pulse length of about the time scale of the

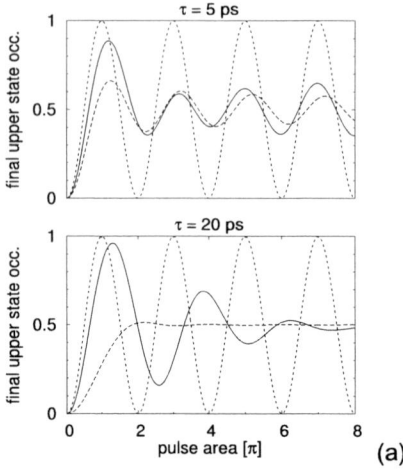
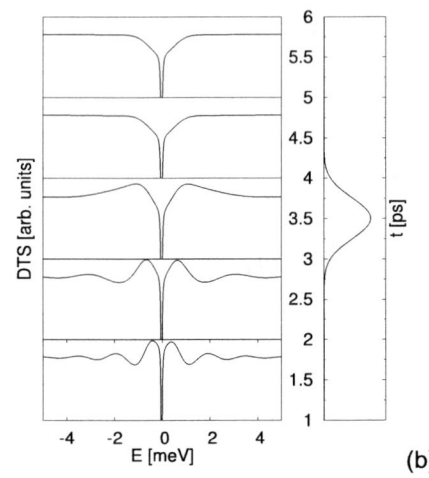

Fig. 9.5. (a) Rabi oscillations of the final upper state occupation of a QD interacting with LA phonons (*solid line*) compared to an artificial two-level system with an exponential pure dephasing ($T_2 = 4$ ps) and no energy relaxation (*dashed line*) and an undamped two-level system (*dotted line*) for two different pulse lengths τ in the pulse area domain ($T = 77$ K). (b) Pump probe DTS spectra of a single QD: (*left*) Spectrum with a probe pulse at the corresponding time; (*right*) pump pulse

fast decay of the polarization ($\tau = 5$ ps) (see Fig. 9.2(a)) and a decreasing dephasing for longer pulses since the pulse then spectrally excites mainly within the ZPL and not the phonon sidebands. Furthermore, the phonons show a pulse area-dependent dephasing and a pulse area renormalization clearly seen for $\tau = 20$ ps [12, 16, 46, 47]. The inclusion of coherent and hot phonons becomes important for very low temperatures, large pulse lengths, and strong confinement where the coupling strength becomes large in order to build up the phonon population as well as cause a significant dynamic detuning [48]. Experimental results on Rabi oscillations in QDs can be found in [49–51].

Pump-Probe/Differential Transmission Spectra

For an electron–phonon coupling diagonal in the electronic states, it is possible to obtain an exact solution for the nonlinear optical response using the cumulant expansion technique (cf. Sect. 9.2.3), up to a limited but in principle arbitrary order in the electric field [28]. This is the proper approach for all wave mixing experiments. Figure 9.5(b) shows a set of differential transmission spectra (DTS) for different delay times between pump and probe pulses, where the probe pulse is assumed to be δ-like and the pump pulse has the shape of a Gaussian. For negative delay times, the typical transmission oscillations due to the optical Stark effect also present in bulk semiconductors [9, 52] can be recognized, while for positive delay times the increased occupation in the upper state causes a bleaching of the QD absorption spectra. Such effects have been found experimentally on a single QD [53, 54]. The

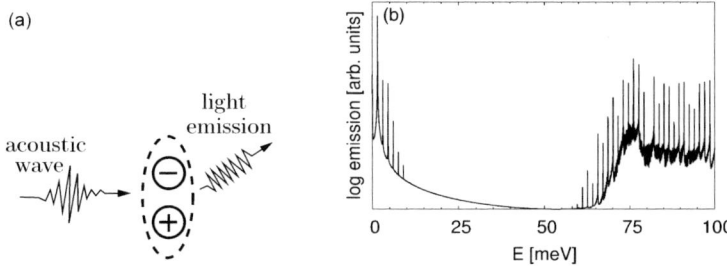

Fig. 9.6. (a) The conversion of acoustic into electromagnetic energy via multiphonon absorption leads to acoustoluminescence. (b) Acoustoluminescence spectrum of a QD excited by a coherent acoustic wave [calculated by K.J. Ahn/F. Milde]

inclusion of electron–phonon interaction in the theory leads to a damping of the spectral oscillations.

Acoustoluminescence

Luminescence under excitation with a strong acoustic wave is called acoustoluminescence (Fig. 9.6(a)). This type of conversion of acoustic into electromagnetic energy was first observed in CdS single crystals and alkali halides [55]. Here, an intersublevel transition in the conduction band of a QD is considered, where the energy gap is bridged by multiphonon absorption. Since the source is a classical coherent acoustic wave, the Hamiltonian of the system is given by the semiclassical version of (9.5):

$$H_{\text{el-ph,semi}} = \sum_{q} \sum_{\lambda,\mu=1}^{2} G_{\lambda\mu}^{q} \Lambda(q) a_{\lambda}^{\dagger} a_{\mu}, \quad (9.24)$$

where $\Lambda(q)$ denotes the amplitude of the acoustic field after its switch-on time and $G_{\lambda\mu}^{q} = s_0 \cdot q D_c F_{\lambda\mu}^{q}$ with the vector of the ion's elongation from the equilibrium position s_0. Figure 9.6(b) shows the emission spectrum ($s_0 = 2\%$ of the lattice constant a_{GaAs}) for a monochromatic acoustic excitation (energy $\omega_0 = 1.76\,\text{meV}$) and an intersublevel gap energy of $\omega_g = 76\,\text{meV}$, exhibiting light emission over a broad THz range. For low energies, the spectrum shows higher order harmonics at frequencies $\omega = (2n+1)\omega_0$ ($n \in \mathbb{N}_0$), while for energies close to the intersublevel energy, hyper-Raman peaks at energies $\omega = \tilde{\omega}_g + 2n\omega_0$ ($n \in \mathbb{N}_0$) around the renormalized transition energy $\tilde{\omega}_g$ are found [56]. Specifically, at the transition and renormalized transition energies, an increased emission of light is found [57, 58].

9.4 Two Coupled Quantum Dots

In the case of two coupled QDs, a model system for coupling on the nanoscale, it is interesting how the interaction manifests itself in the linear and nonlinear dynamics.

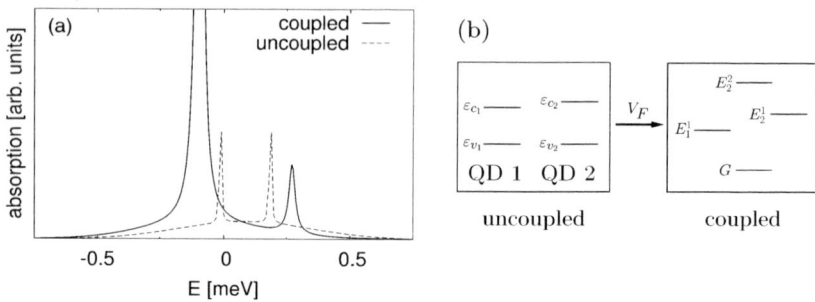

Fig. 9.7. (a) Absorption spectrum of two coupled nonresonant AlAs/GaAs interface QDs (energy gap difference $\Delta E = 0.2\,\text{meV}$) interacting via the Förster coupling ($|V_F| = 0.15\,\text{meV}$). (b) The Förster coupling changes the states from two uncoupled QDs (transition energies $\varepsilon_{c_1} - \varepsilon_{v_1}$ and $\varepsilon_{c_2} - \varepsilon_{v_2}$) to a coupled system (ground state energy G, single excitonic energies E_1^1, E_2^1, and biexcitonic energy E_2^2)

In the calculations, we use parameters suitable for both self-assembled InAs/GaAs QDs in the strong and AlAs/GaAs interface QDs in the weak confinement regime (cf. Sect. 9.2.2), depending on the effects under consideration.

9.4.1 Absorption Spectra

Using the TCL method with the secular and Markovian approximations for the electron–phonon interaction, which is nondiagonal in the excitonic states of the coupled system [59, 60], we can calculate the absorption spectrum of two coupled nonresonant QDs (Fig. 9.7(a) for two interface QDs). Within this scheme, the electron–phonon coupling is treated up to second order (which yields the exact result for diagonal coupling) and the Coulomb coupling to all orders if excitonic states are used. Including the Förster coupling, the previously identical signatures at the uncoupled QD energies change to two different line shapes at the excitonic energies of the coupled system. Specifically, the upper transition is broadened because of the reduced lifetime due to the decay into the lower excitonic state (cf. Sect. 9.4.2), see Fig. 9.7(b).

9.4.2 Excitation Transfer

Figure 9.8(a) shows the upper state occupation transfer dynamics due to the Förster coupling of a pair of resonant InAs/GaAs QDs where one QD is excited locally by a strong ultrafast pulse. The excitation is fully transferred from one QD to the other on a nanosecond scale, provided that there is no dephasing mechanism present in the system; otherwise, the density oscillation between the QDs is damped as seen in Fig. 9.8(b). Here the same system as in Fig. 9.7, which has a stronger Förster coupling than the InAs/GaAs QDs in Fig. 9.8(a), is excited strongly with a long pump pulse, leading to faster transfer times on a picosecond scale and incomplete transfer due to the nonresonance of the QDs.

9 Theory of the Optical Response of Single and Coupled Semiconductor   203

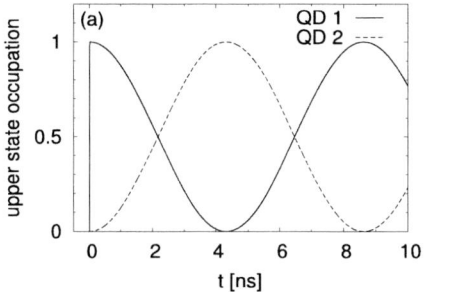
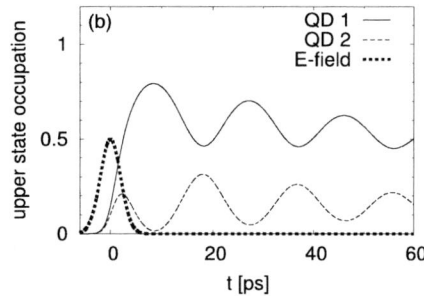

Fig. 9.8. Upper state occupation transfer dynamics of a pair of (**a**) resonant InAs/GaAs QDs interacting via the Förster coupling ($|V_F| = 0.12\,\mu\text{eV}$) for one QD excited locally by a strong ultrafast pulse and (**b**) the nonresonant interface QDs of Fig. 9.7(a) ($|V_F| = 0.15\,\text{meV}$) under nonlinear Gaussian excitation with a long pump pulse including a phenomenological dephasing

9.4.3 Rabi Oscillations

Compared to the last section, where the influence of the coupling on the dynamics was analyzed after the excitation, in this section, we consider the dynamics during a long pulse. Figure 9.9 shows the temporal evolution of the upper state occupations of two resonant InAs/GaAs QDs during excitation with a 200 ps pulse where one QD is excited locally with a strong 10π pump pulse. In Fig. 9.9(a), the second QD is excited weakly due to Förster transfer from the locally excited QD. Including radiative coupling and dephasing (Fig. 9.9(b)), the excitation transfer is much stronger, in addition to a strong damping effect caused by the radiative decay. Finally, Fig. 9.9(c) also includes the biexcitonic shift of the Coulomb interaction, causing the QDs to become nonresonant and thus decreasing the excitation transfer between the two QDs [8].

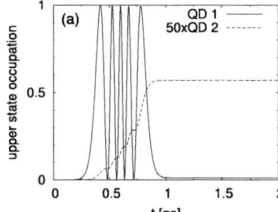
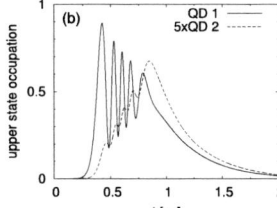
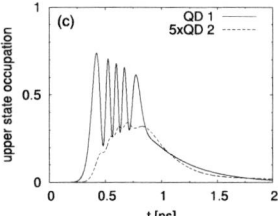

Fig. 9.9. Rabi oscillations of two coupled resonant InAs/GaAs QDs during local excitation of one QD with a 200-ps Gaussian pulse including (**a**) Förster coupling ($|V_F| = 0.12\,\mu\text{eV}$), (**b**) Förster coupling as well as radiative coupling ($|V_{\text{rad}}| = 3.92\,\mu\text{eV}$) and dephasing ($T_2 = 500\,\text{ps}$), and (**c**) Förster coupling, radiative coupling and dephasing, and the biexcitonic shift ($|V_{\text{bs}}| = 1.88\,\mu\text{eV}$) [calculated by K.J. Ahn]

9.4.4 Pump-Probe/Differential Transmission Spectra

In DTS spectra, the Förster coupling and the biexcitonic shift due to the Coulomb interaction can be analyzed separately [8, 17]. Figure 9.10 shows the DTS spectra of a pair of coupled nonresonant QDs including (a) only the biexcitonic shift and (b) only the Förster coupling where the high energetic QD is strongly pumped resonantly. We use parameters corresponding to interface QDs in the weak confinement regime. Here we do not consider phonons, but focus on the Coulomb coupling of the two QDs, thus considering a smaller energy range than in Figs. 9.5/9.7. Therefore, the OSE oscillations are only visible for very large negative delay times (e.g. for $\Delta t = -500$ ps in Fig. 9.10(a,b)), since the oscillations occur on a much broader energy scale. For smaller negative delay times, we can only distinguish between absorptive- and dispersive-like lineshapes. Considering only the biexcitonic shift (Fig. 9.10(a)), at positive delay times the nonlinear pump pulse populates the upper level of the high energetic transition (in Fig. 9.7(b): transition $G \to E_1^2$), causing a bleaching at the $G \to E_1^1$ and an induced absorption at the $E_2^1 \to E_2^2$ transition energies which are spectrally separated due to the biexcitonic shift. The inclusion of the Förster coupling without a biexcitonic shift (Fig. 9.10(b)) shows a dispersive lineshape at negative delay times due to the nonresonance of the QDs (OSE). For positive delay times, the strong π-pulse pumping leads to significant changes in the occupations, resulting in an absorptive lineshape. This effect can be understood within a local field picture (acting from the optically pumped QD on the probed QD), resulting in the population or depopulation of the probed QD after optical excitation. The effect of the Förster coupling is much weaker than that of the biexcitonic shift in this system, confirming recent experimental results [8], since Förster coupling will only have a significant impact on the energies of the states of the coupled system if it is larger than or in the order of the energy gap difference of the QDs.

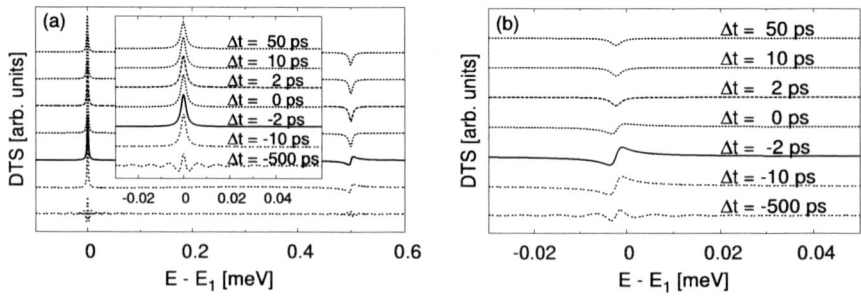

Fig. 9.10. DTS spectra of a pair of nonresonant interface QDs (energy gap difference $\Delta E = 10$ meV) including (**a**) the biexcitonic shift ($|V_{bs}| = 0.5$ meV) and (**b**) Förster coupling ($|V_F| = 0.15$ meV) for different delay times Δt where the high energetic QD is strongly pumped resonantly and the system is probed with respect to the transition of the other QD (E_1)

9.5 Multiple Quantum Dots

The investigations of the optical response of single and coupled QDs in the previous sections can be generalized to more complex systems, e.g. many QDs in an arbitrary configuration as well as ensembles of QDs. In the following, a few examples of the linear and nonlinear response of multiple QDs are considered.

9.5.1 Four-Wave-Mixing: Photon Echo in Quantum Dot Ensembles

Based on recent experiments [61], we present photon echo (PE) results in an ensemble of QDs. Starting from the optical Bloch equations [9] with a radiative dephasing constant γ and considering a local field term in the Hamiltonian (see, e.g., [62, 63]),

$$H_{\mathrm{LF}} = \left(\hbar\omega_{\mathrm{LF}}\langle a_v^\dagger a_c\rangle(t) + \mathrm{c.c.}\right) a_v^\dagger a_c + \mathrm{H.c.}, \quad (9.25)$$

the PE signal at the delay time Δt in third order in the electric field can be derived. Figure 9.11 shows the PE signal for an ensemble of InAs/GaAs QDs for different ratios of the local field energy $\hbar\omega_{\mathrm{LF}}$ to the polarization damping constant $\hbar\gamma$ (the inhomogeneous distribution is assumed to be large compared to both values). The rise in the signal is attributed to local field effects—a longer time Δt between the two pulses gives the local field more time to act back on the QDs, leading to a larger echo signal contribution through the local field, provided that in the same time the dephasing constant does not change.

9.5.2 Absorption of Multiple Coupled Quantum Dots

The cumulant expansion and TCL technique can also be applied to multiple QDs. Figure 9.12(a) shows the absorption spectrum of 14 Förster coupled interface QDs. The 14 QDs are arranged in a circle where each QD couples only to its two nearest neighbors. For the calculation, the spectrum has to be calculated in excitonic states which already include the Förster coupling [60]. The electron–phonon interaction—transformed to these new excitonic operators—is thus nondiagonal and includes relaxation effects between the excitonic states. The ZPL broadening caused by the excitonic relaxation processes can be seen.

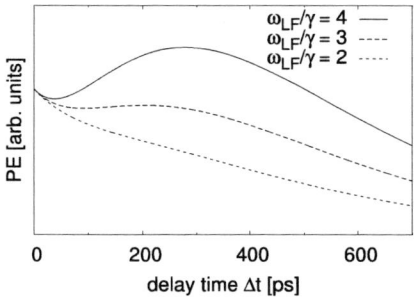

Fig. 9.11. PE signal of an ensemble of InAs/GaAs QDs including local fields for different ratios of the local field energy $\hbar\omega_{\mathrm{LF}}$ to the damping constant $\hbar\gamma$

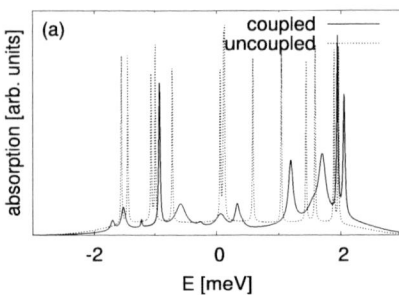 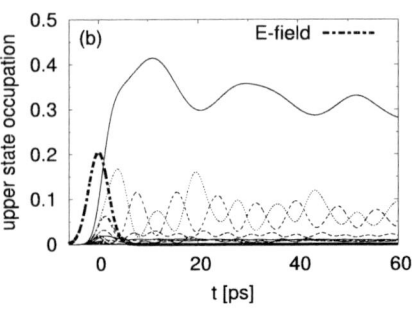

Fig. 9.12. (a) Absorption spectrum of 14 interface QDs with Förster coupling ($|V_F| = 0.3$ meV) and randomly distributed gap energies (distribution range: 4 meV). (b) Upper state occupation transfer dynamics of the same system as in (a) strongly excited with a long pump pulse (*thick line*) including a phenomenological dephasing

9.5.3 Energy Transfer of Multiple Coupled Quantum Dots

As in the case of two coupled QDs (cf. Sect. 9.4.2), it is possible to calculate the temporal evolution of the occupations of many QDs using the correlation expansion. Figure 9.12(b) shows the excitation transfer for the system of 14 QDs considered in the last section. The excitation energy is transferred between pairs of QDs. The decreasing behavior for long times is due to the radiative decay into the lower states.

9.6 Conclusion

This chapter discussed the linear and nonlinear optical dynamics of both single and coupled semiconductor quantum dots.

In the linear regime, the non-Markovian character of the interaction with longitudinal acoustic phonons leads to phonon sidebands observed in absorption as well as resonance fluorescence spectra. In the nonlinear regime, Rabi oscillations show a pulse area dependent dephasing, while differential transmission spectra show optical Stark and bleaching effects as well as a dephasing due to the electron–phonon interaction. The excitation with an acoustic pulse in the intersublevel regime leads via multiphonon absorption to the conversion of acoustic into electromagnetic energy.

For two coupled quantum dots, we discussed the effects of the Coulomb and the radiative coupling on the absorption spectra as well as the excitation transfer and Rabi oscillations. Different parameter regimes, corresponding to different quantum dot prototypes, were investigated with respect to the influence of the Förster and radiative coupling as well as energy renormalization effects. Excitation transfer as well as the character of the excitonic states were analyzed.

Finally, we discussed the optical properties of many quantum dots. While in ensembles, the local field has an effect on photon echo results, an analysis of the excitation transfer shows the effect of Förster coupling on the temporal evolution of the occupations among the different quantum dots.

Acknowledgements

We thank K.J. Ahn, F. Milde, J. Danckwerts, and J. Förstner for the joint work within the Sonderforschungsbereich 296. We also thank V.M. Axt, D. Bimberg, S. Cundiff, A. Hoffmann, A. Schliwa, and R. Zimmermann for fruitful discussions. This project was supported by the DFG via the Sonderforschungsbereich 296.

References

1. D. Bimberg, M. Grundmann, N.N. Ledentsov, *Quantum Dot Heterostructures* (Wiley, Chichester, 1999)
2. P.Y. Yu, M. Cardona, *Fundamentals of Semiconductors* (Springer, Berlin, 1999)
3. N. Baer, S. Schulz, S. Schumacher, P. Gartner, G. Czycholl, F. Jahnke, Optical properties of self-organized wurtzite InN/GaN quantum dots: A combined atomistic tight-binding and full configuration interaction calculation. Appl. Phys. Lett. **87**(23), 231114 (2005)
4. A.J. Williamson, A. Zunger, InAs quantum dots: Predicted electronic structure of free-standing versus GaAs-embedded structures. Phys. Rev. B **59**(24), 15819 (1999)
5. O. Stier, M. Grundmann, D. Bimberg, Electronic and optical properties of strained quantum dots modeled by 8-band k·p theory. Phys. Rev. B **59**(8), 5688 (1999)
6. R. Heitz, O. Stier, I. Mukhametzhanov, A. Madhukar, D. Bimberg, Quantum size effect in self-organized InAs/GaAs quantum dots. Phys. Rev. B **62**(16), 11017 (2000)
7. B. Krummheuer, V.M. Axt, T. Kuhn, Theory of pure dephasing and the resulting absorption line shape in semiconductor quantum dots. Phys. Rev. B **65**(19), 195313 (2002)
8. T. Unold, K. Mueller, C. Lienau, T. Elsaesser, A.D. Wieck, Optical control of excitons in a pair of quantum dots coupled by the dipole–dipole interaction. Phys. Rev. Lett. **94**(13), 137404 (2005)
9. H. Haug, S.W. Koch, *Quantum Theory of the Optical and Electronic Properties of Semiconductors* (World Scientific, Singapore, 2004)
10. A. Thränhardt, S. Kuckenburg, A. Knorr, T. Meier, S.W. Koch, Quantum theory of phonon-assisted exciton formation and luminescence in semiconductor quantum wells. Phys. Rev. B **62**(4), 2706 (2000)
11. G.D. Mahan, *Many-Particle Physics* (Kluwer Academic/Plenum, New York, 2000)
12. J. Förstner, C. Weber, J. Danckwerts, A. Knorr, Phonon-assisted damping of Rabi oscillations in semiconductor quantum dots. Phys. Rev. Lett. **91**(12), 127401 (2003)
13. M.O. Scully, M.S. Zubairy, *Quantum Optics* (Cambridge University Press, Cambridge, 1997)
14. K.J. Ahn, J. Förstner, A. Knorr, Resonance fluorescence of semiconductor quantum dots: Signatures of the electron-phonon interaction. Phys. Rev. B **71**(15), 153309 (2005)
15. F. Rossi, T. Kuhn, Theory of ultrafast phenomena in photoexcited semiconductors. Rev. Mod. Phys. **74**(3), 895 (2002)
16. J. Förstner, C. Weber, J. Danckwerts, A. Knorr, Phonon-induced damping of Rabi oscillations in semiconductor quantum dots. Phys. Status Solidi B **238**(3), 419 (2003)
17. J. Danckwerts, K.J. Ahn, J. Förstner, A. Knorr, Theory of ultrafast nonlinear optics of Coulomb-coupled semiconductor quantum dots: Rabi oscillations and pump-probe spectra. Phys. Rev. B **73**(16), 165318 (2006)
18. T. Förster, Zwischenmolekulare Energiewanderung und Fluoreszenz. Ann. Phys. **6**, 55 (1948)

19. T. Förster, Delocalized excitation and excitation transfer, in *Modern Quantum Chemistry*, ed. by O. Sinanoglu (Academic, New York, 1965), p. 93
20. T. Stauber, R. Zimmermann, Optical absorption in quantum dots: Coupling to longitudinal optical phonons treated exactly. Phys. Rev. B **73**(11), 115303 (2006)
21. M. Kira, S.W. Koch, Quantum-optical spectroscopy of semiconductors. Phys. Rev. A **73**(1), 013813 (2006)
22. I. Waldmüller, J. Förstner, A. Knorr, Self-consistent projection operator theory of inter-subband absorbance in semiconductor quantum wells, in *Nonequilibrium Physics at Short Time Scales*, ed. by K. Morawetz (Springer, Berlin, 2004), p. 251
23. M. Richter, M. Schaarschmidt, A. Knorr, W. Hoyer, J.V. Moloney, E.M. Wright, M. Kira, S.W. Koch, Quantum theory of incoherent THz emission of an interacting electron-ion plasma. Phys. Rev. A **71**(5), 053819 (2005)
24. M. Richter, S. Butscher, M. Schaarschmidt, A. Knorr, Model of thermal terahertz light emission of a two-dimensional electron gas. Phys. Rev. B **75**(11), 115331 (2007)
25. H.-P. Breuer, F. Petruccione, *The Theory of Open Quantum Systems* (Oxford University Press, Oxford, 2002)
26. V.M. Axt, A. Stahl, A dynamics-controlled truncation scheme for the hierarchy of density matrices in semiconductor optics. Z. Phys. B **93**, 195 (1994)
27. S. Butscher, A. Knorr, Theory of strong electron-phonon coupling for ultrafast intersub-band excitations. Phys. Status Solidi B **243**(10), 2423 (2006)
28. S. Mukamel, *Principles of Nonlinear Optical Spectroscopy* (Oxford University Press, New York, 1995)
29. N.G. van Kampen, A cumulant expansion for stochastic linear differential equations. II. Physica **74**, 239 (1974)
30. J. Förstner, K.J. Ahn, J. Danckwerts, M. Schaarschmidt, I. Waldmüller, C. Weber, A. Knorr, Light propagation- and many-particle-induced non-Lorentzian lineshapes in semiconductor nanooptics. Phys. Status Solidi B **234**(1), 155 (2002)
31. R. Zimmermann, E. Runge, Dephasing in quantum dots via electron-phonon interaction, in *Proc. 26th ICPS Edinburgh*, ed. by A.R. Long, J.H. Davies. IOP Conf. Series, vol. 171 (IOP Publishing, Bristol, 2002). Paper M 3.1
32. T. Feldtmann, L. Schneebeli, M. Kira, S.W. Koch, Quantum theory of light emission from a semiconductor quantum dot. Phys. Rev. B **73**(15), 155319 (2006)
33. N. Baer, C. Gies, J. Wiersig, F. Jahnke, Luminescence of a semiconductor quantum dot system. Eur. Phys. J. B **50**(3), 411 (2006)
34. E.A. Muljarov, T. Takagahara, R. Zimmermann, Phonon-induced exciton dephasing in quantum dot molecules. Phys. Rev. Lett. **95**(17), 177405 (2005)
35. E.A. Muljarov, R. Zimmermann, Dephasing in quantum dots: Quadratic coupling to acoustic phonons. Phys. Rev. Lett. **93**(23), 237401 (2004)
36. P. Machnikowski, Change of decoherence scenario and appearance of localization due to reservoir anharmonicity. Phys. Rev. Lett. **96**(14), 140405 (2006)
37. P. Borri, W. Langbein, S. Schneider, U. Woggon, R.L. Sellin, D. Ouyang, D. Bimberg, Ultralong dephasing time in InGaAs quantum dots. Phys. Rev. Lett. **87**(15), 157401 (2001)
38. L. Besombes, K. Kheng, L. Marsal, H. Mariette, Acoustic phonon broadening mechanism in single quantum dot emission. Phys. Rev. B **63**(15), 155307 (2001)
39. I. Favero, G. Cassabois, R. Ferreira, D. Darson, C. Voisin, J. Tignon, C. Delalande, G. Bastard, P. Roussignol, Acoustic phonon sidebands in the emission line of single InAs/GaAs quantum dots. Phys. Rev. B **68**(23), 233301 (2003)
40. A.V. Uskov, A.-P. Jauho, B. Tromborg, J. Mørk, R. Lang, Dephasing times in quantum dots due to elastic LO phonon-carrier collisions. Phys. Rev. Lett. **85**(7), 1516 (2000)

41. P. Machnikowski, L. Jacak, Exciton-LO phonon dynamics in InAs/GaAs quantum dots: Effects of zone-edge phonon damping. Phys. Rev. B **71**(11), 115309 (2005)
42. E.A. Muljarov, R. Zimmermann, Exciton dephasing in quantum dots due to LO-phonon coupling: An exactly solvable model. Phys. Rev. Lett. **98**(18), 187401 (2007)
43. B.R. Mollow, Pure-state analysis of resonant light scattering: Radiative damping, saturation, and multiphoton effects. Phys. Rev. A **12**(5), 1919 (1975)
44. I.I. Rabi, Space quantization in a gyrating magnetic field. Phys. Rev. **51**(8), 652 (1937)
45. I.I. Rabi, J.R. Zacharias, S. Millman, P. Kusch, A new method of measuring nuclear magnetic moment. Phys. Rev. **53**(4), 318 (1938)
46. A. Krügel, V.M. Axt, T. Kuhn, P. Machnikowski, A. Vagov, The role of acoustic phonons for Rabi oscillations in semiconductor quantum dots. Appl. Phys. B **81**, 897 (2005)
47. A. Vagov, M.D. Croitoru, V.M. Axt, T. Kuhn, F.M. Peeters, High pulse area undamping of Rabi oscillations in quantum dots coupled to phonons. Phys. Status Solidi B **243**(10), 2233 (2006)
48. A. Krügel, V.M. Axt, T. Kuhn, Back action of nonequilibrium phonons on the optically induced dynamics in semiconductor quantum dots. Phys. Rev. B **73**(3), 035302 (2006)
49. S. Stufler, P. Ester, A. Zrenner, M. Bichler, Quantum optical properties of a single $In_xGa_{1-x}As$-GaAs quantum dot two-level system. Phys. Rev. B **72**(12), 121301 (2005)
50. A. Zrenner, E. Beham, S. Stufler, F. Findeis, M. Bichler, G. Abstreiter, Coherent properties of a two-level system based on a quantum-dot photodiode. Nature **418**, 612 (2002)
51. P. Borri, W. Langbein, S. Schneider, U. Woggon, R.L. Sellin, D. Ouyang, D. Bimberg, Rabi oscillations in the excitonic ground-state transition of InGaAs quantum dots. Phys. Rev. B **66**(8), 081306 (2002)
52. S.W. Koch, N. Peyghambarian, M. Lindberg, Transient and steady-state optical nonlinearities in semiconductors. J. Phys. C **21**, 5229 (1988)
53. T. Guenther, C. Lienau, T. Elsaesser, M. Glanemann, V.M. Axt, T. Kuhn, S. Eshlaghi, A.D. Wieck, Coherent nonlinear optical response of single quantum dots studied by ultrafast near-field spectroscopy. Phys. Rev. Lett. **89**(5), 057401 (2002)
54. T. Unold, K. Mueller, C. Lienau, T. Elsaesser, A.D. Wieck, Optical Stark effect in a quantum dot: Ultrafast control of single exciton polarizations. Phys. Rev. Lett. **92**(15), 157401 (2004)
55. I.V. Ostrovskii, A.K. Rozhko, V.N. Lysenko, Ultrasonic luminescence of CdS single crystals. Sov. Tech. Phys. Lett. **5**, 377 (1979)
56. P.P. Corso, L. Lo Cascio, F. Persico, Simple vectorial model for the spectrum of a two-level atom in an intense low-frequency field. Phys. Rev. A **58**(2), 1549 (1998)
57. F. Milde, K.J. Ahn, A. Knorr, Theory of quantum dot luminescence from acoustically excited intersubband transitions. Phys. Status Solidi B **243**(10), 2257 (2006)
58. K.J. Ahn, F. Milde, A. Knorr, Phonon-wave-induced resonance fluorescence in semiconductor nanostructures: Acoustoluminescence in the terahertz range. Phys. Rev. Lett. **98**(2), 027401 (2007)
59. T. Renger, R.A. Marcus, On the relation of protein dynamics and exciton relaxation in pigment-protein complexes: An estimation of the spectral density and a theory for the calculation of optical spectra. J. Chem. Phys. **116**(22), 9997 (2002)
60. M. Richter, K.J. Ahn, A. Knorr, A. Schliwa, D. Bimberg, M. El-Amine Madjet, T. Renger, Theory of excitation transfer in coupled nanostructures—from quantum dots to light harvesting complexes. Phys. Status Solidi B **243**(10), 2302 (2006)

61. Y. Mitsumori, A. Hasegawa, M. Sasaki, H. Maruki, F. Minami, Local field effect on Rabi oscillations of excitons localized to quantum islands in a single quantum well. Phys. Rev. B **71**(23), 233305 (2005)
62. A. Knorr, K.-E. Süsse, D.-G. Welsch, Coherent interatomic interaction and cooperative effects in resonance fluorescence. J. Opt. Soc. Am. B **9**(7), 1174 (1992)
63. K.J. Ahn, A. Knorr, Radiative lifetime of quantum confined excitons near interfaces. Phys. Rev. B **68**(16), 161307 (2003)

10

Theory of Nonlinear Transport for Ensembles of Quantum Dots

G. Kießlich, A. Wacker, and E. Schöll

Abstract. This article reviews our work on the description of electronic transport through self-assembled quantum dots. Our main interest is in the effect of Coulomb interaction on quantum dot charging (capacitance-voltage characteristics), on the average current (current-voltage characteristics), and on current fluctuations (quantum shot noise) in quantum dot layers embedded in *pn*- or resonant tunneling devices. Our studies show the particular importance of understanding those interaction mechanisms for future device applications.

10.1 Introduction

Semiconductor quantum dots (QDs) have been at the center of extensive research during the last decade due to their potential importance for applications [1, 2] and basic research [3, 4]. A convenient method to grow QD systems is the Stranski–Krastanow mode of heteroepitaxy, which results in self-assembled arrays of QDs within a layer [5–8]. Such structures typically consist of an inhomogeneous ensemble of QDs which are coupled to each other laterally within the layer of growth, but also stacks of QDs with vertical couplings between adjacent layers are of interest, in particular for the improvement of laser structures [9].

The understanding of electronic transport in such ensembles of QDs, and in particular the role of coupling and interaction phenomena, is an essential prerequisite for the realization of QD devices. Here we review detailed transport models of QDs embedded in *pn*-diode and resonant tunneling structures, giving insight into those interaction mechanisms.

10.2 Coulomb Interaction within a Quantum Dot Layer

Capacitance-voltage measurements are frequently used to gain information on the electronic levels in quantum dot structures. Here we focus on quantum dot layers

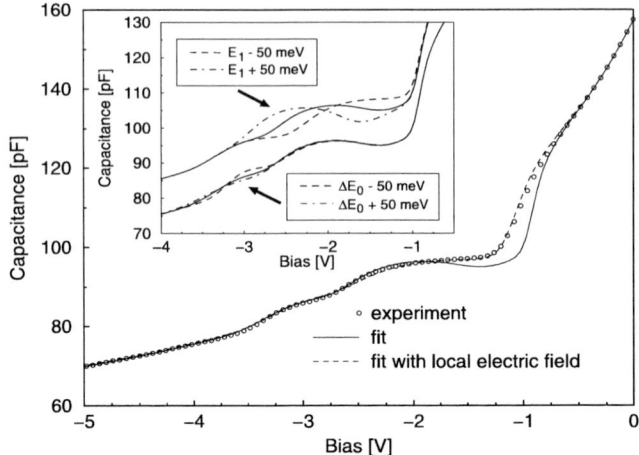

Fig. 10.1. Capacitance as a function of bias for a pn-diode with an embedded layer of quantum dots. Our model can be used to fit the energy levels. The inset demonstrates the sensitivity to varying the energy of the first excited electron state in the quantum dots (from [11])

embedded in a pn-structure [10]. By comparing our simulations with experimental data, we are able to obtain information regarding the energy levels of the quantum dots and their inhomogeneous broadening [11], see Fig. 10.1. In simulating these curves the Coulomb interaction was treated within a mean field model and the energy levels were regarded as randomly distributed with a certain width fitted to the experiment. However, it is important to realize that the width of the distribution results also partly from the discrete charges of the QDs. Due to the interdot interaction, the single particle charging energies increase with increasing occupation of the neighboring QDs, which represents an effective broadening mechanism [12]. In this context the microscopic treatment of screening of Coulomb interaction between the QDs is very important, as the location of bulk charge layers changes with bias in a pn-junction [13]. This demonstrates that level broadening and Coulomb interaction are two features which cannot be treated separately in a quantum dot array. Next to capacitance voltage-measurements these charging effects are also clearly visible in the current-voltage characteristics of $p-i-n$ diodes; see [14] where these effects were studied both theoretically and experimentally.

Similar features are observable in resonant tunneling diodes with an embedded QD layer [15–17]. Here the inhomogeneity of the QD size distribution in connection with Coulomb charging can provide negative differential conductivity [18].

Charging effects in QDs also affect the density N_s of an adjacent two-dimensional electron gas, allowing for a detection of the QD occupation. For the geometry of Fig. 10.2 pronounced bistability in N_s as a function of the bias applied perpendicular to the layers has been observed experimentally [19]. This effect has been suggested for application in future memory devices, where optical switching is of particular interest. Assuming that the generation-recombination kinetics in the QDs is dominated

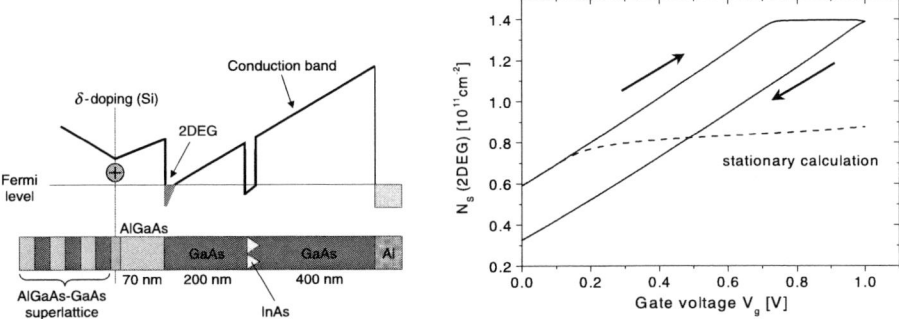

Fig. 10.2. Calculated bistability for the diode structure of [19]. While the stationary density N_S of the 2D electron gas has a unique solution, the slow kinetics of Auger capture and emission provides effective bistability on the time scale of an hour (from [20])

by the Auger effect, the low densities in the regions around the QDs provide an extremely slow kinetics, resulting in an effective bistability for bias sweep up and down of the order of an hour, see the result in Fig. 10.2, which is in excellent agreement with the experiment [20].

A more detailed analysis of the Auger recombination rates for quantum dots spatially separated from free electron states acting as scattering partner was presented in [21]. Again, screening is an important issue, which has to be treated in detail for the complex geometry involved. We could confirm the small rates for the large separation between QDs and free electron gas encountered in the experiment [19]. However, the scattering times can reach the nanosecond range for a separation of 20 nm. This shows that remote electron layers can be responsible for electronic transitions in quantum dots. This effect is of particular importance if the wetting layer, which is usually considered relevant for such processes, is empty. It is essential for modeling QD laser dynamics [22].

10.3 Transport in Quantum Dot Stacks

If two layers of quantum dots are grown on top of each other, the strain field causes strong correlations in the vertical positions [4, 23]. This provides strong electronic coupling between these QDs, so that the states are frequently delocalized. This behavior can be studied by tunneling through such double-dot structures. In this context samples with InAs QDs on InP are very useful, as single stacks can be selectively contacted due to the low QD density [24].

In these structures both tunnel coupling and Coulomb interaction have to be dealt with, providing an excellent paradigm for tunneling in nanosystems. Here a standard tool is the transmission formalism based on Green's function techniques—see, e.g., [25]—which allow for a consistent treatment of level broadening due to the coupling to the leads. However, Coulomb interaction is difficult to include beyond a mean-field approximation within this formalism. Complementary approaches, based on

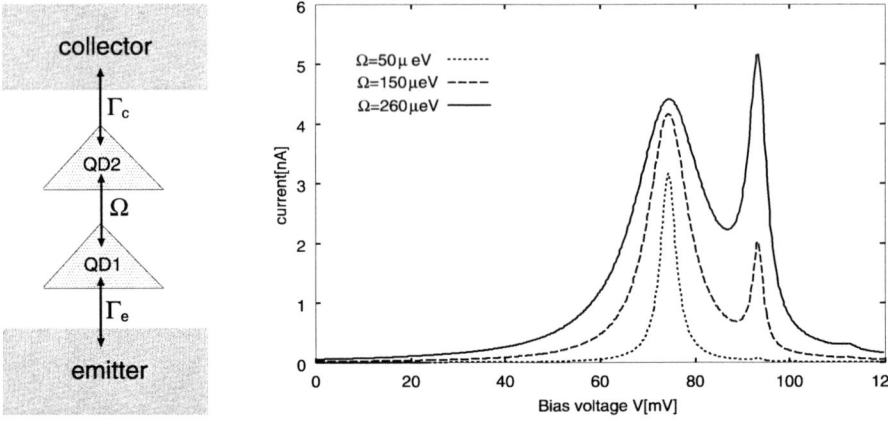

Fig. 10.3. Current-voltage characteristic for a quantum dot stack (see left) for different interdot tunnel coupling matrix elements Ω. The interdot Coulomb interaction ($U = 8$ meV) provides a double peak structure as frequently observed in experiments. Other parameters $\Gamma_e = 17$ μeV, $\Gamma_c = 400$ μeV, $T = 4.2$ K, $\mu_e = 90$ meV, $\mu_c = \mu_e - eV$, QD levels $E_1 = 79.5$ meV $-$ 0.26 eV, $E_2 = 118.7$ meV $- 0.68$ eV; from [32]

the master equation, [26], or quantum rate equations [27, 28] are able to include Coulomb effects, but do not include broadening effects, so that they apply only in the high-temperature or high-bias limit; see [29] for a detailed discussion. Recently, a new approach based on the second-order von Neumann equation was suggested to overcome these limitations [30].

The transport through such coupled QDs exhibits distinct peaks in the current-voltage characteristics when the levels in the respective QDs align [18, 31]. In particular, for asymmetric coupling of the QDs to the leads, we find double peaks due to Coulomb interaction, using a master equation model [32], see Fig. 10.3. A different source of double peaks are phonon replica, which are, however, relatively small [33].

10.4 Current Fluctuations and Shot Noise

The electric current through a nanostructure does not flow smoothly in time. In contrast, the quantization of charge e leads to a granular nature of the current. Assuming a Poissonian process without any correlations or memory for tunneling through a barrier one obtains the well-known result for shot noise [34]

$$S_{\text{Poisson}} = 2e\langle I \rangle, \quad (10.1)$$

which relates the spectral power density

$$S(\omega) = \lim_{T\to\infty} \frac{1}{2T} \left\langle \left| \int_{-T}^{T} dt \left(I(t) - \langle I \rangle \right) e^{i\omega t} \right|^2 \right\rangle \quad (10.2)$$

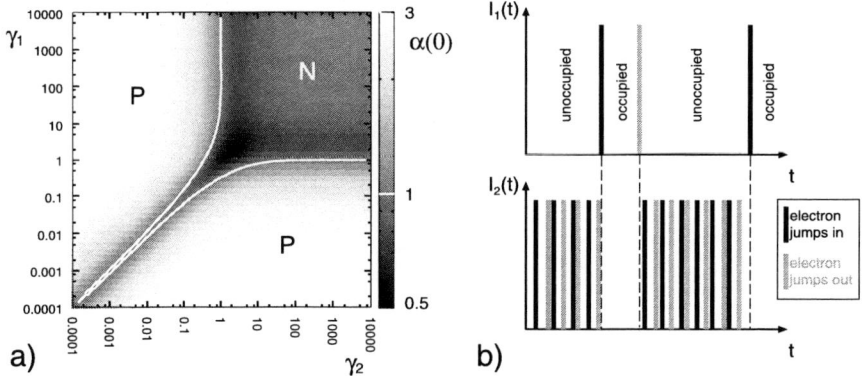

Fig. 10.4. (a) Fano factor $\alpha(\omega \to 0)$ in dependence on the collector couplings γ_1, γ_2 (in units of emitter coupling) for two neighboring QDs. (b) Mechanism for super-Poissonian noise, i.e. $\alpha > 1$, for $\gamma_1 \ll \gamma_2 = 1$. If QD 1 is occupied, tunneling through QD 2 is forbidden due to Coulomb charging. Thus, one observes bunching of electrons in the current through QD 2 (from [41])

to the mean value of the fluctuating current signal $I(t)$. A variety of physical processes like Pauli blocking or Coulomb blockade phenomena can cause deviations from the simple shot noise expression in nanostructures, so that *the noise is the signal* [35] providing information on those processes. The deviation is conveniently quantified by the dimensionless Fano factor [36]:

$$\alpha(\omega) \equiv \frac{S(\omega)}{2e\langle I \rangle} \quad \begin{cases} < 1 & \text{sub-Poissonian noise,} \\ = 1 & \text{Poissonian noise,} \\ > 1 & \text{super-Poissonian noise.} \end{cases} \quad (10.3)$$

See, e.g., [37] for an overview. In order to evaluate the noise spectrum we apply the master equation approach to tunneling through QD systems and use the Wiener–Khinchin theorem to obtain $S(\omega)$, see [29] for details.

For tunneling through a single quantum dot, Pauli blocking prevents a second electron from entering the QD before the first electron has left the QD. This results in correlations in the tunneling processes reducing the noise, so that sub-Poissonian noise occurs, as observed experimentally in resonant tunneling diodes containing lateral arrays of quantum dots [38]. However, the presence of various tunneling resonances at different QD energies provides peaks in the Fano factor whenever a new QD becomes available for transport [39]. In addition, Coulomb interaction between aligned QD energy levels can lead to pronounced features in the noise spectrum, while the average current itself is much less affected [40]. In particular, the temperature dependence of the Fano factor can be used to detect Coulomb correlations between the QD states. In addition to sub-Poissonian noise, super-Poissonian noise is also possible in arrays of laterally coupled QDs, if neighboring QDs exhibit different contact couplings [41]. The reason is that a highly conducting QD is blocked by Coulomb repulsion, if a neighboring QD is temporarily occupied. This induces bunching of tunneling electrons as shown in Fig. 10.4.

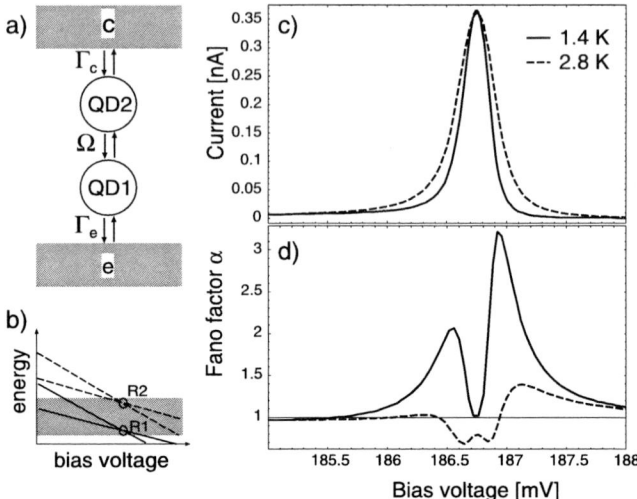

Fig. 10.5. Current and Fano factor $\alpha(0)$ for a stack of two spin-degenerate QDs. Parameters are according to [42]. For details, see [43] or Sect. 4.2.2.2 of [29]

A similar bunching effect can also appear for tunneling through a vertical stack of two quantum dots, see Fig. 10.5(a). Applying the master equation model from [32], we find super-Poissonian noise with a characteristic two-peak structure around the location of the current resonance, see Fig. 10.5(c–d), which is very similar to recent experimental results [42]. Such a behavior relies on the interplay of two resonances at the same bias, see Fig. 10.5(b). Here R1 and R2 refer to the alignment between the single-electron states in both QDs and between two-electron states, respectively. R1 and R2 occur at the same bias, if the intradot charging energies coincide. In addition the Fermi level in the emitter must be high enough such that both the single- and two-particle states in the left QD can be accessed from the emitter.

10.5 Full Counting Statistics and Decoherence in Coupled Quantum Dots

In principle, the transport of discrete charges through conductors is a stochastic process with the number of transferred carriers N in a fixed time interval t_0 being the random variable. The full counting statistics (FCS) deals with the distribution function $P(N, t_0)$ and the associated cumulant-generating function is [45, 46]

$$\exp[-F(\chi)] = \sum_N P(N, t_0) \exp[iN\chi]. \tag{10.4}$$

From this function we can obtain the cumulants $C_k = -(-i\partial_\chi)^k F(\chi)|_{\chi=0}$ which are related to, e.g., the average current $\langle I \rangle = eC_1/t_0$ and to the zero-frequency

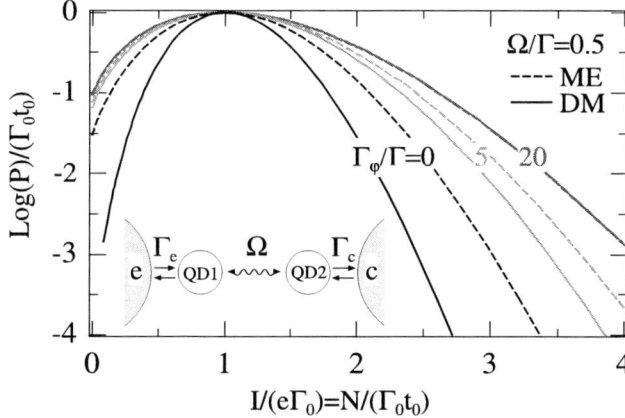

Fig. 10.6. Current statistics for $\Omega/\Gamma = 0.5$ and for various dephasing rates $\Gamma_\varphi/\Gamma = 0, 5, 20$. Dashed lines show the master equation (ME) approach and solid lines show the density matrix (DM) formalism; on-resonance $E_1 = E_2$, symmetric contact coupling: $\Gamma \equiv \Gamma_e = \Gamma_c$. $\Gamma_0 \equiv (2\Gamma\Omega^2)/[4\Omega^2 + \Gamma(\Gamma + \Gamma_\varphi)]$. Inset: Setup of the coupled QD system with (e)mitter and (c)ollector contact and mutual coupling Ω. From [44]

noise $S = 2e^2 C_2/t_0$. The Fano factor is defined as C_2/C_1. The skewness of the distribution of transferred charges is given by the third-order cumulant C_3 [29, 44]. The measurement of C_3 for transport through a single tunnel junction was recently reported [47, 48] confirming that the corresponding charge transfer is a Poissonian process ($C_k = C_1$ for all k). Very recently, Gustavsson et al. [50] presented the first measurement of the FCS in a single QD. They showed that quantum coherence does not play any role in the statistics of charge transfer in agreement with theoretical predictions [51, 52]. In contrast, in serially coupled double QDs—e.g., stacks of QDs as discussed in Sect. 10.3—the superposition between states from both QDs is responsible for prominent coherent effects. In the following we address the question how the FCS is affected by coherence and decoherence.

The setup of the coupled QD system is shown as the inset of Fig. 10.6 [44] and for simplicity noninteracting electrons are considered. The FCS is calculated for fully coherent transport by a density matrix approach (DM) and in a sequential tunneling model by a Pauli master equation (ME) (see [29, 44] for details). The latter approach does not include quantum coherence in terms of nonvanishing off-diagonal elements of a density matrix, but is also not a completely incoherent concept (see below). The resulting distribution functions are shown in Fig. 10.6 (black solid curve for coherent and black dashed curve for sequential tunneling). The maxima in $P(N, t_0)$ are equivalent in those two pictures indicating that the sequential tunneling approach provides a convenient tool for the calculation of average currents in coupled QD systems. However, the detailed statistics of both descriptions differ—in particular the noise of coherent tunneling is smaller, i.e., *quantum coherence suppresses current fluctuations*. The opposite effect can be obtained when Coulomb interaction is in-

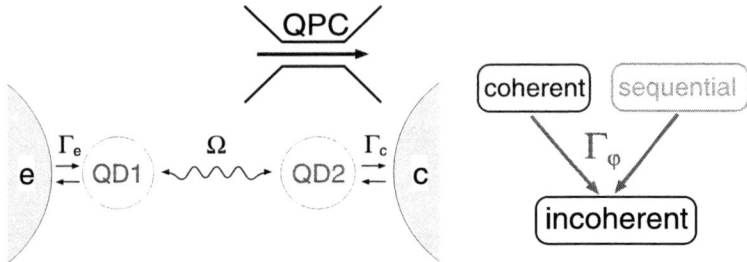

Fig. 10.7. Left: Coupled QD system with capacitively attached quantum point contact (QPC) for charge detection in QD2. Right: scheme of incoherent limit

cluded: *Noise enhancement due to quantum coherence* [49] which provides a robust and simple explanation of the experiment in [42].

What is the effect of decoherence on the FCS? Is sequential tunneling the incoherent limit? To answer those questions we consider a quantum point contact (QPC) capacitively coupled to the QD system (Fig. 10.7, left). Whenever an electron enters QD2, the transmission of the QPC changes.[1] This charge detection leads to decoherence and relaxation of the charge transfer with rate Γ_φ. Due to the coupling to the QPC the resonances are broadened so that the sequential tunneling model also has to be modified. Then, with increasing Γ_φ the statistics of coherent and sequential tunneling tend to merge as shown in Fig. 10.6.

Thus, the coherent and sequential tunneling approaches provide two pictures of electronic transport involving a different degree of quantum coherence. By introducing decoherence in a proper way, both concepts become indistinguishable in the incoherent tunneling limit as sketched in Fig. 10.7 (right).

10.6 Conclusion

The behavior of self-assembled quantum dots ensembles is strongly influenced by the Coulomb interaction between the QDs. The individual charging of QDs leads to a broadening of the single-particle energy level spectrum of the same order as by structural (size) fluctuations. This becomes clearly visible in capacitance or current transport measurements in quantum dot devices. For a quantitative description it is crucial to take into account screening by neighboring electron charged layers, which may also serve as a scattering partner for Auger recombination processes. Also Coulomb blocking leads to distinct peak features in resonant tunneling current-voltage characteristics.

QDs constitute a very interesting system to study the interplay between coherence and scattering in general nanostructure systems. The noise signal can be used to detect different features such as Coulomb correlations or dephasing mechanisms.

---

[1] This is how one can count the electrons to obtain the FCS [50].

References

1. D. Bimberg, M. Grundmann, N. Ledentsov, *Quantum Dot Heterostructures* (Wiley, New York, 1999)
2. V.A. Shchukin, N.N. Ledentsov, D. Bimberg, *Epitaxy of Nanostructures* (Springer, Berlin, 2004)
3. S.M. Reimann, M. Manninen, Rev. Mod. Phys. **74**, 1238 (2002)
4. J. Stangl, V. Holý, G. Bauer, Rev. Mod. Phys. **76**, 725 (2004)
5. D.J. Eaglesham, M. Cerullo, Phys. Rev. Lett. **64**, 1943 (1990)
6. Y.-W. Mo, D.E. Savage, B.S. Swartzentruber, M.G. Lagally, Phys. Rev. Lett. **65**, 1020 (1990)
7. D. Leonard, M. Krishnamurthy, C.M. Reaves, S.P. Denbaars, P.M. Petroff, Appl. Phys. Lett. **63**, 3203 (1993)
8. N. Carlsson, W. Seifert, A. Petersson, P. Castrillo, M.E. Pistol, L. Samuelson, Appl. Phys. Lett. **65**, 3093 (1994)
9. F. Heinrichsdorff, M.-H. Mao, N. Kirstaedter, A. Krost, D. Bimberg, A.O. Kosogov, P. Werner, Appl. Phys. Lett. **71**, 22 (1997)
10. C.M.A. Kapteyn, F. Heinrichsdorff, O. Stier, R. Heitz, M. Grundmann, N.D. Zakharov, D. Bimberg, P. Werner, Phys. Rev. B **60**, 14265 (1999)
11. R. Wetzler, C.M.A. Kapteyn, R. Heitz, A. Wacker, E. Schöll, D. Bimberg, Appl. Phys. Lett. **77**, 1671 (2000)
12. R. Wetzler, A. Wacker, E. Schöll, Phys. Rev. B **68**, 045323 (2003)
13. R. Wetzler, R. Kunert, A. Wacker, E. Schöll, New J. Phys. **6**, 81 (2004)
14. G. Kießlich, A. Wacker, E. Schöll, S.A. Vitusevich, A.E. Belyaev, S.V. Danylyuk, A. Förster, N. Klein, M. Henini, Phys. Rev. B **68**, 125331 (2003)
15. M. Narihiro, G. Yusa, Y. Nakamura, T. Noda, H. Sakaki, Appl. Phys. Lett. **70**, 105 (1997)
16. I.E. Itskevich, T. Ihn, A. Thornton, M. Henini, T.J. Foster, P. Moriarty, A. Nogaret, P.H. Beton, L. Eaves, P.C. Main, Phys. Rev. B **54**, 16401 (1996)
17. I. Hapke-Wurst, U. Zeitler, H.W. Schumacher, R.J. Haug, K. Pierz, F.J. Ahlers, Semicond. Sci. Technol. **14**, L41 (1999)
18. G. Kießlich, A. Wacker, E. Schöll, Physica B **314**, 459 (2002)
19. G. Yusa, H. Sakaki, Appl. Phys. Lett. **70**, 345 (1997)
20. A. Rack, R. Wetzler, A. Wacker, E. Schöll, Phys. Rev. B **66**, 165429 (2002)
21. R. Wetzler, A. Wacker, E. Schöll, J. Appl. Phys. **95**, 7966 (2004)
22. E. Malic, K.J. Ahn, M.J.P. Bormann, P. Hövel, E. Schöll, A. Knorr, M. Kuntz, D. Bimberg, Appl. Phys. Lett. **89**, 101107 (2006)
23. R. Kunert, E. Schöll, Appl. Phys. Lett. **89**, 153103 (2006)
24. T. Bryllert, M. Borgstrom, L.-E. Wernersson, W. Seifert, L. Samuelson, Appl. Phys. Lett. **82**, 2655 (2003)
25. S. Datta, *Electronic Transport in Mesoscopic Systems* (Cambridge University Press, Cambridge, 1995)
26. C.W.J. Beenakker, Phys. Rev. B **44**, 1646 (1991)
27. S.A. Gurvitz, Y.S. Prager, Phys. Rev. B **53**, 15932 (1996)
28. M. Wegewijs, Y. Nazarov, Phys. Rev. B **60**, 14318 (1999)
29. G. Kießlich, Ph.D. thesis, Technische Universität Berlin, 2005, http://nbn-resolving.de/urn:nbn:de:kobv:83-opus-11303
30. J.N. Pedersen, A. Wacker, Phys. Rev. B **72**, 195330 (2005)
31. W.G. van der Wiel, S.D. Franceschi, J.M. Elzerman, T. Fujisawa, S. Tarucha, L.P. Kouwenhoven, Rev. Mod. Phys. **75**, 1 (2003)

32. H. Sprekeler, G. Kießlich, A. Wacker, E. Schöll, Phys. Rev. B **69**, 125328 (2004)
33. C. Gnodtke, G. Kießlich, A. Wacker, E. Schöll, Phys. Rev. B **73**, 115338 (2006)
34. W. Schottky, Ann. Phys. (Leipzig) **57**, 541 (1918)
35. R. Landauer, Nature **392**, 658 (1998)
36. U. Fano, Phys. Rev. **72**, 26 (1947)
37. Y.M. Blanter, M. Büttiker, Phys. Rep. **336**, 1 (2000)
38. A. Nauen, I. Hapke-Wurst, F. Hohls, U. Zeitler, R.J. Haug, K. Pierz, Phys. Rev. B **66**, 161303 (2002)
39. G. Kießlich, A. Wacker, E. Schöll, A. Nauen, F. Hohls, R.J. Haug, Phys. Status Solidi (C) **0**, 1293 (2003)
40. G. Kießlich, A. Wacker, E. Schöll, Phys. Rev. B **68**, 125320 (2003)
41. G. Kießlich, H. Sprekeler, A. Wacker, E. Schöll, Semicond. Sci. Technol. **19**, S37 (2004)
42. P. Barthold, F. Hohls, N. Maire, K. Pierz, R.J. Haug, Phys. Rev. Lett. **96**, 246804 (2006)
43. G. Kießlich, A. Wacker, E. Schöll, in *Proc. 14th Int. Conf. on Nonequilibrium Carrier Dynamics in Semiconductors*, ed. by M. Saraniti, U. Ravaioli. Springer Proc. in Physics, vol. 110 (Springer, Berlin, 2006)
44. G. Kießlich, P. Samuelsson, A. Wacker, E. Schöll, Phys. Rev. B **73**, 033312 (2006)
45. *Quantum Noise in Mesoscopic Physics*, ed. by Y.V. Nazarov (Kluwer Academic, Dordrecht, 2003)
46. W. Belzig, Phys. J. **4**(8/9), 75 (2005)
47. B. Reulet, J. Senzier, D.E. Prober, Phys. Rev. Lett. **91**, 196601 (2003)
48. Y. Bomze, G. Gershon, D. Shovkun, L. Levitov, M. Reznikov, Phys. Rev. Lett. **95**, 176601 (2005)
49. G. Kießlich, E. Schöll, T. Brandes, F. Hohls, R.J. Haug, Phys. Rev. Lett. **99**, 206602 (2007)
50. S. Gustavsson, R. Leturcq, B. Simovič, R. Schleser, T. Ihn, P. Studerus, K. Ensslin, D.C. Driscoll, A.C. Gossard, Phys. Rev. Lett. **96**, 076605 (2006)
51. M.J.M. de Jong, Phys. Rev. B **54**, 8144 (1996)
52. D.A. Bagrets, Y.V. Nazarov, Phys. Rev. B **67**, 085316 (2003)

11
Quantum Dots for Memories

M. Geller and A. Marent

Abstract. This chapter demonstrates the feasibility of a QD-based memory to replace today's Flash and dynamic random access memory (DRAM). A novel memory concept based on QDs is presented, enabling very fast write times below picoseconds, only limited by the charge carrier relaxation time. A thermal activation energy of 710 meV for hole emission from InAs/GaAs QDs across an $Al_{0.9}Ga_{0.1}As$ barrier is determined by using time-resolved capacitance spectroscopy. A hole storage time of 1.6 seconds at room temperature is measured, three orders of magnitude longer than the typical DRAM refresh time. In addition, the dependence of the hole storage time in different III–V QDs on their localization energy is determined and a retention time of more than 10 years in (In)(Ga)Sb/AlAs QDs is predicted. Therefore, a future QD-based memory will show improved performance in comparison to both DRAM and Flash having fast write/read times and good endurance.

11.1 Introduction

Storage and handling of information is one of the fundamental issues of our modern information society. The need for faster processing of an increasing amount of data pushed the information technology over the last decades and forced the semiconductor industry to improve the performance of microelectronic devices. The main strategy is scaling down the feature size. The growing number of components on a chip and the reduction in feature size was predicted by Gordan Moore in 1965 [1], known nowadays as "Moore's Law". Contrary to all expectations, Moore's Law held remarkably well over the last few decades, and microelectronic devices reached a feature size of about 90 nm in 2005. The International Roadmap for Semiconductors (ITRS 2005) [2] predicts a further shrinkage down to 14 nm in 2020. The industry is touching a regime where quantum mechanics dominates the physical properties, without having adequate solutions for the growing number of difficulties to realize such structures.

Considerable effort is devoted to the search for alternative microprocessor and memory technologies. One of the promising options is the use of self-organized materials in future nanoelectronic devices. Self-organized quantum dots (QDs) are especially intriguing because they provide a number of interesting advantages for new generations of memories. Billions of self-organized QDs can be formed simultaneously in a single technological step, allowing massive parallel production in a bottom-up approach and offering an elegant method to create huge ensembles of electronic traps without lithography. They can store just a few or even single charge carriers with a retention time depending on the material combination—possibly up to years at room temperature. With an area density of up to 10^{11} per cm^{-2} an enormous storage density in the order of 1 TBit/inch^2 could be possible, if each single QD would represent one information bit. Furthermore, the carrier relaxation process into QDs is in the order of subpicoseconds [3, 4], an important prerequisite for a very fast write time in such memories. Hence, the implementation of self-organized QDs could lead to a nonvolatile memory with high storage densities combined with a fast read/write access time.

This chapter discusses the possibility of using self-organized QDs in future memory applications. Section 11.2 gives a brief overview of the main semiconductor memories, DRAM and Flash, and presents a possible QD-based memory structure [5]. By using capacitance spectroscopy, the carrier retention time in different QD systems is studied in Sect. 11.3 and a storage time of seconds at room temperature is demonstrated for InGaAs/GaAs QDs with an additional $Al_{0.9}Ga_{0.1}As$ barrier. The retention time in QDs is, hence, already three orders of magnitude longer than the crucial DRAM refresh-time. A further increase of the storage time—up to several years at room temperature (like in a Flash memory)—is predicted in Sect. 11.4.

11.2 Semiconductor Memories

The semiconductor memory industry focuses essentially on two main areas: The dynamic random access memory (DRAM) and the Flash. Both memory concepts have their advantages and disadvantages in speed, endurance, cost and storage time. A memory concept that adds the advantages of a DRAM to a Flash would combine high storage densities, fast read/write access and good endurance in combination with low production cost. In addition, for portable applications like mobile phones and MP3 players, low power consumption in combination with long storage time on the order of years (nonvolatility) is desired. A number of new concepts have been developed, like magnetic random access memory (MRAM) and ferro-electric random access memory (FRAM) [6]. However, the ultimate breakthrough is not in sight.

11.2.1 Dynamic Random Access Memory (DRAM)

DRAM is the main memory in a personal computer and consists of a transistor and a capacitor, Fig. 11.1(a), which stores the information by means of electric charge. The stored charge will be lost typically in a few milliseconds mainly due to leakage and

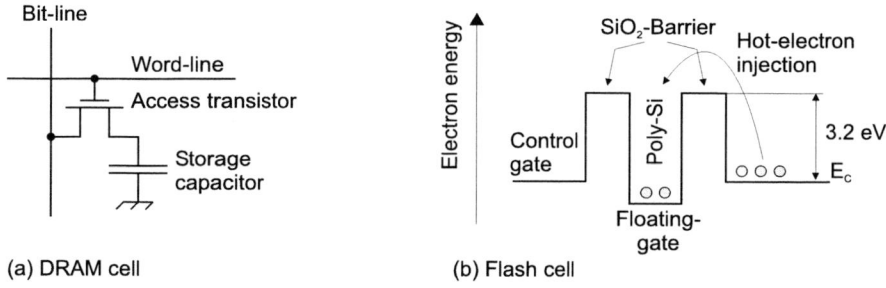

Fig. 11.1. (a) Schematic picture of a single-transistor DRAM cell with a storage capacitor. (b) Schematic bandstructure of a floating-gate Flash memory based on Si as matrix material and SiO$_2$ for the barriers. After [7]

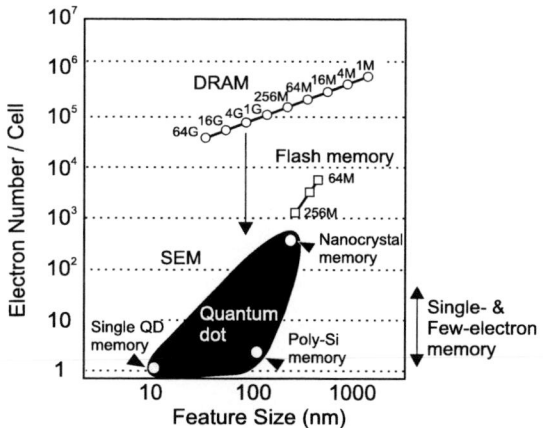

Fig. 11.2. Electron number per memory cell vs. minimum feature size for conventional memories and single- and few-electron memories. After [8]

recombination currents of the capacitor. Hence, dynamic memories require periodic "refreshing": they are volatile.

However, DRAM cells provide a fast read/write access time on the order of $\sim$20 ns and a very good endurance of more than 10^{15} write and erase cycles. On the other hand, a DRAM draws power continuously due to the refresh process and the information is lost when the computer is switched off. One information bit is also stored with a large number of electrons on the order of 10^5, see Fig. 11.2. DRAMs are not suitable for mobile applications, like MP3 players, mobile phones, and personal digital assistants (PDAs), due to the volatility and high power consumption.

11.2.2 Nonvolatile Semiconductor Memories (Flash)

Nonvolatile memories (NVMs) can retain their data for typically more than ten years. The most important nonvolatile memory is the Flash-EEPROM ("Electrically erasable and programmable read-only memory"), in short just Flash. The Flash market is the fastest growing memory market today, due to the increasing number of portable applications. But not only portable applications are ideal for Flash. Electronic products of all types, from microwave ovens to industrial machines, store their operating instructions in Flash memories.

Nowadays Flash is based on a floating-gate structure [7], where the charge carriers are trapped inside a polysilicon structure embedded between two SiO_2 barriers, Fig. 11.1(b). The SiO_2 barriers with a height of ~3.2 eV maintain a storage time of more than 10 years at room temperature. However, these energetically high barriers are also the origin of the two main disadvantages of a Flash cell: the slow write time (on the order of microseconds) and the poor endurance (on the order of 10^6 write/erase cycles). The write process of the information is realized by means of "hot-electron injection", where the electrons have a high kinetic energy to overcome the barrier. This leads to the slow write time and the hot electrons successively destroy the oxide barriers of the floating-gate (poor endurance).

11.2.3 A QD-based Memory Cell

Combining the advantages of both DRAM and Flash is one of the major challenges for the semiconductor memory industry. Such an optimized memory cell should provide long storage times (>10 years) and good endurance (>10^{15} write/erase cycles) in combination with even better read/write access time than a DRAM (<10 ns). In addition, an ultimate memory cell should store one information bit by means of a single charge carrier, either electron or hole.

A possible concept which could fulfill these requirements is to use self-organized QDs in a p–n diode structure [5]. The QDs act as storage units, since they can be charged with electrons or holes representing the "0" (uncharged QDs) and "1" (charged QDs) of an information bit. An emission barrier is needed to store a "1" in such a memory concept. The emission barrier height—which is related to the localization energy of the charge carrier, Fig 11.3(a)—can be varied by varying the material and size of the QDs and the material of the surrounding matrix. Furthermore, a capture barrier is needed to store a "0", Fig. 11.3(a). This barrier protects an

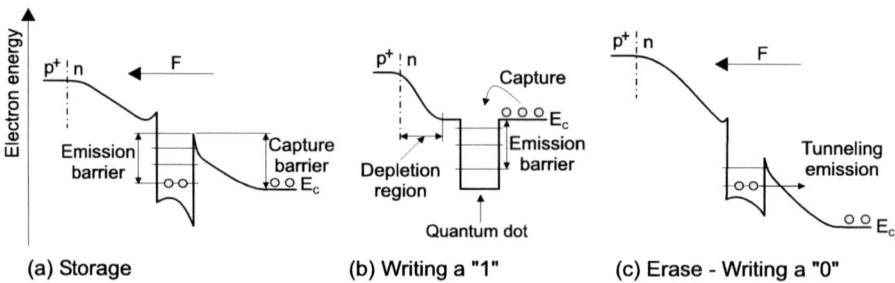

Fig. 11.3. Schematic illustration of the (**a**) storage, (**b**) write, and (**c**) erase process in a possible future QD Flash memory. After [5]

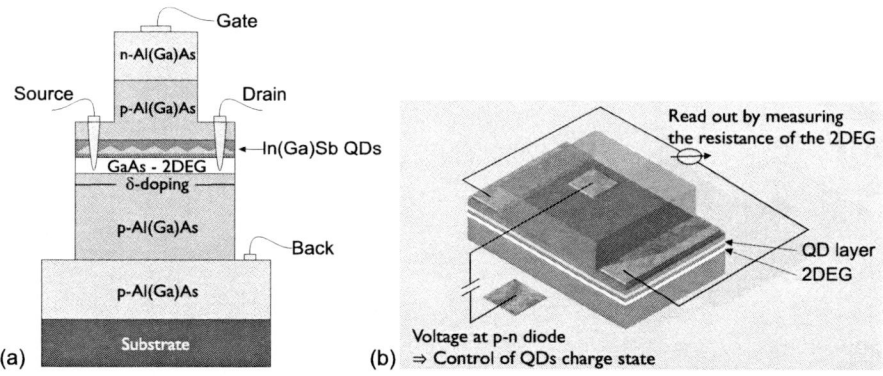

Fig. 11.4. (a) Schematic picture of the layer sequence for a possible QD-based memory. The QDs are located below the p–n junction and a 2DEG is placed below the QD layer to detect the charge state. (b) Schematic picture of the QD-based memory device in a three-dimensional view

empty QD cell from unwanted capture of charge carriers. In this concept, the capture barrier is realized by using the band-bending of a p–n diode within a depletion region.

The major advantage to using a p–n diode is the possibility of tuning the capture barrier's height by an external bias. Therefore, during the write process, Fig. 11.3(b), the capture barrier can be eliminated by reducing the width of the depletion region and the charge carriers can directly relax into the QD states. Fast write operations on the order of the charge carrier relaxation time into the QD states (picoseconds at room temperature [3, 4]) are possible. In addition, a very good endurance of 10^{15} write operations should be feasible. To erase the information, the depletion region is expanded and tunneling in a high electric field occurs, Fig. 11.3(c).

Figure 11.4 shows schematically the device structure of such a QD-based memory, where the QDs are charged with holes to represent an information bit. The doping sequence of the p- and n-regions would be vice versa for an electron storage device.

The QDs are embedded in a n^+–p diode structure below the p–n junction. The distance to the junction is adjusted, such that the QDs are inside the depletion region for zero bias. The write and erase operation can be achieved with an appropriate external bias, applied between gate and back contact. Both contacts are placed on highly p- and n-doped ($>10^{18}$ cm^{-3}) layers. The read-out of the stored information is done by a two-dimensional electron gas (2DEG) below the QD layer. Stored charge carriers inside the QDs reduce the mobility of the 2DEG, hence, a higher resistance is measured between the source/drain contacts. The 2DEG is situated 10–50 nm below the QD layer and is filled with charge carriers, provided by the additional δ-doping.

11.3 Charge Carrier Storage in Quantum Dots

This section discusses experimental results from capacitance spectroscopy measurements on different QD material systems. This time-resolved method allows to derive thermal activation energies and capture cross-sections of electron and hole QD states. In addition, the important storage time at room temperature can be quantified and connected to the localization energy of the charge carriers.

11.3.1 Experimental Technique

Semiconductor QDs can be considered as deep levels (or traps), storing few or single charge carriers for a certain retention time. Time-resolved capacitance spectroscopy and data analysis methods known from deep level transient spectroscopy (DLTS) [9, 10] allow us to investigate the electronic structure of QDs. In addition, the carrier dynamics and the average storage time can be studied in detail. The QDs are usually placed close to the space charge region of a p–n junction or Schottky contact. In conventional DLTS experiments [11–13] the QDs are outside the depletion region during a pulse bias V_p and will be completely filled with charge carriers. After the pulse, the QDs are in the space charge region and release all trapped carriers. The emission of many charge carriers from multiply charged QDs is probed by measuring time-resolved the capacitance of the depletion region, Fig. 11.6(a).

In order to study the QD states in more detail, charge-selective DLTS has been developed, where approximately one charge carrier per QD is captured and emitted, Fig. 11.5. Before the filling pulse, the QDs can already be partly filled with charge carriers. During the filling pulse the Fermi level is adjusted by the applied pulse bias, such that one carrier per QD is captured, Fig. 11.5(b). After the bias pulse, the reverse bias is set to the initial condition and the previously captured carrier is re-emitted. The emission of one charge carrier per QD is probed and narrow peaks appear in the DLTS spectra, which are due to differently charged QDs [14]. Increasing the reverse bias and keeping the pulse bias height fixed enables the observation of thermal emission from the QD ground states up to completely charged QDs.

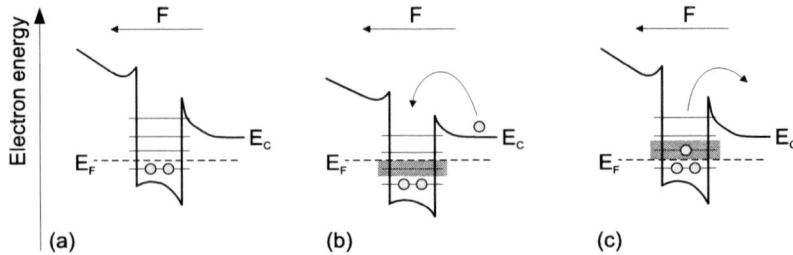

Fig. 11.5. Work cycle of a charge-selective DLTS experiment (**a**) before, (**b**) during and (**c**) after the bias pulse, respectively. The filling pulse V_p is chosen such that the emission or capture of one charge carrier per QD is probed

The thermal emission rate e_{th} is usually given by [9, 10]

$$e_{th} = \gamma T^2 \sigma_\infty \exp(-E_A/kT), \qquad (11.1)$$

where E_A is the thermal activation energy, σ_∞ the apparent capture cross-section for $T = \infty$, and γ a temperature-independent constant. Knowing the emission time constant $1/e_{th}$ for different temperatures enables to derive the thermal activation energy and the capture cross-section. In order to obtain the emission time constant from multiexponential transients the rate window concept is commonly applied.

In the rate window concept the selection of the contribution at a certain reference time constant is done by a simple technique: The DLTS signal at a certain temperature $S(T, t_1, t_2)$ is given by the difference of the capacitance at two times t_1 and t_2, Fig. 11.6(b). The two times t_1 and t_2 define the rate window, which has the reference time constant

$$\tau_{ref} = \frac{t_2 - t_1}{\ln(t_2/t_1)}. \qquad (11.2)$$

Plotting $S(T, t_1, t_2)$ as a function of temperature yields the DLTS spectrum, Fig. 11.6(c). A maximum appears at that temperature, where the emission time constant of the thermally activated process equals the applied reference time constant: $\tau(T_{max}) = \tau_{ref}$. A maximum appears only for a thermally activated process, a temperature-independent tunneling process leads to a constant DLTS signal [15]. For varying τ_{ref} different peak positions T_{max} are obtained. The activation energy can be determined from the slope of an Arrhenius plot of $\ln(T_{max}^2 \tau_{ref})$ as function of T^{-1}. From the y-axis intersection of the extrapolated data the capture cross-section σ_∞

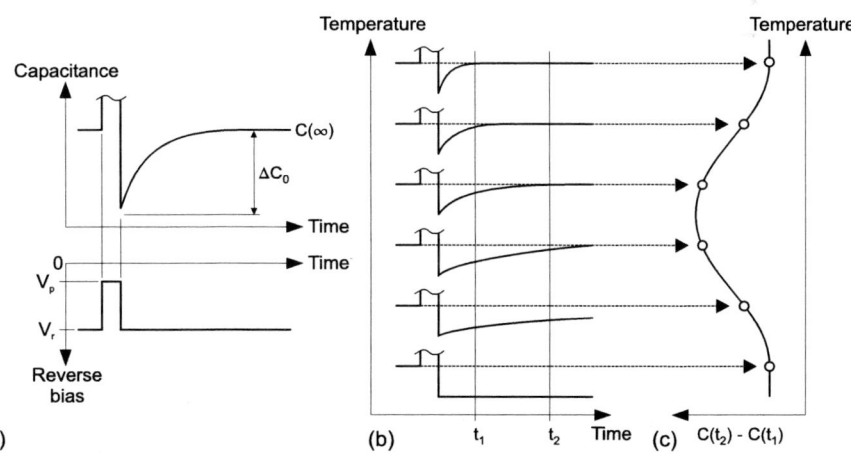

Fig. 11.6. Evaluation of the capacitance transients in a DLTS experiment. The lower part of (**a**) displays the external bias on the device as function of time, the upper part the corresponding capacitance. The evaluation of capacitance transients (**b**) for increasing temperature by a rate window defined by t_1 and t_2 leads to an DLTS plot for a thermally activated process (**c**). After [10]

can be obtained. The activation energy and the capture cross-section can be used to estimate the storage time at room temperature.

11.3.2 Carrier Storage in InGaAs/GaAs Quantum Dots

Emission of electrons and holes from InGaAs/GaAs QDs was observed by Kapteyn et al. [11, 12] using conventional DLTS measurements. The electron/hole emission from InGaAs/GaAs QDs was studied in more detail by using charge-selective DLTS [16] and time-resolved tunneling capacitance measurements (TRTCM) [17]. Two samples, H1 and E1, were investigated to study the hole and electron emission, respectively, where the QD layer was incorporated in the slightly p-doped/n-doped ($\sim 3 \times 10^{16}$ cm^{-3}) region of a n$^+$–p or p$^+$–n diode structure. The QD layer was situated 500 nm/415 nm below the p–n junction for the hole/electron sample. Mesa structures with a diameter of 800 μm and Ohmic contacts were formed by employing standard optical lithography. The results of these measurements are presented in Fig. 11.7.

It turns out that the charge carrier emission process from self-organized QDs in an electric field is always controlled by phonon-assisted tunneling. The influence of the tunneling part, however, depends strongly on the effective mass and the strength of the electric field. The observed hole ground state activation energy $E_A^{H1} = (120 \pm 10)$ meV, hence, underestimates the localization energy. It is the thermal activation part in a phonon-assisted tunneling process: Thermal activation into an excited state and subsequent tunneling through the remaining triangular barrier, cf. schematic pictures in Fig. 11.7. The entire hole localization energy was determined by using the TRTCM method to $E_{\text{loc}}^{H1} = (210 \pm 20)$ meV, in good agreement

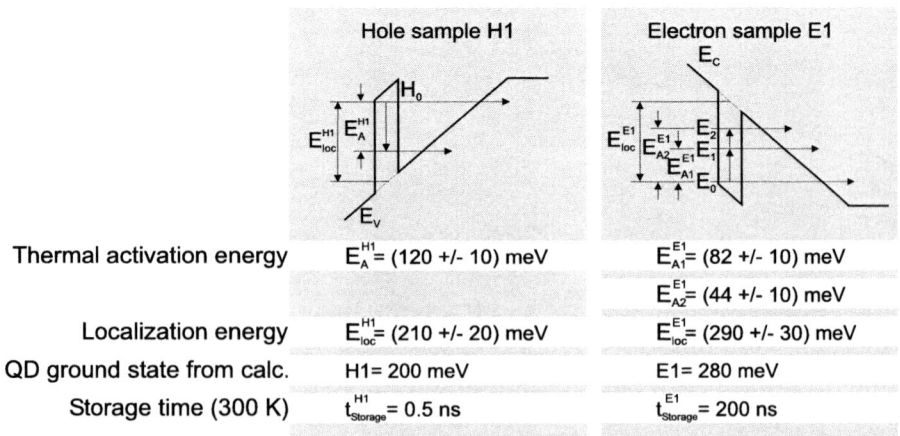

Fig. 11.7. Summary of the results from the charge-selective DLTS [16] and tunneling emission experiments [17] on self-organized InGaAs/GaAs QDs. The calculated values for the electron and hole ground state are taken from [18]

with 8-band $k \cdot p$ theory [18] which predicts a hole ground state localization energy of ~200 meV for QDs with a base length of about 16 nm.

For the electrons, two contributions to the DLTS signal were observed: A DLTS signal with an activation energy of $E_{A1}^{E1} = (82 \pm 10)$ meV, which is in good agreement with the theoretically predicted value for the ground/excited state energy splitting (70 meV). This value was obtained from DLTS [12] and admittance spectroscopy measurements [19]. In addition, a DLTS signal with a smaller activation energy of $E_{A2}^{E1} = (44 \pm 10)$ meV is observed. This value is attributed to the first/second excited state energy splitting and in satisfying agreement with the theoretically predicted value of 50 meV. The ground state localization energy was determined by the TRTCM method to be $E_{loc}^{E1} = (290 \pm 30)$ meV [17]. The theoretical value for the electron localization energy from 8-band $k \cdot p$ calculations is on the order of 280 meV.

The electron/hole storage time at room temperature can be estimated by using (11.1) and the capture cross-section and the localization energy for sample E1/H1, respectively. An average storage time for electrons of about 200 ns and for holes of about 0.5 ns is obtained. This means InAs QDs embedded in a GaAs matrix do not have a sufficiently long storage time to act as a storage unit in future memories.

11.3.3 Hole Storage in GaSb/GaAs Quantum Dots

The storage time can be further increased by changing the material of the QDs and/or the surrounding matrix. A larger difference in the energy bandgap than in InGaAs/GaAs is more promising, e.g. InAs/AlAs QDs. Moreover, large band discontinuities and, hence, strong hole localization is expected in type-II QD heterostructures [20], e.g. GaSb/GaAs or InSb/GaAs. Only holes are confined in GaSb/GaAs QDs, while a repulsive potential barrier exists for the electrons in the conduction band. Type-II material combinations are therefore very attractive for future memory applications.

Hole storage in and emission from GaSb/GaAs QDs was investigated by charge-selective DLTS [14]. The sample was a n^+–p diode structure containing a single layer of GaSb QDs. An area density of about 3×10^{10} cm^{-2}, an average QD height of about 3.5 nm, and an average base width of about 26 nm were determined by structural characterization of uncapped samples grown under identical conditions [21]. The QD layer was placed 500 nm below the n^+–p junction in a slightly p-doped ($p = 3 \times 10^{16}$ cm^{-3}) GaAs region. Mesa structures with a diameter of 800 μm and Ohmic contacts were formed by employing standard optical lithography.

The charge-selective DLTS measurement are discussed in the following. The pulse bias was always set to $V_p = V_r - 0.5$ V, while the reverse bias was increased from 4.5 V to 9.5 V. Narrow peaks appear in the DLTS spectra in Fig. 11.8(a). At $V_r = 4.5$ V the DLTS signal shows a maximum at about 80 K and for increasing reverse bias the DLTS peak shifts to higher temperature. The activation energy—obtained from Arrhenius plots—increases accordingly from 150 meV at 4.5 V to 450 meV at 9.5 V, Fig. 11.8(b).

The maximum activation energy of 450 meV represents the QD hole ground state energy, i.e. the localization energy. The decrease in the activation energy from

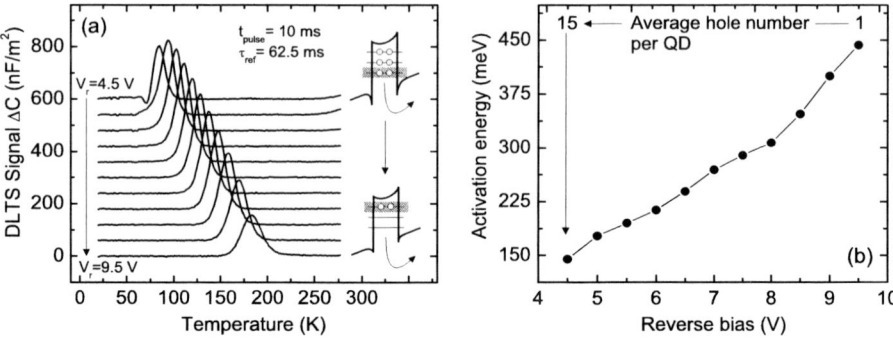

Fig. 11.8. Charge-selective DLTS signal at a reference time constant of $\tau_{\text{ref}} = 62.5$ ms for increasing reverse bias V_r between 4.5 and 9.5 V (**a**). The pulse bias was set to $V_p = V_r - 0.5$ V for all spectra and the pulse width was 10 ms. The dependence of the activation energy on the reverse bias is displayed in panel (**b**). After [14]

450 meV down to 150 meV corresponds to an increase in the average occupation of the QDs; see the schematic insets in Fig. 11.8(a). With increasing amounts of charge in the QDs, state filling lowers the thermal activation barrier. The completely charged QDs are filled with 15 holes up to the Fermi level at the valence band edge, whereas the barrier height is generated by Coulomb charging. In order to compare the storage time for GaSb/GaAs with In(Ga)As/GaAs QDs, the observed emission rates were extrapolated to room temperature using (11.1). A storage time of about 1 μs for localized holes with the ground state energy of 450 meV is estimated, three orders of magnitude longer than the hole storage time in InGaAs/GaAs QDs.

11.3.4 InGaAs/GaAs Quantum Dots with Additional AlGaAs Barrier

This section presents hole storage in InGaAs/GaAs QDs with an additional AlGaAs barrier. The additional AlGaAs barrier increases the activation energies and we observe longer storage time at room temperature. We study two different samples having different AlGaAs barriers. The first contains a 60% $Al_{0.6}Ga_{0.4}As$, the second a 90% $Al_{0.9}Ga_{0.1}As$ barrier below the QD layer. The activation energy in the latter is increased sufficiently to reach a retention time of seconds at room temperature.

Storage Time: Milliseconds at Room Temperature

The first sample is a n^+–p diode structure, grown by molecular beam epitaxy (MBE). It contains a single layer of InGaAs QDs embedded in slightly p-doped GaAs ($p = 2 \times 10^{15}$ cm^{-3}). The QDs are placed 1500 nm below the p–n junction and an additional undoped $Al_{0.6}Ga_{0.4}As$ barrier of 20 nm thickness is situated 7 nm below the QD layer to increase the hole storage time. Again, mesa structures with a diameter of 800 μm and Ohmic contacts were formed.

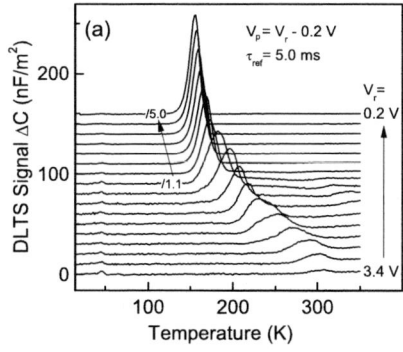 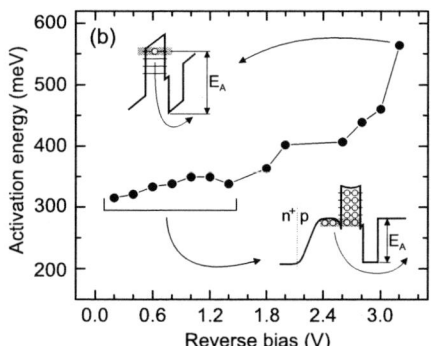

Fig. 11.9. Charge-selective DLTS spectra of thermally activated hole emission from In-GaAs/GaAs QDs with an additional $Al_{0.6}Ga_{0.4}As$ barrier below the QD layer. Spectra below $V_r = 1.6$ V are divided by a factor from 1.1 up to 5 (**a**). (**b**) Dependence of the thermal activation energy on the reverse bias V_r. After [22]

Figure 11.9(a) shows the charge-selective DLTS spectra. The pulse bias height was fixed to 0.2 V for all spectra ($V_p = V_r - 0.2$ V). For a reverse bias above $V_r = 3.2$ V no DLTS signal is visible, as the Fermi level is energetically above the QD states. Therefore, no QD states are occupied. By decreasing the reverse bias the Fermi level reaches the QD ground state at $V_r = 3.2$ V and a peak in the DLTS spectrum appears at 300 K. This peak is related to thermally activated hole emission from the QD ground state across the AlGaAs barrier; see the inset in the upper left corner of Fig. 11.9(b). A further decrease of the reverse bias leads to QD state-filling and emission from higher QD states is observed. A peak in the DLTS appears at that temperature, where the averaged time constant of the thermally activated emission equals the applied reference time constant, cf. (11.2). Therefore, the peak at 300 K for $\tau_{ref} = 5$ ms [$V_r = 3.2$ V in Fig. 11.9(a)] represents an average emission time constant (storage time) of 5 ms for hole emission from the QD ground states across the $Al_{0.6}Ga_{0.4}As$ barrier.

The thermal activation energies are shown in Fig. 11.9(b). The highest value of (560 ± 60) meV at $V_r = 3.2$ V is related to thermal activation from QD hole ground states across the AlGaAs barrier. The decrease in the activation energy corresponds to an increase in the average occupation of the QDs. At $V_r = 1.4$ V the energetic position of the Fermi level is at the valence band edge and no further QD state-filling is possible. As a consequence, between $V_r = 1.4$ V and $V_r = 0.2$ V carrier emission from the valence band edge is probed. The activation energy remains roughly constant with a mean value of 340 meV. This energy represents the energetic height of the $Al_{0.6}Ga_{0.4}As$ barrier.

Storage Time: Seconds at Room Temperature

The second sample is also a n^+–p diode structure, grown by metalorganic chemical vapor depositon (MOCVD). It contains a single layer of InGaAs QDs embedded in

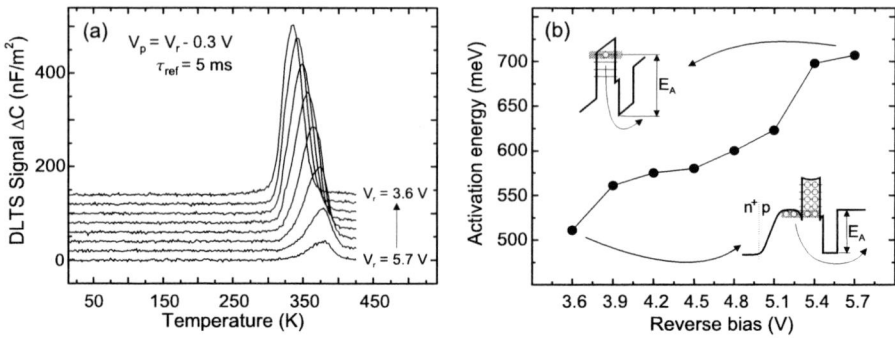

Fig. 11.10. Charge-selective DLTS spectra of thermally activated hole emission from InGaAs/GaAs QDs with an additional $Al_{0.9}Ga_{0.1}As$ barrier below the QD layer (**a**). Dependence of the thermal activation energy on the reverse bias V_r

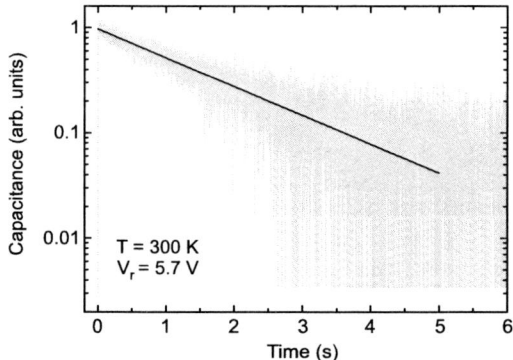

Fig. 11.11. Time-resolved capacitance transient of hole emission from InGaAs/GaAs QDs with an additional $Al_{0.9}Ga_{0.1}As$ barrier (for a reverse bias of $V_r = 5.7$ V). An average hole storage time of 1.6 s at $T = 300$ K is obtained from a linear fit of the transient on a semilogarithmic scale

p-doped GaAs ($p = 3 \times 10^{16}$ cm^{-3}). The QDs are placed 400 nm below the p–n junction and an additional $Al_{0.9}Ga_{0.1}As$ barrier of 20 nm thickness is situated 7 nm below the QD layer.

Figure 11.10(a) shows the charge-selective DLTS spectra. The pulse bias height was fixed here to 0.3 V for all spectra ($V_p = V_r - 0.3$ V). At $V_r = 5.7$ V a peak in the DLTS spectrum appears at 380 K, related to thermally activated hole emission from the QD ground state across the $Al_{0.9}Ga_{0.1}As$ barrier. As mentioned before, the peak at 380 K for $\tau_{ref} = 5$ ms represents an average storage time of 5 ms for hole emission from the QD ground states across the $Al_{0.9}Ga_{0.1}As$ barrier. To obtain the storage time at room temperature, the time-resolved capacitance measurement recorded at 300 K is plotted on a semilogarithmic scale in Fig. 11.11. From a linear fit an average hole storage time of 1.6 s at room temperature is determined.

Furthermore, the thermal activation energies are shown in Fig. 11.10(b). The highest value of (710 ± 40) meV at $V_r = 5.7$ V is related to thermal activation from QD hole ground states across the $Al_{0.9}Ga_{0.1}As$ barrier. The decrease in the activation energy corresponds to an increase in the average occupation of the QDs. At $V_r = 3.6$ V, carrier emission from the valence band edge is probed and the activation energy has a value of 520 meV. This energy now represents the energetic height of the 90% $Al_{0.9}Ga_{0.1}As$ barrier.

11.4 Conclusion and Outlook

This chapter studied carrier emission from different QD systems in order to determine the carrier storage time at room temperature. In addition, the localization energy was obtained by using time-resolved capacitance spectroscopy (DLTS) and related to the hole/electron retention time. The results of the experiments are summarized in Fig. 11.12.

An electron/hole localization energy of 290/210 meV, respectively, was obtained for InGaAs QDs embedded in a GaAs matrix. Based on these values the storage time at $T = 300$ K was estimated to $\sim$200 ns for electrons and $\sim$0.5 ns for holes. Furthermore, the more promising type-II GaSb/GaAs was studied in detail. It was found that these QDs can be charged with up to 15 holes. A ground state activation energy of 450 meV was determined, which accounts for a room temperature emission time on the order of one microsecond. The ground state localization is about twice as large and the retention time at room temperature is about three orders of magnitude longer than in InGaAs/GaAs QDs. Finally, we investigated the hole emission from InGaAs/GaAs QDs with an additional AlGaAs barrier. The activation energy for the QD ground states increases from 210 meV to 560 meV for an additional $Al_{0.6}Ga_{0.4}As$ barrier and the hole storage time is in the order of milliseconds. Using a 90% $Al_{0.9}Ga_{0.1}As$ barrier increases the activation energy accordingly from 210 meV to 710 meV and the hole storage time at room temperature

Material system	Charge carrier typ	Localization energy	Storage time at 300 K
InAs / GaAs [1]	Hole Electron	210 meV 290 meV	~0.5 ns ~200 ns
Ge / Si [2]	Hole	350 meV	~0.1 µs
GaSb / GaAs [3]	Hole	450 meV	~1 µs
InAs / GaAs with $Al_{0.6}Ga_{0.4}As$ barrier [4]	Hole	560 meV	5 ms
InAs / GaAs with $Al_{0.9}Ga_{0.1}As$ barrier	Hole	710 meV	1.6 s

[1] M. Geller et al., PRB **73**, 205331 (2006)
[2] C. M. A. Kapteyn et al., Appl. Phys. Lett. **77**, 4169 (2000)
[3] M. Geller et al., Appl. Phys. Lett. **82**, 2706 (2003)
[4] A. Marent et al., Appl. Phys. Lett. **89**, 072103 (2006)

Fig. 11.12. Summary of the electron/hole storage time (at 300 K) and localization energy for different QD systems

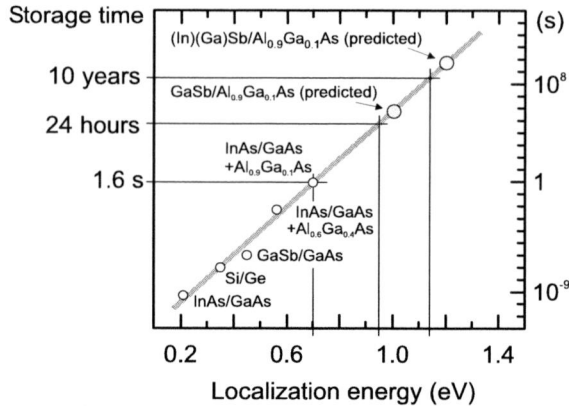

Fig. 11.13. Hole storage time at room temperature versus the localization energy (from Fig. 11.12). The gray line is a fit to the data and corresponds to (11.1)

increases by nine orders of magnitude to 1.6 s. This value is already three orders of magnitude longer than today's DRAM refresh time, which is in the millisecond range.

From the experimental results, a material combination can be predicted to obtain a storage time of more than ten years at room temperature. Figure 11.13 displays the hole storage time in dependence of the localization energy. The gray line is a fit to the data and corresponds to (11.1). This rough estimation gives a retention time of 24 hours for a localization energy of about 0.95 eV. The hole ground state localization energy in GaSb/GaAs QDs is 450 meV and the AlAs/GaAs valence band offset was found to be 550 meV [23], yielding a possible localization energy in GaSb/AlAs QDs of about 1.0 eV. Hence, a retention time of several days at room temperature is possible.

(In)(Ga)Sb QDs in a Al(Ga)As matrix may provide a hole localization of more than 1.2 eV. This localization energy could maintain a storage time on the order of ∼200 years, fulfilling an important prerequisite for a nonvolatile memory. Therefore, based on the experimental results, type-II (In)(Ga)Sb QDs in a Al(Ga)As matrix are considered to be a very promising material combination for a future QD memory.

Acknowledgements

This work was partly funded by the SANDiE Network of Excellence of the European Commission, Contract No. NMP4-CT-2004-500101, the Nanomat project of the European Commission Growth Program, Contract No. G5RD-CT-00545 and SFB 296 of DFG. We acknowledge the support of and helpful discussions with Prof. D. Bimberg, Dr. C. Kapteyn, and E. Stock. Dr. L. Müller-Kirsch, K. Pötschke, D. Feise, A.P. Vasi'ev, E.S. Semenova, Prof. A.E. Zhukov, Prof. V.M. Ustinov are acknowledged for providing the samples.

References

1. G.E. Moore, Electronics **38** (1965)
2. *International Roadmap for Semiconductors* (2005 edn., January 2006)
3. T. Müller, F.F. Schrey, G. Strasser, K. Unterrainer, Appl. Phys. Lett. **83**, 3572 (2003)
4. M. Geller, A. Marent, E. Stock, A.E. Zubkov, I.S. Shulgunova, A.V. Solomonov, D. Bimberg, Appl. Phys. Lett. **89**, 232105 (2006)
5. M. Geller, A. Marent, D. Bimberg, Speicherzelle und Verfahren zum Speichern von Daten. German patent application Nr. 10 2006 059 110.0 (27.10.2006)
6. L. Geppert, IEEE Spectrum **49** (March 2003)
7. S.M. Sze, in *Future Trends in Microelectronics*, ed. by S. Luryi, J. Xu, A. Zaslavsky (Wiley, New York, 1999), p. 291
8. H. Mizuta, Single- and few-electron memories, in *12th Japanese–German IT Forum*, 1998
9. D.V. Lang, J. Appl. Phys. **45**, 3023 (1974)
10. P. Blood, J.W. Orton, *The Electrical Characterization of Semiconductors: Majority Carriers and Electron States* (Academic, London, 1992)
11. C.M.A. Kapteyn, F. Heinrichsdorff, O. Stier, R. Heitz, M. Grundmann, N.D. Zakharov, D. Bimberg, P. Werner, Phys. Rev. B **60**, 14265 (1999)
12. C.M.A. Kapteyn, M. Lion, R. Heitz, D. Bimberg, P.N. Brunkov, B.V. Volovik, S.G. Konnikov, A.R. Kovsh, V.M. Ustinov, Appl. Phys. Lett. **76**, 1573 (2000)
13. C.M.A. Kapteyn, M. Lion, R. Heitz, D. Bimberg, C. Miesner, T. Asperger, G. Abstreiter, Appl. Phys. Lett. **77**, 4169 (2000)
14. M. Geller, C. Kapteyn, L. Müller-Kirsch, R. Heitz, D. Bimberg, Appl. Phys. Lett. **82**, 2706 (2003)
15. C. Kapteyn, *Carrier Emission and Electronic Properties of Self-Organized Semiconductor Quantum Dots* (Mensch & Buch, Berlin, 2001) [Dissertation, Technische Universität Berlin]
16. M. Geller, E. Stock, R.L. Sellin, D. Bimberg, Physica E **32**, 171 (2006)
17. M. Geller, E. Stock, C. Kapteyn, R.L. Sellin, D. Bimberg, Phys. Rev. B **73**, 205331 (2006)
18. O. Stier, M. Grundmann, D. Bimberg, Phys. Rev. B **59**, 5688 (1999)
19. P.N. Brunkov, A.R. Kovsh, V.M. Ustinov, Y.G. Musikhin, N.N. Ledentsov, S.G. Konnikov, A. Polimeni, A. Patanè, P.C. Main, L. Eaves, C.M.A. Kapteyn, J. Electron. Mater. **28**, 486 (1999)
20. F. Hatami, M. Grundmann, N.N. Ledentsov, F. Heinrichsdorff, R. Heitz, J. Böhrer, D. Bimberg, S.S. Ruvimov, P. Werner, V.M. Ustinov, P.S. Kop'ev, Z.I. Alferov, Phys. Rev. B **57**, 4635 (1998)
21. L. Müller-Kirsch, R. Heitz, U.W. Pohl, D. Bimberg, Appl. Phys. Lett. **79**, 1027 (2001)
22. A. Marent, M. Geller, D. Bimberg, A.P. Vasi'ev, E.S. Semenova, A.E. Zhukov, V.M. Ustinov, Appl. Phys. Lett. **89**, 072103 (2006)
23. J. Batey, S.L. Wright, J. Appl. Phys. **59**, 200 (1986)

12

Visible-Bandgap II–VI Quantum Dot Heterostructures

Ilya Akimov, Joachim Puls, Michael Rabe, and Fritz Henneberger

Abstract. The epitaxial growth of self-assembled CdSe/ZnSe quantum dot structures is described. Optical studies on a single-dot level uncover the fundamental electronic excitations of these quasi-zero dimensional structures and their dynamical interactions. The spin finestructure of electron–hole pairs in neutral and charged quantum dots is elaborated. Coherent control of the exciton–biexciton system is experimentally achieved by two-photon excitation. The spin dynamics of holes, electrons, and excitons is investigated. The hyperfine interaction of the electron spin with the nuclear moments of the surrounding lattice atoms is elucidated and a new type of dynamical nuclear polarization is demonstrated. The spin interactions in diluted magnetic quantum dot structures are addressed.

12.1 Introduction

The first zero-dimensional (0D) semiconductor nanostructures were made of II–VI materials. Nanocrystallites aggregated in glass or synthesized by chemical methods are capable of quantum-size energies as large as 1 eV [1, 2]. The epitaxial growth of II–VI quantum dots (QDs) has been tackled only after some delay compared to InAs/GaAs or Si/Ge. In this contribution, we review work on the prototype CdSe/ZnSe heterosystem. Molecular-beam epitaxy (MBE) is used for the fabrication of the structures. The possibility to monitor the growth in real time using different analytical tools and the high flexibility in the choice of the growth parameters make this technique especially suited for explorative studies. Wide-bandgap II–VI QDs stand out by their strong electron-hole (eh) correlation. Using single-dot spectroscopy, we have uncovered the eh-exchange mediated fine structure of the particle states and their magnetic coupling, the coherent quantum dynamics of excitons and biexcitons, as well as the spin relaxation in 0D—all important in the context of using QDs in quantum information processing. Many of these findings have been later reproduced on III–V materials.

12.2 Epitaxial Growth

The lattice mismatch of CdSe on ZnSe(001) is 7.2% and thus very similar to the InAs/GaAs heterosystem for which Stranski–Krastanow (SK) growth was earlier established. The first study for ternary (Zn,Cd)Se on ZnSe revealed 3D localization of the electronic excitations beyond that in standard quantum wells when the Cd content exceeds 30% [3]. The epitaxy of pure CdSe on ZnSe has yielded different, even conflicting results and the mechanisms behind the QD formation as well as their morphology and chemical composition has come under debate [4–14]. Mapping of the Cd distribution of capped structures by transmission electron microscopy (TEM) has merely exposed Cd-enriched regions embedded in a 2D film of ternary (Cd,Zn)Se with lower Cd concentration [15, 16].

Island formation is not necessarily a signature of an equilibrium QD growth mode, but can be simply related to a kinetically controlled and hence incomplete state of the surface. A systematic study of the MBE of CdSe on ZnSe has led us to a specific procedure for the fabrication of QD structures with clear SK morphology and minimized interdiffusion [6, 10, 17–19]. First, ZnSe is grown on [001] oriented GaAs substrate. The optimum temperature for layer-by-layer growth of ZnSe is about $T_G = 310°C$. Deposition of CdSe on ZnSe at this temperature results merely in uncontrollable surface inhomogeneities. However, as revealed by in-situ reflection high-energy electron diffraction (RHEED), coherent Frank–van-der-Merwe growth of CdSe is accomplished when the temperature is sufficiently lowered. In a relatively narrow temperature window ($T_G = 210$–$230°C$), prominent intensity oscillations (Fig. 12.1(a)) appear and the RHEED pattern exhibits sharp diffraction rods from the reconstructed surface, even after five full oscillation periods. Correlating the number and phase of the RHEED oscillations with the material amount deposited, taken from X-ray interferometry, enables monolayer (ML) control of the film thickness by phase-locked epitaxy. Quickly raising the temperature while capping the film with ZnSe forms a quantum well geometry. For more than 3 MLs, pseudomorphic growth is rapidly lost. A decrease of the rod distance in the RHEED pattern and the absence of both X-ray interferences and a measurable photoluminescence (PL) signal signify misfit dislocation-related plastic relaxation. For temperatures above the 2D growth window, increasing surface roughening is indicated by the instant appearance of a strongly enhanced diffuse background in the RHEED pattern. In addition, superstructure reflections displaying surface reconstructions vanish and TEM yields an enhanced (linear) dislocation density of $3 \cdot 10^5 \text{ cm}^{-1}$. The loss of the 2D regime on the low-temperature side is consistent with the reduced mobility of the adatoms, inhibiting surface completion.

The standard SK growth mode where a 2D–3D transition occurs during deposition could not be found for CdSe/ZnSe. However, the as-grown pseudomorphic 2D CdSe film undergoes a striking transformation into a QD morphology under post-growth thermal activation [6], directly verified by the atomic-force microscopy (AFM) images summarized in Fig. 12.1(b). To avoid desorption during activation, the surface is kept Se-rich. The reorganization of the film is also tracked in real time by RHEED, where the otherwise streaky pattern becomes increasingly superimposed

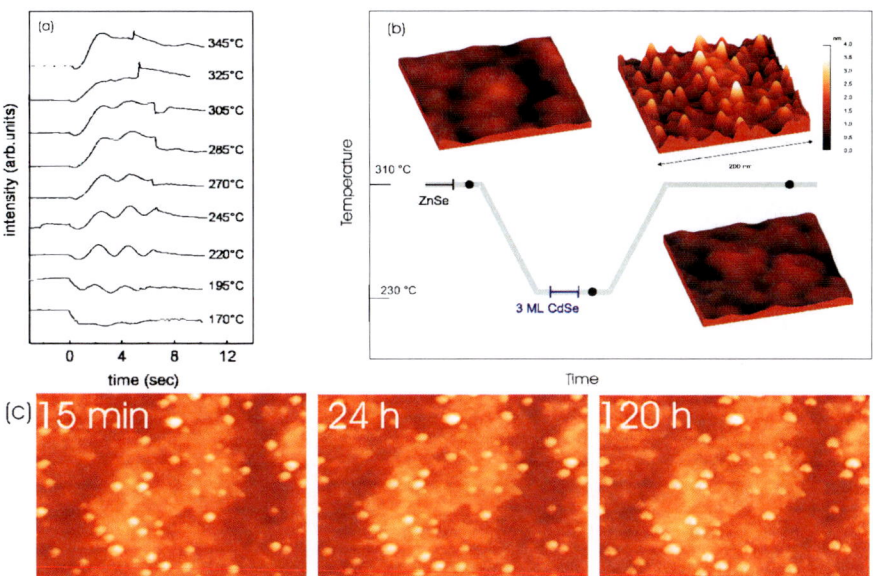

Fig. 12.1. (**a**) Specular beam RHEED intensity oscillations for growth of CdSe on ZnSe at different T_G. (**b**) AFM images revealing the reorganization of a 2D CdSe film into an array of QDs. Top left: ZnSe grown at $T_G = 310°C$; bottom right: CdSe film grown on top at $T_G = 230°C$; top right: after thermal activation at $T_G = 310°C$. (**c**) AFM images of the QDs at certain times after formation taken under ultra-high-vacuum conditions

by distinct diffraction spots [6]. A quantitative analysis of the rod distance, as well as the intensity profile along the diffraction rods, shows that strain relaxation is indeed involved in the island formation [6].

Statistical evaluation of the AFM images—subtracting the long-ranged background variation prior to reorganization by means of a local-mean-value algorithm—provides a Gaussian height distribution of the QDs with 1.6 nm average height and ±0.1 nm standard deviation. The lateral extension of the contrast features is resolution limited, yielding an upper limit of 10 nm for the in-plane QD extension. The average QD density is $(1.1 \pm 0.3) \cdot 10^{11}$ cm^{-2}. From these data, it follows that only a 15–20% fraction of a ML reorganizes in QDs. Artifacts appearing in AFM measurements under ambient air conditions have led to the conjecture that pronounced ripening effects occur in the CdSe/ZnSe heterosystem [7]. Figure 12.1(c) depicts images taken in the MBE apparatus under ultra-high vacuum at room temperature. Neither density nor size changed during a continuous investigation over 120 h. In a long-term experiment, this has been confirmed up to 50 days. Even at an increased temperature of 310°C, only marginal changes are observed. In the standard scenario described by the Gibbs–Thomson equation, ripening is a consequence of an continuous enthalpic gain when the volume of a supercritical nucleus increases. However, theoretical studies on semiconductor SK morphologies have predicted that a minimum of the free energy per atom at a certain island size might exist as a result of energy renormaliza-

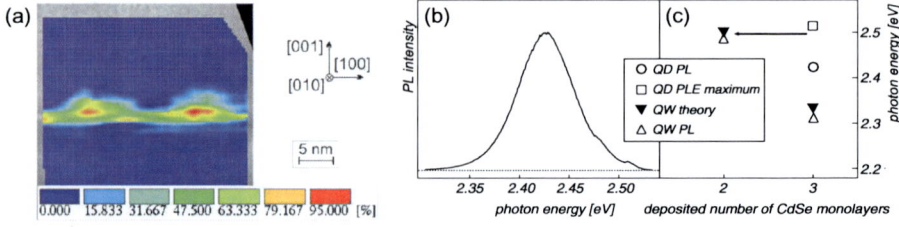

Fig. 12.2. (a) Chemical composition of capped ZnSe/CdSe/ZnSe QD structures evaluated by lattice fringe analysis (CELFA) [19]. The TEM signal is a 10–20 nm depth average which reduces only seemingly the Cd concentration (given by the color scale) in the island center slightly below 100%. (b) PL band of a CdSe QD ensemble at $T = 5$ K. (c) Maxima of PL and PLE of QDs and quantum wells for different number of MLs

tion in the dot/wetting-layer system [20], surface energy contributions from island edges or island-island interaction [21]. The observation of ultra-stable QD assemblages in the CdSe/ZnSe heterosystems is supportive of these predictions.

TEM investigations demonstrate that the SK morphology is maintained after ZnSe overgrowth [19]. The islands are situated on top of a continuous wetting layer and exhibit an extended core of pure CdSe (Fig. 12.2(a)). AFM (of uncapped structures) and TEM yield almost identical island heights, in-plane extensions, and QD density. Therefore, interdiffusion is not a critical issue, neither during activation of the 2D–3D transition nor during overgrowth. Note that the volume size of the QDs is typically two to three orders of magnitude smaller compared to their InAs/GaAs counterparts. This is confirmed by PL and PL excitation (PLE) measurements (Fig. 12.2(b–c)). A useful feature of the thermal activation is that the confinement energy in 2D and 3D can be directly compared through experiments. The PL of the QDs activated from a 3 ML film is high-energy shifted by about 200 meV relative to the 3 ML quantum well. On other SK structures, a low-energy shift is typically found, caused by a confinement weakening in vertical direction. The PL of a 2 ML quantum well is close to the position of the PLE of a 3 ML QD structure suggesting a wetting layer of approximately 2 ML effective width. CdSe/ZnSe QDs fabricated by homoepitaxy on ZnSe substrates along the same recipes have identical properties [22]. Growth on transparent substrates is important in the context of optical and quantum optical devices.

CdSe films of 1 and 2 ML thickness do not undergo the reorganization, confirming that the elastic energy is an essential factor in the QD formation. On the other hand, the thermally activated SK transition takes place just at critical thickness. Contrary to III–V materials, the interplay between strain relaxation and the formation of misfit dislocations is thus a characteristic feature of II–VI QDs. This point has been more explicitly elaborated by covering the 2D film with an amorphous cation layer prior to reorganization and reducing the tendency of dislocation formation [23, 24]. In this way, the growth of CdTe/ZnTe SK QDs succeeded [23]. The thermal activation procedure is also successful for the (CdMn)Se/ZnSe heterosystem up to Mn concentration of about 10% [25]. The in-plane lattice constant of MnSe is closer to

ZnSe so that the mismatch is reduced. In the standard strain scenario, a tendency of forming larger QDs with less area density is expected. In contrast, the QD size is only very little affected, whereas a dramatic decrease of the dot density occurs [25]. That is, incorporation of Mn reduces further the small material amount aggregated in the islands. Therefore, strain is not the only factor that drives the QD formation. Surface processes controlled by the diffusibility of the atomic species must play an important role. This conclusion is supported by the fact that a 3 ML thick CdSe film loses the capability of the 2D–3D transformation when kept at growth temperature for a sufficiently long time [6, 19]. A recent study addressed the kinetics between 2D precursors and the mobile surface adatoms in more detail and showed that the QD density can be engineered by delaying the activation step [26].

12.3 Few-Particles States and Their Fine Structure

12.3.1 Excitons and Biexcitons

The spatially resolved PL from the QD structures decomposes into discrete lines. A typical spectrum under weak excitation is shown in Fig. 12.3(a). The spectral width of the lines of about 40 μeV is resolution limited. For samples grown at enhanced purity conditions, there is no resolvable spectral line drift or blinking effect within typical detection periods of 30 min [27]. Closest to the ground-state emission, the PLE spectrum exhibits broader features due to acoustic and LO-phonons as well as a narrow peak related to the first excited hole state (Fig. 12.3(b)). In experiments addressing coherent dynamics or spin lifetimes, resonant excitation is desired. Direct excitation of the ground-state is faced with enormous stray-light problems. However, the phonon features allow for selection of a particular QD with minimum excess energy. A demonstration is depicted in the inset of Fig. 12.3(b) for the acoustic band: Only one single line is emitted in this case. In what follows we refer to such excitation as "quasi-resonant".

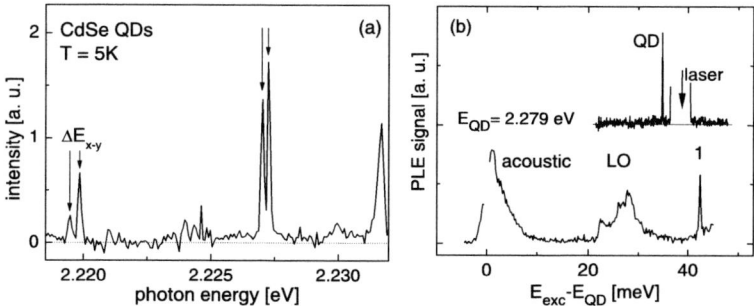

Fig. 12.3. (a) PL spectrum from $150 \times 150\,\text{nm}^2$ mesa prepared by wet chemical etching of a thermally activated CdSe/ZnSe specimen. The mesa is selectively addressed both in excitation and emission in a confocal geometry. **(b)** Typical PLE spectrum of a single emission line in (a). Inset: Secondary emission under excitation in the acoustic phonon feature of a particular QD

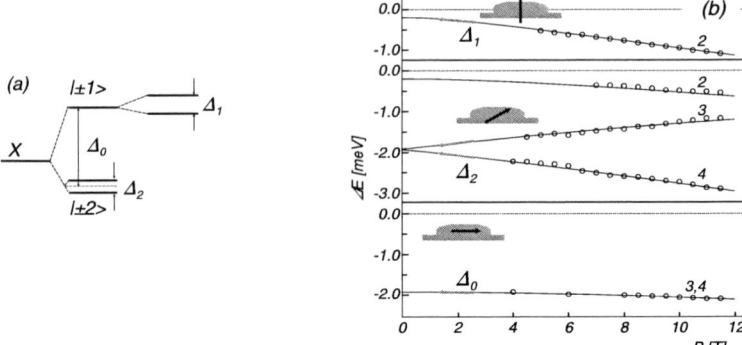

Fig. 12.4. (a) Spin fine structure of the QD exciton. (b) Experimental fan chart in an external magnetic field with different angles relative to the [001] quantization axis as indicated. The measurements are performed on ensembles. The excitation selects resonant QDs with the same high-energy component while the secondary emission of the lower components is recorded

Evidently, the single-dot PL lines exhibit a characteristic double structure [28]. The energy splitting scatters across the QD ensemble from below resolution up to 500 µeV. The $s = 1/2$ electron spin and the $J = 3/2$ heavy-hole angular momentum give rise to four exciton states with total spin projections $|\pm 1\rangle$ and $|\pm 2\rangle$. The eh exchange interaction reorganizes the energy spectrum of these otherwise degenerate states as shown in Fig. 12.4(a). The short-ranged exchange sJ moves the optically active states $|\pm 1\rangle$ away from the dark excitons $|\pm 2\rangle$ by Δ_0. For [001] growth, the D_{2d} symmetry of the zincblende crystal implies that the in-plane directions [110] and [1$\bar{1}$0] are equivalent. In this case, a higher-order sJ^3 exchange term is present that characteristically splits both the radiative and nonradiative doublet ($\Delta_1, \Delta_2 \neq 0$) [29]. However, shape, composition as well as strain anisotropy can reduce the symmetry to C_{2v}. Then, the long-ranged exchange comes into play, which acts only in the subspace of the optically allowed states ($\Delta_2 = 0$). Application of an external magnetic field in tilted geometry allows us to elucidate the true fine structure, because all four exciton states become visible [30]. Interpolating the fan charts in Fig. 12.4(b) down to $B = 0$ provides the zero-field splitting. It turns out that Δ_2 is below the spectral resolution of 20 µeV, while the averages of Δ_0 and Δ_1 are 1.9 meV and 200 µeV, respectively. This finding clearly identifies the D_{2d}–C_{2v} symmetry reduction as the major origin of the exciton PL doublet in SK QDs.

At higher excitation density, the exciton features saturate and new lines related to biexciton emission emerge on the low-energy side. The energy shift is given by the interaction energy ΔE_{XX} of the two excitons constituting the biexciton. This energy is as large as 20–25 meV for CdSe/ZnSe QDs [29, 31]. The complete transition scheme of the exciton-biexciton system is drawn in Fig. 12.5(a). The radiative exciton states $|x/y\rangle \propto |+1\rangle \pm |-1\rangle$ created by the eh exchange are linearly polarized along the principal in-plane axes of the QD. The biexciton ground-state is nondegenerate as both electrons and holes have opposite spin orientation. The spin part of the

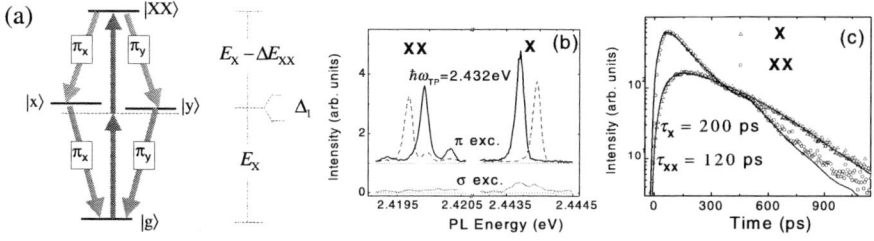

Fig. 12.5. (a) Optical transition scheme of the exciton-biexciton system. (b) Exciton (X) and biexciton (XX) emission lines from a single QD under circularly (σ) and linearly (π) polarized TP excitation. The polarization detection is along the two orthogonal transition dipole axes. (c) Decay transients of biexciton and exciton emission. Solid lines are double-exponential data fits accounting for the apparatus response

wavefunction can be thus written as $|XX\rangle \propto |x\rangle|x\rangle - |y\rangle|y\rangle - |+2\rangle|-2\rangle - |-2\rangle|+2\rangle$. The polarization of the biexciton decay is hence the same as for the exciton left behind. Resonant excitation of the biexciton is achieved by two-photon (TP) absorption where the photons have the energy $\hbar\omega = E_X - \Delta E_{XX}/2$. The cascaded spontaneous emission of a QD subsequent to TP excitation is depicted in Fig. 12.5(b–c) [32]. In full accord with the transition scheme of Fig. 12.5(a), exciton and biexciton features, placed symmetrically to the excitation photon energy, are present. Both features consist of a fine structure doublet of linearly cross-polarized lines, however, with a reversed sequence in the polarization in the exciton and biexciton emission. It is also demonstrated that the TP transition is forbidden in circular excitation polarization. The time-resolved emission shown in Fig. 12.5(c) clearly confirms the existence of an emission cascade, with the biexciton recombining first and a respective rise time for the exciton. The radiative lifetime of the biexciton ($\tau_{XX} = 120$ ps) is about two times shorter than for the exciton ($\tau_X = 200$ ps). This finding is consistent with the fact that the biexciton has two recombination channels and that the exciton and exciton-biexciton transition possess almost the same dipole moment.

12.3.2 Trions in Charged Quantum Dots

Most II–VI semiconductors are naturally n-doped. A certain fraction of the QDs, depending on the stoichiometry regime of the MBE growth, are thus charged by a single electron. The fundamental optical excitation of a singly charged QD is the trion, for negative charge, a state of two electrons and one hole. Contrary to the exciton, the trion is a particle with half-integer spin and, according to the Kramer's theorem, its eigenstates are doublets degenerate in the absence of a magnetic field. In the singlet ground-state, the electron spins are compensated and the total spin is thus given by the hole. The transition schematics of the trion accounting for only heavy-hole states is depicted in Fig. 12.6(a). Measurements in an external magnetic field allow for the unambiguous identification of the trion emission (Fig. 12.6(b–c)) [33]. Unlike the exciton in uncharged QDs, no zero-field splitting remains when the fan charts are

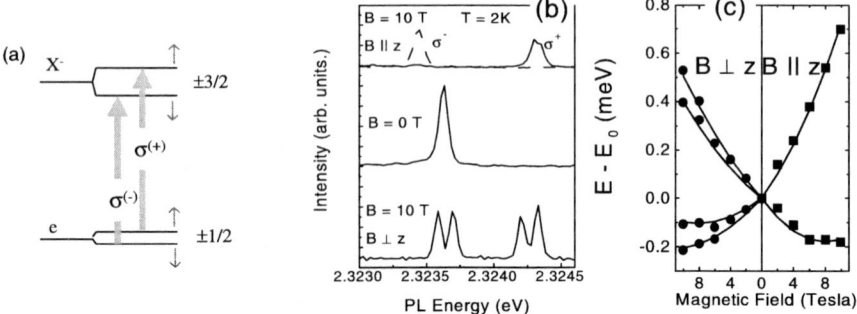

Fig. 12.6. (a) Optical transition scheme of the trion in a negatively charged QD. (b) Single-dot PL spectra at $B = 0$ (*middle*) and $B = 10\,\mathrm{T}$ in Faraday (*top*) and Voigt (*bottom*) geometry. (c) Line position versus magnetic field

extrapolated down to $B = 0$. In a transverse field, all four optical transitions are uncovered. From this finding, it can be further inferred that the hole possesses a noticeable transverse g factor. In D_{2d} symmetry, the heavy hole exhibits no transverse Zeeman interaction linear in the magnetic field. A higher-order coupling $\boldsymbol{J}^3\boldsymbol{B}$ is very weak and generally neglected. In lower symmetry, heavy-light hole mixing appears and, by this, the in-plane g factor of the light hole is incorporated [34]. This picture is confirmed by a study of the polarization properties of the line quartet as a function of the orientation of the transverse field relative to the crystal axes [34]. The g factor is extremely anisotropic and reverses sign between the principal in-plane axes. The largest absolute values observed are about $g_h^\perp \sim 0.3$. The existence of an appreciable transverse g factor of the hole is of essence for protocols in quantum information processing using QD spins.

Heavy-light hole coupling is also manifested by a partial linear polarization degree of the zero-field trion emission. Both this degree as well as the g factor scatter similarly as the energy splitting of the radiative excitons in uncharged QDs across the ensemble. In the studies on trions presented in the following, we have selected QDs where heavy-light hole mixing can be safely ignored. Note that the mixing contributes also to the splitting of the radiative exciton in uncharged QDs, but here can hardly be separated from the exchange contribution.

Consistent with Kramer's theorem, the singlet trion ground-state is not affected by the eh exchange. The first excited state, where one electron occupies the second shell, is built up by a triplet state ($S = 1$) with total spin projections $|\pm 1/2\rangle$, $|\pm 3/2\rangle$, $|\pm 5/2\rangle$ and a singlet state ($S = 0$), considerably high-energy shifted by the electron–electron exchange. As the total electron spin is not compensated in the triplet state, the eh exchange is active here [35, 36]. The isotropic part creates a set of three levels equidistantly spaced by the energy $\tilde{\Delta}_0$ (Fig. 12.7(a)). In the exciton, ignoring the weak contribution from the $s\boldsymbol{J}^3$-term, the anisotropic part couples states that differ in their angular momentum by ± 2. This translates into a coupling of $|\pm\tfrac{3}{2}\rangle$ and $|\mp\tfrac{1}{2}\rangle$ in the trion triplet state producing in lowest order spin wave func-

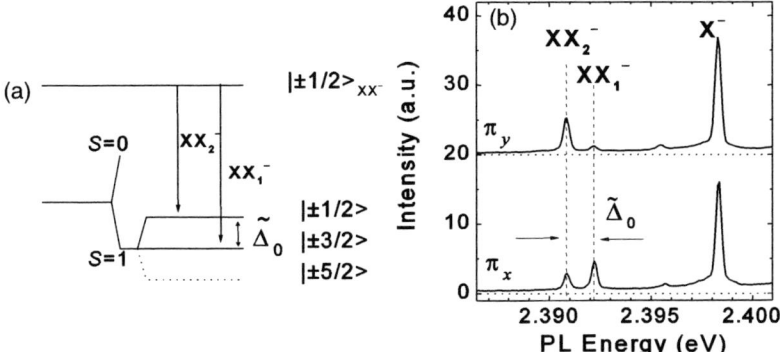

Fig. 12.7. (a) Spin structure of the charged biexciton and the trion triplet state; arrows indicate optically allowed transitions. (b) Emergence of the charged biexciton lines. π_x and π_y refer to linear polarization detection along [110] and [1$\bar{1}$0], respectively

tions of the type $\cos(\alpha)|\pm 3/2\rangle - \sin(\alpha)|\mp 1/2\rangle$ and $\cos(\alpha)|\pm 1/2\rangle + \sin(\alpha)|\mp 3/2\rangle$, $\tan(2\alpha) = \sqrt{2}\tilde{\Delta}_0/\tilde{\Delta}_1$. The respective optical transitions hence become partially linearly polarized. The energies $\tilde{\Delta}_0$ and $\tilde{\Delta}_1$ are different from the exciton as one of the electrons is in an excited state in the trion triplet.

PLE measurements reveal that the triplet state is situated about 80 meV above the singlet ground-state [37–39]. Its direct PL is too weak to be observed. However, the radiative recombination of the charged biexciton can be used as a monitor of the triplet state [40]. In the ground-state of this five-particle complex, according to Pauli's principle, the third electron must occupy the second shell. The most probable transition is recombination of an electron and a hole from the lowest shells hence leaving behind an excited trion in the final state. Owing to the 1/2 spin of the biexciton, its radiative decay obeys the same polarization selection rules as the trion-electron transition. Indeed, at higher pump intensities, the emission from the charged biexciton appears in a characteristic double structure of partial linear polarization (Fig. 12.7(b)). Inspecting a set of QDs, the line separations provide an average $\tilde{\Delta}_0 = 1.5 \pm 0.5$ meV and, from the linear polarization degree as well as from the Zeeman splitting in a longitudinal magnetic field, it follows $\tilde{\Delta}_1/\tilde{\Delta}_0 \sim 0.5$ [36]. This ratio is distinctly larger than for the exciton in uncharged QDs which is caused by the stronger asymmetry of the p-type wave function of the excited electron involved in the exchange. The anisotropy-induced mixing in the trion triplet state allows for a combined electron–hole spin flip by which the spin of the resident electron can be aligned [41]. The magnitude of the anisotropic exchange contribution is thus an important parameter in the context of QD-based spin devices.

12.4 Coherent Control of the Exciton–Biexciton System

Making the identification $|00\rangle = |g\rangle$, $|10\rangle = |x\rangle$, $|01\rangle = |y\rangle$, and $|11\rangle = |b\rangle$, a two-quantum-bit system is built up by the QD states [42]. Coherent optical cou-

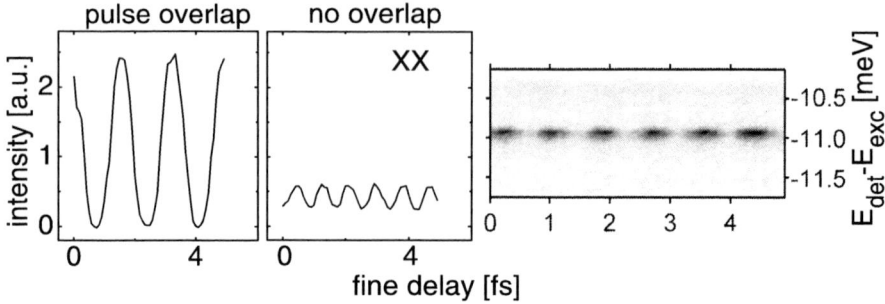

Fig. 12.8. Single-dot TP coherent control of the biexciton monitored by the signal modulation of the biexciton PL line as a function of the time difference produced by the passage of the excitation pulses in the interferometer arms. Left: No pulse delay. Middle: Pulse delay 5 ps. Right: Two-dimensional plot in the photon-energy/time-difference plane, the signal intensity is gray-coded

pling of the ground-state and the biexciton creates an entangled state of the type $a_{00}|00\rangle + a_{11}|11\rangle$. Manipulation of entangled states is a key step in conditional quantum dynamics and logical gates. Using a pair of two ps-pulses with a well-defined phase shift produced in a Michelson interferometer, coherent control of the biexciton by TP excitation is accomplished [43]. Without delay, the emission from the biexciton reflects merely the optical interference of the pulses and the signal is a periodic function of the single phase shift (Fig. 12.8). However, beyond pulse overlap, the behavior changes qualitatively: a doubling of the modulation frequency occurs and the base line drops by a factor of 2, both are characteristic of a TP process. The interferograms in Fig. 12.8 directly visualize the quantum coherence of the biexciton state. Figure 12.8 also shows that the degree of entanglement can be optically adjusted by appropriate choice of the phase shift.

An important advantage of the TP excitation is that the spontaneous emission from the cascaded biexciton–exciton decay is clearly separated from the excitation stray light. The ability to track the QD emission is essential, as it enables one to generate nonclassical light states, e.g., antibunched single-photon emission [44] or entangled photon pairs [45]. Using a second pulse that is resonant to the exciton–biexciton transition, the conversion of the biexciton into an exciton by stimulated emission is accomplished [32]. In this way, the recombination path is strictly defined by the polarization of the stimulation (ST) pulse, in contrast to the spontaneous emission cascade. The result of the stimulation process is a photon as well as an exciton. While the photon can hardly be detected, the stimulated exciton is manifested by the photon that it emits subsequently. The data in Fig. 12.9 directly verify this scenario. While both exciton emission components have equal intensity only for TP excitation, the line polarized along the ST polarization is amplified at the expense of the cross-polarized line when both pulses are present. During pulse delay, the biexciton state is increasingly emptied by spontaneous decay. Consistent with the radiative lifetime, the transition is no longer capable of stimulated emission after about 250 ps.

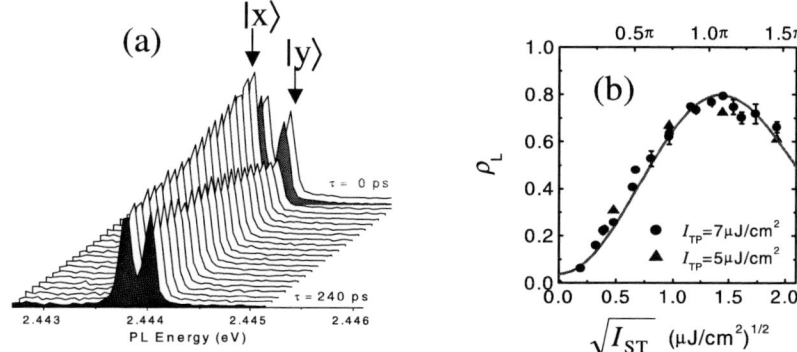

Fig. 12.9. (a) Evolution of the single-dot exciton emission spectrum as a function of the delay τ between TP and ST pulse. The ST polarization is along the polarization of the low-energy line. (b) Induced linear polarization degree versus square-root ST pulse-density (TP-ST pulse delay: 5 ps). The solid line shows a fit to the data based on the numerical solution of the master equation for the two-exciton density matrix. For details see [32]

In a quantum information sense, the stimulated emission corresponds to a disentanglement of the Bell-like state $a_{00}|00\rangle + a_{11}|11\rangle \rightarrow (a_{00}|0\rangle + a_{11}|1\rangle)|0\rangle$. The TP approach is different from the three-pulse arrangement of [46] as it allows for an arbitrary time (within the biexciton lifetime) to pass after the creation of the biexciton to apply the disentangling pulse. In an incoherent regime, the linear polarization degree $\rho_L = (I_x - I_y)/(I_x + I_y)$ is limited to $\rho_L = 0.5$, because the inversion of the exciton and biexciton populations cannot exceed zero. Figure 12.9(b) demonstrates that this limit is clearly overcome and that even the onset of Rabi oscillations is manifested within the experimentally available pulse-density range. The damping of the oscillations is related to an essentially non-Markovian acoustic phonon contribution [47]. The resulting decoherence limits the entanglement of formation that can be achieved by the TP pulse to a value of 0.5. This value given, the experimentally demonstrated degree of disentanglement is 100% [32].

12.5 Spin Relaxation of Excitons, Holes, and Electrons

Use of the spin for information processing has initiated extensive research in recent years. Practical applications require long spin lifetimes to store and to manipulate the spin without losses. In this regard, the spin states in QDs have attracted much interest. Here, the 0D confinement can split the energy levels apart so much that the standard spin relaxation mechanisms controlling the spin dynamics in bulk semiconductors are strongly suppressed.

12.5.1 Exciton Quantum Coherence

As described in Sect. 12.3.1, the optically active exciton is composed of a doublet of linearly cross-polarized eigenstates. Using linear polarization excitation, which is rotated by $\pi/4$ relative the principal axes, a quantum superposition $\Psi = (|x\rangle + |y\rangle)/\sqrt{2}$

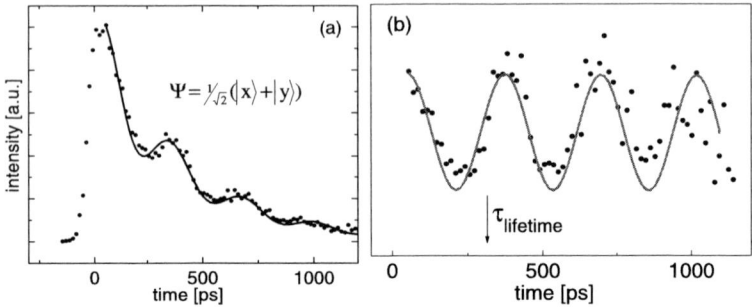

Fig. 12.10. (a) Time-resolved single-dot exciton PL under quasi-resonant excitation (1-LO) with linear polarization along [100] direction. (b) Decay-corrected signal modulation

is prepared [27]. Figure 12.10(a) depicts the time-resolved secondary emission of the exciton subsequent to quasi-resonant excitation.

The coherence of the superposition is signified by quantum beats with a period corresponding to the exchange splitting Δ_1. No beats are present for excitation polarization along the principal axes. The modulation is not 100% because a part of the coherence is spoiled before the exciton has reached the ground-state. The beats are superimposed to the overall signal decay due to depopulation of the exciton states by radiative recombination. If the decay is deducted, it turns out that no measurable damping of the beats occurs within the exciton lifetime (Fig. 12.10(b)). The coherence is limited by the exciton spin lifetime. It can therefore be concluded that the spin lifetime is longer than the lifetime of the exciton itself. This finding is at odds with bulk semiconductors or quantum wells where generally the opposite relation is found.

12.5.2 Hole Spin Lifetime

The substructure of the valence band-edge related to the p-symmetry of the Bloch factor makes the spin lifetime of the hole typically much shorter than for the electron. It is thus interesting whether the confinement-induced reduction of the spin relaxation rate of the exciton in QDs applies also to the hole. As set out in Sect. 12.3.2, the spin of the trion in a negatively charged QD is defined by the hole spin. Resonant excitation of the trion with circularly polarized light allows for the creation of holes with certain spin orientation. Hence, resolving in time the circular polarization degree $\rho_c = (I_+ - I_-)/(I_+ + I_-)$ of the secondary emission subsequent to short-pulse excitation, the dynamics of the hole spin are uncovered. The result of the experiment is summarized in Fig. 12.11 [33].

At low temperature almost no decay of ρ_c is observed. An estimate of the longitudinal spin lifetime yields $\tau_{1,h} > 10$ ns. Data recorded in a longitudinal magnetic field confirm this result [33]. Increasing temperature causes the spin-flip rate to grow with an activation energy corresponding to the energy of the LO-phonon (Fig. 12.11(b)). The transverse g factor of the hole originates from parity-conserving heavy-light

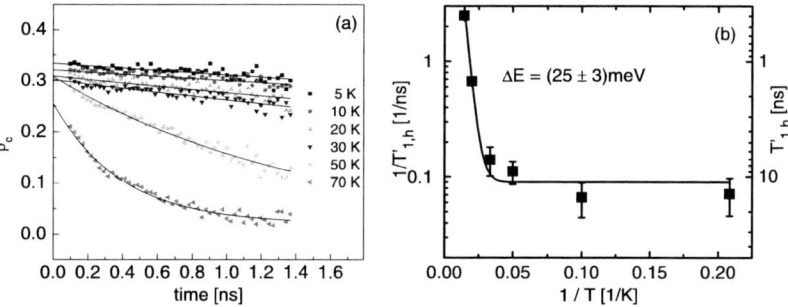

Fig. 12.11. (a) Time dependence of the degree of circular polarization after quasi-resonant σ^+-excitation of X^- at various temperatures. The solid lines are exponential fits to deduce the longitudinal hole spin lifetime $\tau_{1,h}$. (b) Temperature dependence of the rate $1/\tau_{1,h}$. Solid curve is a fit with $[0.1 + 84 \cdot \exp(-\Delta E/k_B T)]\,\text{ns}^{-1}$

hole coupling (Sect. 12.3.2). However, the spin flip requires parity breaking. Then, spin-independent interactions, e.g. with phonons, can induce transitions between the spin states. The very long low-temperature hole spin lifetime signifies that parity mixing is weak, again a result of the strong 0D confinement. Thermal population of the trion-LO-phonon state opens up new scattering channels which makes the spin relaxation faster.

12.5.3 Spin Dynamics of the Resident Electron

The spin of resident carriers in charged QDs is of particular interest as the time of operation is not intrinsically limited by radiative decay. Spin-orbit coupling as the leading process for the electron spin relaxation in the bulk or in quantum wells was predicted to be inefficient in QDs [48, 49]. In strong magnetic fields, when the spin states are sufficiently Zeeman-split, lifetimes in the millisecond range have been indeed observed [50, 51]. At weak or zero magnetic field, the hyperfine interaction of the electron with the nuclear moments of the lattice atoms sets an ultimate limit for the spin lifetime in QDs [52–54].

In the context of the hyperfine interaction, II–VI QDs deserve special attention. Low natural abundance ($a_{Cd} = 25\%$, $a_{Se} = 8\%$), a small value of the hyperfine coupling constant A related to a $1/2$ isospin, and a number of lattice atoms of only $N_L \sim 1000$ generate a dynamic scenario quite different from the widely studied InAs/GaAs QD system.

Spin polarization of the resident electron without application of an external magnetic field can be efficiently achieved by resonant excitation of the trion with circular light polarization [55]. Selective pumping, say with photons of σ^- orientation, addresses only spin-down electrons (cf. Fig. 12.6(a)). The resultant spin-down trions can radiatively recombine via the same channel or, after hole spin-flip, by emitting σ^+ photons and creating spin-up electrons. In this way, an optically driven spin occupation with more spin-up than spin-down electrons is established. As a consequence,

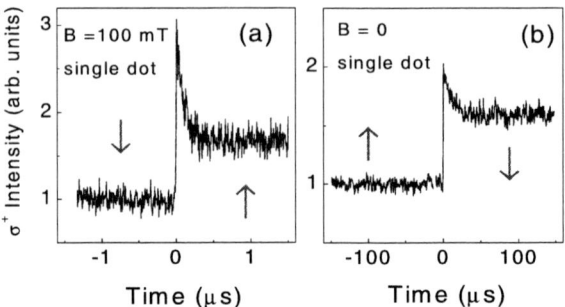

Fig. 12.12. Single-electron spin transients of CdSe QDs recorded via the X^--PL under stationary quasi-resonant excitation with rectangular modulation between σ^+ and σ^- polarization. The rise time is about 10 ns. (**a**) Weak longitudinal magnetic field. (**b**) No external field. Note that both transients are taken at the same optical excitation density

the absorption of σ^- photons is reduced which shows up in a decrease of the X^- PL signal providing a direct optical monitor of the spin population.

Figure 12.12 depicts single-dot spin transients recorded by circular polarization modulation of a stationary excitation power. Time-resolving the transient PL change after the polarization switch provides the electron spin lifetime given in the present experimental configuration with the time needed to establish a new steady-state spin population with reversed orientation. A longitudinal external magnetic field of 100 mT safely outranges the nuclear hyperfine field. An example for this case is given in Fig. 12.12(a) where a time constant of about 100 ns is seen. However, the transients are power dependent and the spin lifetime closely follows, over two orders of magnitude, an inverse-power law [56]. This finding reveals a specific spin-flip scenario where the QD captures an extra charge created in its environment as a concomitant of the direct optical excitation. For example, capture of a hole produces an exciton that radiatively recombines followed by recapture of another electron. Recharging processes of that type limit the lifetime of the electron itself, but also randomize the spin as the latter is undetermined when the QD recovers to single-electron occupation. For lowest excitation, when the X^- PL disappears in the noise floor, the spin lifetime reaches 10 μs, but no offset value is found. The intrinsic spin-orbit time of an isolated QD is thus significantly longer.

The much slower transient at zero field (Fig. 12.12(b)) reflects a situation where the electron spin is in steady-state while the spin lifetime increases slowly due to formation of a nuclear spin polarization (NSP), confirmed by clear asymmetries of the transients with respect to the direction of a very weak external magnetic field [55]. The formation time of the NSP is given by $1/\tau_{\text{NSP}} \sim (A/N_L)^2 \tau_c$ [57] where the correlation time τ_c represents the inverse broadening of the electron energy levels. In bulk semiconductors, because of the huge N_L, τ_{NSP} is much longer than the decay of the NSP caused by the nuclear dipole–dipole interaction (e.g. [60]). Here, the NSP occurs in an equilibrium mode through cooling of the nuclear spin system by an external magnetic field [57]. It has been argued that an external field is not required for QDs, as the hyperfine Knight field seen by the nuclei scaling like A/N_L is sufficiently strong to suppress the dipole–dipole interaction [55, 58]. On InAs/GaAs QDs with a more than 10 times larger Overhauser field, the presence of a NSP is directly visualized by a zero-field splitting of the X^- line with (indirectly measured) formation times in the range of 1 s [58]. Consistent with the two to three orders of magnitude smaller N_L of the present QDs, the transients in Fig. 12.12(b) demonstrate dramatically shorter times of $\tau_{NSP} \sim 10\text{–}100\,\mu s$. Complementary measurements with dark periods in the excitation provide dipole-dipole decay times τ_{dd} of several 100 µs [59]. CdSe/ZnSe QDs, owing to their small volume size, realize a situation of $\tau_{NSP} < \tau_{dd}$ where unlike the standard spin cooling concept a nonequilibrium nuclear polarization is created by the optical electron spin pump at zero external field.

Electron-nuclear spin dynamics of the above type are not restricted to low temperature but have been observed up to 100 K, signifying that the nuclear moments are strongly decoupled from the lattice. A detailed analysis provides maximum electron as well as nuclear spin degrees as large as 0.5 at about 50 K. The formation of the NSP allows for the suppression of the electron spin depolarization by the fluctuating nuclear field. Its observation at practical temperatures is thus an important step toward realizing potential applications.

12.6 Diluted Magnetic Quantum Dots

Diluted magnetic QD structures have the potential of controlling the carrier spin on a nm-length scale through the giant Zeeman effect. Embedding CdSe QDs in a magnetic (Zn,Mn)Se environment is one possibility to advance toward this goal [61, 62]. Truly magnetic QDs are formed in (CdMn)Se/ZnSe structures (Sect. 12.2). Critical issues include the location of the magnetic moments within the heterostructure and how the sp-d coupling is modified by the 0D confinement.

In accord with the bandgap, the PL of the (CdMn)Se/ZnSe QDs undergoes a prominent high-energy shift with increasing Mn concentration. A specific feature of these structures is that the energy required to promote a Mn-3d electron to the first excited shell is generally lower than the energy of the confined exciton. As a consequence, nonradiative energy transfer strongly decreases the QD PL yield by Auger processes. Figure 12.13(a) demonstrates that the yield recovers in a longitudinal external magnetic field where the Auger process becomes increasingly forbidden by spin selection rules. In parallel, the PL band splits in two oppositely circularly polarized Zeeman components (Fig. 12.13(c)). The g factor deduced from the low-field slope is as large as 300 [63]. At a higher magnetic field, because of thermalization, the high-energy band disappears. An exciton spin polarization degree of 90% is already reached at a field of only 0.5 T (Fig. 12.13(b)). Such a field can be supplied by permanent magnets.

(CdMn)Se/ZnSe quantum well structures with the same Mn concentration, fabricated by omitting the thermal activation step, exhibit practically the same g factors. A reduction of the sp-d interaction in QDs—as might be anticipated from the

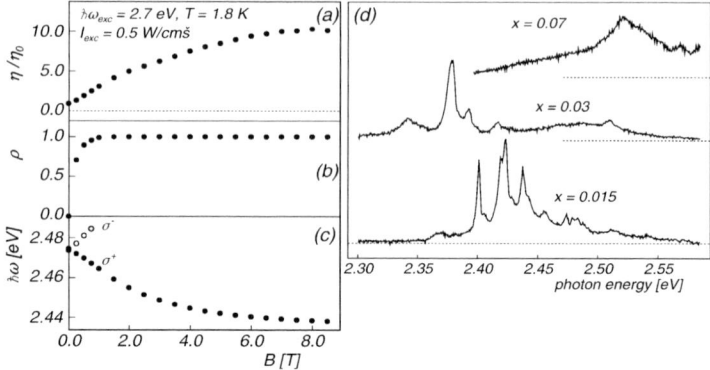

Fig. 12.13. (a–c) Magneto-optical data of (CdMn)Se/ZnSe QD structures as deduced from ensemble measurements. (a) PL yield η measured relative to zero field (η_0). (b) Circular polarization degree ρ. (c) Energy position $\hbar\omega$ of the Zeeman components, all plotted versus magnetic-field strength B. (d) Micro-PL spectra for a mesa size of 200 nm and three different Mn concentrations

k-dependence of the bulk matrix element—is thus not found. A detailed analysis further provides that the Mn ions are homogeneously distributed across the wetting layer and QDs as well as that Zn–Mn interdiffusion does not play a significant role [25]. The magnetic ion spin dissipation times are in the 1-µs range [64]. Their dependence on the Mn concentration, temperature, and magnetic field is indicative of an Orbach process (in a locally heated environment) via Mn pairs. The spin-lattice-relaxation scenario is thus fully analog to the bulk material. A total number of six MLs in the center of the QDs is large enough to produce a sufficient number of Mn pairs. Long-term processes, like 2D spin diffusion or spin transfer between QDs and wetting layer, are obviously not important in the SK morphology of the (CdMn)Se/ZnSe structures.

The spectral line width of the PL from single (CdMn)Se/ZnSe QDs is more than two orders of magnitude larger than for their nonmagnetic CdSe counterparts. The number of lines in Fig. 12.13(d) is consistent with the areal dot density for the respective Mn concentration. The fast energy transfer to internal Mn states excludes magneto-polaron formation. The situation is thus complementary to CdSe/(Zn,Mn)Se [65] and (Cd,Mn)Te/(Cd,Mg)Te QDs [66] where magnetic polarons play a leading role. In the present case, each photo-generated exciton experiences a different Mn spin configuration during its short lifetime. These fluctuations of the magnetic environment are the origin of the large single-dot PL line width. Indeed, calculations of the line width using Gaussian-type fluctuations reproduce the experimental values reasonably well [63]. In the Mn concentration range covered by Fig. 12.13(d), the average number of magnetic ions per QD ranges from 4–17. If this number is reduced down to a single Mn ion, the broad PL decomposes in a set of discrete lines corresponding to the six possible orientations of the 5/2 spin [67].

Acknowledgements

We thank A. Betke, D. Feng, T. Flissikowski, A. Hundt, M. Lowisch, K.V. Kavokin, A. Koudinov, P. Kratzert, F. Kreller, Y. Kusrayev, and H.J. Wünsche for the joint work on the above subjects.

References

1. A.I. Ekimov, A.A. Onushchenko, Sov. Phys. Semicond. **16**, 775 (1982)
2. A. Henglein, in *Elecrochemistry II*, ed. by E. Steckham. Topics Curr. Chem., vol. 143 (Springer, Berlin, 1988)
3. M. Lowisch, M. Rabe, B. Stegemann et al., Phys. Rev. B **54**, R11074 (1996)
4. S.H. Xin, P.D. Wang, A. Yin et al., Appl. Phys. Lett. **69**, 3884 (1996)
5. F. Flack, N. Samarth, V. Nikitin et al., Phys. Rev. B **54**, R17312 (1996)
6. M. Rabe, M. Lowisch, F. Henneberger, J. Cryst. Growth **184/185**, 248 (1998)
7. S. Lee, I. Daruka, C.S. Kim et al., Phys. Rev. Lett. **81**, 3479 (1998)
8. H. Kirmse, R. Schneider, M. Rabe et al., Appl. Phys. Lett. **72**, 1329 (1998)
9. M. Grün, F. Funfrock, P. Schunk et al., Appl. Phys. Lett. **73**, 1343 (1998)
10. P.R. Kratzert, M. Rabe, F. Henneberger, Phys. Rev. Lett. **83**, 239 (1999)
11. K. Arai, T. Hanada, T. Yao, J. Cryst. Growth **214/215**, 703 (2000)
12. C.S. Kim, M. Kim, J.K. Furdyna et al., Phys. Rev. Lett. **85**, 1124 (2000)
13. D. Schikora, S. Schwedhelm, D.J. As et al., Appl. Phys. Lett. **76**, 418 (2000)
14. T. Passow, H. Heinke, J. Falta et al., Appl. Phys. Lett. **77**, 3544 (2000)
15. N. Peranio, A. Rosenauer, D. Gerthsen et al., Phys. Rev. B **61**, 16015 (2000)
16. D. Litvinov, A. Rosenauer, D. Gerthsen et al., Phys. Rev. B **61**, 16819 (2000)
17. P.R. Kratzert, M. Rabe, F. Henneberger, Appl. Surf. Sci. **166**, 32 (2000)
18. P.R. Kratzert, M. Rabe, F. Henneberger, Phys. Stat. Sol. (b) **224**, 179 (2001)
19. D. Litvinov, A. Rosenauer, D. Gerthsen et al., Appl. Phys. Lett. **81**, 640 (2002)
20. L.G. Wang, P. Kratzer, M. Scheffler, Phys. Rev. Lett. **82**, 239 (1999)
21. V. Shchukin, D. Bimberg, Rev. Mod. Phys. **71**, 1125 (1999)
22. S. Sadofev, S. Blumstengel, F. Henneberger, Appl. Phys. Lett. **84**, 3678 (2004)
23. F. Tinjod, B. Gilles, S. Moehl et al., Appl. Phys. Lett. **82**, 4340 (2003)
24. I.-V. Robon, R. André, C. Bougerol et al., Appl. Phys. Lett. **88**, 233103 (2006)
25. P.R. Kratzert, J. Puls, M. Rabe et al., Appl. Phys. Lett. **79**, 2814 (2001)
26. S. Mahapatra, T. Kiessling, E. Margopati et al., Appl. Phys. Lett. **89**, 043102 (2006)
27. T. Flissikowski, M. Lowisch, A. Hundt et al., Phys. Rev. Lett. **86**, 3172 (2001)
28. D. Gammon, E.S. Snow, B.V. Shanabrook et al., Phys. Rev. Lett. **76**, 3005 (1996)
29. V.D. Kulakovskii, G. Bacher, R. Weigand et al., Phys. Rev. Lett. **82**, 1780 (1999)
30. J. Puls, M. Rabe, H.J. Wünsche et al., Phys. Rev. B **60**, R16303 (1999)
31. M. Lowisch, M. Rabe, F. Kreller et al., Appl. Phys. Lett. **74**, 2489 (1999)
32. I.A. Akimov, J.T. Andrews, F. Henneberger, Phys. Rev. Lett. **74**, 067401 (2006)
33. T. Flissikowski, A. Hundt, A.I. Akimov et al., Phys. Rev. B **68**, R161309 (2003)
34. A. Koudinov, I. Akimov, Y. Kusraev et al., Phys. Rev. B **70**, R241305 (2004)
35. K.V. Kavokin, Phys. Stat. Sol. (a) **195**, 592 (2003)
36. I.A. Akimov, K.V. Kavokin, A. Hundt et al., Phys. Rev. B **71**, 075326 (2005)
37. J. Puls, I.A. Akimov, F. Henneberger, Phys. Stat. Sol. **234**, 304 (2002)
38. I.A. Akimov, A. Hundt, T. Flissikowski et al., Physica E **17**, 31 (2003)
39. I.A. Akimov, T. Flissikowski, A. Hundt et al., Phys. Stat. Sol. (a) **201**, 412 (2004)

40. I.A. Akimov, A. Hundt, T. Flissikowski et al., Appl. Phys. Lett. **81**, 4730 (2002)
41. S. Cortez, O. Krebs, S. Laurent et al., Phys. Rev. Lett. **89**, 207401 (2002)
42. D. Gammon, D.G. Steel, Phys. Today **55**, 36 (2002) and references therein
43. T. Flissikowski, I.A. Akimov, A. Betke et al., Phys. Rev. Lett. **92**, 227401 (2004)
44. P. Michler, A. Kiraz, C. Becher et al., Science **290**, 2282 (2000)
45. O. Benson, C. Santori, M. Pelton et al., Phys. Rev. Lett. **84**, 2513 (2000)
46. X. Li, Y. Wu, D. Steel et al., Science **301**, 809 (2003)
47. E.A. Muljarov, R. Zimmermann, Phys. Rev. Lett. **93**, 237401 (2004)
48. A.V. Khaetskii, Y.V. Nazarov, Phys. Rev. B **61**, 12639 (2000)
49. A.V. Khaetskii, Y.V. Nazarov, Phys. Rev. B **64**, 125316 (2001)
50. J.M. Elzerman, R. Hanson, L.H.W. van Beveren et al., Nature **430**, 431 (2004)
51. M. Kroutvar, Y. Ducommun, D. Heiss et al., Nature **432**, 81 (2004)
52. I.A. Merkulov, A.L. Efros, M. Rosen, Phys. Rev. B **65**, 205309 (2002)
53. A.V. Khaetskii, D. Loss, L. Glazman, Phys. Rev. Lett. **88**, 186802 (2002)
54. A.V. Khaetskii, D. Loss, L. Glazman, Phys. Rev. B **67**, 195329 (2003)
55. A.I. Akimov, D. Feng, F. Henneberger, Phys. Rev. Lett. **97**, 056602 (2006)
56. A.I. Akimov, F. Henneberger, Phys. Stat. Sol. (c) **3**, 841 (2006)
57. M.I. Dyakanov, V.I. Perel, in *Optical Orientation*, ed. by F. Meier, B.P. Zakharchenya (North-Holland, Amsterdam, 1984), p. 51
58. C.W. Lai, P. Maletinsky, A. Badolato et al., Phys. Rev. Lett. **96**, 167403 (2006)
59. D. Feng, A.I. Akimov, F. Henneberger, Phys. Rev. Lett. (2007)
60. N.T. Bagrayev, L.S. Vlasenko, R.A. Zhitnikov, Fiz. Tverd. Tela **19**, 3170 (1977)
61. L.V. Titova, J.K. Furdyna, M. Dobrowolska et al., Appl. Phys. Lett. **80**, 1237 (2002)
62. S. Mackowski, S. Lee, J.K. Furdyna et al., Phys. Stat. Sol. (b) **229**, 469 (2002)
63. A. Hundt, J. Puls, F. Henneberger, Phys. Rev. B **69**, R121309 (2004)
64. A. Hundt, J. Puls, A.V. Akimov et al., Phys. Rev. B **72**, 033304 (2005)
65. J. Seufert, G. Bacher, M. Scheibner et al., Phys. Rev. Lett. **88**, 027402 (2002)
66. A.A. Maksimov, G. Bacher, A. McDonald et al., Phys. Rev. B **62**, R7767 (2000)
67. L. Besombes, Y. Léger, L. Maingault et al., Phys. Rev. Lett. **93**, 207403 (2004)

13

Narrow-Gap Nanostructures in Strong Magnetic Fields

T. Tran-Anh and M. von Ortenberg

Abstract. The presented work provides insight into MBE-growth and growth-optimization of HgSe and HgSe:Fe on different buffer/substrate systems as well as its utilization to fabricate zero-gap II–VI semiconductor quantum structures: quantum wells, quantum wires, and quantum dots. The special feature of these nanostructures is the intrinsic population of the quantum states by electrons (as many as 50–500 electrons in a single dot for example), which allows them to be directly investigated by magneto-transport measurements and infrared magneto-resonance spectroscopy in high magnetic fields up to 300 Tesla. The investigation showed strong correlation effects manifested in a 50% increase of the cyclotron mass with respect to that in a bulk structure.

13.1 Introduction

In the last few decades, semiconductor quantum structures have been the subject of intense research. These investigations often focus on III–V and wide-gap II–VI compound semiconductors for laser applications. In this context, the outstanding zero-gap II–IV semiconductor HgSe:Fe is a perfect supplementation, being a Fermi-level pinned system with a high carrier concentration and mobility. This study is devoted to the fabrication and characterization of HgSe/HgSe:Fe quantum structures (quantum wells, quantum wires, and quantum dots). The growth technique used for the investigation is *molecular beam epitaxy* (MBE) with its ability to fully use the advantages offered by ultra-high vacuum technology. By employing different in situ and ex situ analysis tools, we should gain physical insight into epitaxial growth and relaxation phenomena of semiconductor heterostructures. These insights can then be investigated and used to realize and optimize nanostructures. The special properties of HgSe/HgSe:Fe enable the confinement effects in these structures to be studied directly by magneto-transport and magneto-optical measurements using high magnetic fields.

13.2 Materials: HgSe/HgSe:Fe

HgSe crystallizes in the zincblende structure (space symmetry group $F\bar{4}3m$). The material from bulk ingots has been extensively studied with respect to its electronic band structure and macroscopic properties [1–3]. HgSe is one of those zincblende semiconductors with an inverted band structure, i.e. the Γ_6 band lies below the Γ_8 bands and has a negative curvature, while the light-hole Γ_8 band takes over the part as conduction band. Due to degeneration of the Γ_8 bands at $k = 0$, a band gap is literally nonexistent. Consequently, these materials are commonly called zero-gap semiconductors. From experimental results, the main band parameters of HgSe were determined to be $E_g = -274$ meV for the band gap between Γ_6 and Γ_8 bands, and $\Delta_{so} = 383$ meV for the spin-orbit splitting energy at low temperatures. The conduction band was found to be nearly isotropic but considerably nonparabolic [4], thus the effective mass depends on the carrier concentration, which varies in the range from 2×10^{17} cm^{-3} to 1×10^{19} cm^{-3}.

HgSe can be doped with iron with a composition x for Hg$_{1-x}$Fe$_x$Se in the range $0 \leq x \leq 0.14$. In this range, Hg atoms in HgSe are substituted by Fe atoms, leading to a negligibly small modification of the crystal lattice. Nevertheless, the influence of the Fe impurity on the electronic properties of HgSe is dramatic. The free atom electronic configuration of Fe ($Z = 26$) is (Ar)3d^{6}4s^2. The ground iron impurity states are Fe^{2+}3d^6, which locate energetically about 210 meV above the conduction band edge. As a result, they are shifted into the region of the conduction band forming a resonant donor level [5]. This process can be understood as an auto-ionization, where a Fe^{2+} ion delivers one electron to the conduction band, raising the Fermi level, and becomes Fe^{3+}. The process can continue as long as the Fermi level is lower than the impurity level Fe^{2+}3d^6. In this doping range, the free electron concentration, apart from the background, directly reflects the iron concentration. At the point where the Fermi level reaches the iron impurity level (i.e. energetically in balance with it), the iron concentration has a so-called threshold value $n_{Fe}^* = 5 \times 10^{18}$ cm^{-3}, the corresponding composition is $x = 0.0003$. Above this value, the free carrier concentration remains virtually constant, and only a fraction of the resonant donors is ionized. This leads to the formation of an impurity system with coexisting Fe^{2+} and Fe^{3+} charge states, the so-called mixed valence regime [6].

13.3 Fabrication of HgSe/HgSe:Fe Nanostructures

The HgSe:Fe nanostructures were fabricated using the Molecular Beam Epitaxy technique (MBE), which allows precise growth control to monolayer scale accuracy. For experimental growth and subsequent growth optimization, both the in situ monitoring-and-characterization capability and the fast feedback from structural characterization are indispensable. The growth was monitored and analyzed in real time by means of Reflection High-Energy Electron Diffraction (RHEED). The crystal structure of the samples was characterized by High Resolution X-Ray Diffraction (HRXRD). Surface morphology was investigated by Scanning Electron Microscopy (SEM) and Atomic Force Microscopy (AFM).

13.3.1 Quantum Wells

A HgSe/HgSe:Fe quantum well structure itself is a very interesting object for fundamental research. Moreover, the understanding and the control of the growth process of these 2D structures are the foundation for working on the more sophisticated quantum wire and quantum dot structures. Different configurations for different purposes have been investigated.

ZnTe/GaAs

To understand the MBE-growth of HgSe, ZnTe/GaAs (the only buffer/substrate system for HgSe known at the time) was used as a start point. The main characteristic of this system is a very high misfit of $f = -7.4\%$, while the misfit between HgSe and ZnTe is relatively small ($f = +0.3\%$). This means that defects and residual strain in the ZnTe buffer should be suppressed to get HgSe layers of high quality. Because of the high lattice mismatch to GaAs, a ZnTe layer will start to relax after the growth of only few monolayers (MLs) by forming misfit dislocations. With increasing layer thickness, the overall line length of misfit dislocations will increase and reach its saturation value when the layer is fully relaxed (i.e., having its natural lattice structure). From this point on, any further growth will reduce the dislocation density. Thus a sufficient low defect density required for the growth of HgSe films of good crystalline quality can be obtained by growing a very thick ZnTe buffer. The trade off between low defect density and short growth time is usually made at a layer thickness in the range between 2 μm and 4 μm. Detailed experiment setups for the growth and structural characterization of this system were reported in [7].

ZnTe$_{1-x}$Se$_x$/GaSb

This buffer/substrate system was investigated with the aim of improving the crystalline quality of the HgSe layer [8]. By replacing GaAs as substrate with GaSb, the misfit of the system is reduced drastically to $f = -0.11\%$ for ZnTe/GaSb. This allows the growth of ZnTe buffers with a much lower defect density. An even more remarkable improvement could be obtained by fitting the buffer to HgSe using a ternary compound of type ZnTe$_{1-x}$Se$_x$. By varying the Se content $0.016 \leq x \leq 0.041$, the lattice constant of the buffer can be varied between that of GaSb and that of the HgSe layer, thus keeping the misfit at both interfaces between 0% and 0.2%—a nearly lattice-matched system.

ZnTe$_{1-x}$Se$_x$/GaAs with $x = 0 \rightarrow 0.98$

For the growth of HgSe roof-ridge quantum wires, both of the buffer/substrate systems mentioned above have some crucial disadvantages. The process forming the quantum wires requires a limited buffer thickness, which in case of the ZnTe/GaAs system causes a domain-splitting due to the large misfit. The use of the ZnTe$_{1-x}$Se$_x$/GaSb system allows to grow high-quality HgSe quantum wires. However, the high residual carrier concentration of commercially available GaSb-wafers make them unsuitable for many experimental purposes. For these reasons, a graded buffer of $ZnTe_{1-x}Se_x$ with $x = 0 \rightarrow 0.98$ was investigated, which allowed the growth of monocrystalline HgSe on the patterned GaAs substrates [9].

ZnSe/GaAs

For the growth of epitaxial HgSe structures of high crystalline quality, the ZnSe/GaAs buffer/substrate system is rather unsuitable because of the high misfit between HgSe and ZnSe of -6.83%. However, the deformation energy stored in a highly mismatched epitaxial system often turned out to be the driving force for the spontaneous formation of quantum dots. For this purpose, the growth of the ZnSe-buffer on GaAs and the growth of HgSe on this buffer have been investigated [10].

13.3.2 Roof-Ridge Quantum Wires

The principle of the method we employed to create HgSe:Fe roof-ridge quantum wires is shown schematically in Fig. 13.1. The key mechanism is the spontaneous formation of the quantum wires on prestructured substrates. A special surface potential profile is realized by structuring the substrate using the traditional method of lithography and etching. The self-formation of quantum wires takes advantage of the *orientation-dependent growth*, also known as *area-selected growth* or as *self-limiting growth* on nonplanar substrates, where the growth rate depends on the surface orientation due to the crystalline anisotropy.

In this work, substrates for growing quantum wires were all prestructured using conventional structuring techniques, i.e. *photolithography* and *electron beam lithography* followed by wet-chemical etching. Using this method, controllable patterns, with stripes having a top-width of down to 200 nm and a distance between them of down to 1 µm, can be created. The investigation on substrate structuring was reported in detail in [7].

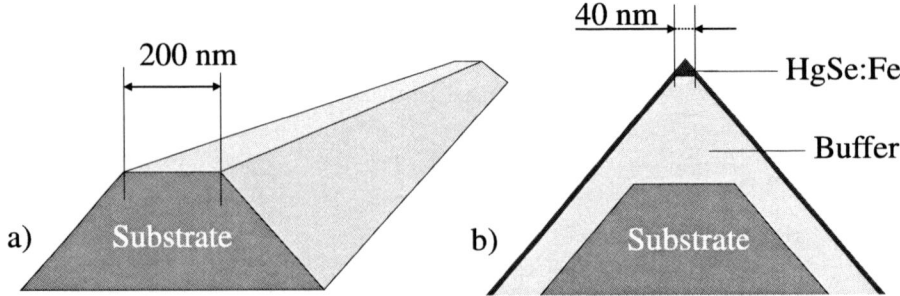

Fig. 13.1. Schematic illustration of the formation of quantum wires on pre-structured substrates [9]: (**a**) the shape of a roof-like bank in mesa patterns on the structured substrate surface before growth; (**b**) cross-section of that bank after growth, roof-ridge-like nano wires are formed

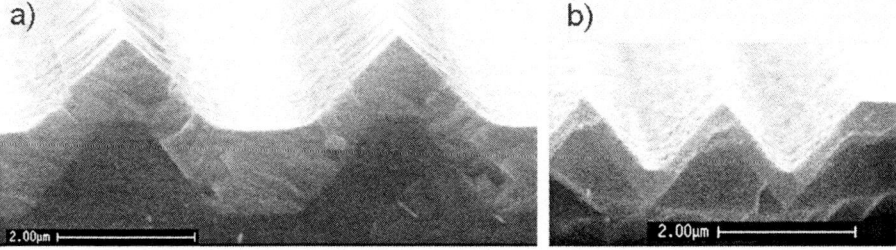

Fig. 13.2. SEM-images of a cleaved ($1\bar{1}0$) cross-section of HgSe:Fe quantum wires grown on a $ZnTe_{1-x}Se_x$-graded buffer on GaAs(001) substrates prestructured using [9]. (**a**) Photolithography and (**b**) electron-beam lithography

The procedure of forming quantum wires began with the overgrowth of the prestructured substrate by the buffer material. The growth rate on the mesa top with [001]-orientation was sufficiently higher than the growth rate on the side-faces giving rise to a decreasing mesa top width. When this width reached the desired value of the quantum wires (about 40 nm), the growth of the buffer was terminated. The process continued by the deposition of HgSe:Fe and finished in a ridge defining the triangular shaped quantum wire sitting on two declined wings of a quasi-wetting layer, as illustrated in Fig. 13.1.

The crucial problem of using the roof-ridge arrangement is the precise growth control, where the buffer should form a template for the subsequent growth of the quantum wires at the roof ridges. To stop the growth of the buffer and the HgSe layer, respectively, at the right time, the actual top-width of the roof has to be known at any time during the growth process. A so-called *mean-profile model* was introduced to achieve this goal [10]. Figure 13.2 shows some images of the roof-ridge quantum wires taken by SEM. These samples were grown on patterned GaAs(001) using a $ZnTe_{1-x}Se_x$-graded buffer [9], as the ZnTe/GaAs system turned out to be not suitable for growing monocrystalline HgSe quantum wires [7]. For applications where the high residual carrier concentration in GaSb is not a problem, GaSb substrates together with a $ZnTe_{0.97}Se_{0.03}$ buffer were employed to produce nearly defect-free HgSe:Fe quantum wires [11].

13.3.3 Quantum Dots

The realization of semiconductor quantum dot systems is a difficult challenge. The critical issues are the size uniformity and the lateral ordering of the quantum dots—besides their crystalline quality. The investigation reported here was devoted to finding mechanisms for in situ formation of HgSe:Fe quantum dots during MBE growth, both on prestructured substrates—so-called *filled-pit quantum dots*, because they embody the potential of flexible pattern design—and on planar substrates that exploit the strain-induced surface roughening phenomenon—so-called *s*elf-assembled quantum dots [12].

Filled-Pit Quantum Dots

The basic idea here is the same for producing roof-ridge quantum wires, i.e. pre-structuring the substrates and exploiting special growth features to form nanoscale objects. The first step was to create nanoscale pits on the surface of the substrate. The pitted surface was then coated with a few ZnSe MLs and subsequently with HgSe:Fe filling the pits until the initial 3D RHEED patterns transformed into streaky patterns hinting at a smooth planar surface (about 10–15 MLs). The transform was due to the higher growth rate in the pits, which can be explained as follows. MBE growth of HgSe:Fe on ZnSe is principally of Stranski–Krastanow type, i.e. the layer grows laterally at the initial stage. This means for a thin HgSe:Fe layer, step edges, kinks, and such special lattice sites are preferred over those on planar surface areas by the adsorbed atoms to incorporate into the crystal phase.

In order to create a template, i.e. a pitted surface structure, for the growth of quantum dots, a special property of GaAs has been exploited: a slow thermal oxide desorption process of GaAs substrates produces dense pits (10^{10} cm^{-2}) with a depth of up to 20 nm and a comparable lateral extension on the surface [13]. The mechanism was found to be chemical with reactions taking place at temperatures above 500°C during the thermal desorption, where GaAs of the substrate consumed locally to form the pits. The effect is even more significant when the epi-ready oxide layer is replaced by a fresh oxide due to a polish etching process.

Stranski–Krastanow Growth

The growth of HgSe/HgSe:Fe on the largely mismatched ZnSe ($f = -6.83\%$) showed very typical characteristics of the Stranski–Krastanow (SK) growth phenomenon. Both a strained ZnSe-buffer ($f_{\text{ZnSe}}^{\text{HgSe}} = f_{\text{GaAs}}^{\text{HgSe}} = -7.08\%$) and a fully relaxed ZnSe-buffer ($f = -6.83\%$) can be used. However, the latter with a larger thickness were preferred to suppress the bad surface quality of the GaAs substrate due to thermal oxide desorption. At the initial growth stage, the ZnSe buffer was fairly wetted by HgSe:Fe manifesting a significant streaky RHEED pattern. A transition to island growth could be recognized by RHEED (transmission spots) after just a few MLs.

The first stage of SK-growth, i.e. the lateral growth of the first few MLs, was performed for all samples under the same optimized growth conditions. The growth, then, continued to a thickness of ca. 30 MLs at a (for HgSe) very high substrate temperature of 140°C to enhance surface processes, making it easier for the system to evolve toward equilibrium (surface with island structure). Figure 13.3 shows a typical RHEED pattern and a typical AFM image of those SK-grown samples. The island density is about 10^{10} cm^{-2}, where two types of islands can be roughly identified: isolated islands with a diameter of about 20–30 nm and a height between 5 nm and 12 nm; and larger islands with a diameter of about 50 nm and a height of about 15–20 nm.

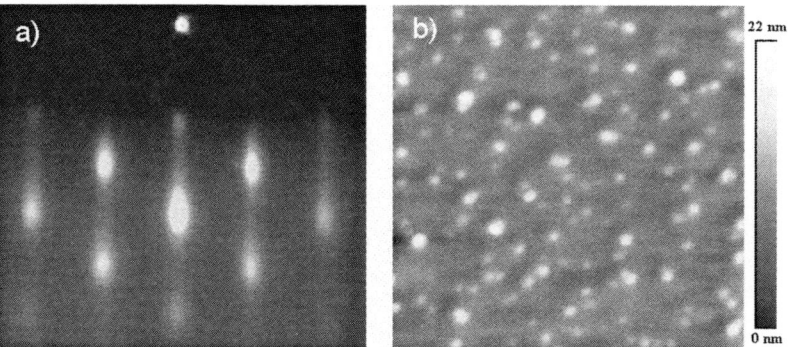

Fig. 13.3. Surface structure of HgSe:Fe grown on ZnSe at a very high growth temperature: (**a**) RHEED pattern and (**b**) AFM image of the corresponding surface [12]

Thermally Activated Surface Reorganization

The results of the SK-growth experiments confirm the theoretical prediction that a thin HgSe:Fe layer (near the critical thickness for the formation of misfit dislocations) on ZnSe with an islanded surface structure is thermodynamically more stable than a planar, pseudomorphic strained layer. It was possible, however, to grow planar HgSe:Fe layers on ZnSe with a thickness of up to a few 10 nm beyond the critical thickness. These structures, especially the fully strained ones, must consequently be kinetically frozen, i.e. thermodynamically metastable. Such lateral frozen layers can be achieved—due to the desorption-limited characteristics of MBE growth of HgSe—at a low substrate temperature and a relatively low impinging rate using growth interruptions.

A thermal annealing should now activate kinetic processes, moving such a metastable pseudomorphic layer towards equilibrium. After the growth of about 4 MLs HgSe:Fe, the pseudomorphic strained layer was heated to 200°C for about 10 min. Although no significant change of the surface structure due to a thermal annealing process in an UHV environment could be recognized by RHEED, a pronounced transition from a planar to an islanded surface structure occurred during annealing under the same Hg beam flux used during the growth.

The dramatic restructuring of the surface of HgSe:Fe occurring during the thermal annealing under Hg-pressure could be documented by RHEED and also confirmed by AFM (see Fig. 13.4). Again, relatively large islands with a lower density can be observed. This may correspond to defects on the surface which enhanced the growth of larger islands having a lateral extension of about 50–60 nm and a height of about 10 nm. More dominating, however, are the highly dense small islands with a basis width of 10–20 nm and a height of 3–6 nm. Numerical integrations yielded an astonishingly large total island volume equivalent to 2.1 MLs.

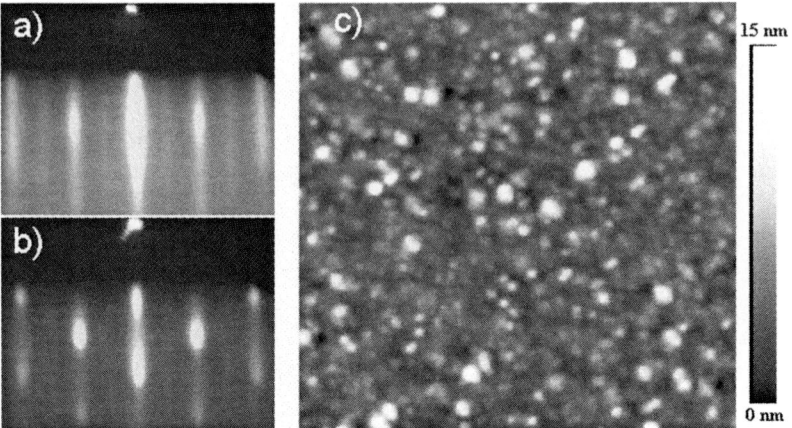

Fig. 13.4. Thermally activated surface reorganization: (**a**) RHEED pattern before thermal annealing; (**b**) RHEED pattern after thermal annealing; (**c**) AFM of the HgSe:Fe surface after thermal annealing [12]

13.4 Electronic Characterization of the HgSe/HgSe:Fe Nano-Structures in Strong Magnetic Fields

The notion of "*quantum structures*" refers to structures having spatial extensions smaller than a microscopic length scale. In such structures, the charge carriers or another quantum-mechanical object being considered are said to be confined. The length scales which can be used as a critical size are, for example, the *Bohr radius* or the magnetic length $l = \sqrt{\hbar/eB}$. The latter implies a very interesting aspect as it can be tuned by variation of the magnetic field B. For magnetic field intensities up into the megagauss regime in routine experiments, l is of the order of nm. This means that the magnetic length is an ideal measure to probe nanostructure potential in semiconductor devices. Indeed, magneto-transport and magneto-optical measurements proved to be powerful methods for studying quantum structures [14, 15].

13.4.1 High-Field Magneto Transport

Due to the intrinsically populated quantum states, electronic properties of HgSe/HgFeSe quantum wells and wires could be characterized using direct resistance measurements [14, 8]. Carrier concentration and mobility were calculated by analyzing the Shubnikov–de Haas (SdH) oscillations. These experiments also gave evidence for the low-dimensional character of the investigated structures.

Figure 13.5 shows the dependence of the transverse magneto-resistance on the magnetic field (up to 12 T) for a HgSe single quantum well structure at 4.2 K. The shift of the Landau levels according to $1/\cos\theta$, where θ is the tilt angle of the external magnetic field relative to the surface normal, corroborates the existence of a two-dimensional quantum system (Q2D).

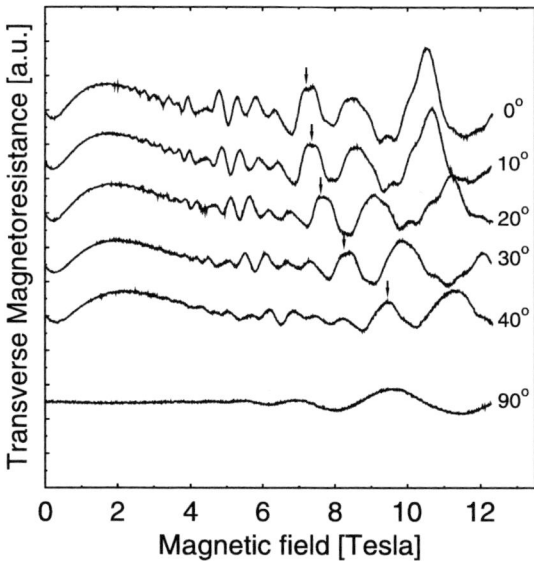

Fig. 13.5. Transverse magneto-resistance of a HgSe single quantum well structure grown on ZnTeSe/GaSb at 4.2 K [8]. The arrows mark the shift of a Landau level with the change of the tilt angle of the external magnetic field relative to the surface normal

Strong SdH-oscillations in transverse magneto-resistance could also observed in quantum wire structures (Q1D) as shown in Fig. 13.6. The interesting fact is that the SdH-oscillation in the Q1D-sample starts only after the magnetic field has reached a threshold value of about 5 Tesla. In contrast, the Q2D reference sample shows homogeneous oscillation starting from very low fields.

This effect can be explained by means of magneto-condensation. For magnetic fields stronger than 5 T, the diameter of the cyclotron orbit $2R_c = l\sqrt{2N+1}$, where N is the Landau quantum number and l is the magnetic length, can fit into the quantum wires. This is no longer the case for magnetic fields below this threshold, and the quantization collapses. From the SdH-oscillation data, the threshold orbit diameter $2R_c^* = 76$ nm could be calculated ($B = 5$ T, $N = 5$), which is comparable with the width of the quantum wires of about 80 nm obtained from scanning electron microscopy.

13.4.2 Infrared Magneto-Resonance Spectroscopy

Since HgSe:Fe is a Fermi-level pinned system with a high carrier concentration, quantum-dot states are intrinsically populated. This means that the localized electrons can be investigated directly by optical intraband excitation [16]. Thus *infrared magneto-resonance spectroscopy* (IR-MRS) has been employed to investigate the

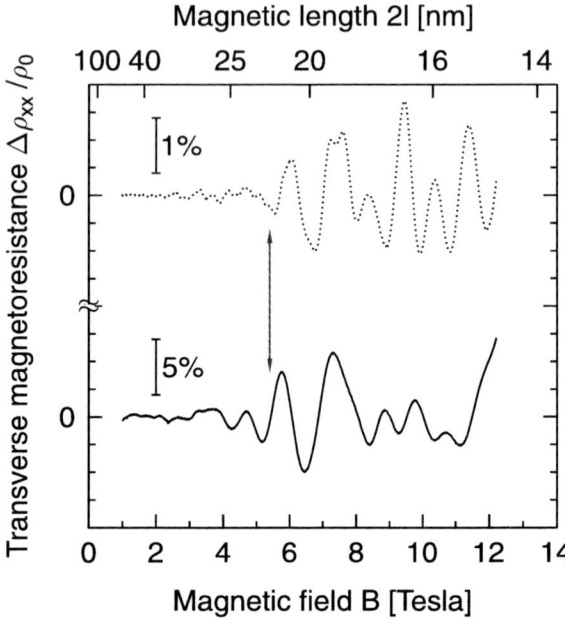

Fig. 13.6. SdH-oscillation of a HgSe quantum wire structure (*upper curve*) in comparison with that of a unpatterned 2D structure (*lower curve*) [7]

electronic properties of the QD-structures obtained from different approaches as described in Sect. 13.3.3. In these experiments, the transmission (or absorption) of radiation by the HgSe:Fe structures in dependence on an external magnetic field was measured. Because of the high number of free carriers, about $5 \times 10^{18}\,\text{cm}^{-3}$, a high radiation energy beyond the plasma edge had to be applied for a reasonable transparency. A high radiation energy, however, also shifts the positions of *cyclotron resonances* to higher magnetic fields. As a radiation source, we used a CO_2-laser in cw-mode (*continuous wave operation mode*) at 117 meV. To enforce cyclotron resonance (CR) for this high radiation energy, high magnetic fields in the megagauss regime were also necessary. The experiments were performed in magnetic fields up to 140 T using the *single-turn coil technique* [17]. A detailed description of the experimental setup can be found in [18].

Figure 13.7 shows the magneto-absorption spectrum of a HgSe:Fe QD-sample grown in the Stranski–Krastanow regime (SK-QD). Typical for samples grown in this regime is a very thick wetting layer—about 20–30 MLs in the case of the presented sample, where the total integrated volume of the islands is equivalent to only about 0.9 MLs. The dominant resonance in the spectrum locates around 60 T, which is expected for a Q2D-structure [19] and thus can be identified as the effect from the wetting layer. The special feature, however, is the additional resonance around 100 T. The magnitude ratio between both resonances is approximately the same as the ratio of the total volume of the islands over that of the wetting layer. This fact

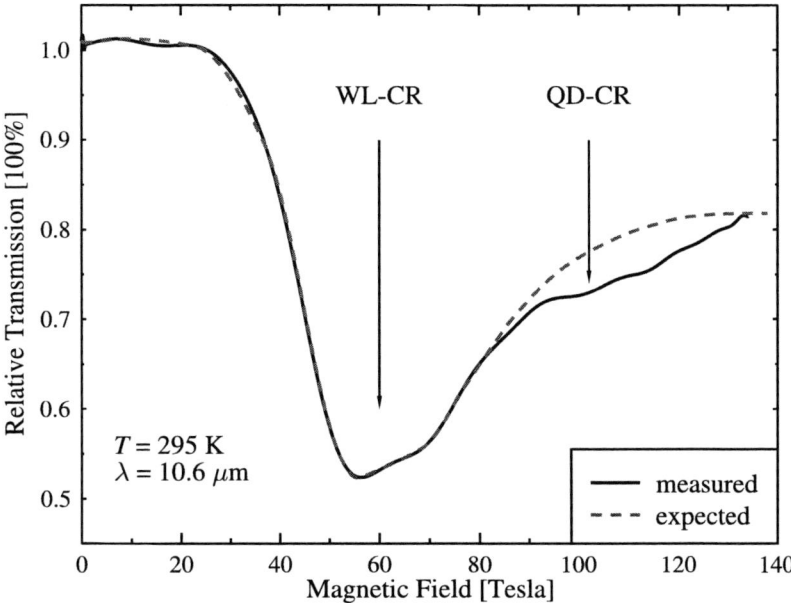

Fig. 13.7. Magneto-absorption spectrum at room temperature of a HgSe:Fe layer grown at 140°C in the Stranski–Krastanow regime [12]. The dashed curve is expected for a 2D layer without islands

strongly hints that the additional absorption may originate from resonance in quantum dots.

With the other methods of growing HgSe:Fe-QDs, namely *pit filling* and *thermally activated surface reorganization*, the thickness of the wetting layer could largely be reduced. Figures 13.8 and 13.9 show cyclotron resonance spectra of HgSe:Fe QD samples fabricated using these two methods. Here, the special resonance at about 100 T can clearly be observed. In these spectra, however, the expected resonance peak of the 2D wetting layer completely disappears. This can be easily explained by looking at the population of the quantum well states. The energy can be estimated as $E_n = (\hbar^2/2m^*)(n\pi/D)^2$, where $n = 1, 2, 3, \ldots$, which means that the energy levels rise with decreasing layer thickness D. With a thickness smaller than a critical value of about 5 nm, the quantum-well states of the system at hand rise higher than the pinned Fermi-level, becoming completely depopulated and thus do not contribute to magneto-absorption.

The resonance at about 100 T corresponds to an effective cyclotron mass of about $0.1 m_0$, where the cyclotron mass in volume is $0.065 m_0$. This indicates an increase by 50% of the cyclotron mass for electrons confined in the quantum dots. The question is what mechanism is behind this phenomenon? For magnetic fields as high as 100 T, the magnetic length already reaches a value of $l = 2.56$ nm. This means that the cyclotron orbit is considerably smaller than the dots, and thus the influence of the side walls might well be negligible. It should be noted that as opposed to most quantum

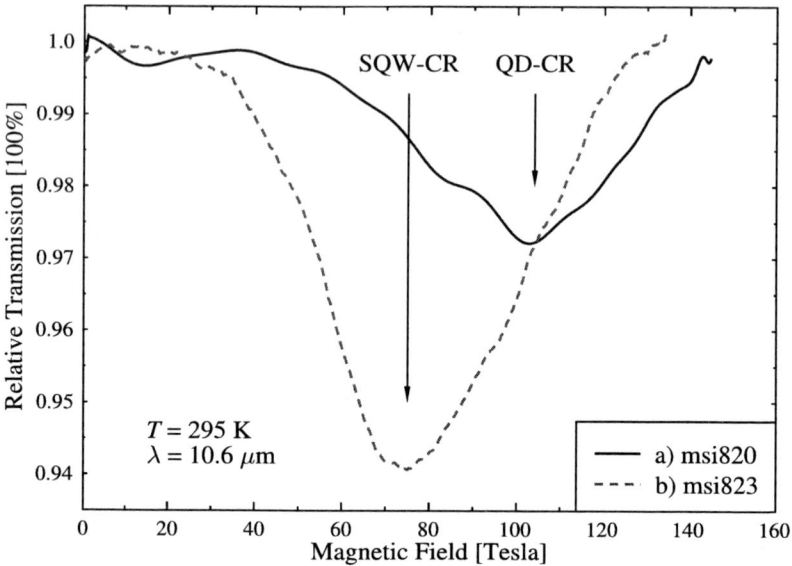

Fig. 13.8. Magneto-absorption spectra at room temperature of HgSe:Fe [12] with a thickness of about 2–5 nm: (**a**) grown on a pitted GaAs substrate and (**b**) a 2D sample for comparison

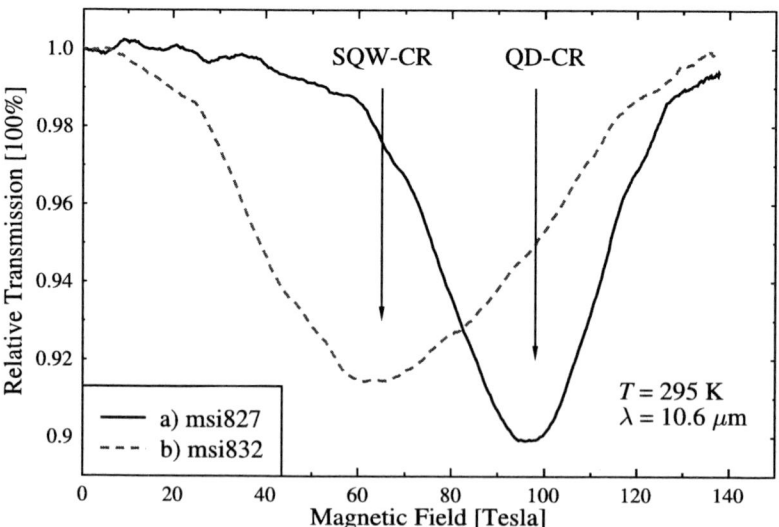

Fig. 13.9. Magneto-absorption spectra at room temperature of HgSe:Fe [12] with (**a**) islanded surface due to thermally activated restructuring, and (**b**) planar surface without thermal annealing (single quantum well)

dot systems, where they are as a rule not populated, there are about 20 simultaneously occupied states in a HgSe:Fe quantum dot of a size of ca. $20 \times 20 \times 10\,\text{nm}^3$.

13.5 Summary

The presented work provided insight into MBE-growth and growth-optimization of HgSe and HgSe:Fe on different buffer/substrate systems as well as its use to fabricate intrinsically populated II–VI semiconductor nanostructures: quantum wells, quantum wires, and quantum dots, respectively. The low-dimensionality of these quantum systems could be verified by magneto-transport measurements and infrared magneto-resonance spectroscopy in high magnetic fields. In particular, an increase by 50% of the cyclotron mass in quantum dots with respect to that in bulk is a very interesting finding for theoretical studies of the structure.

The nearly lattice-matched configuration HgSe:Fe/ZnTe$_{0.97}$Se$_{0.03}$/GaSb is perfect to produce nearly strain- and dislocation-free quantum wells and quantum wires. The easier-to-handle system HgSe:Fe/ZnSe/GaAs requires a relatively thick buffer layer of at least 1 μm and is only suitable for growing quantum wells. For applications where the high residual carrier concentration in GaSb is not tolerable, a graded ZnTe$_{1-x}$Se$_x$-buffer with $x = 0 \to 0.98$ can be used to sufficiently bridge the high lattice mismatch between HgSe and the semi-insulating GaAs, allowing the formation of monocrystalline roof-ridge quantum wires. A substantially different system is HgSe:Fe/ZnSe/GaAs with a very high misfit of $f = -6.83\%$ between HgSe and ZnSe, which can be exploited to induce a spontaneous formation of quantum dots by means of either Stranski–Krastanow growth or thermally activated surface reorganization. Quantum dots can also be formed on prestructured substrates by filling the pits with HgSe:Fe.

References

1. C.R. Whitsett, Phys. Rev. A **138**(3), 829–839 (1965)
2. D.G. Seiler, R.R. Galazka, W.M. Becker, Phys. Rev. B **3**(12), 4274–4284 (1971)
3. M. Dobrowolska, W. Dobrowolski, A. Mycielski, Solid State Commun. **34**, 441–445 (1980)
4. G.B. Wright, A.J. Strauss, T.C. Harman, Phys. Rev. **125**(5), 1534–1536 (1962)
5. A. Mycielski, J. Appl. Phys. **63**(8), 3279–3284 (1988)
6. Z. Wilamowski, Acta Phys. Pol. A **77**(1), 133–145 (1990)
7. H.A. Wißmann, Ph.D. thesis, Institut für Physik der Humboldt-Universität zu Berlin, Shaker Verlag, Aachen, 1999
8. L. Parthier, H. Wißmann, S. Luther, G. Machel, M. Schmidbauer, R. Köhler, M. von Ortenberg, J. Cryst. Growth **175/176**, 642–646 (1997)
9. T. Tran-Anh, H. Wißmann, S. Rogaschewski, M. von Ortenberg, J. Cryst. Growth **214/215**, 40–44 (2000)
10. T. Tran-Anh, Ph.D. thesis, Institut für Physik der Humboldt-Universität zu Berlin, Cuvillier Verlag, Göttingen, 2003

11. H. Wißmann, T. Tran-Anh, S. Rogaschewski, M. von Ortenberg, J. Cryst. Growth **201/202**, 619–622 (1999)
12. T. Tran-Anh, S. Hansel, A. Kirste, H.U. Mueller, M. von Ortenberg, Physica E **20**, 444–448 (2004)
13. M. López López, A. Guillén Cervantes, Z. Rivera Alvarez, I. Hernández Calderón, J. Cryst. Growth **193**, 528–534 (1998)
14. M. von Ortenberg, O. Portugall, N. Puhlmann, H.-U. Müller, S. Luther, M. Barczewski, G. Machel, M. Thiede, Y. Imanaka, Y. Shimamoto, H. Nojiri, N. Miura, D. Schikora, T. Widmer, K. Lischka, Physica B **216**, 384–387 (1996)
15. M. von Ortenberg, K. Uchida, N. Miura, F. Heinrichsdorff, D. Bimberg, Physica B **246/247**, 88–92 (1998)
16. M. von Ortenberg, in *Proc. NGS 10*. IPAP Conf. Ser. **2**, 69–72 (2001)
17. O. Portugall, N. Puhlmann, H.-U. Mueller, M. Barczewski, I. Stolpe, M. von Ortenberg, J. Phys. D: Appl. Phys. **32**(18), 2354–2366 (1999)
18. N. Puhlmann, O. Portugall, M. Barczewski, I. Stolpe, H.U. Mueller, M. von Ortenberg, Physica B **246–247**, 323–327 (1998)
19. O. Portugall, M. Barczewski, M. von Ortenberg, D. Schikora, T. Widmer, H. Lischka, J. Cryst. Growth **184/185**, 1195–1199 (1998)

14

Optical Properties of III–V Quantum Dots

Udo W. Pohl, Sven Rodt, and Axel Hoffmann

Abstract. Results on excitonic properties of few and many particles-complexes confined in self-organized In(Ga)As/GaAs and InGaN/GaN quantum dots (QDs) are highlighted. The renormalization of transition energies in InGaAs QDs is found to be proportional to the number of excitons per dot and in the wetting layer. Resonant Raman scattering on such dots reveals localized TO- and LO-like phonon modes being blue shifted with respect to unstrained InAs bulk modes, and a localized interface mode. The localized modes are largely independent on the structural properties of QDs within different samples. Embedding InAs QDs in an InGaAs well shifts the QDs' emission to lower energies. The reduction of strain is identified as the main reason for this redshift. Binary InAs/GaAs dot ensembles show a distinct formation of subensembles due to self-similar shapes and height variations in steps of integral InAs monolayers. A decreasing number of excited states with decreasing QD size is observed. Spectra of individual dots in such ensembles reveal a biexciton binding energy changing from binding to antibinding as the size of the dots decreases. The trend is well explained by a varying number of bound hole states. Furthermore, a monotonous decrease of the exciton fine-structure splitting with QD size from large values of 0.5 meV to small and even negative values is found, highlighting the effect of piezoelectricity. For nitride structures a clear proof of the quantum-dot nature is provided by resonantly excited time-resolved photoluminescence. Single InGaN QD emission-lines show a pronounced linear polarization, which is attributed to the valence band structure of the wurtzite type nitrides.

14.1 Introduction

Quantum dots (QDs) represent the ultimate limit in charge carrier confinement with discrete atomic-like energy states. The size of the InGaAs and InGaN QDs studied in this chapter is sufficiently small to effect fully quantized electron and hole states. Quantization energies are of the order of kT or larger at room temperature. The energies sensitively depend on size, shape and composition of the QDs (see Chap. 7). Since these quantities show some variation among individual dots within an ensemble comprising many dots, optical spectra detected as a response of the entire ensemble are inhomogeneously broadened due to a superposition of transitions with varying energies. This broadening can be smaller or comparable to the energy separation of transitions between different electronic states of the confined charge carriers. Effects like state filling or phonon coupling can then be well studied by simultaneously probing many dots of the ensemble. The interaction between charge carriers confined to the same dot leads to small splittings and shifts of the energies. Such effects are generally obscured by the inhomogeneous broadening. Spectroscopy of single QDs has therefore been established as a powerful tool to study few-particle interactions. This chapter highlights a number of results of fundamental importance on optical properties of self-organized quantum dots of the prominent materials In(Ga)As and InGaN. While the zincblende GaAs-based dots show piezoelectric effects only due to shear components of the inhomogeneous strain induced by the lattice-mismatch to the matrix [1], wurtzite GaN-based dots exhibit a strong built-in piezo field. The consequences of such effects govern the electronic properties of the dots, which are studied in both materials for dot ensembles and individual dots.

14.2 Confined States and Many-Particle Effects

14.2.1 Renormalization

In a quantum dot the single particle energies of electrons or holes depend almost solely on the QD's structural properties like size, shape, composition and the surrounding material. The single particle picture is, however, no longer sufficient if a QD is occupied by more than one charge carrier, because the Coulomb interaction between the confined particles alters the overall energy of the system. For bulk and quantum well structures this well-known effect is referred to as *renormalization* of the band gap. For fully confined systems like QDs such renormalization is also expected to play a role but to have only minor impact, since the strong confinement dominates the electronic properties of a QD.

Occupation effects in QDs are revealed in photoluminescence spectra recorded as a function of excitation power. The sample presented here was grown using metalorganic vapor phase epitaxy and contains $In_{0.6}Ga_{0.4}As$ QDs in a GaAs matrix, with an $Al_{0.3}Ga_{0.7}As$ barrier separated by a 1 nm GaAs spacer from the bottom side of the QDs. The AlGaAs barrier increases the confinement, giving rise to four excited states besides the QD ground state. Taking the degeneracy $2 \cdot (n + 1)$ of a harmonic potential as a rough estimate for a QD, the maximum overall occupation number of the confined excitonic states will be ~ 30.

Figure 14.1 shows photoluminescence spectra of the QD ensemble for increasing excitation density. For low excitation density the only observable transition is from the QDs' ground state (bottom spectrum) whereas for increasing excitation density four additional peaks appear consecutively due to occupation and radiative recombination from excited states. All peaks obviously shift to lower energy with increasing

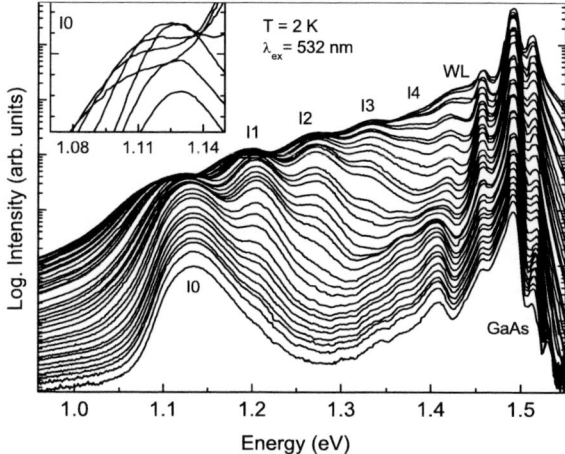

Fig. 14.1. Series of PL spectra for increasing excitation density (*bottom to top*), given on a logarithmic scale. I0: ground state; I1–4: excited states; WL: wetting layer. Inset: Evolution of the ground-state peak energy on a linear scale. From [2]

excitation density. The ground state transition, e.g., exhibits a red shift of as much as 35 meV as illustrated in the inset of Fig. 14.1.

Since the ground state emission can be readily observed for small excitation densities, it's evolution can be easily traced as a function of excitation density. To get information on the true energy position of the excited states, absorption measurements with low excitation density are appropriate. Figure 14.2 shows a photoluminescence excitation measurement of the QD ensemble (dashed line in the bottom panel) together with three photoluminescence spectra for different mean occupations of excitons in the QDs. The occupation numbers $\langle N_X \rangle$ have been determined by comparing the intensity evolution of the different peaks to calculations based on the "master equations for micro states" scheme [3]. The excited states also show an obvious shift to lower emission energy with increasing excitation density, i.e. with increasing occupation (see Fig. 14.2).

To gain insight into the origin of the red shifts observed in Figs. 14.1 and 14.2, we investigated two more QD ensembles embedded into different structures. Sample 2 has a structure comparable to the first sample, but it does not contain an AlGaAs barrier close to the dots. In sample 3 the QDs were overgrown by a 2 nm $In_{0.25}Ga_{0.75}As$ layer to reduce the strain of the QDs and the barrier energy. Figure 14.3 gives a comparison of the photoluminescence spectra of all three samples. Due to the modified potential sample 2 has only three excited states and sample 3 (bottom) has only two. For samples 2 and 3 the excitation density dependence of the energy position of the ground state was determined in the same way as for the first sample. When comparing the results for all three samples one has to keep in mind that the external excitation density differs from the effective excitation of the QDs which depends on the actual sample structure. In order to obtain a sample-independent measure for

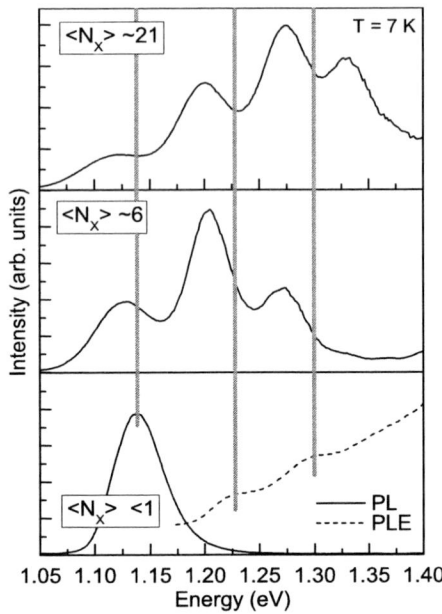

Fig. 14.2. PL spectra (*solid lines*) for an increasing mean occupation $\langle N_X \rangle$ of excitons in the quantum dots. The dashed line in the bottom panel gives a PL excitation spectrum recorded for the $\langle N_X \rangle < 1$ situation at the maximum of the ensemble emission peak. The vertical gray lines mark the energetic positions of the excited states in the 1 exciton limit as given by the PLE curve. From [4]

the excitation density, the measured intensities of the ground state and excited states were compared to calculations based on the "master equations for micro states" [3]. The excitation density can then be expressed in units of excitons per QD and exciton lifetime τ_{rad}, i.e., $\frac{N_X}{N_{QD} \cdot \tau_{rad}}$.

Figure 14.4 compares the red shift of the ground states for the three samples as a function of the effective excitation of the QDs.

Sample 1 exhibits the largest red shift and sample 3 the smallest. The main difference between the samples is the number of confined states. As a consequence an increased red shift is found if more states are bound. This observation can be understood in terms of correlation. The larger the number of bound states, the higher the impact of correlation on the bound excitonic complexes, resulting in a stronger red shift. Besides the number of bound states, the number of onsite charge carriers is also influencing the red shift. The dashed vertical lines in Fig. 14.4 denote the estimated maximum number of confined excitons in the QDs, assuming a degeneracy of $2 \cdot (n + 1)$ of the QD levels. For larger excitation densities the wetting layer will be occupied. From the persistent red shift at further increased excitation density it is concluded that an increase of onsite carriers in the wetting layer also results in a red shift of the ground state transition. This effect is also known for bulk semiconductors and two-dimensional wetting layer systems [5].

14 Optical Properties of III–V Quantum Dots   273

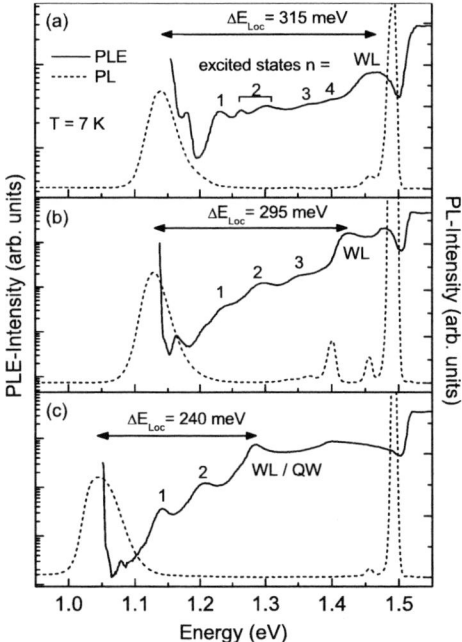

Fig. 14.3. PL (*dashed lines*) and PLE spectra (*solid lines*) of three different QD structures with gradually decreased confining potential from top to bottom panel. (**a**) Sample 1, (**b**) sample 2, and (**c**) sample 3 (see text for description). From [4]

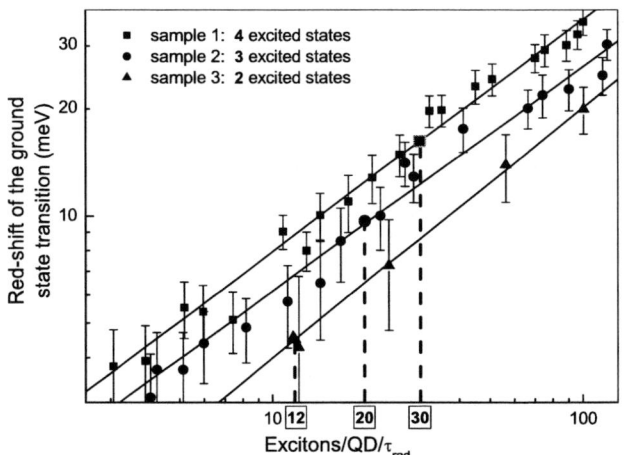

Fig. 14.4. Decrease of the ground state transition energy as a function of the number of excitons per dot and per radiative lifetime, given on a double logarithmic scale. Solid vertical lines mark maximum occupation numbers of excitons in the three QD structures. Further occupation is expected to take place in the wetting layer. From [4]

14.2.2 Phonon Interaction

The interaction with phonons is of great importance for the recombination and dephasing dynamics of confined excitons. In crystals with full translational invariance the number of existing phonons depends solely on the atomic basis. New phonon modes may develop if there are interface boundaries or impurity atoms in the crystal. These modes have mostly a local character, i.e., their spatial distribution is restricted to the interface regions or defect atoms, respectively. As a consequence, there is a large spatial overlap between electronic states related to the interface or defect atom and the respective local phonons, leading to a large interaction among these states. QDs grown in the Stranski–Krastanow growth mode are strained and they have interfaces in all spatial directions, with their bottom interface being the wetting layer. Therefore an interaction between electronic QD states and local phonons is very likely [6]. For a comprehensive presentation five samples with different QD structures have been examined. Samples A to C contain QDs of pure InAs, prepared employing punctuated island growth and a low-temperature GaAs cap prepared using migration-enhanced epitaxy to preserve a pyramidal shape [7, 8]. Sample A consists of a single QD layer, samples B and C contain small InAs seed QDs and either a single active layer of InAs QDs, separated by a 36 monolayer (ML) thick GaAs spacer from the seed QDs, or a five-fold stack of InAs QDs with 45 ML thick GaAs spacers between all layers. Sample D contains a three-fold stack of InAs QD layers, each overgrown by an $In_{0.13}Ga_{0.87}As$ quantum well and a 110 ML thick GaAs spacer [9]. Sample E was prepared using metal-organic vapor phase epitaxy and has a single $In_{0.2}Ga_{0.8}As$ QD layer in GaAs matrix [10]. Luminescence spectra of these QD structures are compiled in Fig. 14.5.

Resonant Raman scattering via the QD ground state was applied for the investigation of localized phonons. Since the separation between Raman response and ground state luminescence equals the respective phonon energy, QD ensembles with small inhomogeneous broadening are best suited for the investigation. Resonant Raman peaks for samples A and E as given in Fig. 14.6 appear at the low-energy side of the respective PL emission, two orders of magnitude weaker in intensity. The excitation

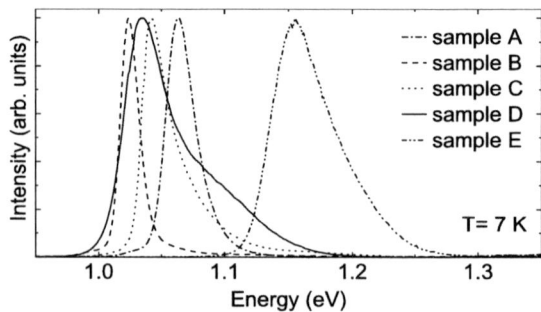

Fig. 14.5. Normalized PL spectra of different In(Ga)As/GaAs QD structures (see text for details), excited nonresonantly into the GaAs matrix with low excitation density. From [11]

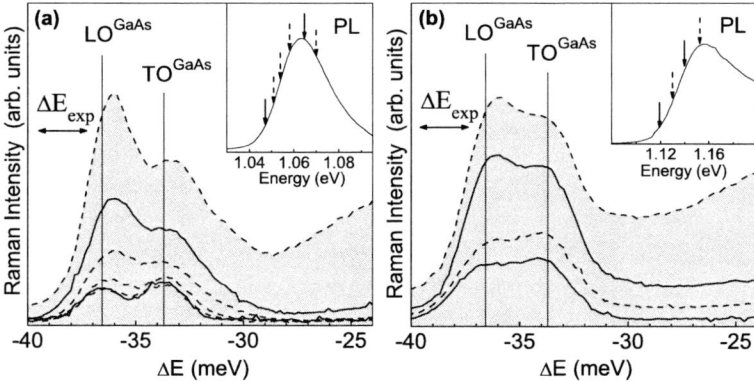

Fig. 14.6. Resonant Raman spectra of single layers of (**a**) InAs QDs and (**b**) $In_{0.2}Ga_{0.8}As$ QDs (samples A and E, respectively). The double arrow denoted ΔE_{exp} indicates the spectral resolution. Insets: PL spectra and excitation energies of the recorded Raman spectra. From [11]

energies indicated by arrows in the insets were chosen in a way that they match some ground state energy within the QD ensemble. The vertical black lines mark LO and TO phonon energies of GaAs at the Γ-point. The intensities of the Raman signals clearly follow the spectral density of the QDs, confirming that the Raman signal is indeed due to resonant scattering via the QDs ground state. At the same time the energies of the Raman peaks deviate from the GaAs LO and TO energies towards smaller values.

The red shift indicates a gradual crossover from Raman coupling with GaAs bulk modes to a coupling with local QD modes as the spectral density and thereby the coupling probability to QDs increases. Figure 14.7 shows the resonant Raman signal as obtained when exciting on the respective QDs' absorption maximum for all QD structures. For all samples the phonon energies are similar and deviate from the GaAs bulk values.

The identification of the confined QD phonon modes is based on a linear chain simulation using the strain modeling described in [12]. The local strain-induced frequency shifts were calculated along a linear path through the center of a QD from discrete 3D strain data and phonon deformation potentials. The calculation shows that the compressive strain as present in the center of the In(Ga)As QDs increases the InAs Γ-point LO phonon energy from the bulk value of 30.3 meV to 33.8 meV, whereas the TO phonon energy is increased from 27.9 meV to 31.5 meV. The experimental LO and TO values as obtained from Fig. 14.7 ($\approx$33 meV (LO) and $\approx$30.5 meV (TO)) are in very good agreement with the simulation. The interface mode cannot be explained solely by strain. Here, the identification is based on the coupling strength and energy position as compared to the local QD modes. The energy laying in between those of Γ-point energies of strained InAs and GaAs is typical for QD interface modes [13]. The high intensity of this mode can be related to the fact that GaAs has a much stronger exciton–phonon coupling than InAs, and

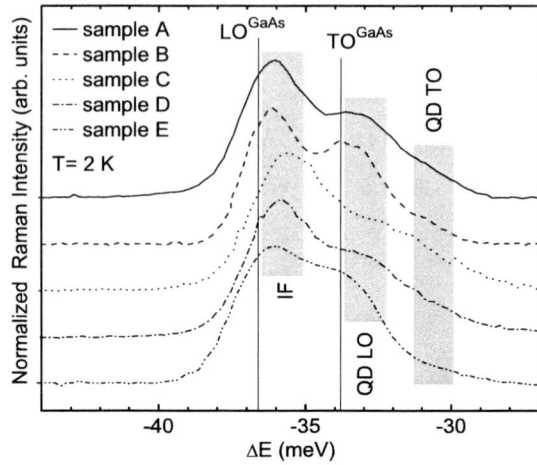

Fig. 14.7. Resonant Raman spectra excited on the respective absorption maximum of the QD ensembles for all five samples. Spectra are normalized and vertically shifted for clarity. The vertical black lines mark the energies of Γ-point GaAs bulk phonons and the gray areas highlight the resonance peaks in the measured spectra. From [11]

that a large part of the mode is located in the GaAs matrix. Hence, the interface mode shows the strongest coupling, followed by the QD LO phonon and, finally, by the QD TO phonon.

14.2.3 Electronic Tuning by Strain Engineering

Tuning the emission of QDs to target wavelengths is of greatest importance for actual devices like lasers and detectors. The emission wavelength of InAs QDs in GaAs matrix is typically below 1.2 μm. Deposition of thicker InAs layers to obtain longer wavelengths generally results in the formation of In-rich clusters and defects within the InAs. Hence, other methods must be applied to reach, e.g., the important telecom range at 1.3 μm. One of the most promising procedures is overgrowing the QDs with an InGaAs quantum well of lower In content. The underlying mechanism of this approach is discussed in this section.

Combined PL and PLE spectra for five MBE-grown samples with In contents in the well ranging from 0 to 25% are given in Fig. 14.8. As the In content increases, the maximum of the QD ground state emission shifts by 120 meV to lower energy. The PLE data provide information on the energies of the excited states in the QDs and of the energy resonances of the InGaAs well. The increasing In content of the well can clearly be seen by the red shift of the well resonances.

The red shift of the QD luminescence with increasing In content of the well originates basically from two effects: The enlargement of the QDs by In migration from the well to the dots, i.e., a partial decomposition of the InGaAs well, and a reduction of the hydrostatic strain inside the dots due to a surrounding of larger

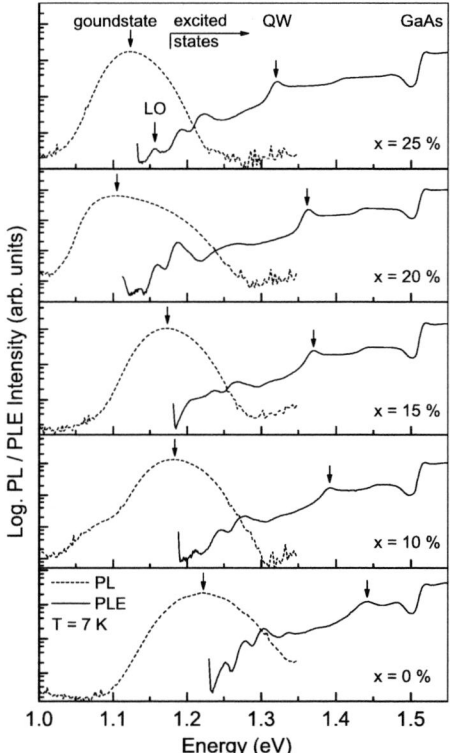

Fig. 14.8. PL (*dashed*) and PLE spectra (*solid lines*) of InAs quantum dots overgrown by an $In_xGa_{1-x}As$ quantum well of varied composition x. PLE spectra were recorded at the respective maxima of the PL spectra marked by arrows above the dashed curves. The QW resonances are highlighted by arrows on top of the PLE spectra. From [9]

lattice constant. The following discussion will render strain reduction as the main effect for the observed red shift of the ground-state transition.

Figure 14.9 combines energy positions of the excited state transitions as a function of In content in the well as obtained from the PLE measurements (Fig. 14.8) and numerical calculations. The calculations have been performed in the framework of 8-band $k \cdot p$ theory including band coupling, correlation, strain, and piezoelectricity [1] (see Chap. 7). An InAs model QD of truncated pyramidal shape and constant size, and a 4 nm thick InGaAs well were assumed for all In contents. The calculations predict the excited states to be rather independent on In content, in agreement with the experimental data. Furthermore, the calculated energies match the measured values very well. The applied theory treats strain in a realistic way, but neglects any In accumulation from the well near the dots. It can hence be concluded that the reduction of strain is the main reason for the red shift of the quantum dot ground state and that a possible change of QD size is of minor importance.

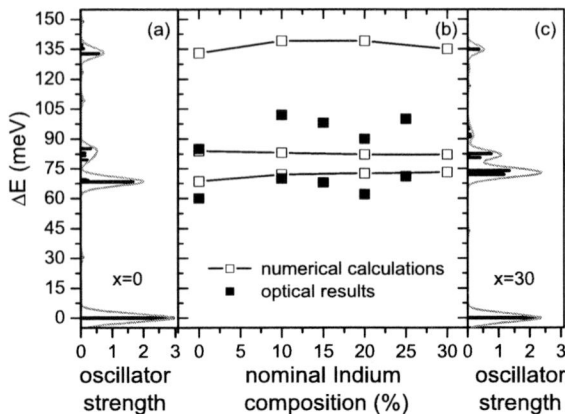

Fig. 14.9. (**b**) Measured (*solid squares*) and calculated energies (*open symbols*) of excited states with respect to the ground state in InAs QDs covered by an $In_xGa_{1-x}As$ well with varied In content x. Panels (**a**) and (**c**) give calculated oscillator strengths and artificially broadened absorption spectra for In compositions of 0 and 30% in the well. From [9]

14.2.4 Multimodal InAs/GaAs Quantum Dots

The quantum-dot ensembles considered so far show a single, inhomogeneously broadened ground state emission peak. Under certain growth conditions such a unimodal distribution of QDs decomposes into a more complex distribution with well-separated peaks in emission spectra. Generally, the differences in such subensembles are related to the size distribution of the QDs; that shows distinctly deviating mean values in the distribution with a spread that is smaller than the differences of the mean values. Coexisting bimodal size distributions have been observed quite frequently for various QD materials. In addition, trimodal distributions were found and models for their formation have been proposed (for a review see, e.g., [14]). For InAs QDs in GaAs matrix, self-organized growth of ensembles showing a *multi*modal distribution of sizes was recently developed and studied in detail [15]. The structural properties of these QD ensembles are well defined and directly linked to the emission energy, leading to spectrally well-separated subensembles. In the following we show that the spectral separation of the subensembles' emission maxima is due to a discrete variation of QD height in steps of InAs monolayers accompanied by a simultaneous increase in base length. As such ensembles allow for a direct correlation of structural and excitonic properties, these QDs represent an ideal model system to unravel the complex interplay of Coulomb interaction and a quantum dot's confining potential.

The PL of InAs/GaAs QDs with a multimodal size distribution is given in Fig. 14.10(a) [16]. A low excitation density <4 mW/cm^2 ensured an average occupation below one exciton per dot, excluding the appearance of excited state transitions. This is confirmed by the PLE data in Fig. 14.10(b) that give a state splitting much larger than the 30–50 meV spacings of the PL peaks.

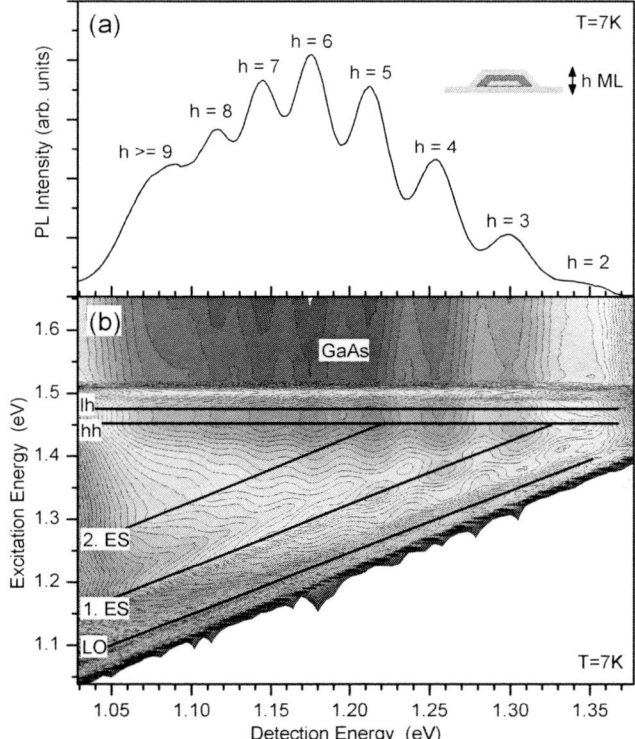

Fig. 14.10. (a) PL spectrum of an InAs QD ensemble with multimodal size distribution. The height of QDs within a subensemble is given on top of the respective peak in MLs. (b) PLE contour plot of the QD ensemble on a logarithmic gray scale. Horizontal black lines mark light-hole (lh) and heavy-hole (hh) resonances of the wetting layer, inclined lines highlight resonances of the first and second excited QD state (ES) as well as the 1-LO phonon transition (LO). From [16]

The PLE spectra reveal a first (second) excited state about 120 (220) meV above the ground state, indicative of a very high InAs content in the dots. The heavy-hole (hh) resonance of the wetting layer at 1.45 eV marks the onset of the continuum that limits the localization in the QDs. The combined representation of the photoluminescence intensity as a function of the detection and excitation energies in the contour plot, Fig. 14.10(b) reveals the evolution of the QDs' energy spectrum as a function of ground state recombination energy E_g. QDs with E_g below 1.22 eV have at least two excited states, whereas smaller QDs have only one (1.22 eV $< E_g <$ 1.32 eV) or, for $E_g >$ 1.32 eV, no excited states at all.

Structural investigations by TEM [15] and XSTM [17] outlined in Chap. 2 showed that the dots consist in fact of pure InAs, have sharp top and bottom interfaces, and steeply inclined side facets. The energy of excitons confined in such truncated pyramidal InAs dots was calculated using the 8-band $\bm{k} \cdot \bm{p}$ model including configuration interaction [18], also discussed in Chap. 7. A 1 ML thick wetting layer is assumed,

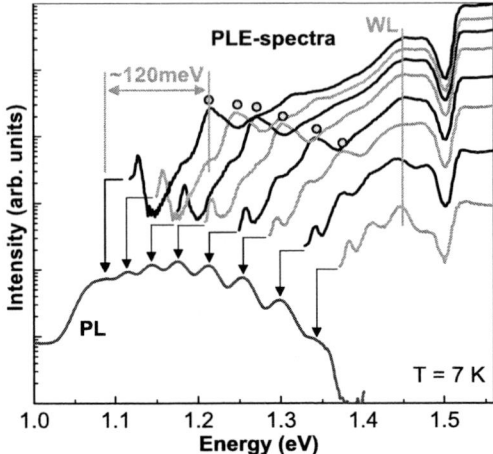

Fig. 14.11. PLE spectra of the multimodal QD ensemble with detection energies being equal to the local PL maxima, as given by arrows on top of the PL curve. The energetic position of the wetting layer is independent of the detection energy. From [4]

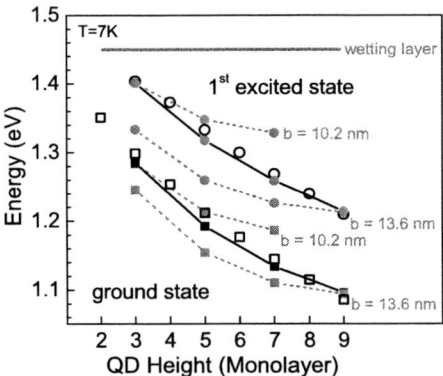

Fig. 14.12. Calculated (*solid symbols*) and experimental (*open symbols*) energies of the exciton ground state (*squares*) and first excited state (*circles*) for truncated pyramidal InAs QDs. Black and gray lines connect calculated energy for varied and constant base length b, respectively. From [2]

in agreement with the heavy- and light-hole energies observed in the PLE spectra. Calculated data are compared to the experimental results (taken from Fig. 14.11) in Fig. 14.12. The energies of the ground and excited state in a small QD of 3 ML height are well reproduced, if a base length of 10.2 nm is used. However, for taller dots this base length results in transition energies that are too large. Similarly energies of a 9 ML high dot are well matched using 13.6 nm base length, but such a base length yields energies too small for smaller dots. Consequently, both height and base length actually increase for larger dots, as proved by the excellent agreement of the calculated energies connected by the black lines in Fig. 14.12. QDs in the ensemble with multimodal size distribution hence show a gradual shell-like increase in volume, giving rise to self-similar dots throughout the whole ensemble.

14.3 Single InAs/GaAs Quantum Dots

QDs are promising building blocks for a number of novel applications like single-photon emitters or qubit registers. For optimal system performance knowledge about the electronic structure of few-particle states and the interrelation of structural and electronic properties is essential for targeted growth. Since the inhomogeneous broadening of a QD ensemble is much larger than the energy differences between recombination energies of different excitonic complexes or exchange energies, single-dot spectroscopy is required to gain insight into the Coulomb-affected few-particle properties.

The following results were obtained using spatially high-resolved cathodoluminescence spectroscopy (CL) with an optically opaque nearfield shadow mask made of Au which is evaporated onto the sample surface to reduce the number of simultaneously probed QDs. The samples under investigation are of multimodal type as presented in the previous section. The dot density is about 4×10^{10} dots cm^{-2}. Applying apertures of ~100 nm in diameter about four QDs are detected simultaneously. Further discrimination is achieved spectroscopically.

14.3.1 Spectral Diffusion

A typical CL spectrum as recorded through a shadow-mask aperture is given in Fig. 14.13(a). A couple of emission lines is spread over an energy range of some 10 meV and the spectral width of the lines is limited by the spectral resolution of the setup. Since the emission lines may originate from different QDs or different excitonic complexes within one QD, the emission spectra of single QDs have to be identified first. For the illustrated case the identification is based on the spectral diffusion [19, 20] as depicted by Fig. 14.13(b). Two sets of lines obviously exhibit the same spectral diffusion pattern in time. The spectral diffusion results from randomly charging and discharging of defects and interface states around the QDs, resulting in a synchronously changing electric field [21]. Since each QD experiences a different electric field in time, all its emission lines show the same, characteristic emission pattern—allowing for a discrimination of single dot spectra, even if they overlap. The magnitude of the spectral diffusion depends not only on the sample structure but also on the way of excitation. For optical excitation (e.g., via a frequency doubled Nd:YAG laser) the spectral diffusion of a particular individual dot was found to be one order of magnitude weaker than for excitation via the electron beam in CL.

The different excitonic transitions within a single-QD spectrum were identified via polarization- and excitation-dependent measurements [22]. Due to the statistical

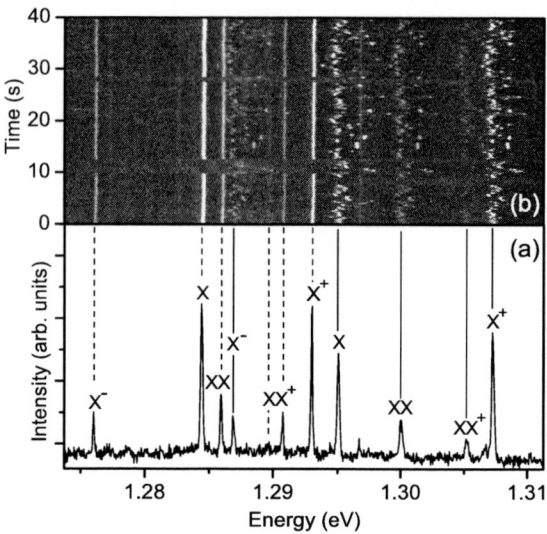

Fig. 14.13. (a) Typical CL spectrum of an InAs QD sample as measured through a shadow mask with 100 nm aperture. $X^{(+/-)}$ and $XX^{(+)}$ denote recombinations of neutral and charged excitons and biexcitons. There is emission from two different QDs as marked by the dashed and solid lines. (b) Temporal evolution of the CL spectra. The series consists of 500 spectra, each being integrated for 80 ms. The intensity is gray-scale coded. From [22]

nature of carrier capture into the dots emission from neutral excitons, biexcitons, as well as from charged excitonic complexes is observed.

14.3.2 Size-Dependent Anisotropic Exchange Interaction

In polarization-dependent measurements four of the emission lines show distinct features which monotonously evolve with QD size as shown in the following. Polarized single-QD spectra from a large QD and a small QD appearing at small and high emission energy, respectively, are displayed in Fig. 14.14. The directions of polarization were chosen to be parallel to [110] and [$\bar{1}$10]. The emissions of the neutral exciton and biexciton split into two lines of equal intensity with reversed order of their polarization with respect to the transition energy. The splitting of both doublets is identical. The other two polarization-dependent lines stem from the recombination of the positively charged biexciton. They switch intensity when the polarizer is rotated by 90°.

Both features originate from the anisotropic part of the exchange interaction among the confined electrons and holes. The anisotropic exchange interaction lifts the degeneracy of the bright exciton state, which is the final state of the biexciton decay and the initial state of the exciton decay, see Fig. 14.15(a). This fine-structure splitting (FSS) leads to equal splittings of the biexciton and exciton transitions, because both the biexciton state XX and the empty QD state "0" have an angular momentum of zero.

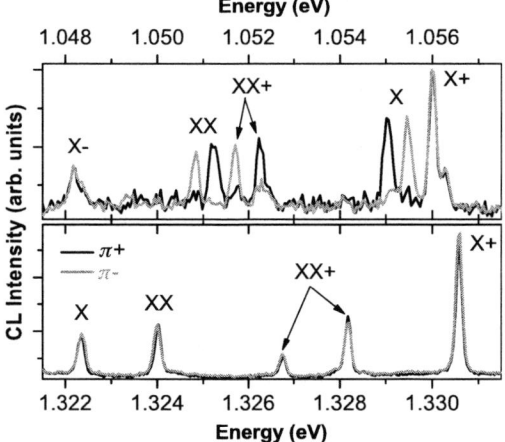

Fig. 14.14. Single dot CL spectra of a single large (*top*) and a small (*bottom*) QD for linear polarization along [$\bar{1}10$] ($\pi+$, black spectra) and [110] ($\pi-$, gray spectra) direction. From [23]

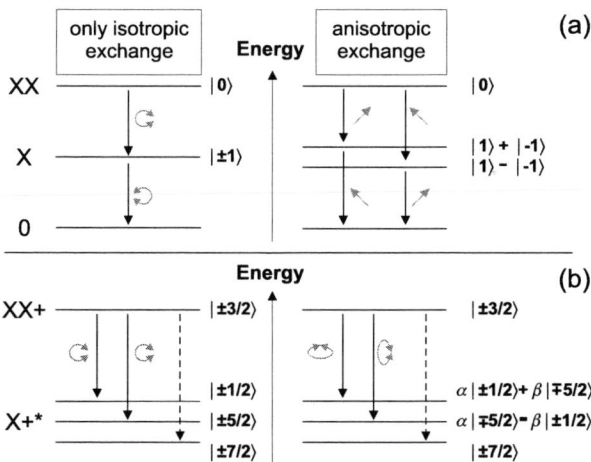

Fig. 14.15. Energy-level scheme of the (**a**) neutral and (**b**) positively charged biexciton and exciton. Solid and dashed vertical arrows signify allowed and forbidden transitions, respectively, with polarizations marked by gray symbols. $|M\rangle$ denotes the total angular momentum of the excitonic complex in the respective configuration. From [23]

In case of a positively charged QD, the anisotropic exchange does not lift the degeneracy of the hot trion states which represent the final states of the XX+ recombination, see Fig. 14.15(b). It effects, however, a mixing of the hot trion states and hence a change of the polarization from circular for pure states to elliptical.

The two single dot spectra given in Fig. 14.14 suggest that the magnitude of exchange interaction is related to the size of QDs. For a systematic analysis a number

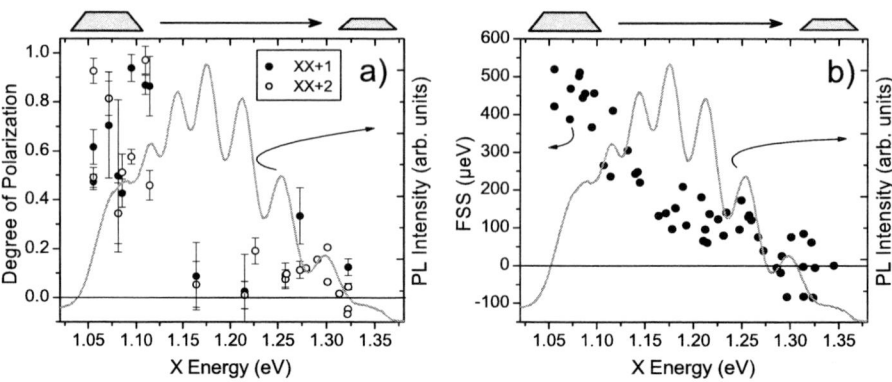

Fig. 14.16. (a) Degree of polarization of the two XX+ emission lines and (b) magnitude of the X fine-structure splitting versus emission energy. The gray spectrum represents the photoluminescence of the QD ensemble. From [23]

of QDs was investigated with respect to the exciton fine-structure splitting and the mixing of the hot trion states. The FSS is defined to be positive, if the X line at lower energy is $\pi+$ polarized, i.e., along the $[\bar{1}10]$ crystal direction. For the mixing degree of the hot trion states the polarization degree of the two XX+ recombination lines defined by $p = \frac{I^{\pi+} - I^{\pi-}}{I^{\pi+} + I^{\pi-}}$ is taken as measure. Figure 14.16 gives the FSS and polarization degree data as measured all over the inhomogeneous broadened QD ensemble peak.

Both graphs show a clear dependence of the respective data on QD size, which is directly connected to the exciton recombination energy for the measured InAs QD ensemble with multimodal size distribution. Hence, in both cases the underlying exchange interaction scales with QD size in a similar way.

The measured data clearly demonstrate the impact of the anisotropic part of the exchange interaction on the confined electrons and holes. Such anisotropy arises in quantum dots, when the symmetry of the confining potential is lower than C_{4v}. Sources for such symmetry lowering are structural anisotropy of the dots, piezoelectricity induced by strain [12, 1], and atomistic symmetry anisotropy [24]. Structural elongation of the studied quantum dots can be ruled out as a main source for the observed exchange effects. TEM measurements of our quantum dots do not show a significant anisotropy ([15], cf. also Chap. 2) and numerical modeling fails to reproduce experimental results [25]. Piezoelectricity provides a reasonable explanation for the observed trend [25]. Its magnitude is proportional to the occurring shear strain in the QDs. Due to the lattice mismatch between GaAs and InAs, the shear strain is larger for larger QDs [12]. Larger QDs with stronger shear strain components hence lead to stronger piezoelectric fields. Consequently such QDs have larger values of the fine-structure splitting, and the XX+ lines show a larger degree of polarization. The importance of strain for the magnitude of the FSS was pointed out earlier [24]. The role of atomistic symmetry anisotropy has not yet been assessed in detail. A major role cannot be excluded in a complete treatment of exchange interaction in quantum dots.

14.3.3 Binding Energies of Excitonic Complexes

Another important aspect of Coulomb interaction is the renormalization of few-particle transition energies acting, e.g., when one electron and hole recombine in the presence of additional charge carriers. Of particular interest for applications is the biexciton binding energy, i.e., the energetic difference between excitonic and biexcitonic recombination. The larger this spacing is, the easier is the spectral selection of each transition. Emitters of polarization-entangled single-photon pairs from excitonic and biexcitonic transitions will benefit from that. For a systematic investigation of the interrelation between excitonic binding energies and structural properties of the QDs, many single QD spectra were recorded all over the inhomogeneously broadened ensemble peak of the multimodal QD sample from Sect. 14.2.4. The resulting binding energies for the biexciton and the two trions are plotted in Fig. 14.17 as a function of the neutral exciton recombination energy.

Obviously, there is a characteristic trend and energy regime for the three excitonic complexes. The negatively charged exciton always has a positive binding energy being almost independent on the neutral exciton's recombination energy. In contrast, the binding energies of the positively charged exciton and the biexciton clearly decrease for increasing exciton recombination energy (decreasing QD size). Moreover, for the biexciton a transition from positive to negative binding energies is observed.

The nonzero binding energies originate from the Coulomb interaction between the confined charge carriers. For understanding the observed energy regimes and

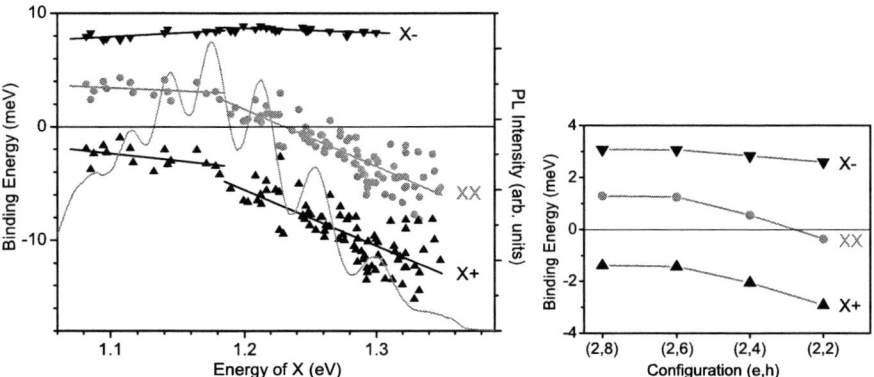

Fig. 14.17. Left: Energy distances between the recombination energies of confined excitonic complexes as a function of the neutral exciton's recombination energy. Right: Calculated energy distances (like on the left) as a function of the number of states included in the configuration interaction calculation (1 particle level with 2 spin orientations = 2 states). From [16]

trends the Coulomb interaction will be discussed in terms of direct Coulomb interaction and correlation.

The energy regimes of the two trions can be explained by the direct Coulomb interaction alone [22]. The binding energy of X- directly depends on the difference between the two direct Coulomb terms $C(e, h)$ and $C(e, e)$. Correspondingly E_b of X+ depends on $C(e, h)$ and $C(h, h)$. Due to the larger effective mass of the holes and the small size of the QDs the wave function of the hole is much stronger localized than that of the electron, which penetrates the barrier to a large extent. Consequently $|C(e, e)| < |C(e, h)| < |C(h, h)|$ and the negative trion has a positive binding energy, while the positive trion has a negative E_b.

For the trends of the binding energies the impact of correlation is to be considered. To model the impact of correlation the number of confined QD states for electrons and holes was varied [22]. The procedure is motivated by the finding presented in Sect. 14.2.4 that the number of confined states decreases for increasing ground state recombination energy as shown in the absorption measurement (Fig. 14.10(b)). In our numerical model the spectrum of bound states is expressed by the number of single particle states for electrons and holes that serve as the basis states in CI calculations. A given set of electron and hole states is called configuration. By comparing calculations for different configurations it becomes obvious that the number of bound hole states is the main factor determining the trend of the binding energies: The larger the number of hole states is, the larger is the binding energy of, e.g., biexcitons. A sample calculation is given in the right panel of Fig. 14.17 where the number of electron states is kept constant at three while the number of hole states was varied from eight for large QDs to two for small QDs. Obviously the trends and energy regimes observed in the experiment are well reproduced—including a binding energy for the X- that is almost independent of the actual configuration. The relevance of correlation for transition energies in QDs was described in Sect. 14.2.1 for many-particle transitions.

14.3.4 Data Storage Using Confined Trions

Besides shining light onto fundamental quantum phenomena, the trion states can be used for information storage. This was recently demonstrated by photocurrent (PC) measurements on InAs/GaAs QDs [26]. By fabricating a p–i–n diode structure with an AlGaAs barrier on the n- (p-) side of the QDs the tunnel rate of holes (electrons) in an electric field is drastically reduced as compared to the opposite charge-carrier type. When exciting the QDs resonantly with monochromatic light a subensemble of QDs will be occupied with excitons. A subsequently applied electric field in reverse direction will lead to an efficient tunneling-out process of one kind of charge carrier. Consequently, the QDs are charged with the opposite one. To reveal the charge state of the QDs and thereby the stored information, photocurrent spectroscopy can be applied. Here, the sample is illuminated with monochromatic light again and the photocurrent is measured as a function of excitation energy. On the energetic position of the formerly neutral QDs a dip appears in the PC spectra: By charging the QDs the energetic position of absorption is shifted by the trion binding energy. Consequently

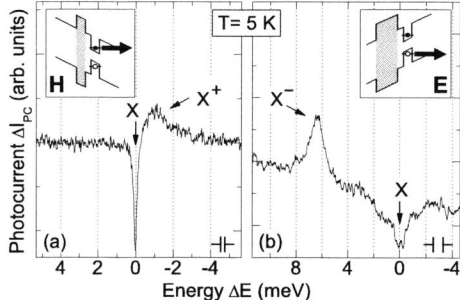

Fig. 14.18. Photocurrent spectra of diode structures with an AlGaAs barrier on (**a**) the n-side for hole storage and (**b**) the p-side for electron storage. The opposite charge carrier escapes the QD by tunneling as depicted by the black arrows in the insets. Symbols in the lower right corners give the spectral resolution. From [26]

the spectral density for absorption is reduced here. On the other hand a peak appears at the energy plus the trion binding energy as the charged QDs increase the spectral density at that energy.

Figure 14.18 gives PC spectra for the storage of holes and electrons, respectively. The shift of the trion resonance with respect to that of the neutral exciton is qualitatively the same as observed in the single QD spectra in the previous section though the QDs are different. The X+ has a negative binding energy while the X- has a positive one with the absolute value being larger than for the X+.

14.3.5 Electronic Tuning by Annealing

For actual single-QD applications it is crucial to optimize the respective electronic property of a QD. An example will be emitters of polarization-entangled single-photon pairs where the fine-structure splitting is to be close to zero. Besides control during growth of the QDs a post-growth (ex-situ) treatment can be applied by thermally annealing the sample. This was previously demonstrated for ensembles of QDs [27, 28] or randomly chosen single QDs [29].

Recently, we have realized an experiment where one and the same QD was subjected to subsequent annealing steps and single-dot measurements [30]. To trace the single QD, mesa structures were fabricated on a sample with a low QD density in a given spectral region. By mapping the luminescence of the mesa in CL the exact position of one QD can be determined and relocated after each annealing step. Figure 14.19 displays such a trace for the as grown QD and two annealing steps of 5 minutes at 710°C and 720°C, respectively.

Figure 14.20 shows the drastic change of a single QD's electronic properties due to the annealing procedure. For clarity, one panel focuses on the binding energies of the excitonic complexes and the other on the excitonic fine-structure splitting. For all excitonic complexes a change of the binding energies occurs. The trend depends on the respective few-particle states. The binding energy of the negatively charged

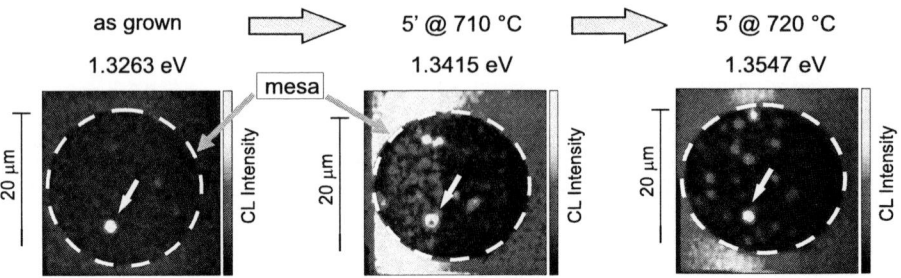

Fig. 14.19. Monochromatic CL images of mesa structures containing InAs/GaAs dots with a low areal density. The images were recorded at the emission energies noted on top. From [30]

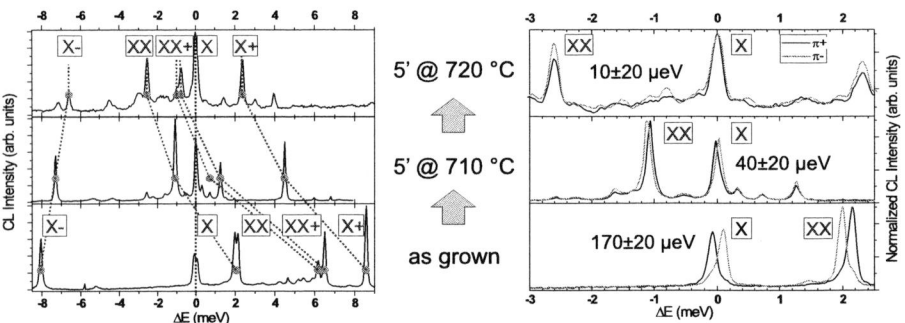

Fig. 14.20. Effect of two consecutive annealing steps on the spectrum of a single QD. 0 meV corresponds to the respective neutral exciton recombination energy. Left: Impact on binding energies. Right: Excitonic fine-structure splitting revealed by polarization dependent measurements. From [30]

exciton decreases for increasing annealing strength whereas all the other complexes become more binding (larger binding energy). The neutral and positively charged biexcitons even show a transition from antibinding to binding.

A drastic reduction is observed for the fine-structure splitting. An initial value of about 170 µeV decreases to less than 20 µeV which represents the resolution limit of the setup. The generation of polarization-entangled photon pairs by annealing-induced tuning should hence be feasible—an important fact for the realization of quantum-cryptography schemes using semiconductor quantum dots.

14.4 Optical Properties of InGaN/GaN Quantum Dots

Over the last couple of years, the study of the optical properties of InGaN/GaN heterostructures has led to great improvements in light-emitting devices that operate in the violet and visible spectral range [31, 32]. Numerous investigations have been carried out to clarify the origin of the luminescence and that of the laser gain in

InGaN-based structures [33, 34]. As a result, localized electron-hole pairs were identified as the microscopic origin of the luminescence in optoelectronic devices. As a consequence, localization centers were deliberately created in the form of quantum dots (QDs). Until now, however, there is no clear picture of the basic processes of the radiative recombination in InGaN QD structures. In nitride QDs, the carriers have large effective masses, high exciton binding energies and they are seriously confined due to the large bandgap differences. As a result, they are found in the QDs' excitonic ground state, even at room temperature, leading to a large overlap of the wave functions of the electron and the hole and, thus, to good luminescence efficiency. A prominent problem of InGaN/GaN heterostructures is that fluctuations of size [35] and composition [36] of InGaN insertions lead to a red-shift of the luminescence due to In-rich regions and also affect the radiative decay. Moreover, the properties of nitride-based nano-objects are significantly affected by the presence of strong built-in electric fields along the (0001) growth direction [37] that are caused by the arrangements of polar atoms in the wurtzite crystal structure. Experiments and theoretical calculations demonstrate that these piezoelectric fields in nitrides are in the range of MV/cm. This is some orders of magnitude larger than in GaAs-based QDs [38, 39]. Such an electric field can no longer be treated as a perturbation of the confined electrons and holes in the QD. In fact, it leads to a Stark shift of the transition energy to smaller energies and to large time constants of the radiative recombination of up to ms [40, 41].

14.4.1 Time-Resolved Studies on Quantum Dot Ensembles

The purpose of this section is to highlight the results of time-resolved luminescence experiments on InGaN/GaN quantum dot ensembles. The investigated InGaN/GaN structure was grown on Si (111) substrate by metal-organic vapor phase epitaxy (MOVPE). The sample consists of a 1.2 µm thick GaN buffer layer grown on top of an AlN nucleation layer, the active InGaN region and a 50 nm thick GaN cap layer. The active region is made up of a superlattice consisting of five layers of 1.5 nm thick InGaN separated by 3 nm thick GaN layers. Vertical electronic coupling between the InGaN layers can be ruled out. Transmission electron microscopy (TEM) measurements indicate lateral fluctuations of the indium concentration inside the InGaN layers and the formation of In-rich nanodomains with sizes between 2.5 and 4 nm. Additionally, a vertical decrease of the In concentration was found. Now the question is if these inhomogeneous layers can be considered as rough QWs or if they rather behave like QDs. In the following we will show, by means of optical experiments, the exact nature of the heterostructure.

Figure 14.21 shows a typical PL spectrum for nonresonant excitation. A broad band originating from the active InGaN region appears between 2.6 eV and 3.3 eV. The PL is spectrally modulated due to an interference effect between the Si/GaN and air/GaN interfaces which provide large refractive index steps causing strong reflection. On the basis of the period of the oscillation we determined the thickness of the structure to be 1.6 µm, in reasonable agreement with the thickness expected from growth-rate calibration. The sharp line around 3.5 eV is attributed to the re-

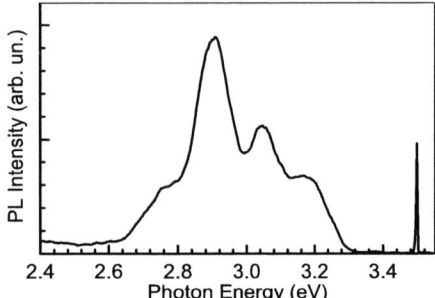

Fig. 14.21. PL spectrum of InGaN/GaN QDs measured under nonresonant excitation. $T = 1.8$ K

combination of bound excitons in the GaN buffer layer. At resonant excitation, the shape of the broad PL does not change, but its intensity does because of the changed excitation condition. PL excitation (PLE) investigations reveal an excitation channel around 3.0 eV, more than 100 meV above the maximum of the PL spectrum ($\sim$2.9 eV) where the detection energy was set. Three possible reasons may account for this Stokes shift: (i) the higher oscillator strength of transitions involving excited subbands in a quantum well (QW) subject to a high piezoelectric field; (ii) the existence of mesoscopic localized states due to long-range disorder in a quantum well; and (iii) QD-like structures that typically exhibit such a PLE spectrum [42]. In the last case, the PLE maximum would correspond to the first excited exciton state of a QD subensemble. This effect is well documented for 3D (with vertical extension) InAs QDs in a GaAs matrix that are proven to have a quasi-ideal δ-function-like density of states [42]. Which of the three explanations is the true scenario can be decided on the basis of resonant and nonresonant time-resolved (TR) PL studies (for clarity, *resonant* excitation means that the excitation energy is equal to the detection energy; *nonresonant* excitation means the excitation energy is larger than the bandgap of the GaN matrix material).

For a QW structure, resonant PL measurements within the continuum of the emission peak should reveal short lifetimes as carriers can relax instantly into QW states of lower energy because of the continuous density of states. Under resonant excitation of an ensemble of QDs with a δ-function-like density of states, the carriers are lifted directly into the ground state and long lifetimes will be observed, similar to nonresonant excitation. For nonresonant excitation, the rise time of the QW luminescence should be affected by carrier migration through small barriers (given by the QW's roughness) into states of minimal energy causing a delay of the evolution of the longer wavelength emission. Concerning the rise time of QD emission for resonant excitation or excitation into excited states, maximal intensity should be reached instantly as, here, the barriers between neighboring dots are high and excited states relax fast into the ground state.

The results of time-resolved PL measurements are presented in Fig. 14.22. The decay after nonresonant and resonant excitation exhibits the same highly nonexpo-

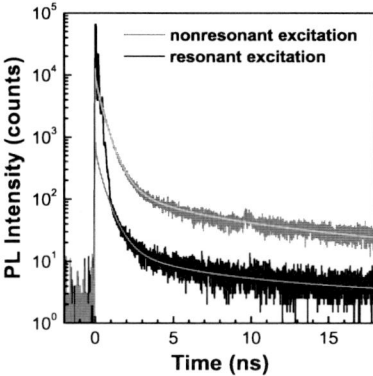

Fig. 14.22. Time-resolved PL under resonant and nonresonant excitation. The detection energy was ≈2.88 eV (430 nm). $T = 1.8$ K

nential behavior and the same time constants [43]. (For resonant excitation, the few maxima and the rapid decay at near zero times reflect the transient behavior of the excitation pulse.) The observed time behavior can be fitted with the help of a stretched exponent β [44, 45], as is typical for disordered media [46]

$$I(t) = I_0 \exp\left(\frac{-t}{\tau^*}\right)^{\beta}.$$

For the parameters we receive the values $\beta = 0.2$ and $\tau^* = 0.4$ ns regardless of the detection energy. The nonexponential decay can be assigned to a disordered ensemble of low-dimensional islands if we assume the occurrence of islands with different shapes. For equal transition energies the values of the electron-hole-overlap are similar in the case of flat 2D islands but may differ considerably for 3D islands. As a consequence, the stretched exponent β is smaller than 1 in the 3D case and equal to 1 in the 2D case. Hence, the value found for β of 0.2 suggests the presence of 3D islands. The observed rise times after nonresonant excitation indicate the presence of QDs rather than QWs. Regardless of the detection energy, the PL signal reaches maximum intensity after less than 7 ps, which is very close to our time resolution (5 ps).

In order to confirm our presumption regarding the QD nature of the InGaN/GaN structures on Si, we performed resonant TR PL investigations in a manner similar to that reported for InGaN/GaN multilayers on sapphire [47]. As mentioned earlier, the continuous function of density of QWs' states leads to fast carrier relaxation into states of minimal energy after resonant excitation significantly above the PL onset. Consequently for TR resonant PL of QWs, we expect a considerable spectral shift of the luminescence peak toward lower energies accompanied by a broadening. With elapsing time, an additional red-shift should become observable, caused by the recombination of nonequilibrium carriers that were initially screening a potential piezoelectric field.

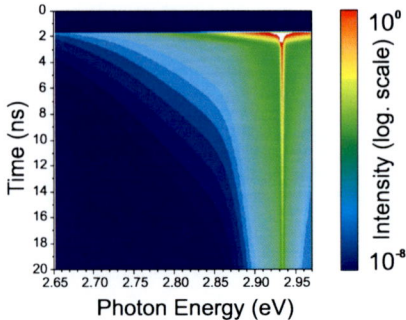

Fig. 14.23. Contour plot of the time-resolved resonant PL spectrum. The white spot is the scattered light of the laser pulse

However, TR resonant PL measurements on our samples do not show such a shift. Figure 14.23 presents the evolution of the PL spectrum after resonant excitation. After the resonant excitation pulse (white color) has decayed the spectral shape of the PL spectrum is governed by the relaxation processes within the InGaN structure. With elapsing time, the shape of the PL peak does not change. Neither a red-shift nor a broadening can be observed. Also the decay time corresponds to that after nonresonant excitation. Consequently, effects of spectral diffusion or piezoelectric screening that are expected for QW structures can be ruled out. In fact, the observed behavior perfectly matches that of resonantly excited QDs with a δ-function-like density of states. Here, no carrier transfer occurs at low temperatures. In independent measurements, a comparison of the spectra of the PL's resonant component, after any period of time after excitation, with the spectrum of the exciting laser pulse's scattered light revealed that their shapes are identical. This is in disagreement with the spectrally asymmetric spectrum developing in Fig. 14.23 after excitation by the symmetric laser pulse. In order to extract the difference between the observed spectrum and the purely resonant part of the luminescence, the spectrum of the laser pulse was normalized to fit the intensity of the PL maximum after any period of time. Subsequently, it was subtracted from the TR spectrum of Fig. 14.23 revealing a well-resolved peak on the low-energy side of the PL maximum. This peak shows up immediately (<5 ps). It is the result of the recombination of ground-state excitons in QDs that were excited via the first excited state and a subsequent fast relaxation. These QDs must be a subensemble of larger QDs with the energy of the first excited state equal to that of the resonantly excited ground state of the major subensemble. The energy difference between the two peaks of about 100 meV provides an estimate of the energy splitting between the QD ground state and the first excited state. About the same value was obtained by PLE experiments (see above).

14.4.2 Single-Dot Spectroscopy

In the previous section, it is demonstrated how the nature of nanoheterostructures can be determined to be quantum-well-like or rather quantum-dot-like by means of

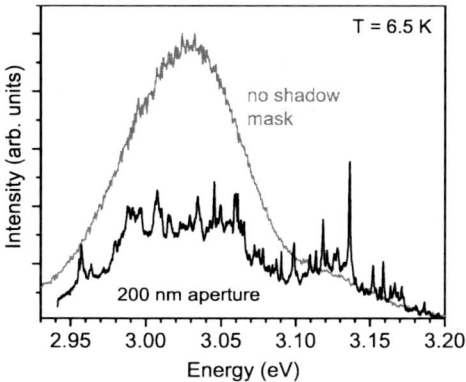

Fig. 14.24. Luminescence spectra of the InGaN QDs detected with (*black curve*) and without (*gray curve*) a shadow mask

time-resolved PL spectroscopy. A direct experimental access to the optical properties of quantum dots is single-dot spectroscopy. For this purpose we studied an InGaN/GaN QD sample grown by MOVPE on n-Si(111) substrate. The sample consisted of a 40 nm thick AlN nucleation layer, a 1 μm thick $Al_{0.05}Ga_{0.95}N$/GaN buffer layer, the active InGaN layer, and a 30 nm thick GaN cap layer. Cross-sectional TEM measurements revealed compositional fluctuations in the InGaN layer with indium-rich domains indicating a spinodal decomposition of the InGaN layer and, thus, a fluctuation-induced formation of the QDs [48]. Their lateral extensions of about 5 nm are comparable to the exciton Bohr radius, thus being small enough to provide not only a strong localization of excitons, but also a confinement of the individual wave functions. The sample was masked by a 70 nm thick metal cap leaving apertures with diameters of about 200 nm for spatially resolved CL investigations.

Spatially high-resolved cathodoluminescence measurements show discrete QD states and demonstrate 3D confinement of carriers in the InGaN QDs [48]: When detected through an aperture of the shadow mask the broad luminescence peak of the QD ensemble decomposes into sharp lines (see Fig. 14.24). The narrowest line observed has a full width at half maximum (FWHM) of 0.48 meV. These lines were observed between 2.8 and 3.2 eV indicating that the entire broad peak originates from QDs.

CL spectra covering several single lines were recorded with short integration times (80–300 ms) to generate a time series (see Fig. 14.25). One series consists of 500 spectra with a fixed integration time. The time series shows slight stochastic variations of the peak energies and intensities caused by fluctuating electric fields and known as spectral diffusion (see Sect. 14.3.1). Groups of up to five lines with the same jitter could be found, indicating the existence of a number of different excitonic complexes in one quantum dot. The typical energetic spread of lines belonging to one QD was found to be 20 meV.

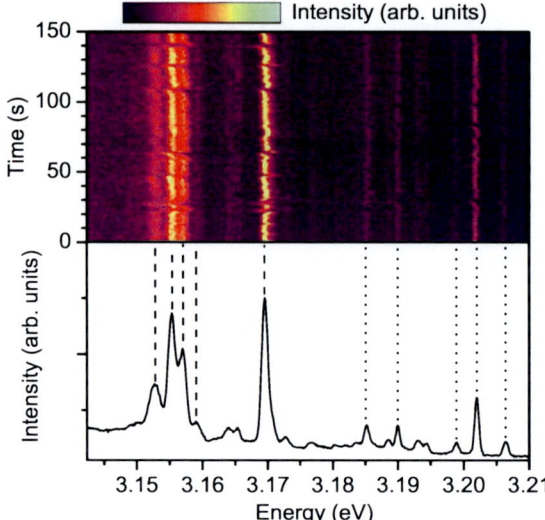

Fig. 14.25. Time series of 500 CL spectra with 300 ms integration time each. Spectral diffusion is visible, allowing us to distinguish lines that originate from the same QD. Dashed lines mark peaks belonging to one QD, dotted lines those of another QD

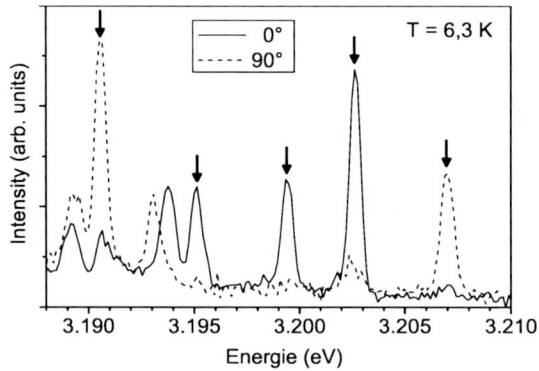

Fig. 14.26. Polarized emission lines of a single InGaN QD (as marked by arrows). The other emission lines belong to different QDs

All lines show a pronounced linear polarization in orthogonal directions (Fig. 14.26). These directions are scattered around the [11$\bar{2}$0] and [1$\bar{1}$00] crystal directions. Both directions were found for each investigated line group. This behavior has not been previously observed in nitrides. In [49], the electronic structure was modeled by realistic 8-band $k \cdot p$ calculations. The model includes piezoelectric and pyroelectric effects as well as crystal-field splitting and spin-orbit interaction. Few-particle states were calculated using a self-consistent Hartree (mean-field) approach. The current parameter set was adjusted to include recently developed

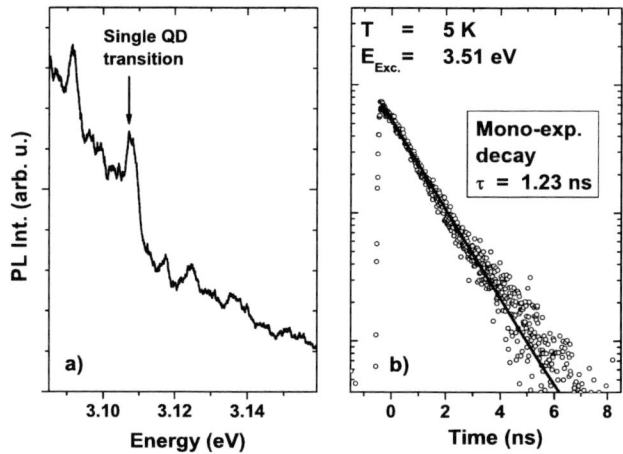

Fig. 14.27. (a) Single-dot spectrum of an InGaN/GaN QD. (b) The monoexponential decay of the marked line was obtained by subtracting the transient behavior of the background

band-dispersion parameters [50] derived from accurate quasiparticle energy calculations [51]. The model structure was chosen to match the TEM data [48]. Both the A- and B-hole states are included in the calculation of the oscillator strengths for the confined exciton, biexciton, and negative trion. Contrary to the heavy-hole-light-hole splitting in InAs QDs, in InGaN QDs the energetic separation of A- and B-holes that is affected by biaxial strain is relatively small ($\approx$7 meV). The calculations predict a linear polarization of all transition lines. The direction of the polarization of each transition depends on whether an A- or a B-hole state is involved in the recombination process, providing an explanation for the polarization of all of the emission lines.

The dynamics of the recombination of excitons in single QDs was investigated using TR micro-PL. Figure 14.27 displays the transient behavior of a single-QD line. The studied line is superimposed on a background signal making up 70% of the PL intensity. This background signal is attributed to luminescence light from the QD ensemble transmitted through the mask and scattered light from other apertures of the mask. To eliminate the transient behavior of the background and isolate the single QD emission line, two transients for each QD line were recorded: one at the maximum of the line and one 5–10 meV away. Since the changes in the PL dynamics of the entire QD ensemble are negligible within energy differences of a few meV, the latter transient can be subtracted from the former to obtain the true PL decay of the single QD state as suggested by Robinson et al. [52]. Contrary to the ensemble luminescence, single QD lines show a monoexponential behavior. Their decay times are between 0.4 and 1.6 ns.

In Fig. 14.28, the observed decay times of emission lines from single QDs are depicted as a function of the transition energy. A large spread of the time constants was observed, even for similar detection energies [53]. Consequently, excitonic QD

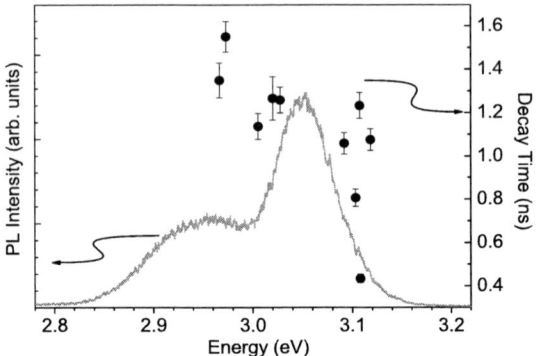

Fig. 14.28. PL spectrum of the InGaN QD ensemble and the time constants of single-QD lines (dots)

states with the same transition energy do not necessarily have the same lifetime. The dynamics of the luminescence are governed by two competing processes: the transition probability, which primarily depends on the overlap of the wave functions of the electron and the hole, and the probability of nonradiative escape. The overlap of the wave functions is affected by size, shape, and composition of the QD. The depth of its localization potential is not given by the absolute indium content alone, but it also depends on the difference in indium concentration between the QD and its adjacent area. Due to the disordered distribution of indium in the investigated sample [48], there is no correlation between the localization energy and the transition energy. Luminescence detected at a particular energy originates from a QD subensemble with equal transition energies. However, this subensemble still consists of QDs with a wide distribution of electron–hole wave-function overlap, resulting directly in a distribution of transition times. Generally, the trend toward shorter time constants occurring for higher transition energies (Fig. 14.28) can be explained by an increasing importance of escape processes. Nevertheless, the transfer probability also differs inside one QD subensemble of equal transition energy. For QDs of a subensemble this leads to different time constants. Thus, the nonexponential PL decay is assigned to the summation of monoexponential decays originating from individual QDs.

14.5 Summary

This chapter presents optical investigations on self-organized In(Ga)As and InGaN QDs. Detailed size-selective and time-resolved investigations demonstrate that particular excitonic properties can be engineered. Both the exciton–phonon and the exciton–photon interactions are sensitive to the actual structural properties of the QDs, realizing exciton dynamics which are limited either by relaxation or by recombination processes. Based on 8-band $\mathbf{k} \cdot \mathbf{p}$ model calculations, the crucial role of the electron-hole overlap and the local charge density, which depend on the size, shape, and composition (profile) of the QDs, is demonstrated.

Many- and few-particle effects are investigated and correlated with the QDs' structural properties. Many-particle effects have been identified for the ground state, when filling the excited states and the wetting layer with spectator charge carriers. Successive quenching and renormalization of PLE resonances allow us to identify transitions corresponding to specific electron and hole levels. The single-particle picture is no longer sufficient if a QD is occupied by more than one charge carrier because renormalization effects appear that strongly influence the optical spectra.

Resonant Raman spectroscopy and line-narrowing photoluminescence measurements on a variety of self-organized In(Ga)As/GaAs QD structures reveal three localized optical phonon modes (QD interface phonon, LO-like phonon of the strained In(Ga)As QD and TO-like QD phonon), being nearly identical for all samples. The interface mode shows the strongest coupling, followed by the QD LO phonon and the QD TO phonon. Thus, the observed localized phonon modes show only very weak dependence on the structural properties of the QDs in the respective sample.

A distinct multimodal modulation of the PL of InAs/GaAs QDs grown by MOVPE is unambiguously attributed to a shell-like growth mode leading to monolayer-steps of the QD height. Based on the optical data we deduce a flat shape with both a structurally and chemically well-defined upper InAs/GaAs interface. The large splitting between the exciton ground and excited state of 120 meV supports the assumption of pure InAs in the QDs.

Single-dot spectroscopy reveals recombination of neutral and charged excitons and biexcitons. The biexciton binding energy strongly depends on the QDs' size. With decreasing size the biexciton complex changes from binding to antibinding. The occurrence of an antibinding biexciton is attributed to the impact of the confining potential, which generates a repulsive effect of the direct Coulomb interaction, and the decreasing number of localized excited states with decreasing QD size, which quenches the impact of correlation.

Polarization-dependent measurements reveal the fine-structure splitting (FSS) of exciton and biexciton emission that monotonously evolves with QD size. We found a variation by one order of magnitude and observed a sign inversion of the FSS. Large values of up to 520 µeV for large QDs were found.

For the nitride structures, the observation of a resonantly excited narrow PL line and sharp CL lines gives clear proof of the QD nature of the luminescence. In addition, the recombination dynamics of excitons localized in a single InGaN QD was compared to the recombination dynamics of the entire QD ensemble. The luminescence decay of single QDs was found to be monoexponential. The nonexponential PL decay of the entire QD ensemble is unambiguous proof that the disorder of the QD system governs the recombination dynamics here.

The single emission lines show a pronounced linear polarization, which is attributed to the valence band structure of the wurtzite type nitrides. This observation reveals a large application potential of nitride QDs in high-operating-temperature single-photon sources. In comparison to InAs QDs, nitride QDs open a new wavelength region in the blue and near-ultraviolet spectral range for single-photon sources.

References

1. O. Stier, M. Grundmann, D. Bimberg, Phys. Rev. B **59**(8), 5688 (1999)
2. F. Guffarth, R. Heitz, A. Schliwa, K. Pötschke, D. Bimberg, Physica E **21**, 326 (2004)
3. M. Grundmann, D. Bimberg, Phys. Rev. B **55**, 9740 (1997)
4. F. Guffarth, *Elektronische Eigenschaften von* In(Ga)As/GaAs *Quantenpunkten* (Mensch & Buch, Berlin, 2003)
5. G. Tränkle, H. Leier, A. Forchel, H. Haug, C. Ell, G. Weimann, Phys. Rev. Lett. **58**, 419 (1987)
6. R. Heitz, H. Born, A. Hoffmann, D. Bimberg, I. Mukhametzhanov, A. Madhukar, Appl. Phys. Lett. **77**, 3746 (2000)
7. I. Mukhametzhanov, R. Heitz, J. Zeng, P. Chen, A. Madhukar, Appl. Phys. Lett. **73**, 1841 (1998)
8. I. Mukhametzhanov, Z. Wie, R. Heitz, A. Madhukar, Appl. Phys. Lett. **75**, 85 (1999)
9. F. Guffarth, R. Heitz, A. Schliwa, O. Stier, N.N. Ledentsov, A.R. Kovsh, V.M. Ustivov, D. Bimberg, Phys. Rev. B **64**, 085305 (2001)
10. R. Sellin, F. Heinrichsdorff, C. Ribbat, M. Grundmann, U.W. Pohl, D. Bimberg, J. Cryst. Growth **221**, 581 (2000)
11. A. Paarmann, F. Guffarth, T. Warming, A. Hoffmann, D. Bimberg, AIP Conf. Proc. **772**, 689 (2005)
12. M. Grundmann, O. Stier, D. Bimberg, Phys. Rev. B **52**, 11969 (1995)
13. Y.A. Pusep, G. Zanelatto, S.W. da Silva, J.C. Galzerani, P.P. Gonzales-Borrero, A.I. Toropov, P. Basmaji, Phys. Rev. B **58**, R1770 (1998)
14. V.A. Shchukin, N.N. Ledentsov, D. Bimberg, *Epitaxy of Nanostructures* (Springer, Heidelberg, 2003)
15. U.W. Pohl, K. Pötschke, A. Schliwa, F. Guffarth, D. Bimberg, N.D. Zakharov, P. Werner, M.B. Lifshits, V.A. Shchukin, D.E. Jesson, Phys. Rev. B **72**, 245332 (2005)
16. S. Rodt, R. Seguin, A. Schliwa, F. Guffarth, K. Pötschke, U.W. Pohl, D. Bimberg, J. Lumin. **122–123**, 735 (2007)
17. R. Timm, H. Eisele, A. Lenz, T.-Y. Kim, F. Streicher, K. Pötschke, U.W. Pohl, D. Bimberg, M. Dähne, Physica E **32**, 25 (2006)
18. R. Heitz, F. Guffarth, K. Pötschke, A. Schliwa, D. Bimberg, N.D. Zakharov, P. Werner, Phys. Rev. B **74**, 045325 (2005)
19. S.A. Empedocles, D.J. Norris, M.G. Bawendi, Phys. Rev. Lett. **77**, 3873 (1996)
20. P. Castrillo, D. Hessman, M.-E. Pistol, J.A. Prieto, C. Pryorand, L. Samuelson, Jpn. J. Appl. Phys., Part 1 **36**, 4188 (1997)
21. V. Türck, S. Rodt, O. Stier, R. Heitz, R. Engelhardt, U.W. Pohl, D. Bimberg, R. Steingrüber, Phys. Rev. B **61**(15), 9944 (2000)
22. S. Rodt, A. Schliwa, K. Pötschke, F. Guffarth, D. Bimberg, Phys. Rev. B **71**, 155325 (2005)
23. R. Seguin, S. Rodt, A. Schliwa, K. Pötschke, U.W. Pohl, D. Bimberg, Phys. Stat. Sol. (b) **243**, 3937 (2006)
24. G. Bester, S. Nair, A. Zunger, Phys. Rev. B **67**, R161306 (2006)
25. R. Seguin, A. Schliwa, S. Rodt, K. Pötschke, U.W. Pohl, D. Bimberg, Phys. Rev. Lett. **95**, 257402 (2005)
26. W. Wieczorek, T. Warming, M. Geller, D. Bimberg, G.E. Cirlin, A.E. Zhukov, V.M. Ustinov, Appl. Phys. Lett. **88**, 182107 (2006)
27. W. Langbein, P. Borri, U. Woggon, V. Stavarache, D. Reuter, A.D. Wieck, Phys. Rev. B **69**, 161301 (2004)

28. A.I. Tartakovskii, M.N. Makhonin, I.R. Sellers, J. Cahill, A.D. Andreev, D.M. Whittaker, J.-P.R. Wells, A.M. Fox, D.J. Mowbray, M.S. Skolnick, K.M. Groom, M.J. Steer, H.Y. Liu, M. Hopkinson, Phys. Rev. B **70**, 193303 (2004)
29. J. Young, R.M. Stevenson, A.J. Shields, P. Atkinson, K. Cooper, D.A. Ritchie, K.M. Groom, A.I. Tartakovskii, M.S. Skolnick, Phys. Rev. B **72**, 113305 (2005)
30. R. Seguin, A. Schliwa, T.D. Germann, S. Rodt, K. Pötschke, A. Strittmatter, U.W. Pohl, D. Bimberg, M. Winkelnkemper, T. Hammerschmidt, P. Kratzer, Appl. Phys. Lett. **89**, 263109 (2006)
31. H. Morkoc, *Nitride Semiconductors and Devices* (Springer, Berlin, 1999)
32. S. Nakamura, Science **281**, 956 (1998)
33. S. Chichibu, T. Azuhata, T. Sota, S. Nakamura, Appl. Phys. Lett. **69**, 4188 (1996)
34. Y. Narukawa, Y. Kavakami, M. Funato, S. Fujita, S. Nakamura, Appl. Phys. Lett. **70**, 981 (1993)
35. N.N. Ledentsov, Z.I. Alferov, I.L. Krestnikov, W.V. Lundin, A.V. Sakharov, I.P. Soshnikov, A.F. Tsatsul'nikov, D. Bimberg, A. Hoffmann, Compound Semicond. **5**, 61 (1999)
36. O. Moriwaki, T. Someya, K. Tachibana, S. Ishida, Y. Arakawa, Appl. Phys. Lett. **76**, 2361 (2000)
37. F. Bernadini, V. Fiorentini, D. Vanderbilt, Phys. Rev. B **62**, 15851 (2001)
38. A.D. Andreev, E.P. O'Reilly, Appl. Phys. Lett. **79**, 521 (2001)
39. V. Ranjan, G. Allan, C. Priester, C. Delerue, Phys. Rev. B **68**, 115305 (2003)
40. T. Bretagnon, P. Lefebvre, P. Valvin, R. Bardoux, T. Guillet, T. Taliercio, B. Gil, N. Grandjean, F. Sermond, B. Damilano, A. Dussaigne, J. Massies, Phys. Rev. B **73**, 113304 (2006)
41. J. Simon, N.T. Pelekanos, C. Adelmann, E. Martinez-Guerrero, R. Andre, B. Daudin, L.S. Dang, H. Mariette, Phys. Rev. B **68**, 035312 (2003)
42. N.N. Ledentsov, M. Grundmann, N. Kirstaedter, O. Schmidt, R. Heitz, J. Böhrer, D. Bimberg, V.M. Ustinov, V.A. Shchukin, P.S. Kopev, Z.I. Alferov, S.S. Ruvimov, A.O. Kosogov, P. Werner, U. Richter, U. Gösele, J. Heydenreich, Solid State Electron. **40**, 785 (1996)
43. I.L. Krestnikov, M. Strassburg, A. Strittmatter, N.N. Ledentsov, J. Christen, A. Hoffmann, D. Bimberg, Jpn. J. Appl. Phys. **42**, L1057 (2003)
44. M. Pophristic, F.H. Long, C. Tran, I.T. Ferguson, R.F. Karlicek Jr., Appl. Phys. Lett. **73**, 3550 (1998)
45. M. Pophristic, F.H. Long, C. Tran, I.T. Ferguson, R.F. Karlicek Jr., J. Appl. Phys. **86**, 1114 (1999)
46. H. Scher, M.F. Shlesinger, J.T. Bender, Phys. Today **26**, 24 (1991)
47. I.L. Krestnikov, N.N. Ledentsov, A. Hoffmann, D. Bimberg, A.V. Sakharov, W.V. Lundin, A.F. Tsatsul'nikov, A.S. Usikov, Z.I. Alferov, Y.G. Musikhin, D. Gerthsen, Phys. Rev. B **66**, 155310 (2002)
48. R. Seguin, S. Rodt, A. Strittmatter, L. Reißmann, T. Bartel, A. Hoffmann, D. Bimberg, E. Hahn, D. Gerthsen, Appl. Phys. Lett. **84**, 4023 (2004)
49. M. Winkelnkemper, R. Seguin, S. Rodt, A. Schliwa, L. Reissmann, A. Strittmatter, A. Hoffmann, D. Bimberg, J. Appl. Phys. **101**, 113708 (2007)
50. P. Rinke, M. Scheffler, A. Qteish, M. Winkelnkemper, D. Bimberg, J. Neugebauer, Appl. Phys. Lett. **89**(16), 161919 (2006)
51. P. Rinke, A. Qteish, J. Neugebauer, C. Freysoldt, M. Scheffler, New J. Phys. **7**, 126 (2005)
52. J.W. Robinson, J.H. Rice, A. Jarjour, J.D. Smith, R.A. Taylor, R.A. Oliver, G.A.D. Briggs, M.J. Kappers, C.J. Humphreys, Y. Arakawa, Appl. Phys. Lett. **83**, 2674 (2003)
53. T. Bartel, M. Dworzak, M. Strassburg, A. Hoffmann, A. Strittmatter, D. Bimberg, Appl. Phys. Lett. **85**, 1946 (2004)

15

Ultrafast Coherent Spectroscopy of Single Semiconductor Quantum Dots

Christoph Lienau and Thomas Elsaesser

Abstract. This chapter summarizes our recent work—performed within the project B6 of the Sonderforschungsbereich 296—on combining ultrafast spectroscopy and near-field microscopy to probe the nonlinear optical response of a single quantum dot and of a pair of dipole-coupled quantum dots on a femtosecond time scale. We demonstrate coherent control of both amplitude and phase of the coherent quantum dot polarization by studying Rabi oscillations and the optical Stark effect in an individual interface quantum dot. By probing Rabi oscillations in a pair of laterally coupled interface quantum dots, we identify couplings between excitonic dipole moments and reveal the microscopic origin of these couplings. Our results show that although semiconductor quantum dots resemble in many respects atomic systems, Coulomb many-body interactions can contribute significantly to their optical nonlinearities on ultrashort time scales. This is important for realizing potentially scalable nonlocal quantum gates in chains of dipole-coupled dots, but also means that decoherence phenomena induced by many-body interactions need to be carefully controlled.

15.1 Introduction

The experimental implementation of quantum information processing (QIP) relies on identifying, coherently manipulating, coupling and detecting elementary excitations of individual quantum systems. All these operations need to be performed on a time scale much shorter than the decoherence time of the quantum system. This extremely challenging task has attracted the interest of an increasing number of researchers in all areas of science. Implementations of quantum logic operations are currently explored in a wide range of different quantum systems [1], e.g., nuclear magnetic spins in liquids and solids [2, 3], ions in traps [4–7], atoms in microwave resonators [8], optical lattices [9], photonic band gap materials [10], Josephson junctions [11, 12] or photons in quantum-optical systems [13, 14]. The complexity of this endeavor is quite clearly demonstrated by the fact that, despite the outstanding progress in this field over the last few years, the most complex quantum calculation performed to date is the factorization of the number 15 [3].

A particularly attractive approach to realizing all-solid-state quantum information processing relies on using charge or spin excitations of semiconductor quantum dots (QDs) as quantum bits. In QDs, electron and hole wave functions are localized in all three spatial dimensions on a nanometer length scale due to growth-induced nanoscale variations of the semiconductor composition. This makes QDs interesting model systems for exploring the basic physics of quasi-zero-dimensional quantum confinement as well as interesting for building novel optical devices for information processing.

The optical and electronic properties of semiconductor QDs have been intensely studied during the last decade. Due to the pronounced and so far unavoidable growth-induced inhomogeneous broadening in ensembles of semiconductor QDs, the recent development of single QD spectroscopy has provided a wealth of new information [15]. It is now understood that sufficiently confined QDs resemble in many respects atomic systems, showing atomic-like densities of states [16–18], a shell-like absorption spectrum [19] and—at low temperatures—comparatively long dephasing times of up to 1 ns [20, 21]. In addition, the nanometer spatial extent of the electron wave function in QDs gives rise to excitonic dipole moments of 10–100 Debye, much larger than those of atomic systems. This strong coupling to light makes charge excitations of single quantum bits interesting for quantum information processing. Ultrafast light pulses with pulse durations in the 100-fs range allow for generating and manipulating exciton excitations of single QDs on a subpicosecond time scale. With such ultrafast coherent carrier control, dephasing times in the 100-ps to 1-ns range are comparably long, making in principle up to 10,000 coherent manipulations possible before decoherence destroys the quantum information stored in excitonic quantum bits [21]. Another important consequence of the large excitonic dipole moments are comparatively strong dipolar interactions between adjacent quantum dots [22, 23]. Those interactions give rise to a nonlocal coupling between adjacent excitonic quantum bits, an important prerequisite for implementing scalable quantum gates. Consequently, different ideas for realizations of such gates have been proposed theoretically in recent years [22–25].

Such perspectives have triggered a research effort toward coherent control of excitonic excitations in semiconductor quantum dots. Initial successful experiments have shown coherent control on excited state transitions in the weak excitation regime [26–28] before Rabi oscillations could be demonstrated on different quantum dot systems [29–33]. All these experiments have so far revealed a finite damping of Rabi oscillations, which has been attributed either to excitation-induced dephasing due to Coulomb interactions among charge excitations [29, 34] or to exciton–phonon coupling [35, 36]. Most recently, an all-optical two-bit quantum logic gate was demonstrated using the exciton and biexciton transitions of a single quantum dot [37].

Here we present our recent experimental work on coherent control of excitonic excitations in quantum dots. We discuss a novel nano-optical technique [34] for probing optical nonlinearities of single quantum dots on ultrafast time scales [34, 38]. Coherent control of both amplitude [39] and phase [40] of the coherent exciton polarization in a single interface quantum dot is demonstrated and interactions between permanent excitonic dipole moments in a pair of neighboring quantum dots are resolved by analyzing Rabi oscillations in their nonlinear optical response [39].

In Sect. 15.2 we summarize the relevant properties of the investigated samples. In Sect. 15.3 we describe our experimental techniques. Results on coherent control of single quantum dots are given and discussed in Sect. 15.4. In Sect. 15.5 we present first results on dipolar couplings between two quantum dots. In Sect. 15.6 we give some conclusions.

15.2 Interface Quantum Dots

An important QD model system are thin semiconductor quantum wells (QWs). In quantum wells, local monolayer height fluctuations at the interfaces (interface roughness) and fluctuations of the alloy composition (alloy disorder) are unavoidable (Fig. 15.1(a)). The resulting disordered potential leads to the localization of excitons in single "interface" QDs with a confinement energy of about 10 meV (Fig. 15.1(b)). This disorder gives rise to a pronounced inhomogeneous broadening of far-field optical spectra. In experiments with high spatial and spectral resolution, however, the smooth, inhomogeneously broadened photoluminescence (PL) spectra break up into narrow emission spikes from a few localized excitons [16–18, 41–43].

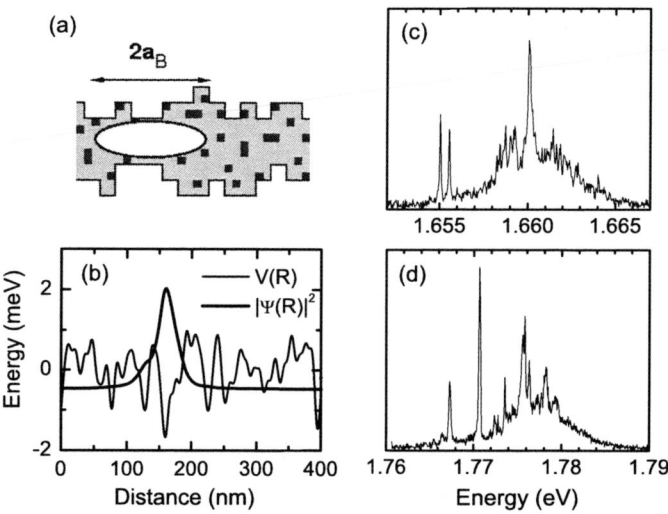

Fig. 15.1. (a) Disorder in quantum wells arises from spatial fluctuations of the local quantum well thickness (interface roughness) and of the quantum well composition (alloy disorder). (b) Schematic illustration of the effective disorder potential $V(R)$ and of a localized excitonic center-of-mass wave function $|\Psi(R)|^2$. (c)–(d) Representative near-field PL spectra ($T = 12$ K) of (c) a 5.1 nm thick and (d) a 3.3 nm thick (100) GaAs QW

The linear optical properties of interface QDs resemble in many aspects those of atomic systems. At low temperatures, the excitonic lines display a narrow homogeneous line width of 30–50 µeV, in agreement with measured dephasing times of 20–30 ps. The QDs show a discrete absorption spectrum [18] and often a fine structure splitting due to the spatial asymmetry of the monolayer islands. The temperature dependence of the exciton line width and the fine structure splitting has been thoroughly investigated [18, 44]. The correlation length of the disordered potential and thus also the center-of-mass wave function of localized excitons in QDs typically extends over several tens of nm, as known from near-field autocorrelation spectroscopy [42]. This large extension of the excitonic wave function results in large QD dipole moments of 50–100 Debye and a particularly strong coupling of these excitons to light [45, 46]. This makes interface QDs a particularly interesting model system for nonlinear spectroscopy of single QDs.

In this work, we investigate a sample consisting of 12 single QW layers of different thicknesses grown on a (100) GaAs substrate. The QW layers are separated by AlAs/GaAs short-period superlattice barriers, each formed by nine AlAs and GaAs layers with a total thickness of 23.8 nm. Here, we study the top seven QWs with thicknesses of 3.3–7.1 nm. The layers are buried at distances between 40 and 211 nm below the surface. Growth interruptions of 10 s at each interface lead to a large correlation length of the QW disorder potential and to the formation of interface QDs. The growth interruptions are kept short in order to avoid a monolayer splitting of the macroscopic PL spectra and to minimize the incorporation of impurities at the interfaces.

In Fig. 15.1(c) and (d) representative low temperature ($T = 12$ K) near-field PL spectra are shown for the 3.3 and 5.1 nm thick (100) GaAs QW. The spectra clearly reveal the emission from excitons localized in interface QDs. The line width of the sharp resonances is limited by the spectral resolution of 100 µeV. The spectra are recorded at an excitation intensity of 110 nW, corresponding to an average excitation density well below one exciton per monolayer island. For excitation powers between 1 and 500 nW, we find a linear intensity dependence and an excitation-independent shape of the emission spectra, indicating negligible contributions from biexcitons and charged excitons.

In addition to the sharp localized exciton emission, at higher energies these spectra display a spectrally broad background emission from more delocalized excitons in QW continuum states [42]. This is a disadvantage for QIP applications, as it may be difficult to avoid the uncontrolled population of such delocalized exciton states when ultrashort and thus spectrally broadband pulses are used for optical excitation. Yet, we will demonstrate in the following that such problems can be reduced by careful spectral shaping of the excitation pulses. Important properties of interface QDs are the excellent interface quality of the (100) GaAs quantum wells and the strong reduction of piezoelectric and strain fields. In the investigated samples, the energetic positions of the sharp exciton emission lines remains unchanged over many hours and we observe no signs of a spectral diffusion of the exciton lines.

15.3 Coherent Spectroscopy of Interface Quantum Dots: Experimental Technique

In the experiments on interface QDs, we read out quantum information from a single QD by directly probing the transient nonlinear optical spectrum of ground-state exciton transitions with subpicosecond time resolution. Our experimental concept is outlined in Fig. 15.2(a). We use spectrally broad femtosecond laser pulses which are centered around the excitonic QW absorption resonance and coupled into a near-field fiber probe to probe the optical QD nonlinearity. As a near-field probe we use an uncoated etched single mode optical fiber taper with a cone angle of about 30° [47]. With such probes we reach—in an illumination/collection geometry—a spatial resolution of about 150 nm, i.e., about $\lambda/5$ [41]. This high spatial resolution together with their large collection efficiency makes such uncoated fiber probes particularly well suited for semiconductor nanospectroscopy. Experimentally, we find that for GaAs samples about 1% of the light coupled into the fiber is collected in this illumination/collection geometry.

In the pump-probe experiments, the probe laser light reflected from the QW sample is collected by the same fiber probe, dispersed in an 0.5 m monochromator and then detected with a high-sensitivity liquid-nitrogen cooled CCD camera. This steady-state reflectivity spectrum $R_0(\omega_{det})$ contains weak, spectrally narrow resonances from single QD transitions (Fig. 15.2(a)).

The interaction with a second pump pulse now affects the QD spectrum and thus gives rise to a modified probe reflectivity $R(\omega_{det})$. Differential probe reflectivity spectra $\Delta R(\omega_{det}, \Delta t)/R_0 = [R(\omega_{det}, \Delta t) - R_0(\omega_{det})]/R_0(\omega_{det})$ are recorded at a fixed spatial position of the near-field tip as a function of the time delay Δt between pump and probe pulses. To probe the nonlinear optical response from single QDs, the high spatial resolution of the near-field technique is needed for two reasons. First, the

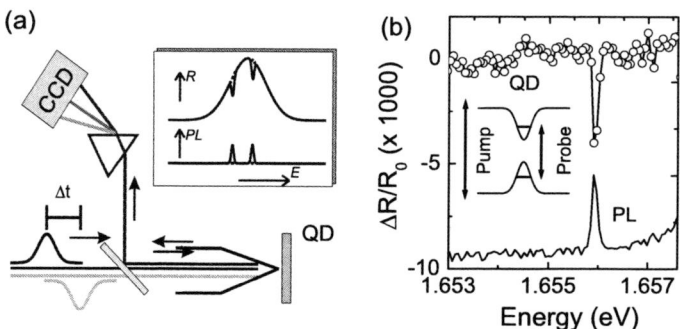

Fig. 15.2. (a) Schematic illustration of the experimental setup and of near-field PL and reflectivity spectra of a QD sample. (b) Near-field PL spectrum of a single QD (*solid line*) and differential reflectivity spectrum $\Delta R/R_0$ at $\Delta t = 30$ ps. PL and ΔR are recorded with identical pump pulses centered at 1.675 eV, exciting electron–hole pairs in 2D continuum states. The 100 nW probe pulses of 19 meV bandwidth are centered at 1.655 eV, around the QD absorption resonance. Inset: Schematic energy diagram

combined spatial and spectral resolution allows us to isolate single QD resonances (Fig. 15.1). Second, the relative amplitude of the QD resonance in $R_0(\omega_{det})$ scales, in first approximation, inversely proportional to the square of the spatial resolution.

We assume for simplicity that the QD absorption spectrum can be modeled as that of an ideal two-level system (TLS) and that the incident laser power is homogeneously distributed over an area A (the areal resolution of the microscope). Then the incident power is $P = I_0 A$ and the absorbed power is $P_{QD} = I_0 \sigma_{QD}$, with I_0 being the incident intensity and σ_{QD} the QD absorption cross section. For an ideal TLS

$$\sigma_{QD}(\omega) = \frac{\omega \mu_{QD}^2}{nc\epsilon_0 \hbar} \frac{\gamma}{\gamma^2 + (\omega - \omega_{QD}^2)}, \quad (15.1)$$

where $\omega_{QD} = 2\pi c / \lambda_{QD}$ denotes QD resonance frequency, ω the laser frequency, μ_{QD} the QD dipole moment, n the refractive index and $\gamma = 1/T_2$ the dephasing rate of the QD polarization. Improving the resolution from 1 µm to 100 nm increases the weak nonlinear QD signal by two orders of magnitude. For values typical for our experiments, $A = (250\,\text{nm})^2$, $\mu_{QD} = 60\,\text{D}$, $\gamma = (30\,\text{ps}^{-1})$, $n = 3.5$, $\lambda_{QD} = 750\,\text{nm}$, we estimate $\sigma_{QD}/A \simeq 0.04$, certainly measurable with the sensitivity of our setup.

The near-field PL and differential reflectivity ΔR spectrum of a single QD are compared in Fig. 15.2(b). To record the ΔR spectrum we use 100-fs pump pulses derived from a 80 MHz repetition rate Ti:sapphire laser. The pump pulses in this experiment create less than five electron–hole pairs in QW states, corresponding to an excitation density of $5 \cdot 10^9\,\text{cm}^{-2}$. Relaxation of these extra carriers into the QD bleaches the QD absorption and this bleaching is probed with 1 fJ probe pulses centered around the QW absorption resonance. Figure 15.2(b) depicts a differential reflectivity spectrum $\Delta R(E_{det})$ at a time delay of 30 ps in the low-energy region of the 5.1 nm QW absorption spectrum. It displays a single spectrally sharp resonance at exactly the same spectral position E_{QD} as the simultaneously recorded near-field PL spectrum. The large amplitude of the signal of $5 \cdot 10^{-3}$ is consistent with a spatial resolution of the experiment of 200–250 nm. Two-dimensional spatial scans indicate a resolution of 230 nm, partly limited by the QD-to-surface distance.

To better understand the image contrast in these pump-probe experiments, we compare in Fig. 15.3 differential reflectivity $\Delta R(E_{det})$ and PL spectra recorded under similar excitation conditions for single localized excitons in five different QWs buried at distances of 95–211 nm below the surface. We very clearly observe a transition between a dispersion-like and an absorption-like line shape as the QW-to-surface distance is varied. This behavior of the QD line shape can be understood in the framework of a local oscillator model as caused by the interference between the electric probe laser field $E_R(t)$ reflected from the sample surface and the field $E_{QD}(t)$ emitted from the QD in back direction. Our experiment works in the following way (Fig. 15.4). A fraction $E_R(t)$ of the probe laser transmitted through the near-field probe is reflected from the sample surface and coupled back into the near field fiber probe. The probe field $E_T(t)$, transmitted into the semiconductor, induces a polarization $P_{QD}(t) = \int dt' \chi_{QD}(t') E_T(t - t')$ of the QD located at a distance d below the sample surface. Here, $E_T(t)$ and χ_{QD} denote the probe field interacting with the

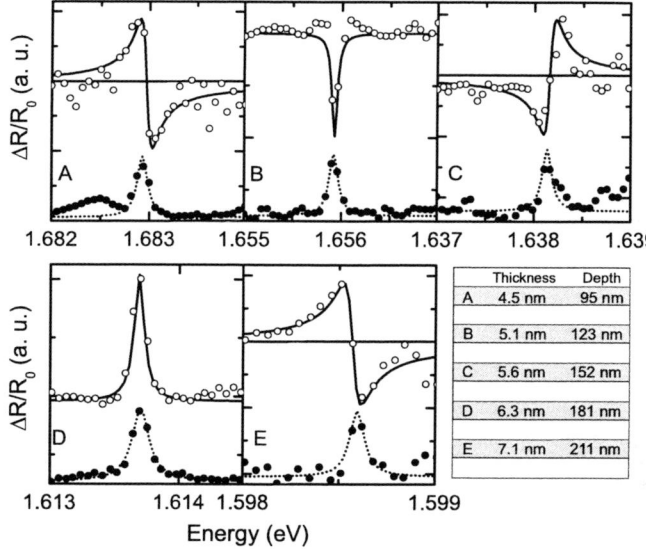

Fig. 15.3. Differential reflectivity spectra (*open circles*) of five interface QD located at different depths of 95–210 nm below the sample surface (see inset). The differential reflectivity spectra are compared to simultaneously recorded PL spectra. Note the transition between dispersive and absorptive line shapes

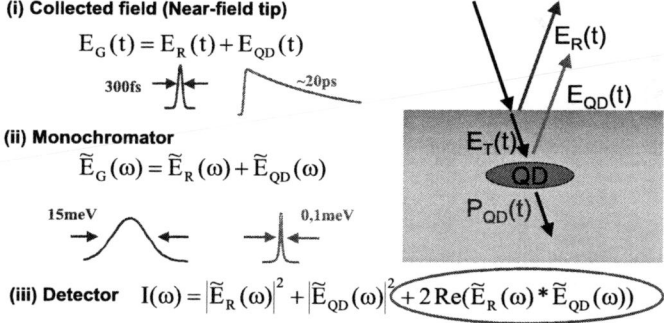

Fig. 15.4. Heterodyne detection of coherent QD polarizations. A femtosecond probe laser is coupled through a near-field fiber probe. A large fraction E_R of the probe laser is directly reflected the sample surface into the fiber probe. The transmitted probe light E_T is induces a QD dot polarization P_{QD} and the fraction E_{QD} of the electric field re-emitted from the QD is collected by the near-field probe. E_R and E_{QD} are spectrally dispersed in a monochromator and interfere on the CCD detector. This heterodyne detection scheme greatly enhances the weak QD field

QD and the QD susceptibility, respectively. The QD polarization re-emits an electric field and a fraction of this field, $E_{QD}(t)$ is locally collected by the near-field probe where it interferes with $E_R(t)$.

The time-integrated reflectivity $R(\omega)$ detected behind the monochromator is proportional to $|\tilde{E}_{QD}(\omega) + \tilde{E}_R(\omega)|^2 \simeq |\tilde{E}_R(\omega)|^2 + 2\text{Re}[\tilde{E}_R^*(\omega)\tilde{E}_{QD}(\omega)]$, where $\tilde{E}(\omega)$ denotes the Fourier transform of the field $E(t)$. Here, the finite monochromator resolution and the weak contribution from $|\tilde{E}_{QD}|^2$ has been neglected. Excitation by the pump laser affects the QD polarization and thus results in a change of the QD reflectivity. The differential reflectivity $\Delta R(\omega, \Delta t)$ represents the spectral interferogram of $\tilde{E}_R$ and $\tilde{E}_{QD}$:

$$\Delta R(\omega, \Delta t) \propto \text{Re}\{\tilde{E}_R^*(\omega)[\tilde{E}_{QD}(\omega, \Delta t) - \tilde{E}_{QD,0}(\omega)]\}. \quad (15.2)$$

The spectral shape of this interferogram evidently depends on the QD polarization dynamics and on the phase delay between $E_{QD}(t)$ and $E_R(t)$. Treating the QD for simplicity as a point dipole and the near-field tip as a point-like emitter, the phase delay and thus the spectral shape of this interferogram depends on the distance between QD and the near field tip. This interference effect is nicely seen in Fig. 15.3 and explains the transition between absorptive and dispersive line shapes. Since the QDs are buried below the surface by more than 50 nm, the near-field terms of the QD dipole emission can be neglected since they decay on a typical length scale of $\lambda/(2\pi n) \simeq 35$ nm ($n \simeq 3.5$—refractive index). Based on an optical path of $4\pi nd/\lambda$, we estimate a phase change of $\pi/2$ for a change in QD-sample distance of 28 nm. This is in quite good agreement with the results of Fig. 15.3. We consider this convincing evidence for the validity of the phenomenological local oscillator model described earlier. Clearly a detailed analysis of these data using, e.g., a Green function solution of Maxwell's equations for a realistic experimental geometry, is desirable for a more quantitative comparison between experiment and theory.

15.4 Coherent Control in Single Interface Quantum Dots

In this section, we describe experiments probing the coherent polarization dynamics of a single interface QD induced by impulsive excitation with ultrafast light pulses. Specifically, three different topics are addressed. First, we ask the fundamental questions: To what extent does the ultrafast nonlinear optical response of a single QD resemble that of an atomic system and how do many-body Coulomb interactions—often governing optical nonlinearities of higher-dimensional systems such as quantum wells and wires—affect the QD polarization dynamics? Then we demonstrate coherent control of the phase of the QD polarization by probing the optical Stark effect in a single QD and coherent control of the polarization amplitude by probing Rabi oscillations in single QD.

15.4.1 Ultrafast Optical Nonlinearities of Single Interface Quantum Dots

To study the effects of many-body interactions on the QD nonlinearities, we perform a quasi-two-color pump-probe experiment, exciting the QD sample in the QW

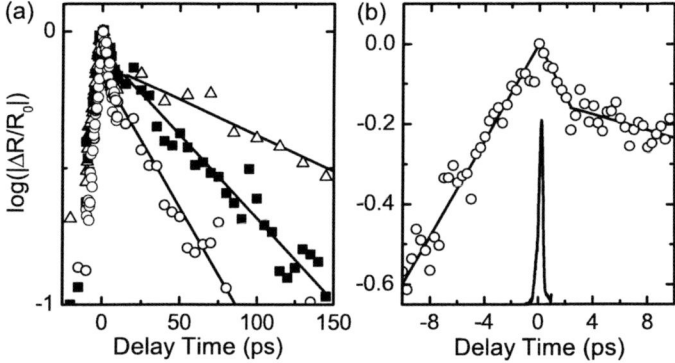

Fig. 15.5. (a) Temporal dynamics of $\Delta R/R_0$ for three different QD resonances (logarithmic ordinate scale). All decays are biexponential with a slow decay time varying between 30 and 150 ps. (b) Early time $\Delta R/R_0$ dynamics of a single QD resonance. A slow rise of $\Delta R/R_0$ is observed at negative time delays. The time resolution of the experiment is 150 fs, as indicated by the cross-correlation measurement (solid line around $\Delta t = 0$)

absorption continuum with 100 fs pump-pulses with a pulse energy of 1.5 fJ. These pulses create carriers in QW states and the resulting change in the QD spectrum is probed. The dynamics of the spectrally resolved reflectivity change measured on different QD resonances is shown in Fig. 15.5. After a picosecond rise of the signal at negative delay times (probe precedes pump) one finds a partial decay with a time constant of about 6 ps, followed by a much slower decay with time constants of 50–150 ps, depending on the specific QD resonance investigated.

The nonlinearities observed at sufficiently long positive Δt are easily understood on the basis of a simple two-level model for the QD nonlinearity. The pump pulse creates a nonequilibrium distribution of electron–hole pairs in QW continuum states. Subsequent trapping of these carriers gives rise to a bleaching of the QD absorption and a concomitant change of the QD reflectivity ΔR. Hence, the decay time of ΔR reflects the lifetime of the individual exciton state probed. Following an earlier conjecture [18], the QD population decay is mainly dominated by radiative recombination, i.e. $\tau_{rad} \simeq \tau_{QD}$. We can then estimate the dipole moment μ_{QD} of the individual QD using [46, 48]:

$$\frac{1}{\tau_{rad}} = n \frac{\omega^3 \cdot \mu_{QD}^2}{3\pi \epsilon_0 \hbar c^3}. \quad (15.3)$$

We estimate dipole moments μ_{QD} of 50–85 Debye for τ_{rad} between 150 and 50 ps. These values are in rather good agreement with previous estimates [29, 48]. They exceed those of atomic systems by more than an order of magnitude and reflect the large spatial extension of the exciton center-of-mass wave function in these QDs. Near-field autocorrelation spectra indicate an exciton localization length of about 40–50 nm [49]. Due to the statistical nature of the disorder potential, the exciton localization length and thus the dipole moment and radiative recombination rate varies

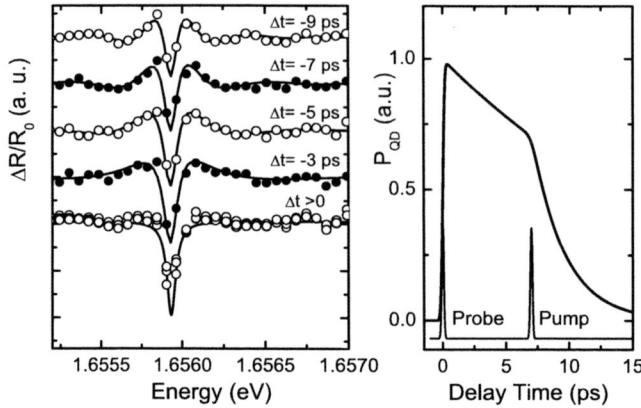

Fig. 15.6. (a) Near-field $\Delta R/R_0$ spectra (*circles*) at different delay times Δt. The spectra at $\Delta t < 0$ display pronounced spectral oscillations around the excitonic resonance. The solid lines show simulated spectra for the perturbed free induction decay of the coherent QD polarization assuming $T_2 = 15$ ps. (b) Dynamics of $P_{QD}(t)$ extracted from the time dependent near-field $\Delta R/R_0$ spectra

quite strongly from QD to QD, as seen in Fig. 15.5(a). Theoretical models of localized excitons in disordered quantum wells [49, 50] yield comparable results.

The dynamics of the QD reflectivity on a time scale of less than 10 ps, however, are quite different from what is expected for an ideal atomic system. The time evolution shows an 8-ps rise at negative delay times, much slower than the 150-fs cross-correlation of pump and probe pulses. A biexponential decay is found at positive delays, where the slow component reflects the exciton lifetime as discussed earlier. The fast decay time of about 6 ps is similar for all different QDs. The spectral characteristics of the differential reflectivity are markedly different at positive and negative delays. At negative delays, pronounced spectrally symmetric oscillations around the excitonic resonance are observed (see Fig. 15.6). Their oscillation period decreases with increasing negative time delay. At large positive delays, the spectra show a bleaching of the QD resonance [34].

This complex behavior reflects directly the coherent polarization dynamics of the excitonic QD excitation. To account for this behavior, one has to consistently describe the dynamics of the field $E_{QD}(t)$ radiated from the coherent QD polarization $P_{QD}(t)$. We phenomenologically describe the QD as an effective two-level system with a ground (no-exciton) state $|0\rangle$, and an excited one-exciton state $|1\rangle$. Then the quantum state of the two-level system is given as a coherent superposition $|\psi(t)\rangle = c_0(t)|0\rangle + c_1(t)|1\rangle$. Within the density matrix formalism, $P_{QD}(t)$ is given as $P_{QD}(t) = \mu_{QD}^* \rho_{01} + \text{c.c.}$, where the microscopic QD polarization $\rho_{01} = \langle c_0^* c_1 \rangle$, μ_{QD} denotes the QD dipole moment and $\langle \ldots \rangle$ the ensemble average [51]. Then, the well-known Bloch equations hold and ρ_{01} obeys the equation of motion

$$\frac{\partial}{\partial t}\rho_{01}(t) = -i\omega_{QD}\rho_{01}(t) + i(1 - 2n_{QD})\omega_R - \gamma\rho_{01}(t), \qquad (15.4)$$

with exciton energy ω_{QD}, dephasing rate γ, exciton population n_{QD} and generalized Rabi frequency ω_R.

In the absence of a pump laser, the resonant probe laser impulsively excites a coherent QD polarization which then decays with the dephasing rate γ (free induction decay). The re-emitted field interferes with the reflected probe laser field, giving rise to a Lorentzian QD line shape in $R_0(\omega)$. The fact that we observe a line width that is limited by our monochromator resolution of about 60 µeV gives a lower limit for the QD dephasing time of $T_2 = 1/\gamma > 15$ ps.

The transient spectral oscillations around the QD exciton resonance at negative time delays indicate that this free induction decay of $P_{QD}(t)$ is perturbed by the presence of the pump laser. In semiconductors, such oscillations have so far only been observed for higher dimensional system, e.g., studies of transient QW nonlinearities [52, 53]. In our experiments, the off-resonant pump does not directly interact with the QD dipole but creates electron–hole pairs (density n_{QW}) in the QW continuum. Many-body interactions between such carriers and the exciton that are mediated via the Coulomb interaction perturb the free induction decay of $P_{QD}(t)$.

The spectra at $\Delta t < 0$ are quantitatively described by assuming that an excitation-induced dephasing [54], i.e., an increase in γ due to the interaction between ρ_{01} and n_{QW} is the leading contribution to the QD nonlinearity at early times. Coulomb scattering between the QD dipole and the initial nonequilibrium carrier distribution in the QW causes this additional fast damping of ρ_{01}. In the frequency domain, this excitation-induced dephasing leads to oscillatory structures in the spectrum with a period determined by the time delay between probe and pump. The solid lines in Fig. 15.6(a) are calculated from (15.4) by assuming that the probe-induced QD polarization $P_{QD}(t)$ decays initially with an effective dephasing time $T_2 = 15$ ps, decreasing to $T_{EID} = 3$ ps after the arrival of the pump laser (Fig. 15.6(b)). Such an excitation-induced dephasing model accounts quantitatively for the transient oscillations and this analysis allows us to extract the QD polarization dynamics. A detailed theoretical analysis of the data was performed on the basis of the semiconductor Bloch equations in the mean-field approximation [34] and gives strong support for this interpretation.

Beyond the density-dependent dephasing processes leading to the perturbed free-induction decay, excitation of electrons and holes into continuum states gives rise to pump-probe signals at positive delay times, in particular the fast decay of the differential reflectivity at early delay times (Fig. 15.5(b)). Carriers initially populating continuum states relax into localized QD states, i.e., the population n_{QW} of QW states decays on a time scale of about 3 ps. Thus, the initial fast differential reflectivity decay reflects the transition from a QD nonlinearity dominated by excitation-induced dephasing to a nonlinearity dominated by exciton bleaching due to the population relaxation into the QDs.

These results highlight two important features of interface quantum dots. First, the coherence of the excitonic QD polarization persists for more than 10 ps even after resonant excitation with spectrally broadband femtosecond pulses. This decoherence time is two orders of magnitude larger than the duration of the excitation pulses. On the other hand, Coulomb many-body interactions may contribute significantly to their optical nonlinearities on ultrashort time scales if an additional exciton population in quantum well continuum states is created during the optical excitation process. Such many-body interactions have to be taken into account as important additional dephasing mechanisms.

15.4.2 Rabi Oscillations in a Quantum Dot

In this section, we demonstrate coherent control of both the amplitude and phase of the coherent QD polarization on an ultrafast time scale. Coherent control of the population of a generic two-level system with finite electronic dipole moment μ can be achieved by resonant impulsive excitation with light pulses much shorter in duration than the decoherence time of the microscopic polarization ρ_{01}. Neglecting for simplicity the finite decoherence time, the excited state population after the interaction with the excitation laser is given as $n_1 = \sin^2(\theta/2)$, with θ being the pulse area

$$\theta = \frac{\mu \cdot \hat{\varepsilon}}{\hbar} \int_{-\infty}^{\infty} |E(t)| \, dt, \tag{15.5}$$

where $E(t)$ denotes the time-dependent electric field of the excitation laser and $\hat{\varepsilon}$ its polarization direction. Thus, for weak excitation pulses the excited state population first increases linearly with increasing pulse intensity until it reaches a value of $n = 1$ for $\theta = \pi$, i.e., until the two-level system is fully inverted. Further increase in the pulse intensity induces stimulated emission from the excited state back to ground state and thus a decrease of excited state population. After interaction with a light pulse of area $\theta = 2\pi$ the excited state population reaches again $n = 0$, i.e. the system is back in its original state. For higher excitation, the population shows the well-known Rabi oscillations. This simplified picture only holds for resonant excitation and negligible decoherence, of course. Thus, the study of QD Rabi oscillations should generally give insight into the decoherence of quantum systems in the strong excitation regime.

For resonant excitation of only a single excitonic transition, we use spectrally tailored optical pulses with a spectral width of about 1 meV and a pulse duration of about 1.5 ps. We tune these pump pulses to a specific quantum dot resonance and probe the induced optical nonlinearity with collinearly polarized pulses of 15-meV spectral width and 200 fs duration. The idea of the experiment is to read out the transient exciton population by probing the induced absorption on the exciton–biexciton transition. Since each confined electron state in the QD can be occupied with two electrons of opposite spin orientation, two distinguishable single exciton states with orthogonal polarization can be optically excited. Simultaneous excitation of both exciton states results in a transient population of the bound biexciton state [Fig. 15.7(a)]. In interface QDs, the biexciton energy is normally slightly smaller (1–4 meV) than the sum of the two exciton energies due to the Coulombic interaction between the two excitons. Since the monolayer islands in interface quantum dots are slightly elongated along the [−110] direction [18], the energetic degeneracy of the two single-exciton states is lifted (by typically less than 100 μeV) [18]

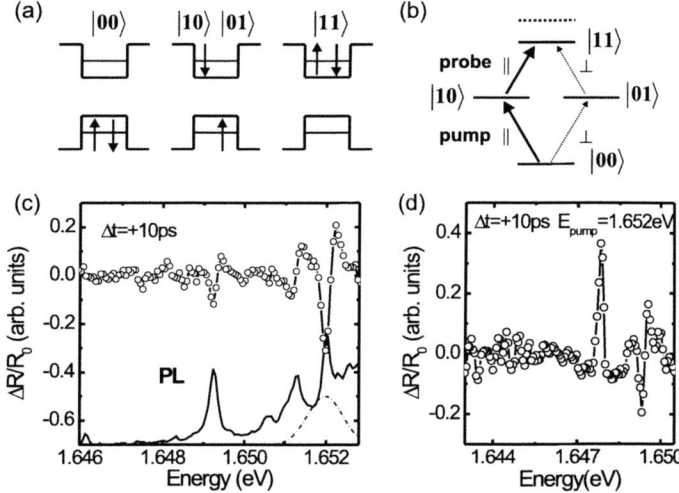

Fig. 15.7. (a) Schematics of exciton ground ($|00\rangle$), one-exciton ($|10\rangle$, and $|01\rangle$), and biexciton states $|11\rangle$ in a QD. (b) Excitation-level diagram in an interface QD and optical selection rules for pump and probe laser. (c) PL and ΔR for above-band gap excitation. In the studies of biexciton nonlinearities, the pump laser (*dashed line*) is tuned to the exciton resonance at 1.652 eV. (d) Pump-induced biexciton nonlinearity. The time delay between pump (at 1.652 eV) and probe laser is $\Delta t = 10$ ps

and one finds linear polarization selection rules for the exciton and biexciton transitions [18, 55]. The energy level structure of the QD states and the polarization selection rules can be summarized in a four-level system for the (no-exciton) crystal ground state $|00\rangle$, two exciton states with orthogonal polarization $|10\rangle$ and $|01\rangle$ and the biexciton state $|11\rangle$ [Fig. 15.7(b)]. Optical excitation of the $|00\rangle \rightarrow |10\rangle$ single exciton transition gives rise to excited state absorption on the $|10\rangle \rightarrow |11\rangle$ transition. This excited state absorption is not present in the absence of single exciton excitation.

In the experiment we observe, after single exciton excitation at 1.652 eV [Fig. 15.7(c)], a new transition at 1.648 eV, red-shifted by 4 meV from the single exciton transition which is assigned to the exciton–biexciton $X \rightarrow XX$ transition [Fig. 15.7(d)]. The dynamics of the pump-probe signal on the $X \rightarrow XX$ transition is consistent with this assignment [Fig. 15.8(a)]: There is no biexciton nonlinearity at negative delay times (probe precedes pump), i.e., in the absence of $|10\rangle$ exciton excitation. Around time zero the $X \rightarrow XX$ signal rises within the time resolution of the experiment and then decays exponentially on a 40-ps time scale with the radiative single exciton lifetime, demonstrating that the amplitude of the induced $X \rightarrow XX$ reflectivity change is a direct measure of the transient $|10\rangle$ exciton population generated by the pump pulse. The effect of the pump power on this biexciton nonlinearity is shown in Fig. 15.8(b) where the pump-probe signal ΔR_{XX} for a delay time of $\Delta t = 10$ ps is plotted vs. the maximum field strength $E_{pu} \propto \sqrt{P_{pu}}$ of

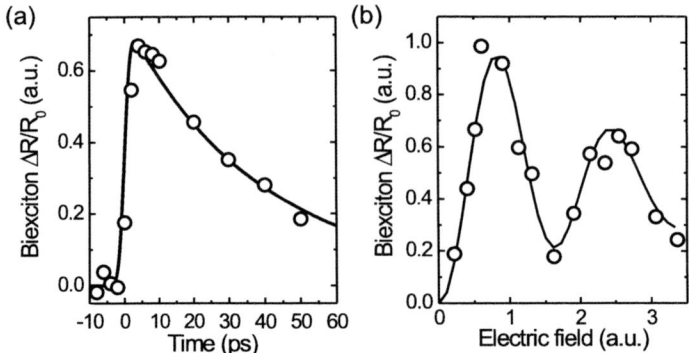

Fig. 15.8. (a) Temporal dynamics of the biexciton nonlinearity at 1.648 eV. The excitation conditions are as shown in Fig. 15.7. The 40-ps decay at $\Delta t > 0$ reflects the exciton lifetime. (b) Rabi oscillation in a single interface QD. Magnitude of the biexcition nonlinearity as function field amplitude of the pump laser at $\Delta t = 10$ ps

the pump pulse. The magnitude of the differential biexciton nonlinearity ΔR_{XX} displays pronounced oscillations, giving clear evidence for Rabi oscillations on a single ground-state exciton transition in a single interface QD. Despite the clarity of these oscillations, the experiment also shows that interface quantum dots are not an ideal two-level system. The biexcitonic nonlinearity at the second maximum, corresponding to a 3π excitation pulse, is about 1/3 smaller than that at the first maximum— corresponding to a π excitation. This unwanted damping of the Rabi oscillations is caused again by excitation-induced dephasing as an additional source of decoherence.

The field dependence of the biexciton nonlinearity, $\Delta R_{XX}(E_{pu})$, is well reproduced within the framework of Optical Bloch Equations of a two-level system with an intensity-dependent dephasing rate $\gamma = 1/T_2 + \gamma_1 \cdot E_{pu}^2$ [Fig. 15.8(b)]. Good agreement between experiment and model is achieved by assuming a dipole moment of 60 D, similar to those previously measured. The microscopic physics underlying this excitation-induced dephasing is similar to that reported for above-band gap excitation of QW continuum states in Sect. 15.4.1. Excitation by the picosecond pump pulse not only drives the desired single exciton transition but also creates coherent polarizations and incoherent populations in the QD environment. Since our pump-probe signals are accumulated over a large number of typically 10^8 laser pulses, these unwanted excitations give rise to fluctuating electric fields acting on the QD and, thus, to decoherence of the ensemble-averaged QD polarization. Our experimental results are well reproduced by assuming that the QD dephasing rate increases from less than $(15\,\text{ps})^{-1}$ (an upper limit given by our finite monochromator resolution) to about $(6\,\text{ps})^{-1}$ for a pulse area of 3π. In summary, these experiments show coherent control of the population of a single QD exciton through the demonstration of Rabi oscillations.

15.4.3 Optical Stark Effect: Ultrafast Control of Single Exciton Polarizations

Full coherent control over a single exciton excitation requires not only control over population—or more precisely the amplitude of the microscopic polarization ρ_{01}—but also control over the polarization phase $\phi = \arctan(\mathrm{Im}(\rho_{01})/\mathrm{Re}(\rho_{01}))$. In a Bloch sphere representation, which is often used to visualize the quantum dynamics of quasi-two-level systems, the momentary polarization ρ_{01} is represented as a three-dimensional vector $\rho_B = [\mathrm{Re}(\rho_{01}), \mathrm{Im}(\rho_{01}), 1/2(n_1 - n_0)]$, with $n_i = \langle c_i^* c_i \rangle$ describing the population of state $|i\rangle$. In this representation, polarization control thus means controlling the azimuthal angle ϕ in the $[\mathrm{Re}(\rho_{01}), \mathrm{Im}(\rho_{01})]$-plane.

Here we demonstrate control of the relative phase between the driving laser and the excitonic polarization by making use of the optical Stark effect (OSE). The OSE is one of the fundamental coherent light-matter interactions describing the light-induced shift ("dressing") of energy levels in the presence of nonresonant laser fields. In atomic systems the OSE is well known and, for weak excitation, well described by optical Bloch equations for independent two-level systems [56, 57]. In higher-dimensional semiconductors, e.g., quantum wells, however, the polarization dynamics induced by nonresonant light fields is much more complex than in atomic systems and often dominated by Coulomb-mediated many-body interactions [58–61]. Effects such as exciton–exciton interaction, biexciton formation or higher-order Coulomb correlations may affect the magnitude of the energy shift, the exciton oscillator strength and may even reverse the sign of the shift [60, 62–65]. Here, we report the first experimental study of the OSE in a single quasi-zero-dimensional semiconductor quantum dot [40].

Figure 15.9 compares the PL (solid line) from a single QD and the $\Delta R(\omega, \Delta t = 30\,\mathrm{ps})/R_0$ spectrum for above-band gap excitation of QW continuum states (solid circles). The absorptive ΔR spectrum reflects the bleaching of the QD resonance as described earlier. For below-band gap excitation, however, we observe for weak excitation ($P_{\mathrm{pu}} < 0.2\,\mathrm{\mu W}$) and negative delay times, $\Delta t = -4\,\mathrm{ps}$ (probe precedes pump), a dispersive line shape centered around ω_{QD}. With increasing excitation power, we find a drastic change in the line shape of $\Delta R(\omega)$ [Fig. 15.10(a)]: The signal maximum shifts slightly toward higher energies and an increasing number of spectral oscillations is observed, in particular on the high-energy side of the QD resonance. This change in line shape occurs together with a saturation of the strength of the nonlinear signal ΔR_{m}, taken as the difference between minimum and maximum of $\Delta R(\omega)$ [Fig. 15.10(b)]. As we will show in the following, this characteristic change in line shape allows us to extract the phase shift $\Delta\phi$ of the QD polarization due to the interaction with the off-resonant pump laser from a Bloch equation simulation. The extracted phase shift $\Delta\phi$ is plotted as a function of excitation power in Fig. 15.10(c).

To ensure that we are indeed probing only a light-induced shift of the exciton resonance, we also plot the time evolution of the QD nonlinearity $\Delta R_{\mathrm{m}}(\Delta t)/R_0$ (Fig. 15.11). It is important that the signal vanishes completely for positive delay times $\Delta t > 0$ (pump precedes probe) and rises around $\Delta t = 0$ within the time resolution of our experiment of 250 fs. For $\Delta t < 0$, $\Delta R(\Delta t)$ decays with a time constant of $\tau_{\mathrm{d}} = 8\,\mathrm{ps}$.

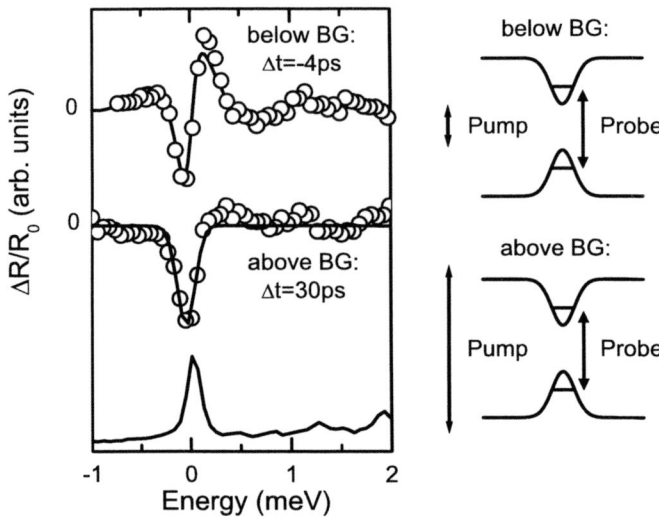

Fig. 15.9. (Left) PL spectrum of a single QD resonance $\omega_{QD} = 1.6503\,\text{eV}$ and differential reflectivity $\Delta R(\omega)/R_0$ for above-band gap excitation at $\Delta t = 30\,\text{ps}$ and for below-band gap excitation at $\Delta t = -4\,\text{ps}$ with 2-ps pulses at $1.647\,\text{eV}$ (bandwidth $\sigma = 0.8\,\text{meV}$) in the weak excitation limit ($P_{pu} = 0.12\,\mu\text{W}$). Solid lines: Bloch equation model. (Right) Schematic excitation diagram

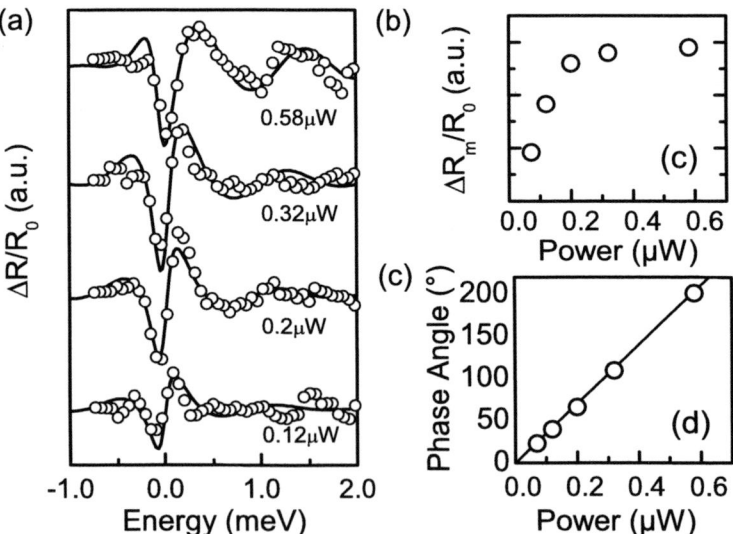

Fig. 15.10. (a) Optical Stark Effect in a single QD. Differential reflectivity spectra $\Delta R(\omega)/R_0$ for below-band gap excitation at $\Delta t = -4\,\text{ps}$ with 2-ps pulses at $1.647\,\text{eV}$ (bandwidth $\sigma = 0.8\,\text{meV}$) for excitation powers between 0.12 and $0.58\,\mu\text{W}$. Solid lines: Bloch equation model. (b) Variation of the signal magnitude $\Delta R(\omega_{QD})/R_0$ with pump power. (c) Phase shift of the QD polarization vs. pump power

15 Ultrafast Coherent Spectroscopy of Single Semiconductor Quantum Dots 317

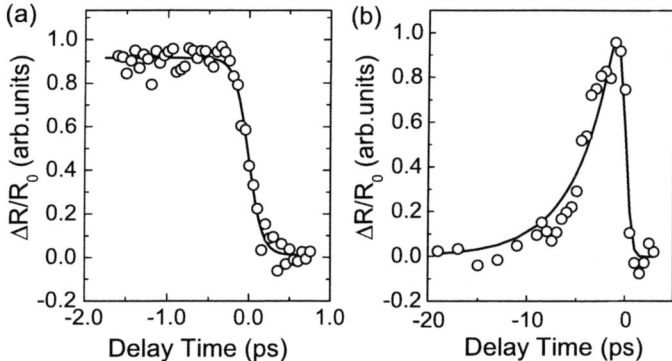

Fig. 15.11. Time evolution of $\Delta R(\Delta t)/R_0$ for a single QD at $\omega_{QD} = 1.6544\,\text{eV}$. Here 200-fs pump pulses with a power of 58 µW were centered at 1.640 eV. $\Delta R_m(\Delta t > 0)$ vanishes and the signal for $\Delta t < 0$ decays on a ps time scale. (**a**) 3-ps time scale. (**b**) 25-ps time scale

The dispersive ΔR line shape observed in Fig. 15.9(a) for below-band gap excitation and small time delays is the signature of the OSE in the *weak excitation limit* [58]. It reflects a transient light-induced blue shift of the QD exciton resonance. In the presence of an AC electric field of frequency ω_p, the transition frequency of a two-level system shifts by $\Delta\omega_0(t) = \sqrt{[(\omega_0 - \omega_p)^2 + \Omega_R(t)^2]} + \omega_p - \omega_0$. Here, ω_0 is the transition frequency without external field, $\Omega_R(t) = \mu \cdot E_p(t)/\hbar$ is the Rabi frequency, μ the transition dipole moment and $E_p(t) \cdot \cos(\omega_p t)$ the (pump) AC electric field. The blue shift $\Delta\omega_0(t)$ of the QD absorption resonance results in a dispersive $\Delta R(\omega)/R_0$ line shape, which can be approximated as $\Delta R(\omega)/R_0 \propto \Delta\omega_{0,\text{max}} \cdot \partial\alpha(\omega)/\partial\omega$, where $\alpha(\omega)$ is the QD absorption spectrum and $\Delta\omega_{0,\text{max}}$ is the maximum blue shift. Thus, in the weak excitation limit, the amplitude of the $\Delta R(\omega)/R_0$ signal is expected to increase linearly with increasing pump power, without change of the line shape. The spectra of Fig. 15.10(a), taken with pump powers $\leq 0.2\,\mu\text{W}$, exactly display this behavior. For such pump powers, the Rabi frequency has a maximum value of $\Omega_{R,\text{max}} = 1.75\,\text{meV} \simeq 5\Delta\omega_{0,\text{max}}$.

The origin of this transient blue shift becomes clear from an analysis of the optical Bloch equations. We describe the QD as a two-level system with a radiative lifetime of $T_1 = 100\,\text{ps}$ corresponding to a dipole moment $\mu = 50$ Debye [34, 45, 46]. A dephasing time of $T_2 = 8\,\text{ps}$ is assumed to account for our finite monochromator resolution. It is important to stress that since we know both power and duration τ_p of the pump pulses and the spatial resolution of about 250 nm, the electric field of the pump laser is estimated to within a factor of 2 and no free parameters enter the simulation.

The calculated dynamics of the QD polarization in the weak excitation limit are displayed in the rotating frame in Fig. 15.12(a). The probe field resonant to the exciton line changes the QD population and drives a coherent polarization oscillating at the QD resonance frequency ω_{QD}. This polarization is 90° phase-shifted with respect to the probe field ($\text{Re}[P_{QD}] = 0$). During the presence of the pump

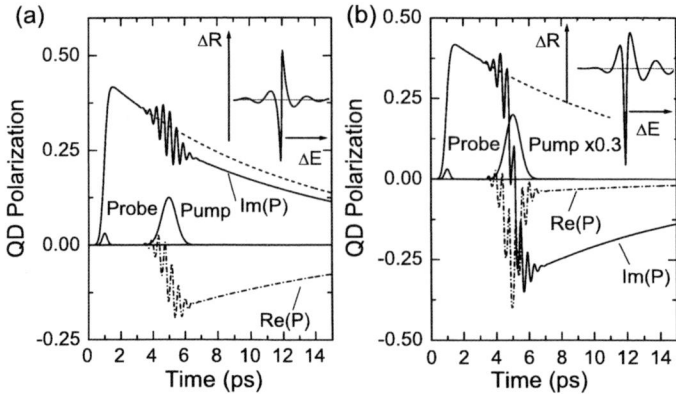

Fig. 15.12. Bloch equation simulation of the single QD optical Stark effect. Shown is the time-dependent QD polarization P_{QD} in the rotating frame with (*solid line*) and without (*dashed line*) pump laser. Nonlinear ΔR spectra are given in the inset. (**a**) Weak excitation limit. (**b**) Strong excitation limit. The chosen pump power corresponds to a phase shift $\Delta\phi = 172°$

pulse, the polarization is externally driven, leading to oscillations at the detuning frequency $\omega_{det} = \omega_0 - \omega_p$. After the interaction, the polarization is phase-shifted by $\Delta\phi \approx \int \Delta\omega_0(t)\,dt$. It is this shift $\Delta\phi$ of the QD polarization which changes the product $E_{pr}(\omega) \cdot E_{QD}(\omega)$ of the complex electric fields and therefore the line shape. Fourier-transformation of the polarization dynamics gives directly the dispersive line shape of the $\Delta R(\omega)$ spectrum in the weak excitation limit, $\Delta\phi < 40°$, at early delay times (Fig. 15.9; inset in Fig. 15.12(a)).

The simulation also reproduces the time-dependent data shown in Fig. 15.11. Evidently, a vanishing nonlinearity at positive time delays is predicted by the Bloch model, since then the pump laser interacts with the sample before the excitonic polarization is created. At negative delays, the OSE nonlinearity is expected to decay with the dephasing time of the polarization. The fact that we reproduce both predictions of this simple model experimentally is quite striking. In particular we find, within our signal-to-noise ratio, no measurable nonlinearity at $\Delta t > 0$. This indicates that we are indeed probing a pure light-induced shift of the resonance and that nonlinearities induced by real carriers generated by one- or two-photon absorption in the surrounding of the QD obviously play a negligible role [40]. This conclusion is strongly supported by recording transient nonlinear spectra at different negative delay times between 0 and -10 ps. Here, pronounced spectral oscillations are observed which are quantitatively fit by the Bloch equation. Thus, even under femtosecond excitation, the nonlinear response of the interface QD for below-band gap excitation is very close to that of an isolated atomic system and it appears that the excitonic QD excitation is only weakly influenced by the complex solid state environment. To be precise, one should note that from our experiments one cannot directly tell whether the 8-ps decay at negative delay times reflects the polarization dephasing time. We are spectrally resolving the QD nonlinearity with a monochromator with about 100 μeV resolution and this finite resolution puts an upper limit of slightly less than 10 ps to the measurable decay. Thus the 8-ps decay is close to our instrument resolution and gives a lower limit only for the excitonic decoherence rate.

For higher electric fields of the pump pulse, the weak excitation limit of the OSE nonlinearity is no longer valid. Experimentally, one finds additional features in the transient reflectivity spectra [Fig. 15.10(a), traces for pump intensities of 0.32 and 0.58 μW]. These spectral oscillations are a direct consequence of the interaction of the QD polarization with the strong pump field. The pump laser induces pronounced large amplitude oscillations of QD polarization at the detuning frequency during the presence of the pump laser. This is illustrated in Fig. 15.12(b) showing the solution of Bloch equations for strong excitation with $\Omega_R = 6\,\text{meV}$ ($\omega_{\text{det}} = -10\,\text{meV}$). A large phase shift $\Delta\phi$ of 172° of the QD polarization results from this interaction and the nonlinear ΔR spectrum shows additional oscillatory structures on the high-energy side, as found in the experiment. This large amplitude phase rotation corresponds to the observation of *gain* on the resonance of a single QD. In the Bloch sphere representation this phase rotation basically reflects a nutation-like motion of the Bloch vector, resulting in a change in azimuthal angle after the interaction. A comparison between experimental spectra and simulation [solid lines in Fig. 15.10(a)] allows us to quantify the phase shift $\Delta\phi$ experienced by the QD polarization. In Fig. 15.10(c) we plot $\Delta\phi$ obtained from the simulation of the data in Fig. 15.10(a) as a function of the pump power P_{pu}. We find a linear increase in $\Delta\phi$ with P_p. This means that the light shift also increases linearly in our experiment, despite the saturation of ΔR_m. This linear increase in the polarization phase $\Delta\phi$ is somewhat analogous to the pulse area theorem for Rabi oscillations of the population of a two-level system when driven with a resonant pulse. Currently, we can quantitatively measure the phase shift with an accuracy of about 10° and achieve phase rotations of as much as 200°. Control of the exciton density, on the other hand, has been established above by the observation of Rabi oscillations when varying the pulse area of a resonant excitation pulse. The result show that a sequence of a resonant and an off-resonant laser pulse gives full control over both amplitude *and* phase of the coherent excitonic polarization. In particular, we can switch the QD from absorption to gain within about 1 ps.

15.5 Coupling Two Quantum Dots via the Dipole–Dipole Interaction

Coherent control of excitonic transitions in a single QD, as demonstrated in the last section, is an essential prerequisite for exploring excitonic couplings between adjacent dots and attempting to implement potentially scalable two-qubit operations. Over the last few years, different microscopic coupling schemes have been proposed to theoretically achieve such implementations, among them couplings via photonic or plasmonic nanoresonator modes, via optical phonon wavepackets or through dipolar interactions. In particular, ultrafast optical realizations of two-qubit operations in dipole-coupled QDs have been studied theoretically in some detail [22, 23, 66–68].

So far, experimental studies of the proposed ideas have not been reported. This is partly due to a lack of suitable experimental methods. Since the strength of the dipole–dipole interaction depends strongly on both the geometric arrangment (orientation and separation of the dipoles) and on the microscopic interaction mechanism (permanent dipole couplings, van der Waals dispersion forces, Förster dipole energy transfer, etc.), studies of single nanosystems and/or ordered and homogeneous nanoarrays are often needed to resolve such couplings. Such experiments are scarce and have so far investigated a pair of molecules in an organic crystal [69] or the light-harvesting-2 complex [70] with steady-state techniques. Here, we demonstrate that combining high spatial resolution with time-resolving optical techniques allows us to separate different couplings through their individual real-time dynamics and control nanosystems on ultrashort time scales [39].

To probe dipole interactions between two individual QDs, we go back to the experimental situation depicted in Fig. 15.7. In Sect. 15.4.2 we discussed experiments demonstrating coherent population control of the QD resonance at an energy of 1.652 eV (QD A). In these experiments, we probed the pump-induced biexcitonic nonlinearity of this QD. The broad spectral bandwidth of the femtosecond probe pulses enables us to simultaneously probe the pump-induced nonlinear optical response of the other QDs detected at this NSOM tip position. We focus on the optical nonlinearity of the neighboring QD resonance at 1.649 eV (QD B) and study now the effect of a single-exciton excitation of QD A on the optical nonlinearity of this QD. Nonlinear spectra ΔR_B of QD B recorded with resonant excitation of QD A are displayed in Fig. 15.13(a). The excitation conditions are identical to those in Fig. 15.7(c) with an excitation pulse area of $\theta = 0.75\pi$.

Now, optical nonlinearities are observed at both positive *and* negative Δt, the latter being evident from the noninstantaneous rise of the signal in Fig. 15.13(b). In contrast to the absorptive line shape in Fig. 15.7(b), the nonlinear spectra display a time-independent dispersive line shape, reflecting a transient *blue* shift of the exciton resonance which does not change much with time delay. From the amplitude and shape of the nonlinear spectra we deduce a line shift of 30 ± 15 μeV around zero time delay. As seen in Fig. 15.13(b), the time evolution of $\Delta R_{B,m}$, defined as the difference between maximum and minimum of $\Delta R_B(\omega)$, is very different from that observed at the biexciton resonance. At negative time delays, $\Delta R_{B,m}(\Delta t)$ shows a rise with a time constant of about 6 ps, followed by a slight dip and a slower decay on a time scale of more than 100 ps. The change of $\Delta R_{B,m}$ with the excitation field displays clear Rabi oscillations.

To discuss these results, we stress the following observations: (i) Dispersive line shapes, caused by a transient blue shift of the QD resonance, are observed at all time delays and we find no signature of absorptive ΔR changes which would reflect pump-induced changes of the exciton population of QD B. This indicates that the observed nonlinearity is not due to an exciton relaxation between QD A and B. (ii) The presence of a strong laser field gives rise to transient excitonic line shifts via the optical Stark effect (OSE). However, as shown in Sect. 15.4.3, the OSE leads to optical nonlinearities at negative time delays ($\Delta t < 0$) only. Also, for a pump frequency above the exciton resonance, a red-shift of the QD line is expected, in contradiction

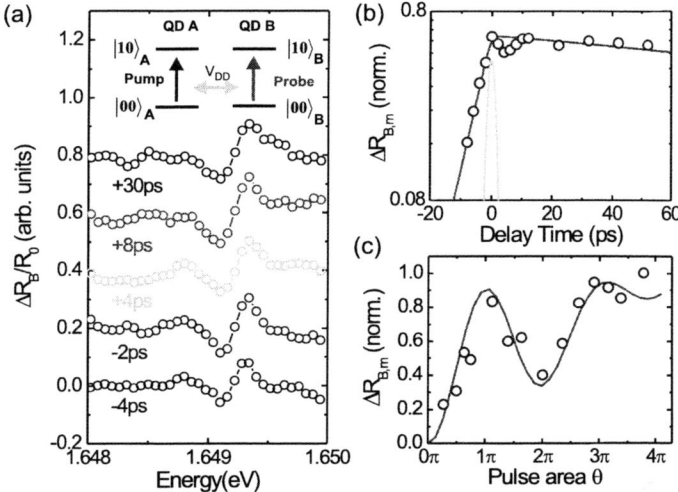

Fig. 15.13. (a) Nonlinear ΔR spectra of quantum dot B for resonant single-exciton excitation of QD A at 1.652 eV as a function of time delay Δt. The pulse area of the 2-ps excitation pulses is $\theta \simeq 0.75\pi$. Inset: Excitonic $|00\rangle \rightarrow |10\rangle$ transitions in QD A and QD B coupled through V_{DD}. (b) Time dynamics of $\Delta R_B(\Delta t)$. The excitation conditions are the same as in (a) and the time resolution of the experiment is indicated (*thin solid line*). (c) Rabi oscillation in a coupled QD. Magnitude of $\Delta R_B(\Delta t = 10\,\text{ps})$ as a function of the field amplitude of the pump laser. The solid line shows a simulation based on an optical Bloch equation model

with our present findings. (iii) There is a clear correlation between the pulse-area dependence of ΔR_{XX} in Fig. 15.8(b) and of ΔR_B in Fig. 15.13(c).

The data in Fig. 15.13 thus reflect an electronic coupling between the QDs A and B. The most likely candidate for such an interaction is a dipole–dipole coupling between both QDs. Theoretical studies [22, 23, 66, 67] indicate that two different mechanisms can contribute: resonant Förster energy transfer and direct Coulomb interaction between permanent excitonic dipole moments. For two quantum dots separated by less than the wavelength of light, pulsed optical excitation of one QD leads to the re-emission of a transient electric field which can be reabsorbed by the second QD, thus (Förster) transferring the excitation. The interaction Hamiltonian $H_F = V_F p_A p_B^* + \text{c.c.}$ includes the coupling $V_F \propto \mu_A \mu_B / R_{AB}^3$ between coherent excitonic polarizations $p_i(t) = |10\rangle_i \langle 01|_i + \text{c.c.}$ in QDs A and B. The coupling strength is determined by the transition dipole moments $\mu_i = |\langle 00|M_i|10\rangle_i|$ (M_i: dipole operator) and the QD separation R_{AB}. In the strong coupling limit, $V_F/\hbar$ is larger than the detuning $\Delta\omega = \omega_A - \omega_B$ between the QD resonances and the dephasing rate $1/T_2$, leading to entangled states of the coupled system and cooperative effects in its radiative decay [67, 69, 71]. In the weak coupling limit, $V_F \ll \hbar\Delta\omega, \hbar/T_2$, the interaction induces a population relaxation between the coupled states [72].

The direct dipole interaction H_D on the other hand involves permanent excitonic dipole moments and thus interaction between the exciton populations $n_i = |10\rangle_i \langle 10|_i$

with $H_D = V_D n_A n_B$ and $V_D \propto d_A d_B / R_{AB}^3$. Here, d_i represents the permanent dipole moment originating from a shift of the electron and hole charge distributions in the exciton. This interaction leads to a biexcitonic energy shift V_D in case that both QDs are excited [22].

To examine these two interaction mechanisms, nonlinear optical spectra are calculated from the time evolution of the density matrix in rotating wave approximation. Here, the QDs are treated as effective two-level systems (states $|00\rangle_i$, $|10\rangle_i$), interacting with the pump and probe fields and coupled via the dipole–dipole interaction. Most of the parameters of these calculations, such as ω_i, μ_i, $T_{2,i}$, and electric field profiles of the lasers are quantitatively known. The basic unknown is the mechanism and strength of the dipole–dipole interaction.

For the Förster mechanism, the time-evolution of the spectra depends critically on the ratios of V_F, $\hbar\Delta\omega$, and $\hbar/T_2$. In our case, typical interdot distances are limited by the finite exciton size to about 20 nm, giving $V_F \simeq 30\,\mu eV$ for $\mu = 60\,D$. Therefore $V_F \leq \hbar/T_2$ (0.1 meV) $\ll \hbar\Delta\omega$ (3 meV), i.e., we are in the weak coupling limit. At negative delay times $\Delta t < 0$, $\Delta R_{B,m}$ is due to the optical Stark effect induced by the pump field with a dispersive line shape reflecting a red shift of the exciton line, and a rise of $\Delta R_{B,m}(\Delta t < 0)$ with $T_{2,B}$ (Fig. 15.14(a)). At $\Delta t > 0$, the Förster mechanism induces exciton population relaxation between both QDs, resulting in absorptive line shapes. The decay of $\Delta R_{B,m}(\Delta t > 0)$ reflects both the exciton lifetime $T_{1,A} \simeq 40\,ps$ and the exciton transfer rate which scales as $\Gamma_F \propto V_F^2 T_2 [1 + (\Delta\omega T_2)^2]^{-1}$ [72]. Although the excitation field dependence of $\Delta R_{B,m}(\Delta t = 10\,ps)$ (Fig. 15.14(c)) displays Rabi oscillations, the line shapes and the temporal dynamics of $\Delta R_{B,m}$ are in disagreement with the experiment. Also the amplitude of $\Delta R_{B,m}$ is much smaller than in the experiment. We infer that dipole coupling via the Förster mechanism is of minor importance for our QDs.

For a direct dipole interaction H_D between permanent excitonic dipole moments, excitation of QD A transiently shifts the energy of QD B by V_D. The sign of this shift depends on the sign of V_D and, thus, a blue shift occurs for parallel dipoles d_A and d_B. For a shift smaller than the homogeneous exciton line width, the coupling results in a dispersive shape of ΔR (Fig. 15.14(b)). Both direct dipole coupling and OSE contribute to the line shifts at $\Delta t < 0$ and net blue shifts are observed if the Coulomb coupling is stronger than the OSE. The signal at $\Delta t < 0$ rises with $T_{2,B}$. For $\Delta t > 0$, $\Delta R_{B,m}$ decays exponentially with the exciton lifetime $T_{1,A}$, as there is no population transfer between the dots. The amplitude of $\Delta R_{B,m}$ monitors the exciton population in QD A and the intensity dependence of the pump-induced Rabi oscillation (Fig. 15.14(b)) is thus similar to that found in the single exciton manipulation experiments.

The experimental line shapes and Rabi oscillations are in good agreement with the direct coupling model. The calculated decay of $\Delta R_{B,m}$, however, is faster. This discrepancy may reflect signal contributions from more delocalized excitonic transitions in the environment of QD A [42]. Such states have smaller dipole moments and thus longer radiative lifetimes. Their presence may also lead to finite dipole shifts that persist on time scales longer than $T_{1,A}$. This notion is supported by finding experimentally a finite optical nonlinearity from QD B when the excitation pulse is

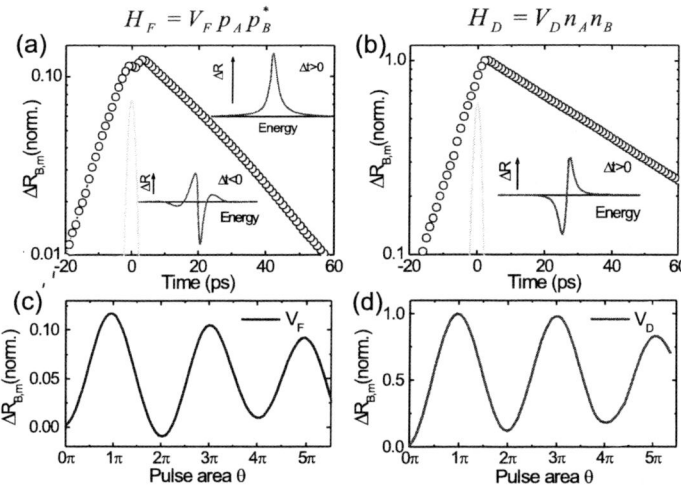

Fig. 15.14. (a) Simulation of optical nonlinearities of two QDs coupled by Förster energy transfer for excitation conditions similar to Fig. 15.13. The nonlinear spectra (inset) display an absorptive line shape at $\Delta t > 0$ and dispersive red-shifted line shape at $\Delta t < 0$. (b) Coupling via permanent excitonic dipole moments. For $V_D > 0$, the nonlinear spectra (inset) reflect a blue shift of the exciton line at all time delays. Pump-induced Rabi oscillations ($\Delta t = 10$ ps) for (c) Förster and (d) direct dipole coupling

slightly detuned from the resonance of QD A. For such a nonresonant excitation, however, Rabi oscillations are not observed. This indicates that the direct coupling between permanent excitonic dipole moments is the dominant interaction mechanism. Apart from the permanent dipoles, the Coulomb interaction between excitons in QD A and B may lead to induced charge rearrangements which lower the energy (formation of distant biexcitons). The absence of a redshift in the experiment points to a dominance of dipole repulsion over such correlation effects.

It is interesting to ask whether the weak Förster coupling is a general property of this class of QD samples. The energy statistics of the localized exciton states are heavily influenced by level repulsion effects [42], resulting in finite energy splittings between excitons in neighboring QDs. Such splittings are typically 1–3 meV and thus stronger than the dipole coupling. Thus, it is quite unlikely to find near-resonance situations between adjacent QDs and Förster coupling is expected to be weak in general.

15.6 Summary and Conclusions

We introduced a novel technique—ultrafast near-field optical spectroscopy—to probe the nonlinear optical response of single semiconductor quantum dots. We used this technique to demonstrate coherent control over amplitude and phase of the excitonic QD polarization. Rabi oscillations of up to 4π are induced and probed by ultrashort light pulses. It appears that even in relatively weakly confined interface quantum dots, the ultrafast polarization dynamics are in many respects similar to those of an atomic system, yet with an enhanced dipole moment. Only when interacting with strong excitation pulses with an area of order 2π, excitation-induced dephasing due to Coulomb-mediated many-body interactions is limiting the visibility of Rabi oscillations. One may expect that using quantum dots with larger confinement energies may reduce excitation-induced dephasing. Yet so far the experiments on Rabi oscillations in more strongly confined self-assembled quantum dots seem to indicate that other factors, such as enhanced exciton–phonon coupling, may be important additional decoherence sources. Certainly, the microscopic origin of exciton decoherence in single quantum dots will be the topic of much additional experimental and theoretical work in the near future.

We introduced an experimental technique that probes transient optical nonlinearities in a broad spectral range and thus is particularly well suited to study excitonic couplings. This allowed us to demonstrate coupling between permanent excitonic dipole moments in a pair of adjacent quantum dots. The coupling strength of about 30 μeV is still about one order of magnitude too small to implement a nonlocal conditional quantum gate as proposed in [22]. An increase in coupling should readily be achievable by applying moderate lateral electric fields and two-qubit gating times of a few picoseconds seem feasible [23].

Recent progress in nanofabrication allows for manufacturing linear arrays of vertically and laterally stacked quantum dots with well-defined interdot distances. Such systems may permit us to go beyond two-qubit operations toward scalable qubit arrays, even though statistical variations of the coupling parameters within such arrays and excitation-induced decoherence still pose major technological challenges. Either energy-selective addressing (with inherently limited scalability) or cellular-automaton schemes with globally applied multicolor pulse sequences may be used for encoding information in such arrays. The now-established real-time probing of many-body interactions between individual solid-state nanostructures will certainly be of key importance for future progress in this area.

Acknowledgment

We thank Francesca Intonti, Tobias Guenther, Kerstin Mueller, and Thomas Unold for their important contributions to the work reviewed here. High-quality semiconductor samples were provided by Soheyla Eshlaghi and Andreas D. Wieck (Ruhr-Universität Bochum). We are very grateful for theoretical support from and stimulating discussion with Markus Glaneman, Vollrath Martin Axt and Tilmann Kuhn (Universität Münster), Andreas Knorr (Technische Universität Berlin), and Roland Zimmermann (Humboldt-Universität Berlin). Financial support by the Deutsche Forschungsgemeinschaft (SFB296) and the European Union through the SQID program is gratefully acknowledged.

References

1. M.A. Nielsen, I.L. Chuang, *Quantum Computation and Quantum Information* (Cambridge University Press, Cambridge, 2000)
2. N. Gershenfeld, I.L. Chuang, Bulk spin-resonance quantum computation. Science **275**, 350 (1997)
3. L.M.K. Vandersypen, M. Steffen, G. Breyta, C.S. Yannoni, M.H. Sherwood, I.L. Chuang, Experimental realization of Shor's quantum factoring algorithm using nuclear magnetic resonance. Nature **414**, 883 (2001)
4. J.C. Cirac, P. Zoller, Quantum computations with cold trapped ions. Phys. Rev. Lett. **74**, 4091 (1995)
5. A. Sorensen, K. Molmer, Quantum computation with ions in thermal motion. Phys. Rev. Lett. **82**, 1971 (1999)
6. S. Gulde, M. Riebe, G.P.T. Lancaster, C. Becher, J. Eschner, H. Häffner, F. Schmidt-Kaler, I.L. Chuang, R. Blatt, Implementation of the Deutsch-Jozsa algorithm on an ion-trap quantum computer. Nature **421**, 48 (2003)
7. F. Schmidt-Kaler, H. Häffner, M. Riebe, S. Gulde, G.P.T. Lancaster, T. Deuschle, C. Becher, C.F. Roos, J. Eschner, R. Blatt, Realization of the Cirac-Zoller controlled-NOT quantum gate. Nature **422**, 408 (2003)
8. Q.A. Turchette, C.J. Hood, W. Lange, H. Mabuchi, H.J. Kimble, Measurement of conditional phase shifts for quantum logic. Phys. Rev. Lett. **75**, 4710 (1995)
9. G.K. Brennen, C.M. Caves, I.H. Deutsch, Quantum logic gates in optical lattices. Phys. Rev. Lett. **82**, 1060 (1999)
10. M. Woldeyohannes, S. John, Coherent control of spontaneous emission near a photonic band edge: A qubit for quantum computation. Phys. Rev. A **60**, 5046 (1999)
11. J.E. Mooij, T.P. Orlando, L. Levitov, L. Tian, C.H. van der Wal, S. Lloyd, Josephson persistent-current qubit. Science **285**, 1036 (1999)
12. T. Yamamoto, Y.A. Pashkin, O. Astafiev, Y. Nakamura, J.S. Tsai, Demonstration of conditional gate operation using superconducting charge qubits. Nature **425**, 941 (2003)
13. J.D. Franson, Cooperative enhancement of optical quantum gates. Phys. Rev. Lett. **78**, 3852 (1997)
14. E. Knill, R. Laflamme, G.J. Milburn, A scheme for efficient quantum computation with linear optics. Nature **409**, 46 (2001)
15. A. Zrenner, A close look on single quantum dots. J. Chem. Phys. **112**, 7790 (2000)
16. K. Brunner, G. Abstreiter, G. Böhm, G. Tränkle, G. Weimann, Sharp-line photoluminescence and two-photon absorption of zero-dimensional biexcitons in a GaAs/AlGaAs structure. Phys. Rev. Lett. **73**, 1138 (1994)
17. H.F. Hess, E. Betzig, T.D. Harris, L.N. Pfeiffer, K.W. West, Near-field spectroscopy of the quantum constituents of a luminescent system. Science **264**, 1740 (1994)
18. D. Gammon, E.S. Snow, B.V. Shanabrook, D.S. Katzer, D. Park, Fine structure splitting in the optical spectra of single GaAs quantum dots. Phys. Rev. Lett. **76**, 3005 (1996)
19. P. Hawrylak, G.A. Narvaez, M. Bayer, A. Forchel, Excitonic absorption in a quantum dot. Phys. Rev. Lett. **85**, 389 (2000)
20. P. Borri, W. Langbein, S. Schneider, U. Woggon, R.L. Sellin, D. Ouyang, D. Bimberg, Ultralong dephasing time in InGaAs quantum dots. Phys. Rev. Lett. **87**, 157401 (2001)
21. M. Bayer, A. Forchel, Temperature dependence of the exciton homogeneous linewidth in In0.60Ga0.40As/GaAs self-assembled quantum dots. Phys. Rev. B **65**, 041308 (2002)
22. E. Biolatti, R. Iotti, P. Zanardi, F. Rossi, Quantum information processing with semiconductor macroatoms. Phys. Rev. Lett. **85**, 5647 (2000)

23. E. Biolatti, I. D'Amico, P. Zanardi, F. Rossi, Electro-optical properties of semiconductor quantum dots: Application to quantum information processing. Phys. Rev. B **65**, 075306 (2002)
24. P. Chen, C. Piermarocchi, L.J. Sham, Control of exciton dynamics in nanodots for quantum operations. Phys. Rev. Lett. **87**, 067401 (2001)
25. C. Piermarocchi, P. Chen, Y.S. Dale, L.J. Sham, Theory of fast quantum control of exciton dynamics in semiconductor quantum dots. Phys. Rev. B **65**, 075307 (2002)
26. N.H. Bonadeo, G. Chen, D. Gammon, D.S. Katzer, D. Park, D.G. Steel, Nonlinear nano-optics: probing one exciton at a time. Phys. Rev. Lett. **81**, 2759 (1998)
27. Y. Toda, T. Sugimoto, M. Nishioka, Y. Arakawa, Near-field coherent excitation spectroscopy of InGaAs/GaAs self-assembled quantum dots. Appl. Phys. Lett. **76**, 3887 (2000)
28. T. Flissikowski, A. Hundt, M. Lowisch, M. Rabe, F. Henneberger, Photon beats from a single semiconductor quantum dot. Phys. Rev. Lett. **86**, 3172 (2001)
29. T.H. Stievater, X. Li, D.G. Steel, D. Gammon, D.S. Katzer, D. Park, C. Piermarocchi, L.J. Sham, Rabi oscillations of excitons in single quantum dots. Phys. Rev. Lett. **87**, 133603 (2001)
30. H. Kamada, H. Gotoh, J. Temmyo, T. Takagahara, H. Ando, Exciton Rabi oscillation in a single quantum dot. Phys. Rev. Lett. **87**, 246401 (2001)
31. H. Htoon, T. Takagahara, D. Kulik, O. Baklenov, A.L. Holmes Jr., C.K. Shih, Interplay of Rabi oscillations and quantum interference in semiconductor quantum dots. Phys. Rev. Lett. **88**, 087401 (2002)
32. A. Zrenner, E. Beham, S. Stuffer, F. Findeis, M. Bichler, G. Abstreiter, Coherent properties of a two-level system based on a quantum-dot photodiode. Nature **418**, 612 (2002)
33. L. Besombes, J.J. Baumberg, J. Motohisa, Coherent spectroscopy of optically gated charged single InGaAs quantum dots. Phys. Rev. Lett. **90**, 257402 (2003)
34. T. Guenther, C. Lienau, T. Elsaesser, M. Glanemann, V.M. Axt, T. Kuhn, S. Eshlaghi, A.D. Wieck, Coherent nonlinear optical response of single quantum dots studied by ultrafast near-field spectroscopy. Phys. Rev. Lett. **89**, 057401 (2002), Erratum Phys. Rev. Lett. **89**, 179901 (2002)
35. A. Vagov, V.M. Axt, T. Kuhn, Electron-phonon dynamics in optically excited quantum dots: Exact solution for multiple ultrashort laser pulses. Phys. Rev. B **66**, 165312 (2002)
36. J. Förstner, C. Weber, J. Danckwerts, A. Knorr, Phonon-assisted damping of Rabi oscillations in semiconductor quantum dots. Phys. Rev. Lett. **91**, 127401 (2003)
37. X. Li, Y. Wu, D. Steel, D. Gammon, T.H. Stievater, D.S. Katzer, D. Park, C. Piermarocchi, L.J. Sham, An all-optical quantum gate in a semiconductor quantum dot. Science **301**, 809 (2003)
38. T. Guenther, C. Lienau, T. Elsaesser, M. Glanemann, V.M. Axt, T. Kuhn, Guenther et al. Reply. Phys. Rev. Lett. **90**, 139702 (2003)
39. T. Unold, K. Mueller, C. Lienau, T. Elsaesser, A.D. Wieck, Optical control of excitons in a pair of quantum dots coupled by the dipole–dipole interaction. Phys. Rev. Lett. **94**, 137404 (2005)
40. T. Unold, K. Mueller, C. Lienau, T. Elsaesser, A.D. Wieck, Optical Stark effect in a quantum dot: ultrafast control of single exciton polarizations. Phys. Rev. Lett. **92**, 157401 (2004)
41. F. Intonti, V. Emiliani, C. Lienau, T. Elsaesser, R. Nötzel, K.H. Ploog, Near-field optical spectroscopy of localized and delocalized excitons in a single GaAs quantum wire. Phys. Rev. B **63**, 075313 (2001)

42. F. Intonti, V. Emiliani, C. Lienau, T. Elsaesser, V. Savona, E. Runge, R. Zimmermann, R. Nötzel, K.H. Ploog, Quantum mechanical repulsion of exciton levels in a disordered quantum well. Phys. Rev. Lett. **87**, 076801 (2001)
43. V. Emiliani, F. Intonti, C. Lienau, T. Elsaesser, R. Nötzel, K.H. Ploog, Near-field optical imaging and spectroscopy of a coupled quantum wire-dot structure. Phys. Rev. B **64**, 155316 (2001)
44. D. Gammon, E.S. Snow, B.V. Shanabook, D.S. Katzer, D. Park, Homogeneous liniwidths in the optical spectrum of a single gallium arsenide quantum dot. Science **273**, 87 (1996)
45. J.R. Guest, T.H. Stievater, X. Li, J. Cheng, D.G. Steel, D. Gammon, D.S. Katzer, D. Park, C. Ell, A. Thränhardt, G. Khitrova, H.M. Gibbs, Measurement of optical absorption by a single quantum dot exciton. Phys. Rev. B **65**, 241310(R) (2002)
46. A. Thränhardt, C. Ell, G. Khitrova, H.M. Gibbs, Relation between dipole moment and radiative lifetime in interface quantum dots. Phys. Rev. B **65**, 035327 (2002)
47. P. Lambelet, A. Sayah, M. Pfeffer, C. Philipona, F. Marquis-Weible, Chemically etched fiber tips for near-field optical microscopy: A process for smoother tips. Appl. Opt. **37**, 7289 (1998)
48. L.C. Andreani, G. Panzarini, J.M. Gerard, Strong-coupling regime for quantum boxes in pillar microcavities: Theory. Phys. Rev. B **60**, 13276 (1999)
49. C. Lienau, F. Intonti, T. Guenther, T. Elsaesser, V. Savona, R. Zimmermann, E. Runge, Near-field autocorrelation spectroscopy of disordered semiconductor quantum wells. Phys. Rev. B **69**, 085302 (2004)
50. V. Savona, E. Runge, R. Zimmermann, F. Intonti, V. Emiliani, C. Lienau, T. Elsaesser, Level repulsion of localized excitons in disordered quantum wells. Phys. Stat. Sol. (a) **190**, 625 (2002)
51. H. Haug, S.W. Koch, *Quantum Theory of the Optical and Electronic Properties of Semiconductors*, 2nd edn. (World Scientific, Singapore, 1994)
52. B. Fluegel, N. Peyghambarian, G. Olbright, M. Lindberg, S.W. Koch, M. Joffre, D. Hulin, A. Migus, A. Antonetti, Femtosecond studies of coherent transients in semiconductors. Phys. Rev. Lett. **59**, 2588 (1987)
53. J.P. Sokoloff, M. Joffre, B. Fluegel, D. Hulin, M. Lindberg, S.W. Koch, A. Migus, A. Antonetti, N. Peyghambarian, Transient oscillations in the vicinity of excitons and in the band of semiconductors. Phys. Rev. B **38**, 7615 (1988)
54. H. Wang, K. Ferrio, D.G. Steel, Y.Z. Hu, R. Binder, S.W. Koch, Transient nonlinear optical response from excitation induced dephasing in GaAs. Phys. Rev. Lett. **71**, 1261 (1993)
55. A. Bertoni, P. Bordone, R. Brunetti, C. Jacoboni, S. Reggiani, Quantum logic gates based on coherent electron transport in quantum wires. Phys. Rev. Lett. **84**, 5912 (2000)
56. B.R. Mollow, Stimulated emission and absorption near resonance for driven systems. Phys. Rev. A **5**, 2217 (1972)
57. H. Häffner, S. Gulde, M. Riebe, G. Lancaster, C. Becher, J. Eschner, F. Schmidt-Kaler, R. Blatt, Precision measurement and compensation of optical Stark shifts for an ion-trap quantum processor. Phys. Rev. Lett. **90**, 143602 (2003)
58. A. Mysyrowicz, D. Hulin, A. Antonetti, A. Migus, W.T. Masselink, H. Morkoç, "Dressed excitons" in a multiple-quantum-well structure: Evidence for an optical Stark effect with femtosecond response time. Phys. Rev. Lett. **56**, 2478 (1986)
59. W.H. Knox, D.S. Chemla, D.A.B. Miller, J.B. Stark, S. Schmitt-Rink, Femtosecond ac Stark effect in semiconductor quantum wells: Extreme low- and high-intensity limits. Phys. Rev. Lett. **62**, 1189–1192 (1989)
60. C. Sieh, T. Meier, F. Jahnke, A. Knorr, S.W. Koch, P. Brick, M. Hübner, C. Ell, J. Prineas, G. Khitrova, H.M. Gibbs, Coulomb memory signatures in the excitonic optical Stark effect. Phys. Rev. Lett. **82**, 3112 (1999)

61. M. Saba, F. Quochi, C. Ciuti, D. Martin, J.-L. Staehli, B. Deveaud, A. Mura, G. Bongiovanni, Direct observation of the excitonic ac Stark splitting in a quantum well. Phys. Rev. B **62**, R16322 (2000)
62. S. Schmitt-Rink, D.S. Chemla, Collective excitations and the dynamical Stark effect in a coherently driven exciton system. Phys. Rev. Lett. **57**, 2752 (1986)
63. M. Combescot, R. Combescot, Excitonic Stark shift: A coupling to "semivirtual" biexcitons. Phys. Rev. Lett. **61**, 117 (1988)
64. S. Schmitt-Rink, D.S. Chemla, H. Haug, Nonequilibrium theory of the optical Stark effect and spectral hole burning in semiconductors. Phys. Rev. B **37**, 941 (1988)
65. C. Ell, J.F. Müller, K.E. Sayed, H. Haug, Influence of many-body interactions on the excitonic optical Stark effect. Phys. Rev. Lett. **62**, 304 (1989)
66. L. Quiroga, N.F. Johnson, Entangled Bell and Greenberger-Horne-Zeilinger states of excitons in coupled quantum dots. Phys. Rev. Lett. **83**, 2270 (1999)
67. B.W. Lovett, J.H. Reina, A. Nazir, G.A.D. Briggs, Optical schemes for quantum computation in quantum dot molecules. Phys. Rev. B **68**, 205319 (2003)
68. A. Nazir, B.W. Lovett, S.D. Barrett, J.H. Reina, G.A.D. Briggs, Anticrossings in Förster coupled quantum dots. Phys. Rev. B **71**, 045334 (2005)
69. C. Hettich, C. Schmitt, J. Zitzmann, S. Kuhn, I. Gerhardt, V. Sandoghdar, Nanometer resolution and coherent optical dipole coupling of two individual molecules. Science **298**, 385 (2002)
70. C. Hofmann, M. Ketelaars, M. Matsushita, H. Michel, T.J. Aartsma, J. Köhler, Single-molecule study of the electronic couplings in a circular array of molecules: Light-harvesting-2 complex from rhodospirillum molischianum. Phys. Rev. Lett. **90**, 013004 (2003)
71. R.H. Dicke, Coherence in spontaneous radiation processes. Phys. Rev. **93**, 99 (1954)
72. J.A. Leegwater, Coherent versus incoherent energy transfer and trapping in photosynthetic antenna complexes. J. Phys. Chem. **100**, 14403 (1996)

16

Single-Photon Generation from Single Quantum Dots

Matthias Scholz, Thomas Aichele, and Oliver Benson

Abstract. Single photons from single quantum dots have already found their way into many important realizations in quantum information processing. In this chapter, we review several demonstrations using a deterministic single-photon source based on InP/GaInP quantum dots. Single-photon generation is described and verified by auto-correlation measurements. Additionally, photon pairs and triplets from multi-exciton cascades are observed. Using a Michelson interferometer, an efficient separation of photon pairs can be achieved allowing multiplexing on the single-photon level. We also show the generation of non-classical states of light from electrically pumped quantum dots. The applicability of our sources in quantum information experiments is demonstrated in quantum key distribution and in a realization of a two-qubit Deutsch-Jozsa algorithm which uses different degrees of freedom of a single photon. By encoding the quantum information in appropriate bases, we implement passive error correction in the presence of phase fluctuations.

16.1 Introduction

Recently, considerable effort has been put into exploiting light for quantum information processing. Photons are ideal tools to transmit quantum information over large distances. In 1984, Bennett and Brassard proposed a secret key-distribution protocol [4] that uses the single-particle character of a photon to avoid any possible eavesdropping on an encoded message (for a review see [26]). Additionally, the implementation of efficient quantum gates based on photons and linear optics was proposed [34, 25] and demonstrated [23, 32]. Proposals describe the role of photons as the ideal information carriers in larger networks [67] between processing nodes of stationary qubits, like ions [14, 16, 57, 37], atoms [31, 15], quantum dots [41, 72], and Josephson qubits [47, 52].

Linear optics applications in quantum information processing require the reliable deterministic generation of single- or few-photon states. However, due to their bosonic character, photons tend to appear in bunches. Thus, classical light sources provide a broad photon number distribution, as depicted for thermal and laser light in

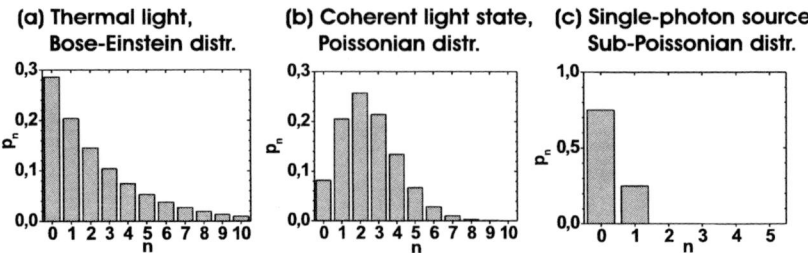

Fig. 16.1. Photon number distributions of (**a**) thermal light, (**b**) a coherent state, and (**c**) a single-photon source with 25% efficiency

Figs. 16.1(a) and (b), respectively. This characteristic hinders the implementation of classical sources particularly in quantum cryptographic systems since an eavesdropper may gain partial information with a beam splitter attack. Similar obstacles occur for linear optics quantum computation where photonic quantum gates [34], quantum repeaters [10], and quantum teleportation [9] require the preparation of single- or few-photon states *on demand* in order to obtain reliability and high efficiency. Although an ideal nonclassical single-photon source emits a sub-Poissonian photon number distribution with exactly one photon at a time, real sources have inevitable losses due to scattering and absorption which lead to typical photon number distributions as shown in Fig. 16.1(c).

A promising process for single-photon generation is the spontaneous emission from a single quantum emitter. Numerous emitters have been used to demonstrate single-photon emission [49]. Single atoms or ions are the most fundamental systems. They have been trapped and coupled to optical resonators to obtain single-mode emission [35, 30]. Other systems capable of single-photon generation are single molecules and single nanocrystals [11, 42, 45]. However, their drawback is a susceptibility to photo-bleaching and blinking [50]. Stable alternatives are nitrogen-vacancy defect centers in diamond [36, 6], but they show broad optical spectra together with comparably long lifetimes ($\approx$12 ns).

Here, we focus on single-photon generation from self-assembled single quantum dots. Most experiments with quantum dots have to be conducted at cryogenic temperature in order to reduce electron–phonon interaction and thermal ionization. High count rates can be obtained due to their short transition lifetimes, and their spectral lines are nearly lifetime-limited with material systems covering the ultraviolet, visible, and infrared spectrum. Quantum dots are even more attractive because of the possibility of electrical excitation [73] and their implementation in integrated photonic structures [3].

Section 16.2 provides an introduction to the radiative properties of single quantum dots and covers the characterization and measurement of single-photon states. Section 16.3 shows decay cascades in single quantum dots. Section 16.4 highlights the topic of electrically excited quantum dots. Section 16.5 demonstrates the application of optically excited photons in a multiplexed quantum cryptography and a quantum computing setup. Section 16.6 concludes with a short summary.

16.2 Single Quantum Dots as Single-Photon Emitters

16.2.1 Photon Statistics of Single-Photon Emitters

Single-photon emitters exhibit light statistics that cannot be obtained from classical sources or lasers. The standard method to test single-photon emission is the measurement of intensity correlations: The normally ordered second-order coherence function $g^{(2)}(t_1, t_2)$ detects these intensity correlations of a light field at two distinct times and is proportional to the joint probability to detect one photon at time $t = 0$ and another at $t = \tau$. For stationary fields, it takes the form

$$g^{(2)}(\tau = t_1 - t_2) = \frac{\langle : \hat{I}(0)\hat{I}(\tau) : \rangle}{\langle \hat{I}(0) \rangle^2}. \tag{16.1}$$

This function has several characteristic properties: It tends to a value of unity for large times since each random process is assumed to become uncorrelated on a sufficiently long timescale. It can further be shown [43] that $g^{(2)}(0) \geq 1$ and $g^{(2)}(0) \geq g^{(2)}(\tau)$ for all classical fields which prohibits values smaller than unity.

The case $g^{(2)}(0) > 1$ is characteristic for thermal light sources: This bunching means an increased probability to detect a second photon soon after a first one (Fig. 16.2(a)). For coherent light fields, such as laser light well above the threshold, $g^{(2)}(\tau) = 1$ for all τ which indicates a Poissonian photon number distribution and photons arriving randomly (Fig. 16.2(b)).

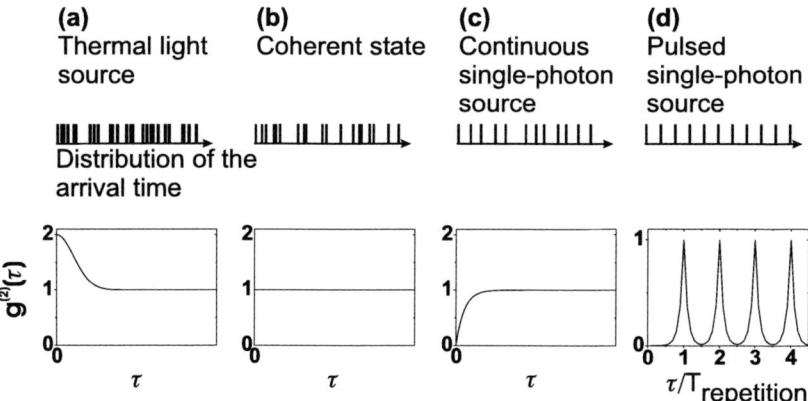

Fig. 16.2. Top: illustrative distribution of the photon arrival time. Bottom: second-order coherence function $g^{(2)}(\tau)$ of (**a**) a thermal light source (e.g., a light bulb), (**b**) coherent light (laser light), (**c**) a continuously driven single-photon source, and (**d**) a pulsed single-photon source

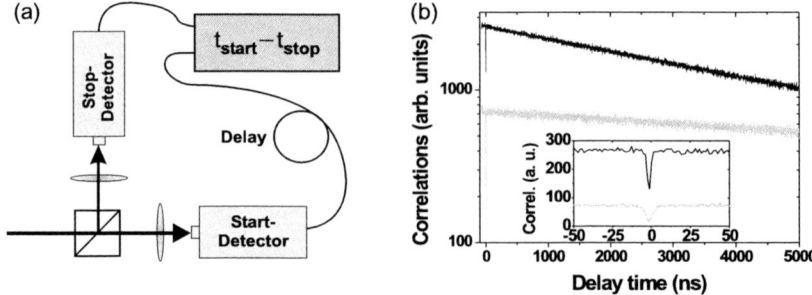

Fig. 16.3. (a) Scheme of the Hanbury Brown–Twiss setup. (b) Correlation measurement of a spectral line of a single quantum dot over a timescale that is large compared to the average time between detection events (7.7 μs for the black and 20 μs for the gray curve). The logarithmic scale emphasizes the exponential behavior. The dip at delay time $\tau = 0$ (see zoom in the inset) indicates single-photon emission

For single-photon sources, however, $g^{(2)}(0) < 1$ is possible (Fig. 16.2(c)) since the probability of detecting a second photon soon after a first detection event is reduced compared to an independent process. This effect is called antibunching and is reserved to nonclassical states with sub-Poissonian statistics. For Fock states $|n\rangle$, $g^{(2)}(0) = 1 - 1/n$ can be obtained, and in the special case of a single-photon state ($n = 1$), this yields $g^{(2)}(0) = 0$. Intermediate values are possible for statistical mixtures of one- and two-photon states. In the case of a pulsed source, the second-order coherence function shows a comb-like structure. The missing peak at $\tau = 0$ indicates the generation of only a single photon per pulse (Fig. 16.2(d)).

The straightforward approach to measure the second-order coherence function is to note the times of detector clicks and to compute the correlation function according to (16.1). However, this method is limited to timescales larger than the detectors' dead time (≈ 50 ns for the avalanche photo detector modules used in our setup [62]). In order to overcome this restriction, a Hanbury Brown–Twiss arrangement is chosen [28] as depicted in Fig. 16.3(a), consisting of two photo detectors monitoring the output of a 50:50 beam splitter. With this setup, the second detector can be armed right after the detection event of the first. For small count rates, one can neglect the case where the first detector is already armed while the second is still dead. Losses, like photons leaving the wrong beam splitter output or undetected photons, simply lead to a global decrease of the measured nonnormalized correlation function.

16.2.2 Micro-Photoluminescence

Optical investigation of quantum dots is usually performed in a micro-photoluminescence setup which combines excellent spatial resolution with high detection efficiency. Samples with low densities of 10^8–10^{11} dots/cm^2 are required to isolate a single dot. In our setup, the sample is held at 4 K inside a continuous-flow liquid Helium cryostat (Fig. 16.4). The dots are excited either pulsed (Ti:Sa, pulse width 400 fs, repetition rate 76 MHz, frequency-doubled to 400 nm) or continuously (Nd:YVO$_4$,

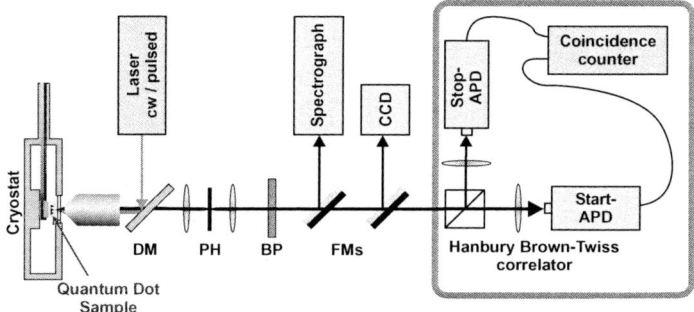

Fig. 16.4. Micro-PL setup (FM: mirrors on flip mounts, DM: dichroic mirror, PH: pinhole, BP: narrow bandpass filter, APD: avalanche photo diode)

532 nm). The microscope system (NA = 0.75) reaches a lateral resolution of 500 nm, and further spatial and spectral filtering selects a single transition of a single quantum dot. The photoluminescence can be imaged on a CCD camera, on a spectrograph, or can be sent to our Hanbury Brown–Twiss setup with a time resolution of about 800 ps to prove single-photon emission.

16.2.3 Single Photons from InP Quantum Dots

The sample used in our experiment was grown by metal-organic vapor phase epitaxy. An aluminum mirror was evaporated on the sample which was then thinned down to 400 nm and glued on a Si-substrate (see Fig. 16.5(a)). Figure 16.5(b) depicts the spectrum of a single InP quantum dot on the sample under weak excitation showing only one dominant spectral line. Its linear dependence on the excitation power identifies it as an exciton emission. A 1-nm bandpass filter was used to further reduce the background.

Figure 16.6 shows the corresponding cw correlation measurement at a count rate of 1.1×10^5 counts/s and 10 K. The function is modeled as a convolution of the expected shape of the ideal correlation function $g^{(2)}(\tau) = 1 - \exp(-\gamma \tau)$ and a

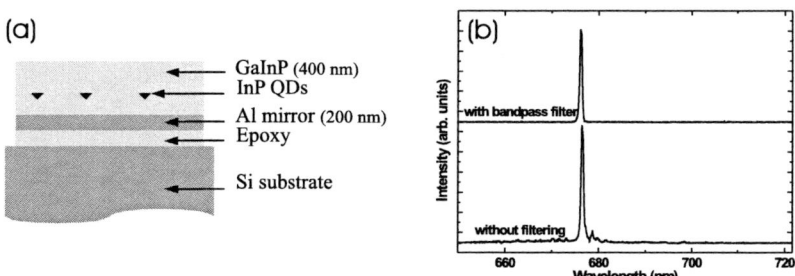

Fig. 16.5. (a) Structure of the InP/GaInP sample. (b) PL spectra from a single InP QD. An offset was added to separate the graphs

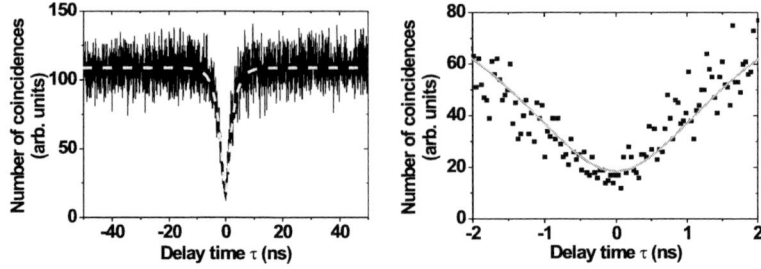

Fig. 16.6. $g^{(2)}$ function at continuous excitation. The fit function corresponds to an ideal single-photon source with limited time resolution. The right image shows a zoom into the central dip region

Gaussian distribution according to a 800-ps time resolution of the detectors. The width of the antibunching dip (2.3 ns) depends on both the transition lifetime and the excitation timescale. Equivalent measurements at pulsed excitation show a vanishing peak at zero delay (Fig. 16.7(a)) which again proves single-photon generation. With higher temperature, phonon interactions lead to an increasingly incoherent background with other spectral lines overlapping the filter transmission window (Fig. 16.7(b)). However, a characteristic antibunching was observed up to 50 K [75].

16.3 Multiphoton Emission from Single Quantum Dots

The potential of single semiconductor quantum dots as emitters in photonic devices is not limited to the generation of single photons on demand. Recently, entangled photon pairs were generated from a single quantum dot, an advance [64] that makes use of polarization correlations in a biexciton–exciton cascade [5]. But even without entanglement formation, multiphoton cascades can find applications in quantum communication experiments as shown in the following chapter.

Here, intensity cross-correlations between different quantum dot transitions in InP quantum dots are measured. Similar exciton–biexciton cross-correlation measurements have also been reported on InAs quantum dots [48, 54, 56] and II–VI

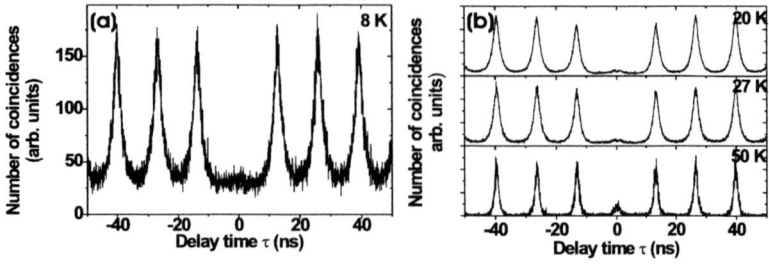

Fig. 16.7. $g^{(2)}$ function of a single InP QD at pulsed excitation: (**a**) at 8 K and (**b**) between 20 and 50 K

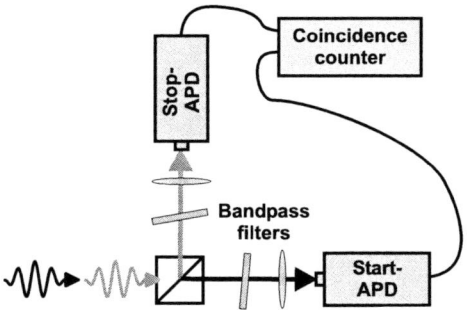

Fig. 16.8. Modified Hanbury Brown–Twiss setup for cross-correlation experiments. The different gray scales of the two beams represent two spectral lines selected by the bandpass filters

quantum dots [68, 17]. These experiments serve several purposes: First, they are important tools to identify the nature of the investigated spectral lines, such as those resulting from an exciton, biexciton, triexciton, or emerging from the same or different quantum dots. Second, they give information about the decay and excitation timescales in multiphoton cascades. Finally, polarization resolved cross-correlations form a first step toward the observation of entangled photon pairs.

In order to detect correlations between different transitions, a modified second-order coherence function is considered. This cross-correlation function is defined equivalently to (16.1), but with the photon number operators assigned to different field modes α and β:

$$g^{(2)}_{\alpha\beta}(\tau) = \frac{\langle : \hat{I}_\alpha(t)\hat{I}_\beta(t+\tau) : \rangle}{\langle \hat{I}_\alpha(t)\rangle\langle \hat{I}_\beta(t)\rangle}. \quad (16.2)$$

In the following experiments, the two modes represent spectral lines of two quantum dot transitions. In order to distinguish this cross-correlation from the second-order coherence function, the latter is also referred to as the autocorrelation function.

In the experiment, the two spectral lines are filtered individually for each photo detector of the Hanbury Brown–Twiss setup. This was realized by placing narrow bandpass filters directly in front of each avalanche photo detector, as sketched in Fig. 16.8. In this way, the resulting correlation function will show an asymmetry with respect to the time origin, as start and stop detection events now arise from different processes and a change in sign of the time axis is in accord with an effective exchange of start and stop detector.

In order to make a first estimate about the origin of the distinct spectral lines, we investigated their different scaling with the excitation intensity. Figure 16.9 shows photoluminescence spectra of a quantum dot, taken at various excitation intensities at 8 K. In the following, the reference excitation power density P_0 was kept constant at $1\,\text{nW}/\mu\text{m}^2$. The spectral behavior in Fig. 16.9 is typical for an InP dot emitting in this energy range in terms of line spacing and power dependence. At low excitation power density, a single sharp emission line at 686.3 nm (1.8155 eV) is present in the spectrum (X_1). As the excitation power is increased, a second line (X_2) appears

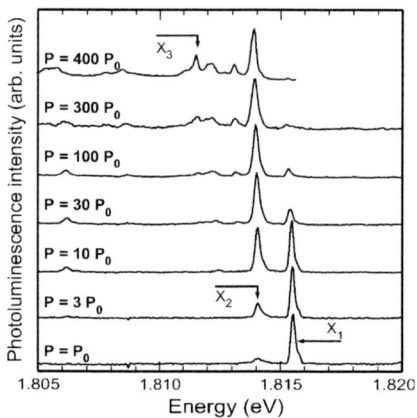

Fig. 16.9. Power-dependent spectroscopy of a single InP quantum dot showing the lines used in the correlation measurements. The excitation intensity is given as a multiple of $P_0 = 1\,\text{nW}/\mu\text{m}^2$. Lines X_1, X_2, and X_3 are assigned to different excitations, as described in the text

about 0.6 nm (1.5 meV) beside the exciton emission. When further increasing the excitation power, additional lines appear. The integrated photoluminescence intensity of X_1 increases linearly with the excitation intensity, whereas X_2 shows a quadratic dependence. This behavior is a good indication of excitonic and biexcitonic emission, respectively. The lines appearing at high excitation power density, such as X_3, are attributed to a multiexciton of higher complexity. In particular, X_3 is assumed to originate from a triexciton, as will be proven later.

For such a complex excitation as the triexciton, it is necessary to invoke additional states to the single-particle ground states of the quantum dot. Figure 16.10(a) shows two of the possible triexciton decays together with a simplified decay chain of a triexciton in Fig. 16.10(b). The triexciton X_3 recombines to an excited biexciton X_2^* that rapidly relaxes to the biexciton ground state X_2 which in turn recombines via the exciton X_1 to the empty ground state G of the quantum dot [18].

After characterization and pre-identification of the different spectral lines, additional information can be gained by performing cross-correlation measurements between these emission lines. Figure 16.11(a) shows the cross-correlations of the exciton and biexciton line of that dot at different cw excitation power densities [53]. A strong asymmetric behavior is observed: At positive times, when the detection of a biexciton photon starts, the correlation measurement and the detection of an exciton photon stops it, photon bunching occurs, as here the detection of the starting biexciton photon projects the quantum dot into the exciton state which now has an increased probability of recombining shortly thereafter. On the other hand, if the correlation measurement is started by the exciton photon, which prepares the dot in the ground state, and stopped by the biexciton photon (negative times in Fig. 16.11(a)), a certain time is needed until the dot is re-excited. In this measurement, effectively the recycling time of the quantum dot is observed which explains the strong anti-

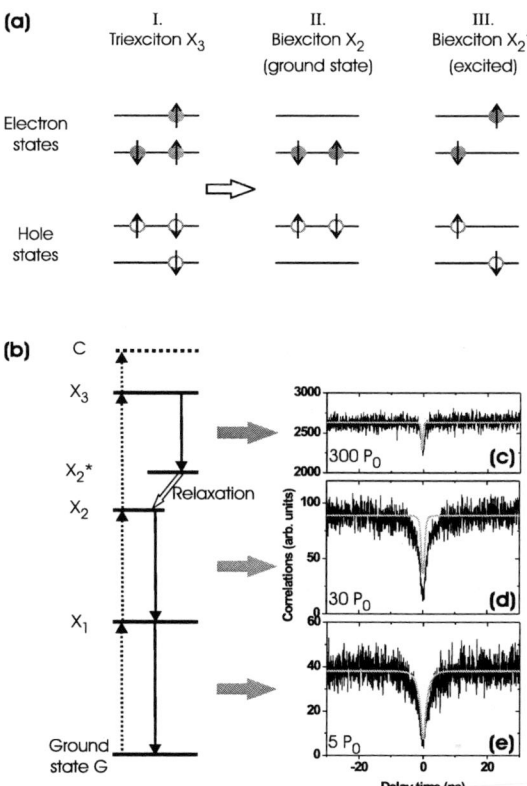

Fig. 16.10. Illustration of the multiexciton cascade in quantum dots. (**a**) Occupation of electron and hole states in the decay of the triexciton state X_3 to the biexciton ground state X_2 and to an excited state X_2^*. (**b**) Decay cascade model used in the discussions and the rate equation approach. Dashed arrows indicate excitation, solid arrows radiative decay, and the open arrow a nonradiative relaxation. The dashed state C symbolizes an effective cut-off state as explained in the text

bunching for negative times. The population of the biexciton state is dependent on the laser power, and the excitation time decreases when the laser power increases. Similar measurements have also been performed on another quantum dot (depicted in Fig. 16.11(c)). Although its timescales are similar, it shows a more pronounced bunching peak.

In the same way, the cross-correlation of the biexciton emission with the triexciton emission was measured. This is shown in Fig. 16.11(b). Its behavior is similar to the exciton–biexciton case, but with different timescales apparent.

The presence of the combined bunching/antibunching shape is a unique hint for observing a decay cascade of two adjacent states. In contrast, the cross-correlation function of spectral lines of two independent transitions (e.g., from two different quantum dots) would show no (anti-)correlations at all. It can be concluded that there

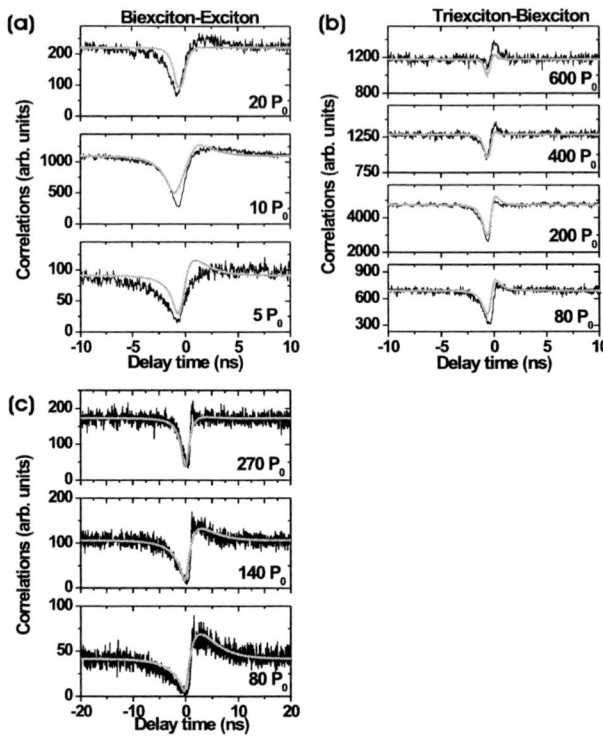

Fig. 16.11. Measured cross-correlation functions (**a**) between the exciton and biexciton line and (**b**) between the biexciton and triexciton line of the quantum dot that was also used for Fig. 16.9 at different excitation intensities ($P_0 = 1\,\text{nW}/\mu\text{m}^2$). (**c**) Exciton–biexciton cross-correlations of a second quantum dot

is a three-photon cascaded emission from the triexciton via biexciton and exciton to the quantum dot ground state. Together with the information of the different scaling of the spectral lines with excitation power, this justifies the previous assignments to these lines.

In order to support the interpretation of the obtained correlation data, the photon cascade was analyzed using a common rate model [48, 54]. The rate equations correspond to the scheme shown in Fig. 16.10(b) where only two transition types account for the dynamics of the excitonic states: spontaneous radiative decay and re-excitation at a rate proportional to the excitation power. As the excitation is performed above the quantum dot continuum, we should also take into account the relaxation of the charge carriers into the multiexciton states, as well as the relaxation of the excited biexciton after the triexciton decay. But as this process happens on a much faster timescale (several 10 ps [69]) than the state lifetimes ($\approx$ ns), it is neglected in this consideration. The according rate equation ansatz then reads:

$$\frac{d}{dt}\mathbf{n}(t) = \begin{pmatrix} -\gamma_E & \gamma_1 & 0 & 0 & 0 \\ \gamma_E & -\gamma_E - \gamma_1 & \gamma_2 & 0 & 0 \\ 0 & \gamma_E & -\gamma_E - \gamma_2 & \gamma_3 & 0 \\ 0 & 0 & \gamma_E & -\gamma_E - \gamma_3 & \gamma_C \\ 0 & 0 & 0 & \gamma_E & -\gamma_C \end{pmatrix} \mathbf{n}(t),$$

with $\mathbf{n}(t) = (n_G(t), n_1(t), n_2(t), n_3(t), n_C(t))$ and $\gamma_i = \tau_i^{-1}$. Here, n_1, n_2, and n_3 represent the populations of the exciton, biexciton, and triexciton, respectively, with corresponding decay times τ_1, τ_2, and τ_3. n_G is the population of the empty ground state, and τ_E^{-1} is the excitation rate. In order to truncate the ladder of states connected by rates in this model, an effective cutoff state with population n_C and lifetime τ_C was introduced. This accounts for population and depopulation of all higher excited states via excitation and radiative decay, respectively.

This rate equation can be solved analytically [1]. The general solution is a sum of decaying exponentials with different time constants. The initial conditions are defined by the transition that forms the start event in the Hanbury Brown–Twiss measurement which prepares the quantum dot in the next lower state α, so that $n_\alpha(0) = 1$ and $n_{\gamma \neq \alpha}(0) = 0$. On the other hand, the detection of a photon from the stop transition dictates the shape of the cross-correlation function, as $g_{\alpha\beta}^{(2)}(t) \propto n_\beta(t)$. Therefore it is clear that the cross-correlation function on the positive and negative side is described by two completely different functions with a possible discontinuity at $\tau = 0$. In the experiment (Fig. 16.11), this discontinuity is washed out, due to the finite time resolution of the detectors. Due to this smoothing of the experimental data, the minima in the graphs are shifted towards the antibunching side, as well.

This model was used to describe the auto- and cross-correlation data in Figs. 16.10 and 16.11. The results are shown as gray lines in the graphs. The lifetimes of exciton, biexciton, and triexciton were taken from independent measurements. The excitation rate was chosen to optimally fit the correlation functions in these figures, but was kept linear to the experimental excitation power P throughout the graphs. In this way, apart from a one-time initialization of the experimentally inaccessible values τ_3, τ_C, and τ_E, the normalization was the only fit parameter in all graphs. No vertical offset was used to compensate the lift of the antibunching dips. Apparently, the model describes the experimental data very well. Minor deviations can be explained by the long-term variation of the excitation power due to a spatial drift of the sample and by the presence of additional states neglected in this model.

16.4 Realization of the Ultimate Limit of a Light Emitting Diode

As described in the previous section, photoluminescence from single semiconductor quantum dots [8] has become an important player to allow the demanded deterministic emission of single-photon states [45, 55]. It even opens the gates to coherent mapping of information onto stationary solid-state quantum bit systems for information processing [20] due to its narrow line width. However, electrical excitation is

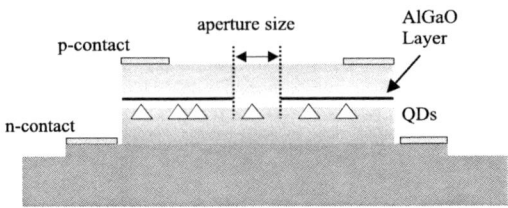

Fig. 16.12. Schematic cross-section of the device structure

required for the fabrication of integrated compact devices and to avoid large-scale external pump sources. Early schemes used a simultaneous Coulomb-blockade for electrons and holes in a semiconductor triple quantum well nanostructure [33]. Other realizations embedded single self-organized quantum dots in pin-diode structures [73, 22, 74, 21, 59]. Their emission can be enhanced and directed into distinctive modes by the growth of Bragg reflectors [46]. But the desired ultimate control in integrated ready-to-go single-photon sources for quantum communication includes the deterministic injection of a single electron and a single hole into a quantum dot since it avoids the need for spectral or spatial filtering of the emission.

It has been shown that it is possible to selectively pump a single QD in a pin-diode and to obtain a pure emission spectrum with only a single exciton line. This promising approach for single-photon generation based on electrical pumping uses a micron-size aluminum oxide aperture to restrict the current flow to a single dot [74, 21, 39, 40]. Thus, electrical excitation of more than one dot was significantly suppressed, and nonclassical statistics of the electroluminescence was measured. But that work could not finally prove the injection of a single electron and a single hole into the pin-junction that generates sub-Poissonian statistics without external filtering. The structure of our device has been described elsewhere [39, 40, 27] and will be just briefly sketched (Fig. 16.12). It is grown on semi-insulating (100) epiready GaAs substrates using a Riber-32P MBE system. The light-emitting diode (LED) consists of an undoped GaAs layer with InAs quantum dots of low density inserted, a 60 nm thick aperture layer of high aluminum content AlGaAs, and p- and n-type GaAs electrical contact layers. The low quantum dot density of 10^8 cm^{-2} was obtained in the Stranski–Krastanow mode by deposition of 1.8 ML of InAs. Cylindrical mesas were processed by inductively coupled plasma reactive ion etching, and selective oxidation of the high aluminum content AlGaAs layers led to submicron-size oxide current apertures. Subsequent Si_3N_4 deposition allowed Au/Pt/Ti and Au/Au–Ge/Ni metalization to form p- and n-contacts, respectively.

In order to probe the electrical excitation of only a single quantum dot, we measured the electroluminescence spectrum at an injection current of 870 pA and a bias voltage of 1.65 V (Fig. 16.13(a)). It reveals just one single line, and emission of the wetting layer is also completely suppressed. For experiments related to entangled pair generation using biexciton–exciton decays [5], emission from an uncharged dot is requisite. Via high-resolution spectroscopy, we could determine a fine structure splitting of 55 μeV due to electron–hole exchange interaction [60, 61].

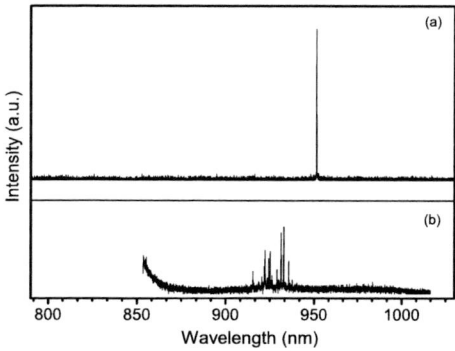

Fig. 16.13. (a) Electroluminscence spectrum at a current of 870 pA and 1.65 V bias voltage. (b) Micro-photoluminescence with a laser spot size of 2 μm

The existence of a splitting proves the electroluminescence to originate from an exciton rather than a trion state and will allow further experiments towards entanglement generation with this device. Although many dots may still be excited with the electroluminescence of residual dots being blocked by a shadow mask, microphotoluminescence of a few dots (Fig. 16.13(b)) at 10 K exhibited a set of several discrete lines. Thus, the oxide aperture above the dot is transparent for near-infrared light, and the absence of light emission other than from the exciton decay in the electroluminescence spectrum clearly demonstrates the pumping of indeed only a single quantum dot.

In Fig. 16.14, six electroluminescence spectra are depicted to further characterize the emission of our device at increasing injection current. At around 1 nA, corresponding to an injection of 5 electrons/ns, the exciton intensity saturates. With the typical exciton lifetime of 1 ns [44, 29, 66], this gives a remarkable injection efficiency of about 20% since the capture time for carriers is in the picosecond regime.

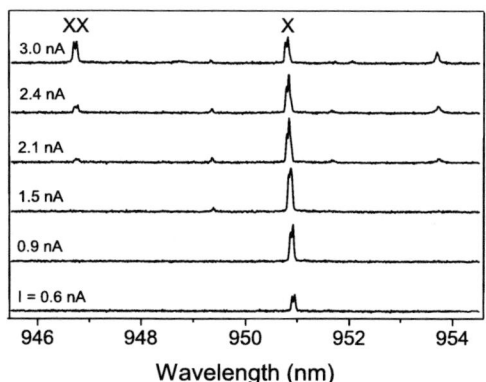

Fig. 16.14. Electroluminescence spectra at different current injection. For clarity an offset is added to each spectrum

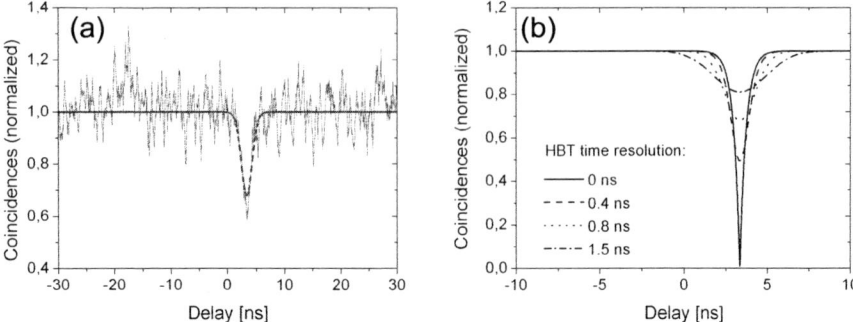

Fig. 16.15. (a) Autocorrelation measurement under continuous wave current injection (0.9 nA, 1.65 V) at 10 K. No spectral filtering was used to isolate a single transition in a single quantum dot. The dashed line shows a fit function corresponding to a HBT time resolution of 800 ps as described in the text. (b) Impact of the HBT time resolution on the measured correlation function

This is two orders of magnitude better than in previously reported structures [73, 22, 74, 21]. At higher currents, two additional lines appear. Both lines originate from the same dot since they obey the same spectral jitter. The high-energy peak is assigned to the biexciton decay due to its super-linear dependence on the injection current. The third line is unpolarized and does not show any splitting; thus, it can be attributed to the trion. As already mentioned, the emission from uncharged dots paves the way for on-demand generation of entangled photon pairs from our device which was not possible in earlier realizations [21].

Given these extremely clean emission features and excellent carrier control, we further exploited the use of our device as a single-photon source. Nonclassical photon statistics is characterized by the second-order intensity correlation function via a Hanbury Brown and Twiss setup (HBT) [28] consisting of a 50:50 beam splitter and two avalanche photo diodes (APDs). In contrast to all previous HBT setups related to quantum dot experiments, no spatial or spectral filtering of the exciton line was needed (except a 10-nm FWHM band pass filter centered at 953 nm in front of one APD to avoid cross-talk between the detectors).

The result of our autocorrelation measurement is depicted in Fig. 16.15. In order to interpret our experimental data, we use the rate model described in [1]. Although an ideal single-photon source shows an antibunching dip of the correlation function down to zero, the limited time resolution of our HBT setup must be taken into account. Thus, the ideal function $g^{(2)}(\Delta t) = 1 - \exp(-\Delta t/\tau)$ was convoluted with a Gaussian (FWHM: 800 ps) and yields a dip depth of around 65%. The time resolution of 800 ps was independently determined by an autocorrelation measurement of ultra-short laser pulses. The convoluted fit shown in Fig. 16.15 agrees well with our experimental data and proves that our quantum dot LED represents indeed an ideal electrically pumped single-photon source.

16.5 Applications in Quantum Information Processing

16.5.1 Quantum Key Distribution

For quantum communication protocols, a single-photon source with high repetition rate and collection efficiency is desired. Besides passive elements like mirrors and solid immersion lenses, resonant techniques exploiting the Purcell effect can greatly enhance the emission rate [24, 63] if the source relies on the decay of an excited state.

A well-established technique from classical communication is *multiplexing*. Each signal is marked with a physical label like its wavelength and identified at the receiver using filters tuned to the carrier frequencies. For single photons, their wavelength can be used as their distinguishing label and polarization to encode quantum information [2]. If the two photons from a biexciton-exciton cascade pass a Michelson interferometer, constructive and destructive interference will be observed at the two output ports by proper choice of the relative path difference (Fig. 16.16). These photons are fed into an optical fiber each, and one of them is delayed by half the repetition rate of the excitation laser. Recombination at a beam splitter then leads to a photon stream with a doubled count rate. This is also reflected by autocorrelation measurements with a halved spacing in the peak structure (see Fig. 16.16(d)) [2].

Multiplexing can enhance the bandwidth of quantum communication systems which use single-photon sources based on spontaneous emission. In our experiment, we implemented the BB84 protocol [4, 26] where the exact photon wavelength is unimportant. Quantum cryptography has been realized with single-photon states

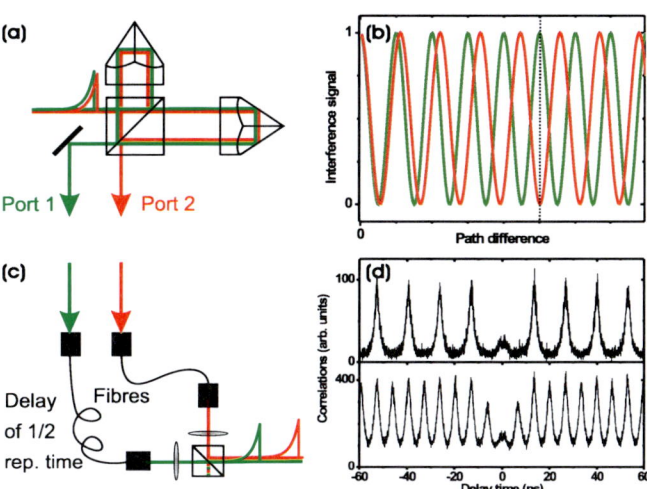

Fig. 16.16. (a) Michelson interferometer with two output ports as a single-photon add/drop filter. (b) Interference pattern for two distinct wavelengths versus the relative arm length. (c) Merging the separated photons with one path delayed by half the excitation repetition rate. (d) Intensity correlation of the exciton spectral line and the multiplexed signal

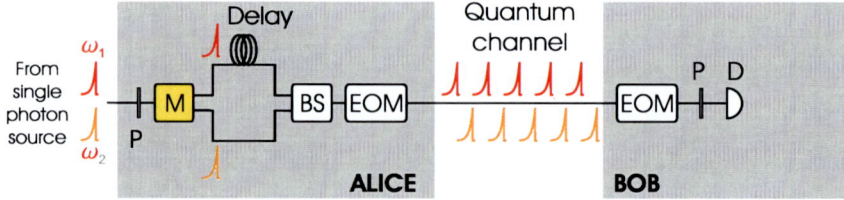

Fig. 16.17. Possible implementation of the multiplexer using the BB84 protocol. For a detailed description of the setup see text

Fig. 16.18. Visualization of the quantum key distribution. Alice's original data (**a**), a photograph taken out of our lab window in Berlin, is encrypted and sent to Bob (**b**). After decryption with his key, Bob obtains image (**c**)

from diamond defect centers [7] and single quantum dots [70]. Figure 16.17 shows an implementation of interferometric multiplexing in the BB84 protocol using a cascaded photon source.

Behind an exciton–biexciton add/drop filter, Alice randomly prepares the photons' polarization. Bob's detection consists of a second EOM, an analyzing polarizer, and an APD. In our experiment, all EOM and APD operation is performed automatically by a LabVIEW program. A visualization of a successful secret transmission of data is depicted in Fig. 16.18. After exchange of the key, Alice encrypts her data by applying an XOR operation between every image bit and the sequence of random bits (the secret key) which she has previously shared with Bob by performing the BB84 protocol. The encoded image (shown in Fig. 16.18(b)) is transmitted to Bob over a distance of 1 m with 30 bits/s. Another XOR operation by Bob reveals the original image with an error rate of 5.5%.

16.5.2 Quantum Computing

Knill et al. proposed the realization of a universal set of quantum gates for quantum computation, using linear optical elements and single-photon detectors [34]. This scheme requires triggered, indistinguishable single photons. Two-photon quantum gates (c-NOT gates) and basic quantum algorithms have been realized using photon pairs created by parametric down-conversion [71, 51]. Here, we report the on-demand operation of a two-qubit Deutsch–Jozsa algorithm [19] using a deterministic single-photon source. Previous all-optical experiments were restricted to emulate the Deutsch–Jozsa algorithm with attenuated classical laser pulses [65].

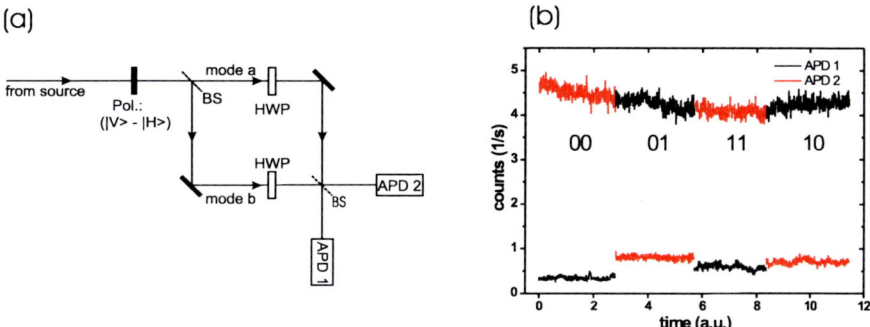

Fig. 16.19. (**a**) Setup of the Deutsch–Jozsa experiment. Pol: polarizer, BS: beam splitter, HWP: half-wave plate. (**b**) Detector count rates of APD 1 and 2 for the four different combinations of the HWPs inside the interferometer: 00 no HWP, 10: HWP in mode a, 01: HWP in b, 11: HWP in a and b

An intuitive and often-cited interpretation compares this algorithm to a "coin tossing" game: While a classical observer has to check both sides of a coin in order to tell if it is fair, i.e. if its two sides are different (head and tail), or unfair with two equal sides (head–head or tail–tail), quantum computation can perform this task with a single shot. In the simplest case, two qubits are needed to store query and answer, respectively.

Here, we apply the proposal of [12] to implement two qubits using different degrees of freedom of a single photon (see also [58]). The first qubit corresponds to the which-way information of the photon path inside a Mach–Zehnder interferometer, namely the spatial modes a and b (Fig. 16.19(a)). The second qubit is implemented via the photon's polarization state. With these definitions of our two qubits, we are able to construct the proper states for a Deutsch–Jozsa algorithm. The initial state is realized by sending a photon into a linear superposition of the two spatial modes using a nonpolarizing 50:50 beam splitter after preparing it in a superposition of horizontal and vertical polarization.

In a next step, a unitary transformation is realized by selectively adding a half-wave plate (HWP) to each mode. A HWP in only one of the two paths corresponds to a balanced situation representing fair coins in the above example. In contrast, if either no HWP or two HWPs in both paths are used, this corresponds to the constant situations, illustrated by the two possible false coins. Physically, the difference between these two situations is an additional local phase difference between the two paths, 0 for the constant and π for the balanced case. Recombining the two interferometer arms on a second 50:50 beam splitter implements a final Hadamard gate on the first qubit. The two interferometer outputs are monitored with one APD each. Constructive interference at the top (bottom) output indicates the balanced (constant) situation which can thus be distinguished by only one detection event of a single photon. Figure 16.19 shows the outcome of many detection events (albeit only one would be enough) for the four possible combinations. In these measurements an overall suc-

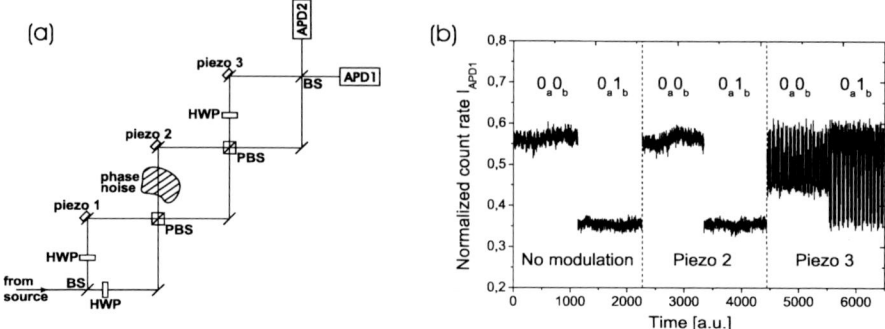

Fig. 16.20. (a) Extended setup. (P)BS: (polarizing) beam splitter. (b) Detector count rates for a constant (00) and a balanced (01) HWP combination. Left graph: No modulation of the piezos, middle graph: modulation on piezo 2, right graph: modulation on piezo 3

cess probability of 79% was achieved. This was mainly limited by a slight temporal and spatial mode mismatch in the interferometer.

A fundamental problem of quantum computation is dephasing, i.e. the tendency of a qubit to lose or change its quantum information due to coupling to the classical environment. Solutions are active quantum error correction [13] or the use of decoherence-free subspaces of the system's Hilbert space [38]. In our experiment, a variant of the setup insensitive to phase noise has been achieved by extending the setup to the scheme in Fig. 16.20(a). Horizontal and vertical polarization are separated and merged at polarizing beam splitters (PBS) so that the modes, that interfere at the final beam splitter, travel along the same path. Thus, phase noise in the central interferometer does not cause perturbations of the algorithm, but only leads to a global phase change [58].

In our experiment, we simulate phase noise by modulating piezoelectric mirror mounts. The left graph of Fig. 16.20(b) displays the detection signals for two combinations of the HWPs in the first interferometer (one constant and one balanced), without any artificial noise. Modulating piezo 2 in the central interferometer does also not affect the count rates, as expected (middle graph). For comparison, phase noise induced by piezo 3 in the final interferometer changes the APD signals substantially (right graph). The same happens when modulating piezo 1 (not shown). Thus, using this extended experimental setup, the robustness of the algorithm against phase noise could be proven when information is properly encoded in an unaffected superposition of polarization states.

16.6 Outlook

In this chapter, we reviewed single-photon generation via photoluminescence from single InP QDs as well as via electroluminescence from InGaAs quantum dots and its application to quantum information processing. We demonstrated single-photon statistics and cross-correlations of various transitions from multi-excitonic states in the visible spectrum around 690 nm. We described a proof-of-principle experiment that shows how multiphoton generation from single QDs may find applications in quantum cryptography devices to enhance their maximum bandwidth. In this experiment, multiplexing on the single-photon level and its implementation in the BB84 quantum key distribution protocol has been realized for the first time. We described the implementation of a Deutsch–Jozsa algorithm with single photons from single quantum dots and showed prospects for the electrical excitation of quantum dots and further integration of photonic devices.

Our experiments prove that today's single-photon sources based on QDs have reached the level of sophistication to be ready-to-use nonclassical light sources. The *on-demand* character of the emission is extremely useful in all-solid-state implementations of quantum information devices. Single QDs will therefore play an important role in the field of quantum cryptography and quantum computation, but also for realizing interfaces in larger quantum networks.

References

1. T. Aichele, Ph.D. thesis, Humboldt-Universität zu Berlin, 2005
2. T. Aichele, G. Reinaudi, O. Benson, Phys. Rev. B **70**, 235329 (2004)
3. A. Badolato, K. Hennessy, M. Atatre, J. Dreiser, E. Hu, P.M. Petroff, A. Imamoğlu, Science **308**, 1158 (2005)
4. C.M. Bennett, G. Brassard, in *Proc. IEEE Conference on Computers, Systems, and Signal Processing in Bangalore, India* (IEEE, New York, 1984), p. 175
5. O. Benson, C. Santori, M. Pelton, Y. Yamamoto, Phys. Rev. Lett. **84**, 2513 (2000)
6. A. Beveratos, S. Kühn, R. Brouri, T. Gacoin, J.-P. Poizat, P. Grangier, Eur. Phys. J. D **18**, 191 (2002)
7. A. Beveratos, R. Brouri, T. Gacoin, A. Villing, J.-P. Poizat, P. Grangier, Phys. Rev. Lett. **89**, 187901 (2002)
8. D. Bimberg, M. Grundmann, N.N. Ledentsov, *Quantum Dot Heterostructures* (Wiley, Chichester, 1998)
9. D. Bouwmeester, J.-W. Pan, K. Mattle, M. Eibl, H. Weinfurter, A. Zeilinger, Nature **390**, 575 (1997)
10. H.-J. Briegel, W. Dür, J.I. Cirac, P. Zoller, Phys. Rev. Lett. **81**, 5932 (1998)
11. C. Brunel, B. Lounis, P. Tamarat, M. Orrit, Phys. Rev. Lett. **83**, 2722 (1999)
12. N.J. Cerf, C. Adami, P.G. Kwiat, Phys. Rev. A **57**, 1477 (1998)
13. J. Chiaverini, D. Leibfried, T. Schaetz, M.D. Barrett, R.B. Blakestad, J. Britton, W. Itano, J.D. Jost, E. Knill, C. Langer et al., Nature **432**, 602 (2004)
14. J.I. Cirac, P. Zoller, Phys. Rev. Lett. **74**, 4091 (1995)
15. J.I. Cirac, P. Zoller, Nature **404**, 579 (2000)
16. J.I. Cirac, P. Zoller, Phys. Today **57**, 38 (2004)
17. C. Couteau, S. Moehl, F. Tinjod, J.M. Gérard, K. Kheng, H. Mariette, J.A. Gaj, R. Romestain, J.-P. Poizat, Appl. Phys. Lett. **85**, 6251 (2004)
18. E. Dekel, D. Gershoni, E. Ehrenfreund, J.M. Garcia, P.M. Petroff, Phys. Rev. B **61**, 11009 (2000)
19. D. Deutsch, R. Jozsa, Proc. R. Soc. Lond. A **439**, 553 (1992)

20. L.-M. Duan, M.D. Lukin, J.I. Cirac, P. Zoller, Nature **414**, 413 (2001)
21. D.J.P. Ellis, A.J. Bennett, A.J. Shields, P. Atkinson, D.A. Ritchie, Appl. Phys. Lett. **88**, 133509 (2006)
22. A. Fiore, J.X. Chen, M. Ilegems, Appl. Phys. Lett. **81**, 1756 (2002)
23. S. Gasparoni, J.-W. Pan, P. Walther, T. Rudolph, A. Zeilinger, Phys. Rev. Lett. **93**, 020504 (2004)
24. J.M. Gérard, B. Sermage, B. Gayral, B. Legrand, E. Costard, V. Thierry-Mieg, Phys. Rev. Lett. **81**, 1110 (1998)
25. D. Gottesmann, I.L. Chuang, Nature **402**, 390 (1999)
26. N. Gisin, G. Ribordy, W. Tittel, H. Zbinden, Rev. Mod. Phys. **74**, 145 (2002)
27. V.A. Haisler, F. Hopfer, R.L. Sellin, A. Lochmann, K. Fleischer, N. Esser, W. Richter, N.N. Ledentsov, D. Bimberg, C. Möller, N. Grote, Appl. Phys. Lett. **81**, 2544 (2002)
28. R. Hanbury Brown, R.Q. Twiss, Nature **178**, 1046 (1956)
29. R. Heitz, A. Kalburge, Q. Xie, M. Grundmann, P. Chen, A. Hoffmann, A. Madhukar, D. Bimberg, Phys. Rev. B **57**, 9050 (1998)
30. M. Keller, B. Lange, K. Hayasaka, W. Lange, H. Walther, Nature **431**, 1075 (2004)
31. D. Kielpinski, C. Monroe, D.J. Wineland, Nature **417**, 709 (2002)
32. N. Kiesel, C. Schmid, U. Weber, R. Ursin, H. Weinfurter, Phys. Rev. Lett. **95**, 210505 (2005)
33. J. Kim, O. Benson, H. Kan, Y. Yamamoto, Nature **397**, 500 (1999)
34. E. Knill, R. Laflamme, G.J. Milburn, Nature **409**, 46 (2001)
35. A. Kuhn, M. Hennrich, G. Rempe, Phys. Rev. Lett. **89**, 067901 (2002)
36. C. Kurtsiefer, S. Mayer, P. Zarda, H. Weinfurter, Phys. Rev. Lett. **85**, 290 (2000)
37. D. Leibfried, B. Demarco, V. Meyer, D. Lucas, M. Barrett, J. Britton, W.M. Itano, B. Jelenkovic, C. Langer, T. Rosenband et al., Science **422**, 412 (2003)
38. D.A. Lidar, I.L. Chuang, K.B. Whaley, Phys. Rev. Lett. **81**, 2594 (1998)
39. A. Lochmann, E. Stock, O. Schulz, F. Hopfer, D. Bimberg, V.A. Haisler, A.I. Totopov, A.K. Bakarov, A.K. Kalagin, Electron. Lett. **42**, 774 (2006)
40. A. Lochmann, E. Stock, O. Schulz, F. Hopfer, D. Bimberg, V.A. Haisler, A.I. Toropov, A.K. Bakarov, M. Scholz, S. Bttner, O. Benson, Phys. Stat. Sol. (c) **4**, 547 (2006)
41. D. Loss, D.P. DiVicenzo, Phys. Rev. A **57**, 120 (1998)
42. B. Lounis, W.E. Moerner, Nature **407**, 491 (2000)
43. L. Mandel, E. Wolf, *Optical Coherence and Quantum Optics* (Cambridge University Press, Cambridge, 1995)
44. P. Michler, *Single Quantum Dots* (Springer, Berlin, 2003)
45. P. Michler, A. Imamoğlu, M.D. Mason, P.J. Carson, G.F. Strouse, S.K. Buratto, Nature **406**, 968 (2000)
46. C. Monat, B. Alloing, C. Zinoni, L.H. Li, A. Fiore, Nano Lett. **6**, 1464 (2006)
47. J.E. Mooij, T.P. Orlando, L. Levitov, L. Tian, C.H. van der Wal, S. Lloyd, Science **285**, 1036 (1999)
48. E. Moreau, I. Robert, L. Manin, V. Thierry-Mieg, J.M. Gérard, I. Abram, Phys. Rev. Lett. **87**, 183601 (2001)
49. *New Journal of Physics* (special issue on single-photon sources) **6** (2004)
50. M. Nirmal, B.O. Dabbousi, M.G. Bawendi, J.J. Macklin, J.K. Trautman, T.D. Harris, L.E. Brus, Nature **383**, 802 (1996)
51. J.L. O'Brien, G.J. Pryde, A.G. White, T.C. Ralph, D. Branning, Nature **426**, 264 (2003)
52. Y.A. Pashkin, T. Yamamoto, O. Astafiev, Y. Nakamura, D.V. Averin, J.S. Tsai, Nature **421**, 823 (2003)
53. J. Persson, T. Aichele, V. Zwiller, L. Samuelson, O. Benson, Phys. Rev. B **69**, 233314 (2004)

54. D.V. Regelman, U. Mizrahi, D. Gershoni, E. Ehrenfreund, W.V. Schoenfeld, P.M. Petroff, Phys. Rev. Lett. **87**, 257401 (2001)
55. C. Santori, M. Pelton, G. Solomon, Y. Dale, Y. Yamamoto, Phys. Rev. Lett. **86**, 1502 (2000)
56. C. Santori, D. Fattal, M. Pelton, G.S. Solomon, Y. Yamamoto, Phys. Rev. B **66**, 045308 (2002)
57. F. Schmidt-Kaler, H. Häffner, M. Riebe, S. Gulde, G.P.T. Lancaster, T. Deuschle, C. Becher, C.F. Roos, J. Eschner, R. Blatt, Nature **422**, 408 (2003)
58. M. Scholz, T. Aichele, S. Ramelow, O. Benson, Phys. Rev. Lett. **96**, 180501 (2006)
59. R. Schmidt, U. Scholz, M. Vitzethum, R. Fix, C. Metzner, P. Kailuweit, D. Reuter, A. Wieck, M.C. Hubner, S. Stufler et al., Appl. Phys. Lett. **88**, 121115 (2006)
60. R. Seguin, A. Schliwa, S. Rodt, K. Pötschke, U.W. Pohl, D. Bimberg, Phys. Rev. Lett. **95**, 257402 (2005)
61. R. Seguin, A. Schliwa, T.D. Germann, S. Rodt, K. Pötschke, A. Strittmatter, U.W. Pohl, D. Bimberg, M. Winkelnkemper, T. Hammerschmidt, P. Kratzer, Appl. Phys. Lett. **89**, 263109 (2006)
62. *Single-Photon Counting Module—SPCM-AQR Series Specifications*. Laser Components GmbH, Germany, 2004
63. G. Solomon, M. Pelton, Y. Yamamoto, Phys. Rev. Lett. **86**, 3903 (2001)
64. R.M. Stevenson, R.J. Young, P. Atkinson, K. Cooper, D.A. Ritchie, A.J. Shields, Nature **439**, 179 (2006)
65. S. Takeuchi, Phys. Rev. A **62**, 032301 (2000)
66. R.M. Thompson, R.M. Stevenson, A.J. Shields, I. Farrer, C.J. Lobo, D.A. Ritchie, M.L. Leadbeater, M. Pepper, Phys. Rev. B **64**, 201302 (2001)
67. L. Tian, P. Rabl, R. Blatt, P. Zoller, Phys. Rev. Lett. **92**, 247902 (2004)
68. S.M. Ulrich, S. Strauf, P. Michler, G. Bacher, A. Forchel, Appl. Phys. Lett. **83**, 1848 (2003)
69. M. Vollmer, E.J. Mayer, W.W. Rhle, A. Kurtenbach, K. Eberl, Phys. Rev. B **54**, R17292 (1996)
70. E. Waks, K. Inoue, C. Santori, D. Fattal, J. Vučković, G.S. Solomon, Y. Yamamoto, Nature **420**, 762 (2002)
71. P. Walther, K.J. Resch, T. Rudolph, E. Schenck, H. Weinfurter, V. Vedral, M. Aspelmeyer, A. Zeilinger, Nature **434**, 169 (2005)
72. W.G. van der Wiel, S.D. Franceschi, J.M. Elzerman, T. Fujisawa, S. Tarucha, L.P. Kouwenhoven, Rev. Mod. Phys. **75**, 1 (2003)
73. Z. Yuan, B.E. Kardynal, R.M. Stevenson, A.J. Shields, C.J. Lobo, K. Copper, N.S. Beattie, D.A. Ritchie, M. Pepper, Science **295**, 102 (2002)
74. C. Zinoni, B. Alloing, C. Paranthoen, A. Fiore, Appl. Phys. Lett. **85**, 2178 (2004)
75. V. Zwiller, T. Aichele, W. Seifert, J. Persson, O. Benson, Appl. Phys. Lett. **82**, 1509 (2003)

Index

{137} facets 125
(2 × 4) reconstruction of InP (001) 73
2D (monolayer-high) islands 25
2D platelets 17
2D–3D transition 43, 56
2D-island growth 74
3D islands 25
3D strained islands 27
3D-island mode 88

ab initio 3
ab-initio DFT-GW calculation 72
ab initio molecular dynamics 3
acoustic phonons 165
acoustoluminescence 201
activation energy 227
adatom diffusion 5
adatom sea 25, 28, 54
adatoms 14, 24
analytical techniques 98
annealing 159, 287
anomalous coarsening 15
anti-correlation 34
antibunching 332
anticorrelated growth 29, 32
antimony surfactants 133
area-selected growth 258
Arrhenius 229
aspect ratio 126, 155, 157
atomic structure 123
atomic-force microscopy 238
attachment 3, 4
attachment process 25
attachment-limited kinetics 28
Auger effect 213

barrier 9, 33
bath 195, 196
BB84 protocol 343

biexciton 139, 241, 282, 285
biexciton binding energy 157, 285
biexcitonic shift 193, 203, 204
bimodal island size distributions 15
bimodal size distribution 44, 58
binding energies of excitonic complexes 285
binding energy 286, 287
bistability 212
Bloch equations 310
Bohr radius 262
bond-counting ansatz 90
bright exciton state 282
bunching 215
buried islands 29

capacitance spectroscopy 226
capacitance-voltage characteristics 211
capping 127, 130
carrier concentration 81
carrier relaxation 222
cascaded emission 338
(CdMn)Se 251
CdSe 238
charge-selective DLTS 226
charged excitonic complexes 282
chemical composition 107
chemical potential 10, 14, 25, 28, 29
chemical reactions 3
chemically sensitive reflection 107, 118
cleavage 124
cleavage surfaces 128
coalescence 28
coalescence on impact 34
coherent control 245, 308
composition 70, 73, 74, 116
composition profiles 49, 157
configuration interaction (CI) method 139
confined excitonic states 270
confined states 286

352 Index

confinement potential 190
constrained equilibrium 14
constraint equilibrium 4
constraint surface equilibrium 22
constraint thermodynamic equilibrium 34
constraint thermodynamics 29, 32
continuum description 3
continuum elasticity theory 3, 34, 166
continuum theory 3
contrast 128
correlated growth 32
correlation 140, 143, 157, 272, 286
correlation expansion 194, 195, 197, 198, 206
correlation of structural and excitonic properties 278
Coulomb charging 230
Coulomb interaction 139, 193, 202–204, 211, 286
coupled quantum dots 182
crater 133
critical layer thickness 44
critical nucleus 33
critical thickness 134
cross-sectional scanning tunneling microscopy 123
cumulant expansion 174, 180, 196, 200, 205
cumulant-generating function 216
current-voltage characteristics 211
cyclotron resonance 264

dark-field image 107, 114, 115, 118
decoherence 216
decoherence-free subspaces 346
deep level transient spectroscopy (DLTS) 226
deep levels 226
defect-reduction technique 2, 35
deformation potential coupling 165, 191, 192, 197
dense array 27, 28, 34
dense metastable arrays 17
density functional theory 3, 11, 34, 169
density matrix 194, 196
density matrix approach 217
density matrix formalism 194
density-functional theory 4, 5, 9, 13
dephasing 165
deposition 3, 23, 24, 26

desorption 6, 9
detachment 3, 4, 23
detachment process 25
Deutsch–Jozsa algorithm 344
DFT 14, 18
differential transmission spectra, *see* DTS spectra
diffraction contrast 98, 100, 107, 116
diffusion 3, 4, 6, 9, 23, 25
diffusion length 74, 76
dilute array of islands 19
diluted magnetic quantum dots 251
dimer vacancies 88
dimer vacancy line 88
dipole approximation 192
dipole–dipole interaction 319
doping 79
DRAM 222
drift 25
DTS spectra 200, 204
DWELL structure 61

effective elastic constants 169
effective mass 190
effective mass approximation 190
eh exchange 242, 244
eight-band $k \cdot p$ theory 139, 140
eight-band $\mathbf{k} \cdot \mathbf{p}$ theory 190
elastic anisotropy 30, 112
elastic constants 167
elastic energy relief 12
elastic relaxation energy 19
elastically soft directions 30
elasticity theory 11
electric field 79, 80
electrical pumping 340
electron beam lithography 258
electron diffraction 98
electron microscopy 98
electron p-state splitting 139
electron–electron exchange 244
electron–phonon coupling 191, 194–197, 199–202
electron-acoustic phonon interaction 172
electronic coupling 29
ellipsometry 68, 71, 77
emission spectrum 201
energy barrier for diffusion 18
energy barriers 6

energy-dispersive X-ray spectroscopy 98, 102
ensemble evolution kinetics 25, 42, 53
entangled photons 287
entry
  subentry 1
envelope function 190
epioptics 68
epitaxial growth 238
equilibrium 3
equilibrium island shape 4
equilibrium shape 11, 13
equilibrium volume 19
evaporation 3
evolution of an island ensemble 3
exchange interaction 282
excitation-density dependence 270
excitation-induced dephasing 311
excited states 270, 271, 279
exciton 139, 165, 241, 282
exciton quantum coherence 247
exciton–biexciton cross-correlation 334
exciton–biexciton transition 246
excitonic states 202, 205

Fabry–Perot 81
facets 125, 128, 130
Fano factor 215
few-particle effects 281
filled-pit quantum dots 259
fine-structure splitting 282, 288, 340
finite element method 98, 102
Flash 222
floating-gate 224
fluctuations of the magnetic environment 252
Fock states 332
Förster coupling 193, 202, 204, 205
Förster energy transfer 321
FRAM 222
Frank–van-der-Merwe 19
free-standing islands 11
free-standing QD 33
full counting statistics 216

GaAs 11
GaAs homoepitaxy 6
GaSb 134
Ga(Sb,As) 113, 114, 116–119

GaSb quantum dots 134
GaSb/GaAs quantum dots 56
Ge distribution 99, 100, 102
Ge/Si 29
global surface equilibrium 4
growth 19
growth interruption 130, 133
growth parameters 46
growth rate 70, 74, 82

Hanbury Brown–Twiss 332
harmonic bath 196
harmonic disc model 190
harmonic oscillator model 190
Hartree–Fock factorization 194
heavy-light hole mixing 244
Heisenberg picture 194
heterodyne detection 307
high index substrates 109
high resolution TEM 100, 117
high-angle annular dark-field STEM 118, 119
high-index facets 14, 34
high-index surfaces 13
high-resolution microscopy 98
high-resolution transmission electron microscopy 24
hole spin 248
homoepitaxy of GaAs on a GaAs(001) 5
hot trion 283
hot-electron injection 224
hydrostatic strain 276
hyperfine interaction 249

IBM, *see* independent boson model
III/V quantum dot 98
image charge 142
(In,Ga)As 103, 104, 107, 109, 116, 117, 119
In distribution 104–106
In migration 105, 106
In segregation 106
in-situ analysis 67
in-situ strain monitoring 69
InAs 11, 125
InAs quantum dots 16, 124, 127
InAs/GaAs 5, 29
InAs/GaAs quantum dots 50
inclined inheritance 112, 113
independent boson model 172, 174, 196–198

indium content 129, 132
infrared magneto-resonance spectroscopy 263
InGaAs 103, 131
InGaAs quantum dots 131
InGaAs/GaAs quantum dots 42
In(Ga)As/GaAs quantum dots 270
InGaN/GaN quantum dots 270
inhomogeneous broadening 212
interaction picture 195, 196
interface quantum dots 303
inverse ripening 26
IR-MRS 263
irrelevant density matrix 195

kinetic Monte Carlo 3, 6, 8, 9, 17, 23, 34
kinetically controlled islands 24

LA phonons 191, 197–199
linear electro-optic effect 79
linear optics quantum computation 330
Liouvillian 196
liquid phase epitaxy 99
LO phonon 192, 197, 275
local field 205
local phonon modes 274
local surface equilibrium 4
localization 279
localization energy 224
localized exciton 304
longitudinal acoustic phonons, *see* LA phonons
longitudinal optical phonons, *see* LO phonons
low-index facets 14

macroscopic polarization 194
many-particle 139
Markovian approximation 199, 202
mass transfer 57
master equation model 214
MBE 125
megagauss regime 264
memories 221
metal-organic vapor phase epitaxy 42, 46, 333
metastable state 22, 26, 29
micro-photoluminescence 332
microscopic polarization 193

minimum energy per atom 22
mobile phones 222
MOCVD 104, 131, 134
modeling 139
molecular beam epitaxy 109, 237
molecular dynamics 3
Mollow splitting 199
monolayer 51, 68
monolayer height-variation 278
monolayer oscillations 74, 76
Monte Carlo 3
Monte Carlo simulations 90
Moore's Law 221
MP3 players 222
MRAM 222
multi-phonon absorption 201
multibody potentials 34
multilayer 109
multimodal size distribution 51, 278, 281
multiplexing 343
multiscale modeling 3, 34
multisheet array 29

Nakajima–Zwanzig formalism 195
nanoengineering 2
nanopatterning 87
nanospectroscopy 305
nanovoid 132
near-field fiber probe 305
negative differential conductivity 212
negatively charged exciton 285
non-exponential decay 291
non-Markovian 195, 197–199
nonresonant excitation 290
nonvolatile memories 223
nuclear spin polarization 250
nucleation 3, 6, 8, 16, 17, 24, 33
nucleation kinetics 34
nucleation problem 103

occupation effects 270
occupation number 271
optical properties 269
optical Stark effect 199, 200, 204, 315
orientation-dependent growth 258
oscillations 68
OSE, *see* optical Stark effect
Ostwald ripening 22, 28, 34, 84

Index 355

p–n diode 224
PE, see photon echo
personal computer 222
phonon amplitude 171
phonon interaction 274
phonon replica 214
phonon sidebands 197–200
phonons 165, 275
photolithography 258
photoluminescence 114
photon echo 205
photons 329
piezoelectric field 290
piezoelectricity 139, 284
plastic relaxation 114
pn-diode 211
Poissonian process 214
polar coupling 192, 197
polarization 282
polarization degree 284
polarized emission 282, 294
positive chemical potential gradient 26
positively charged exciton 285
potential-energy surfaces 5
principle of STM 82
pure dephasing 197–199
pyramidal shape 125

QD lateral ordering 103, 104, 109, 110
QD morphology 105
quadratic coupling model 177
quantitative HRTEM 98, 117
quantum coherence 217
quantum computing 344
quantum dot growth 41
quantum dots 2, 4, 9, 12, 87, 103, 107, 113, 123, 139, 221, 237
quantum information processing (QIP) 301
quantum structures 262
quantum wire 2
quantum-dot laser 60
quantum-dot overgrowth 47, 276
quantum-dot shape 52
quantum-dot size 280
quantum-dot stack 48

Rabi oscillations 199, 203, 312
radiative coupling 203
radiative decay 203

radiative recombination 309
Raman scattering 274
RAS spectra 72
rate window 227
recombination dynamics 295
reconstruction 7–10, 13, 68, 72
red shift 271, 272, 276
reflectance 68, 69
reflectance anisotropy spectroscopy 68, 71
refractive index 69, 81
relevant density matrix 195
renormalization 270
resident carriers 249
resonance fluorescence 196
resonant excitation 290
resonant tunneling structures 211
reversed truncated cone 132
RF, see resonance fluorescence
ring-like structures 133
ripening 3, 22, 25

scanning tunneling microscopy 69, 82, 92, 123
schematic and principle of the MOVPE-STM 82
Schottky model 80
Schrödinger picture 194
second Born approximation 194
second-order coherence function 331
secret key-distribution protocol 329
secular approximation 202
segregation 129
self-assembled quantum dots 259
self-assembled single quantum dots 330
self-limitation of growth 98
self-limiting growth 258
self-organization 2
sequential tunneling 217
setup of a typical RAS-system 72
shape 52, 123
shape transition 14
shear strain 284
shift of quantum-dot emission 276
shot noise 214
(Si,Ge)/Si 98, 99
side facets 20, 125
single quantum dots 281, 292
single-dot spectroscopy 281
single-particle orbitals 140

single-particle states 139
single-photon source 330
single-turn coil technique 264
size 123
size distribution 278
skewness 217
Slater determinants 140
sp-d coupling 251
spectral diffusion 281, 293
spectral power density 214
spin relaxation 247
spin-lattice-relaxation 252
state filling 230
step-edge 74, 76, 79
step-flow growth 74, 77
steps 10
stimulated emission 246
STM 123, 124
stoichiometry 123, 129, 130, 134
Stokes shift 290
storage of holes and electrons 287
strain 125, 128, 130, 139
strain energy distribution 102, 112
strain field 23, 99, 116
strain reduction 277
strain relaxation 11, 14, 45
strain relaxation simulations 129
strain sensitive 116
strain tensor 167
strain-induced renormalization of the surface energy 21
strain-renormalized surface energy 25, 27
Stranski–Krastanow 87, 99, 103, 109
Stranski–Krastanow growth 4, 19, 41
stress 88, 167
stretched exponent 291
strong coupling 195
structural anisotropy 284
structural changes 127
structural characterisation 98
subensemble 51, 53, 278
substrate orientation 109, 111
substrate-mediated elastic interaction 20
suppression of nucleation 98
surface energies 13
surface equilibrium 19
surface free energies 4
surface reconstructions 5
surface stress 4, 19

symmetry 145
symmetry lowering 284
synchrotron radiation 98
system–bath interaction 195, 196

TCL theory 195, 196, 202, 205
TEM 240
thermal emission 226
thermodynamic driving force 10
thermodynamically controlled islands 24
thermodynamically stable ensembles of the islands 22
three-dimensional arrays of QDs 98, 109, 110
three-dimensional strain field 100
time convolutionless theory, *see* TCL theory
time-resolved PL 290
time-resolved spectroscopy 289
TO phonon 275
transmission electron microscopy 97, 98, 100, 107, 113–116
tree-dimensional arrays of QDs 109
trion 243, 286
triplet state 244, 245
truncated pyramid 53, 131
truncated pyramidal shape 128
truncation 130
tunneling 228
two-dimensional electron gas (2DEG) 225
two-dimensional islands 19
two-dimensional nucleation 102
two-level system 306
two-quantum-bit system 245
type II band lineup 56
type-II band alignment 134

ultimate memory 224

VCSEL 61
vertical alignment 48
vertical correlation 34
vertical ordering 110, 113
vertically anticorrelated arrangement 30
vertically correlated arrangement 30
vertically correlated array 29
virtual and real phonon-assisted transition 176
Volmer–Weber 19

wetting layer 4, 10, 15, 20, 22, 28, 42, 56, 104
wetting layer depletion 102
wetting layer thickness 105
Wiener–Khinchin theorem 215

X-ray diffraction 97, 98, 100
X-ray diffuse scattering 98, 104–106, 110

XSTM 123, 127, 131, 134

Z-contrast 98, 118
zero-dimensional 103
zero-phonon line 197, 200
ZnSe 238
ZPL, *see* zero-phonon line